Methods in Enzymology

Volume 375
CHROMATIN AND CHROMATIN REMODELING ENZYMES
Part A

METHODS IN ENZYMOLOGY

EDITORS-IN-CHIEF

John N. Abelson   Melvin I. Simon

DIVISION OF BIOLOGY
CALIFORNIA INSTITUTE OF TECHNOLOGY
PASADENA, CALIFORNIA

FOUNDING EDITORS

Sidney P. Colowick and Nathan O. Kaplan

Methods in Enzymology

Volume 375

Chromatin and Chromatin Remodeling Enzymes

Part A

EDITED BY

C. David Allis

THE ROCKEFELLER UNIVERSITY
NEW YORK, NEW YORK

Carl Wu

NATIONAL CANCER INSTITUTE
BETHESDA, MARYLAND

ELSEVIER
ACADEMIC
PRESS

AMSTERDAM • BOSTON • HEIDELBERG • LONDON
NEW YORK • OXFORD • PARIS • SAN DIEGO
SAN FRANCISCO • SINGAPORE • SYDNEY • TOKYO
Academic Press is an imprint of Elsevier

Elsevier Academic Press
525 B Street, Suite 1900, San Diego, California 92101-4495, USA
84 Theobald's Road, London WC1X 8RR, UK

This book is printed on acid-free paper. ∞

For all information on all Academic Press Publications
visit our Web site at www.academicpress.com

ISBN: 0-12-182779-8

PRINTED IN THE UNITED STATES OF AMERICA
03 04 05 06 07 08   9 8 7 6 5 4 3 2 1

Table of Contents

Section I. Histone Bioinformatics

Section II. Biochemistry of Histones, Nucleosomes, and Chromatin

Section III. Molecular Cytology of Chromatin Functions

Contributors to Volume 375

Article numbers are in parentheses and following the names of contributors. Affiliations listed are current.

CHAD ALEXANDER (3), *The University of Tennessee-Oak Ridge Graduate School of Genome Science and Technology, Oak Ridge National Laboratory, Life Sciences Division, Oak Ridge, Tennessee 37831-8080*

GENEVIÈVE ALMOUZNI (8), *Institut Curie, Section de Recherche, F-75248, Paris Cedex 05, France*

SATOSHI ANDO (18), *Department of Molecular Life Science, School of Medicine, Tokai University, Kanagawa 259-1193, Japan*

YUNHE BAO (2), *Department of Biochemistry and Molecular Biology, Colorado State University, Fort Collins, Colorado 80523-1870*

BLAINE BARTHOLOMEW (13), *Department of Biochemistry & Molecular Biology, Southern Illinois University School of Medicine, Carbondale, Illinois 62901-4413*

DAVID P. BAZETT-JONES (28), *Programme in Cell Biology, Hospital for Sick Children, Toronto, Ontario M5G 1X8, Canada*

ANDREW S. BELMONT (23), *Department of Cell and Structural Biology, University of Illinois at Urbana-Champaign, Urbana, Illinois 61801*

LEISE BERVEN (16), *Children's Medical Research Institute, Westmead, New South Wales 2415, Australia*

YEHUDIT BIRGER (21), *National Cancer Institute, National Institutes of Health, Bethesda, Maryland 20892*

HINRICH BOEGER (11), *Department of Structural Biology, Stanford University School of Medicine, Stanford, California 94305*

WILLIAM M. BONNER (5), *Laboratory of Molecular Pharmacology, National Cancer Institute, Bethesda, Maryland 20892*

MICHAEL BRUNO (14), *Division of Gene Regulation and Expression, The Wellcome Trust Biocentre, Department of Biochemistry, University of Dundee, Dundee, DD1 5EH, Scotland, United Kingdom.*

GERARD J. BUNICK (3), *Life Sciences Division, Oak Ridge National Laboratory, Oak Ridge, Tennessee 37831-8080*

MICHAEL BUSTIN (21), *National Cancer Institute, National Institutes of Health, Bethesda, Maryland 20892*

ANNE E. CARPENTER (23), *Whitehead Institute for Biomedical Research, Cambridge, Massachusetts 02142*

GUSTAVO CARRERO (26), *Department of Mathematical and Statistical Sciences, Faculty of Science, University of Alberta, Edmonton, Alberta T6G 2E1, Canada*

DAVID CARTER (29), *Laboratory of Chromatin and Gene Expression, Babraham Institute, Cambridge CB2 4AT, United Kingdom*

FRÉDÉRIC CATEZ (21), *National Cancer Institute, National Institutes of Health, Bethesda, Maryland 20892*

LYUBOMIRA CHAKALOVA (29), *Laboratory of Chromatin and Gene Expression, Babraham Institute, Cambridge CB2 4AT, United Kingdom*

SRINIVAS CHAKRAVARTHY (2), *Department of Biochemistry and Molecular Biology, Colorado State University, Fort Collins, Colorado 80523-1870*

LAKSHMI N. CHANGOLKAR (15), *Department of Animal Biology, School of Veterinary Medicine, University of Pennsylvania, Philadelphia, Pennsylvania 19104*

LISA ANN CIRILLO (9), *Department of Cell Biology, Neurobiology, and Anatomy, Medical College of Washington, Milwaukee, Wisconsin 53149*

PETER R. COOK (24), *The Sir William Dunn School of Pathology, University of Oxford, Oxford OX1 3RE, United Kingdom*

ELLEN CRAWFORD (26), *Department of Oncology, Faculty of Medicine, University of Alberta and Cross Cancer Institute, Edmonton, Alberta T6G 2E1, Canada*

WOUTER DE LAAT (30), *Department of Cell Biology, ErasmusMC, 3015 GE Rotterdam, The Netherlands*

GERDA DE VRIES (26), *Department of Mathematical and Statistical Sciences, Faculty of Science, University of Alberta, Edmonton, Alberta T6G 2E1, Canada*

GRAHAM DELLAIRE (28), *Programme in Cell Biology, Hospital for Sick Children, Toronto, Ontario M5G 1X8, Canada*

JOHN D. DILLER (10), *Department of Biochemistry and Molecular Biology, Center for Gene Regulation, The Pennsylvania State University, University Park, Pennsylvania 16802*

CHARLES E. DUCKER (10), *Department of Biochemistry and Molecular Biology, Center for Gene Regulation, The Pennsylvania State University, University Park, Pennsylvania 16802*

PAMELA N. DYER (2), *Department of Biochemistry and Molecular Biology, Colorado State University, Fort Collins, Colorado 80523-1870*

RAJI S. EDAYATHUMANGALAM (2), *Department of Biochemistry and Molecular Biology, Colorado State University, Fort Collins, Colorado 80523-1870*

THOMAS G. FAZZIO (6), *Fred Hutchinson Cancer Research Center, Seattle, Washington 98109-1024*

ANDREW FLAUS (14), *Division of Gene Regulation and Expression, The Wellcome Trust Biocentre, Department of Biochemistry, University of Dundee, Dundee, DD1 5EH, Scotland, United Kingdom.*

PETER FRASER (29), *Laboratory of Chromatin and Gene Expression, Babraham Institute, Cambridge CB2 4AT, United Kingdom*

SUSAN M. GASSER (22), *Department of Molecular Biology, University of Geneva, 1211 Geneva 4, Switzerland*

STANISLAW A. GORSKI (25), *National Cancer Institute, National Institutes of Health, Bethesda, Maryland 20892*

JOACHIM GRIESENBECK (11), *Department of Structural Biology, Stanford University School of Medicine, Stanford, California 94305*

FRANK GROSVELD (30), *Department of Cell Biology, ErasmusMC, 3015 GE Rotterdam, The Netherlands*

B. LEIF HANSON (3), *The University of Tennessee-Oak Ridge Graduate School of Genome Science and Technology, Life Sciences Divison, Oak Ridge National Laboratory, Oak Ridge, Tennessee 37831-8080*

JOEL M. HARP (3), *Department of Biochemistry and Center for Structural Biology, Vanderbilt University, Nashville, Tennessee 37232-8725*

KEIJI HASHIMOTO (17), *Core Research for Evolutional Science and Technology, Saitama 332-0012, Japan*

JEFFREY J. HAYES (12), *Department of Biochemistry and Biophysics, University of Rochester Medical Center, Rochester, New York 14642*

FLORENCE HEDIGER (22), *Department of Molecular Biology, University of Geneva, 1211 Geneva 4, Switzerland*

MICHAEL J. HENDZEL (26), *Department of Oncology, University of Alberta and Cross Cancer Instutite, Edmonton, Alberta T6G 1Z2, Canada*

MIKI HIEDA (24), *Sir William Dunn School of Pathology, University of Oxford, Oxford OX1 3RE, United Kingdom*

STEFAN R. KASSABOV (13), *Department of Biochemistry & Molecular Biology, Southern Illinois University School of Medicine, Carbondale, Illinois 62901-4413*

HIROSHI KIMURA (24), *Horizontal Medical Research Organization, School of Medicine, Kyoto University, Kyoto 606-8510, Japan*

ROGER D. KORNBERG (11), *Department of Structural Biology, Stanford University School of Medicine, Stanford, California 94305*

DAVID LANDSMAN (1) *National Center for Biotechnology Information, National Library of Medicine, National Institutes of Health, Bethesda, Maryland 20894*

PAUL J. LAYBOURN (7), *Department of Biochemistry and Molecular Biology, Colorado State University, Fort Collins, Colorado 80523-1870*

JAE-HWAN LIM (21), *National Cancer Institute, National Institutes of Health, Bethesda, Maryland 20892*

KAROLIN LUGER (2), *Department of Biochemistry and Molecular Biology, Colorado State University, Fort Collins, Colorado 80523-1870*

JAMES G. MCNALLY (27), *Laboratory of Receptor Biology and Gene Expression, National Cancer Institute, National Institutes of Health, Bethesda, Maryland 20892*

TOM MISTELI (25) *National Cancer Institute, National Institutes of Health, Bethesda, Maryland 20892*

CRAIG A. MIZZEN (19), *Department of Cell & Structural Biology, University of Illinois at Urbana-Champaign, Urbana, Illinois 61801*

SETSUO MORISHITA (17), *Department of Molecular Biology, School of Science, Nagoya University, Nagoya 464-8601, Japan*

UMA M. MUTHURAJAN (2), *Department of Biochemistry and Molecular Biology, Colorado State University, Fort Collins, Colorado 80523-1870*

FRANK R. NEUMANN (22), *Department of Molecular Biology, University of Geneva, 1211 Geneva 4, Switzerland*

ROZALIA NISMAN (28), *Programme in Cell Biology, Hospital for Sick Children, Toronto, Ontario M5G 1X8, Canada*

TOM OWEN-HUGHES (14), *Division of Gene Regulation and Expression, The Wellcome Trust Biocentre, Department of Biochemistry, University of Dundee, Dundee, DD1 5EH Scotland, United Kingdom.*

JOHN R. PEHRSON (15), *Department of Animal Biology, School of Veterinary Medicine, University of Pennsylvania, Philadelphia, Pennsylvania 19104*

CRAIG L. PETERSON (4) *University of Massachusetts Medical School, Worchester, Massachusetts 01605*

ROBERT D. PHAIR (25), *BioInformatics Services, Rockville, Maryland 20854*

DUANE R. PILCH (5), *Laboratory of Molecular Pharmacology, National Cancer Institute, Bethesda, Maryland 20892*

YURI V. POSTNIKOV (21), *National Cancer Institute, National Institutes of Health, Bethesda, Maryland 20892*

DANNY RANGASAMY (16), *The John Curtin School of Medical Research, Australian National University, Canberra, Australia Capital Territory 2601, Australia*

DOMINIQUE RAY-GALLET (8), *Institut Curie, Section de Recherche, F-75248, Paris Cedex 05, France*

CHRISTOPHE REDON (5), *Laboratory of Molecular Pharmacology, National Cancer Institute, Bethesda, Maryland 20892*

RAYMOND REEVES (20), *School of Molecular Biosciences, Biochemistry/Biophysics, Washington State University, Pullman, Washington 99164-4660*

PATRICIA RIDGWAY (16), *The John Curtin School of Medical Research, Australian National University, Canberra, Australian Capital Territory 2601, Australia*

CHUN RUAN (10), *Department of Biochemistry and Molecular Biology, Center for Gene Regulation, The Pennsylvania State University, University Park, Pennsylvania 16802*

OLGA A. SEDELNIKOVA (5), *Laboratory of Molecular Pharmacology, National Cancer Institute, Bethesda, Maryland 20892*

MICHAEL A. SHOGREN-KNAAK (4), *University of Massachusetts Medical School, Worchester, Massachusetss 01605*

ROBERT T. SIMPSON (10), *Department of Biochemistry and Molecular Biology, Center for Gene Regulation, The Pennsylvania State University, University Park, Pennsylvania 16802*

ERIK SPLINTER (30), *Department of Cell Biology, ErasmusMC, 3015 GE Rotterdam, The Netherlands*

DIANA A. STAVREVA (27), *Laboratory of Receptor Biology and Gene Expression, National Cancer Institute, National Institutes of Health, Bethesda, Maryland 20892*

J. SETH STRATTAN (11), *Department of Structural Biology, Stanford University School of Medicine, Stanford, California 94305*

STEVEN A. SULLIVAN (1), *National Center for Biotechnology Information, National Library of Medicine, National Institutes of Health, Bethesda, Maryland 20894*

ULRICA SVENSSON (16), *The John Curtin School of Medical Research, Australian National University, Canberra, Australian Capital Territory 2601, Australia*

ANGELA TADDEI (22), *Department of Molecular Biology, University of Geneva, 1211 Geneva 4, Switzerland*

JOHN TH'NG (26), *Northwestern Ontario Regional Cancer Centre, Thunder Bay, Ontario P7A 7T1, Canada*

DAVID JOHN TREMETHICK (16), *The John Curtin School of Medical Research, Australian National University, Canberra, Australian Capital Territory 2601, Australia*

TOSHIO TSUKIYAMA (6), *Fred Hutchinson Cancer Research Center, Seattle, Washington 98109-1024*

JAY C. VARY, JR. (6), *Molecular and Cellular Biology Program, University of Washington, Seattle, Washington 98195*

CINDY L. WHITE (2), *Department of Biochemistry and Molecular Biology, Colorado State University, Fort Collins, Colorado 80523-1870*

SRIWAN WONGWISANSRI (7), *Department of Biochemistry and Molecular Biology, Colorado State University, Fort Collins, Colorado 80523-1870*

KINYA YODA (17, 18), *Bioscience and Bio-technology Center, Nagoya University, Nagoya, 464-8601, Japan*

KENNETH S. ZARET (9), *Cell and Developmental Biology Program, Fox Chase Cancer Center, Philadelphia, Pennsylvania 19111*

CHUNYANG ZHENG (12), *Department of Biochemistry and Biophysics, University of Rochester Medical Center, Rochester, New York 14642*

Preface

A central challenge of the post-genomic era is to understand how the 30,000 to 40,000 unique genes in the human genome are selectively expressed or silenced to coordinate cellular growth and differentiation. The packaging of eukaryotic genomes in a complex of DNA, histones, and nonhistone proteins called chromatin provides a surprisingly sophisticated system that plays a critical role in controlling the flow of genetic information. This packaging system has evolved to index our genomes such that certain genes become readily accessible to the transcription machinery, while other genes are reversibly silenced. Moreover, chromatin-based mechanisms of gene regulation, often involving domains of covalent modifications of DNA and histones, can be inherited from one generation to the next. The heritability of chromatin states in the absence of DNA mutation has contributed greatly to the current excitement in the field of epigenetics.

The past 5 years have witnessed an explosion of new research on chromatin biology and biochemistry. Chromatin structure and function are now widely recognized as being critical to regulating gene expression, maintaining genomic stability, and ensuring faithful chromosome transmission. Moreover, links between chromatin metabolism and disease are beginning to emerge. The identification of altered DNA methylation and histone acetylase activity in human cancers, the use of histone deacetylase inhibitors in the treatment of leukemia, and the tumor suppressor activities of ATP-dependent chromatin remodeling enzymes are examples that likely represent just the tip of the iceberg.

As such, the field is attracting new investigators who enter with little first hand experience with the standard assays used to dissect chromatin structure and function. In addition, even seasoned veterans are overwhelmed by the rapid introduction of new chromatin technologies. Accordingly, we sought to bring together a useful "go-to" set of chromatin-based methods that would update and complement two previous publications in this series, Volume 170 (Nucleosomes) and Volume 304 (Chromatin). While many of the classic protocols in those volumes remain as timely now as when they were written, it is our hope the present series will fill in the gaps for the next several years.

This 3-volume set of *Methods in Enzymology* provides nearly one hundred procedures covering the full range of tools—bioinformatics, structural biology, biophysics, biochemistry, genetics, and cell biology—employed in chromatin research. Volume 375 includes a histone database, methods for preparation of histones, histone variants, modified histones and defined chromatin segments, protocols for nucleosome reconstitution and analysis, and cytological methods for imaging chromatin functions *in vivo*. Volume 376 includes electron microscopy and biophysical protocols for visualizing chromatin and detecting chromatin interactions, enzymological assays for histone modifying enzymes, and immunochemical protocols for the *in situ* detection of histone modifications and chromatin proteins. Volume 377 includes genetic assays of histones and chromatin regulators, methods for the preparation and analysis of histone modifying and ATP-dependent chromatin remodeling enzymes, and assays for transcription and DNA repair on chromatin templates. We are exceedingly grateful to the very large number of colleagues representing the field's leading laboratories, who have taken the time and effort to make their technical expertise available in this series.

Finally, we wish to take the opportunity to remember Vincent Allfrey, Andrei Mirzabekov, Harold Weintraub, Abraham Worcel, and especially Alan Wolffe, co-editor of Volume 304 (Chromatin). All of these individuals had key roles in shaping the chromatin field into what it is today.

C. DAVID ALLIS
CARL WU

Editors' Note: Additional methods can be found in Methods in Enzymology, Vol. 371 (RNA Polymerases and Associated Factors, Part D) Section III Chromatin, *Sankar L. Adhya and Susan Garges, Editors.*

METHODS IN ENZYMOLOGY

Section I

Histone Bioinformatics

[1] Mining Core Histone Sequences from Public Protein Databases

By STEVEN A. SULLIVAN and DAVID LANDSMAN

Introduction

Constructing an online database of histones and histone fold-containing proteins has allowed our group to analyze histone sequence variation in some detail.[1,2] Here, we describe how we have inventoried core histone protein sequences as part of this project. The issues involved in such an undertaking are for the most part not unique to histone sequences. Our methods and observations should be broadly applicable to studies of protein families that are highly represented in public sequence databases.

Considerations

Our initial goal was to collect as many reported histone sequences as we could find. Among the considerations that came into play were the following.

1. *Sourcing of sequences.* Several excellent public sequence repositories make protein sequences available to researchers. We relied on the protein database maintained by the National Center for Biotechnology Information (NCBI), which is updated frequently and has been compiled from worldwide sources, including Swiss-Prot,[3] the Protein Information Resource (PIR),[4] the Protein Research Foundation (PRF) (http://www.prf.or.jp/en/), the Protein Data Bank (PDB),[5] and translations from annotated coding regions in GenBank[6] and RefSeq,[7] a curated, nonredundant set of sequences.

[1] S. Sullivan, D. W. Sink, K. L. Trout, I. Makalowska, P. M. Taylor, A. D. Baxevanis, and D. Landsman, *Nucleic Acids Res.* **30,** 341 (2002).

[2] S. A. Sullivan and D. Landsman, *Proteins* **52,** 454 (2003).

[3] B. Boeckmann, A. Bairoch, R. Apweiler, M. C. Blatter, A. Estreicher, E. Gasteiger, M. J. Martin, K. Michoud, C. O'Donovan, I. Phan, S. Pilbout, and M. Schneider, *Nucleic Acids Res.* **31,** 365 (2003).

[4] C. H. Wu, L. S. Yeh, H. Huang, L. Arminski, J. Castro-Alvear, Y. Chen, Z. Hu, P. Kourtesis, R. S. Ledley, B. E. Suzek, C. R. Vinayaka, J. Zhang, and W. C. Barker, *Nucleic Acids Res.* **31,** 345 (2003).

[5] J. Westbrook, Z. Feng, L. Chen, H. Yang, and H. M. Berman, *Nucleic Acids Res.* **31,** 489 (2003).

METHODS IN ENZYMOLOGY, VOL. 375

2. *Sequence-harvesting tools.* In general, a sequence database search is a similarity search of either the actual sequence data or its annotation. We find that both must be targeted in order to maximize the sequence harvest, because sequence-based searches alone can miss small or ambiguous sequence fragments that have been deposited in the public databases, and text-based searches can miss "cryptic" histones, that is, those with inadequate or incorrect annotation.

For text-based searches of sequence annotation we used the Entrez search engine at the NCBI Web site (http://www.ncbi.nlm.nih.gov/Entrez). For sequence-based searching we used several varieties of the popular Basic Local Alignment Search Tool (BLAST) pairwise alignment algorithm. The most commonly used sequence similarity search tools find "hits" based on pairwise alignments of each sequence in the database to either the query sequence alone, for example, in the case of BLAST, or a query profile derived from a previously aligned set of similar sequences, for example, in the case of PSI-BLAST or HMMER.[8,9] The latter tools are better at finding highly divergent members of a protein family but can be expected to return false positives, requiring further filtering of results. PSI-BLAST is actually a hybrid tool that performs one round of standard BLAST, using a user-supplied query sequence, and then builds a profile from the alignment of the initial BLAST results, which becomes the query for the next round of BLAST. The process is reiterated until "convergence" is reached, that is, until no more new matches are found above the cutoff score. Ideally this should take fewer than 10 iterations, but convergence can be elusive when the query sequence matches a diverse and perhaps distantly related set of proteins. This was more difficult to interpret with searches for nonhistone proteins containing the histone fold than for harvesting core histone sequences. With the latter we found that seven iterations were sufficient to reach either convergence or the point at which all the "new" hits appeared by other criteria to be false positives. PSI-BLAST routinely returned a small number of true-positive matches to the query sequences that gapped protein BLAST (BLASTPGP) had missed.

Reasonably fast BLASTPGP and PSI-BLAST servers are available at the NCBI Web site (http://www.ncbi.nlm.nih.gov/BLAST). One advantage of the NCBI Web site PSI-BLAST implementation over a command-line

[6] D. A. Benson, I. Karsch-Mizrachi, D. J. Lipman, J. Ostell, and D. L. Wheeler, *Nucleic Acids Res.* **31,** 23 (2003).
[7] K. D. Pruitt, T. Tatusova, and D. R. Maglott, *Nucleic Acids Res.* **31,** 34 (2003).
[8] S. F. Altschul, T. L. Madden, A. A. Schaffer, J. Zhang, Z. Zhang, W. Miller, and D. J. Lipman, *Nucleic Acids Res.* **25,** 3389 (1997).
[9] S. R. Eddy, *Bioinformatics* **14,** 755 (1998).

version is that the user can edit each set of aligned sequences before it is used to generate a profile. This can redirect a diverging sequence search back toward convergence. Unfortunately, however, it can also happen that a valid match from one iteration falls below the noise cutoff in the next, and in the WWW-based implementation, that match is lost. Therefore we ran PSI-BLAST (and BLASTPGP) from the command line in a UNIX environment, which allowed us to save the results from all of the iterations into one file for subsequent text parsing. It also allows considerable flexibility in setting other BLAST options. Most default values were adequate for typical BLAST searches, but we commonly increased the number of displayed description lines and alignments (the $-b$ and $-v$ options) to 3000, to ensure retrieval of all the possible hits for subsequent filtering steps.

3. *Query sequences.* Histones are ancient proteins, found in all known eukaryote lineages as well as some archaeal microbes. Using a single query sequence, there is the possibility that some valid hits might be missed because of the sequence divergence and extreme biodiversity of the histones, even using a profile-generating protocol. To maximize the identification of eukaryote core histones from the protein databases, we "bracketed" the kingdom evolutionarily by using core histone sequences from human and yeast as search queries. This proved important for the more divergent histones, H2A and H2B, but less so for the more conserved histones, H4 and H3. For example, queries with human or yeast H4 or H3 returned almost the same sets of true-positive hits. In H3 searches, the most common outliers requiring taxonomic bracketing to capture were sequence fragments from protists, and members of the centromeric H3 subclass (data not shown).

4. *Sequence redundancy.* Sequence redundancy is the bane of most database searches. In most cases, redundant sequences in a large public sequence repository such as GenBank are often the same sequence from the same organism, automatically harvested from different databases, rather than originating from discrete sequencing projects in different laboratories. Thus, Web-based sequence similarity search tools, such as PSI-BLAST at the NCBI Web site, tend to present results in a convenient, nonredundant fashion, with sequence identifiers of identical sequences grouped together with an anchored sequence. To populate the histone database, however, we required every sequence in FASTA format (i.e., each record consisting of only a unique definition line and a sequence), one reason being that homologous histones display remarkable degrees of sequence identity, rather than mere similarity, across species. It is not uncommon that fully "redundant" histone sequences in the public database derive from more than one species. We wanted to start with a set in which such identical sequences are properly resolved. Because we

were attempting an exhaustive search, the well-intentioned nonredundancy of the public databases was, for us, an obstacle. Our strategy was to extract all the unique sequence identifiers from the BLAST outputs (in the case of NCBI records, the unique identifier is the GI number found at the beginning of the sequence definition line of a FASTA-formatted record) into a file, and use this file to generate a corresponding library of FASTA records. NCBI Entrez on the World Wide Web can take a file of GI numbers as input for batch retrieval of records; alternatively, we used the SEALS software suite to perform such retrievals in a UNIX environment.[10] SEALS has a tool, fauniq, for reducing a set of redundant FASTA sequence records to a nonredundant format, on the basis of either definition-line identifiers such as the GI number or on the sequence itself. This tool proved invaluable for filtering BLAST outputs to remove GI-based redundancies and for generating nonredundant sequence sets for alignment and variation analysis.

5. *Fragmentary, ambiguous, and frameshifted sequences.* Some sequences in the public databases are less than full-length; for example, a few records annotated as "histone H3" consist of only two or three amino acid residues. As sequences shorten, their detection becomes more difficult using typical "flavors" of BLAST when querying a large database because they become less distinguishable from chance hits. This problem is compounded if, as is the case with histones, the protein features segments of low sequence complexity, or if the fragment records contain ambiguous ("X") residues. To capture sequence fragments, we first divided the full-length query sequence into overlapping segments, with a segment window of 20 residues, sampled at intervals of 10 residues along the length. This was easily done with the SEALS fenestrate command. We then used these segments as queries against the public database in a modified gapped BLAST search optimized to capture short, nearly exact matches (a search option that is also available at the NCBI Website cited earlier). For these searches, low-complexity filters were turned off. The combined results of all the "window BLASTs" for a query sequence were made nonredundant with respect to GI number.

Frameshifted sequences (either authentic or erroneous) can pose a similar problem, depending on the size of the frameshifted region. Putative frameshifts are easily identified by visual inspection of multiple alignments of query results, for example, using the popular CLUSTAL X program,[11] where they manifest as sudden and extensive loss of sequence similarity.

[10] D. R. Walker and E. V. Koonin, *Proc. Int. Conf. Intell. Syst. Mol. Biol.* **5**, 333 (1997).
[11] J. D. Thompson, T. J. Gibson, F. Plewniak, F. Jeanmougin, and D. G. Higgins, *Nucleic Acids Res.* **25**, 4876 (1997).

To verify a frameshift, assuming access to the genomic DNA or cDNA record for the protein (which are often, but not always, available in public databases), one should translate the DNA in all frames and add those conceptual translations to the alignment; the correct frames will be visually evident in a true frameshift. Several tools exist on the Web for doing such translations; we commonly use the one at the ExPASy (Expert Protein Analysis) Web site: http://us.expasy.org/tools/dna.html. A translation tool is also available in the SEALS package.

Comparison of Search Strategies

There are many available variations on the basic BLAST search protocol. We investigated several parameters for their effects in the identification of histone H3 sequences. Histone H3 is a moderately diverse histone class, with more than half of the known full-length sequences displaying >80% identity in their histone fold domains; this figure falls between those for the more highly conserved H4 class and the more diverse H2A and H2B classes.[2] The H3 class comprises two subclasses that are markedly distinct in sequence and in function: replication-dependent H3 (the major H3) and centromeric H3. There is also a third, replication-independent H3.3 class, although its sequence is only marginally divergent from that of the major H3.

We first compiled a redundant reference set of H3 sequences, using a variety of BLAST- and Entrez-based searches, to include as many probable H3 sequence records as we could find in the NCBI protein database. This set was manually reviewed to eliminate false positives, yielding a final set of 1742 good candidate H3 sequences from all three subclasses. We then compared the results of different individual BLAST and Entrez search strategies with the reference set, to determine the efficiency (percentage of hits that are true positives, i.e., that are also found in the reference set) and the success (percentage of the reference set found by the search method). The results are shown in Table I. Entrez searches of eukaryotic sequence record annotation used the queries "H3" or "histon." BLAST parameters that we varied were: query sequence BLAST flavor (gapped BLAST versus gapped PSI-BLAST versus gapped BLAST for short, nearly exact window matches); query sequence (human versus yeast); database size (all versus the eukaryotic subset); and SEG low-complexity filtering (off versus on).

The Entrez results indicate that almost 20% of H3 sequences in the public database are cryptic, lacking specific annotation as H3 histones. The search results for "histon" as a query term recovered 95% of the reference sequences, with a trade-off of many more false positives, as one

TABLE I
COMPARISON OF SEARCH STRATEGIES FOR H3 HISTONE SEQUENCES[a]

Reference H3 set	Unique GI 1742	H3 1742	Success (%)	Efficiency (%)
Entrez "eukaryota[ORGN]"	1,143,461	1742	100.0	0.2
Entrez "H3"	3303	1452	83.4	44.0
Entrez "histon"	9297	1653	94.9	17.8
Entrez "eukaryota[ORGN] AND H3"	2703	1452	83.4	53.7
Entrez "eukaryota[ORGN] AND histon"	7453	1653	94.9	22.2
BLASTPGP H3human	1747	1719	98.7	98.4
BLASTPGP H3human+seg	1747	1719	98.7	98.4
BLASTPGP H3human+eukgi	1754	1722	98.9	98.2
BLASTPGP H3human+eukgi+seg	1754	1722	98.9	98.2
BLASTPGP H3yeast	1777	1718	98.6	96.7
BLASTPGP H3yeast+seg	1777	1718	98.6	96.7
BLASTPGP H3yeast+eukgi	1780	1718	98.6	96.5
BLASTPGP H3yeast+eukgi+seg	1780	1718	98.6	96.5
PSIBLASTPGP H3human	1897	1726	99.1	91.0
PSIBLASTPGP H3human+seg	1897	1726	99.1	91.0
PSIBLASTPGP H3human+eukgi	1949	1727	99.1	88.6
PSIBLASTPGP H3human+eukgi+seg	1949	1727	99.1	88.6
PSIBLASTPGP H3yeast	2011	1726	99.1	85.8
PSIBLASTPGP H3yeast+seg	2011	1726	99.1	85.8
PSIBLASTPGP H3yeast+eukgi	2077	1727	99.1	83.1
PSIBLASTPGP H3yeast+eukgi+seg	2077	1727	99.1	83.1
WINBLASTPGP H3human	69,678	1730	99.3	2.5
WINBLASTPGP H3human+eukgi	60,821	1732	99.4	2.8
WINBLASTPGP H3human+eukgi+seg	1697	1646	94.5	97.0
WINBLASTPGP H3yeast	70,864	1730	99.3	2.4
WINBLASTPGP H3yeast+eukgi	63,949	1730	99.3	2.7
WINBLASTPGP H3yeast+eukgi+seg	1788	1646	94.5	92.1

[a] Entrez queries of the NCBI protein database were conducted from the NCBI Web site www.ncbi.nlm.nih.gov/Entrez. BLAST searches using human or yeast histone H3 sequences were performed from the command line in a UNIX environment: BLASTPGP, gapped protein BLAST; PSIBLASTPGP, interated gapped protein BLAST using profiles; WINBLASTPGP, gapped protein BLAST for short, nearly exact matches, using sequence windows as queries; eukgi, search restricted to sequences from eukaryotes; seg, SEG filtering of low-complexity regions enabled. All results were compared with a curated reference_H3_set of sequences. Column headers: unique GI, number of unique sequence records retrieved; H3, number of retrieved unique GIs shared with the reference set; efficiency, percent H3/unique GI; success, percent H3/reference set.

would expect. The "histon" query also captured all of the true-positive "H3" query results (data not shown).

Any of the BLAST-based strategies was sufficient to capture at least 94% of the reference set from the public databases. The best combination of efficiency and success was achieved using gapped BLAST. The effects of differences in query sequence, database size, and filtering were minor compared with the difference between using BLAST, PSI-BLAST, or windowed BLAST, because the latter two BLAST implementations return far more false positives while increasing the success rate only marginally. Low-entropy sequence filtering appeared to make no difference whatsoever except in the case of windowed searches, in which the query sequence was divided into overlapping segments 20 residues each in length, with several gapped BLAST parameters altered to facilitate finding short, nearly exact matches to the query segments. Using the low-complexity filter here vastly increased efficiency by greatly reducing false positives, although success suffered in comparison with nonfiltered strategies, reflecting the presence of short, often basic low-complexity regions that are a hallmark of core histone sequences.

Unfortunately, as these results show, no single method captures all the relevant sequence records. A combination of strategies was the only way to achieve 100% success. However, the results of our comparison suggest a rational way to mine the maximum number of histone sequence records of a class from a database. The first step is to perform a single-round gapped BLAST search, making sure that the options for "number of descriptions" and "number of alignments" returned are set high (e.g., several thousand each). This should return most of the true positives with high efficiency. This set should be inspected carefully, using a variety of tools including text-search of the definition lines, multiple alignment, and further BLAST searches with a different query sequence, to remove false positives. The resulting validated set becomes most useful in subsequent searches employing other strategies, such as PSI-BLAST or text-based searches. The validated set can be used to subtract known positives from subsequent search results, using difference-finding tools such as the SEALS fanot command, which finds the logical exclusion of two sets of FASTA records or definition lines. This leaves a much shorter list of candidates from the new search results to be examined for new true positives. As these are identified they are added to the validated set, increasing its usefulness as a filter. This search strategy has also served us well in harvesting histone H4, H2A, and H2B sequences, and should work for any well-conserved class of protein sequences.

Histone Sequence Variants

Histone variants have been divided into "homomorphous" and "hetero-morphous" categories.[12,13] Homomorphous variants have relatively minor sequence differences and require high-resolution separation methods to distinguish them biochemically (reviewed in von Holt et al.[14]). They are found in all four core histone classes, and are presumed to be functionally identical. Heteromorphous variants are readily distinguished by conventional biochemical separation methods and tend to be distinct from other histones in their class with respect to function and/or spatiotemporal localization, as well as sequence. The distinction between the two categories of variants is not rigid—for example, the ostensibly homomorphous H3.3 appears to be functionally distinct from the major H3—and may become less so as the functions of more variants are experimentally tested. In clustering trees made from multiple sequence alignments of each histone class, hetero-morphous variants tend to form biodiverse clades distinct from the major form, indicating early branching off from major histones, whereas homo-morphous variants tend to comingle with the major form in clades that are more strongly delineated by phylogeny than by any other factor, suggesting the variants arose after the founding speciation event (data not shown; see also Thatcher and Gorovsky[15]). For all core histone classes, sequence align-ments show clear distinctions between metazoan, plant, fungal, and various basal eukaryote subclasses. Distinct subclasses within the metazoan sequences are also common (e.g., insect or echinoderm sequences). Nomen-clature is only occasionally helpful in classifying histone variants. It is not standardized, and thus "H3.2" in one species may not be similar to "H3.2" in another. The only other constant among aligned histone sequences appar-ent in Figs. 1–4, is that there tends to be less variation in the α-helical regions of the histone fold, than in the interhelical loops and the N- and C-terminal regions flanking the histone fold. This pattern of variation is common in other α helix-containing protein families.

H2A

The H2A class is the most diverse of the four core histone classes, both functionally and in terms of sequence, comprising four subclasses of known or putative functional variants in addition to typical phylogeny-based

[12] M. H. West and W. M. Bonner, *Biochemistry* **19,** 3238 (1980).

[13] J. Ausio, D. W. Abbott, X. Wang, and S. C. Moore, *Biochem. Cell Biol.* **79,** 693 (2001).

[14] C. von Holt, W. F. Brandt, H. J. Greyling, G. G. Lindsey, J. D. Retief, J. D. Rodrigues, S. Schwager, and B. T. Sewell, *Methods Enzymol.* **170,** 431 (1989).

[15] T. H. Thatcher and M. A. Gorovsky, *Nucleic Acids Res.* **22,** 174 (1994).

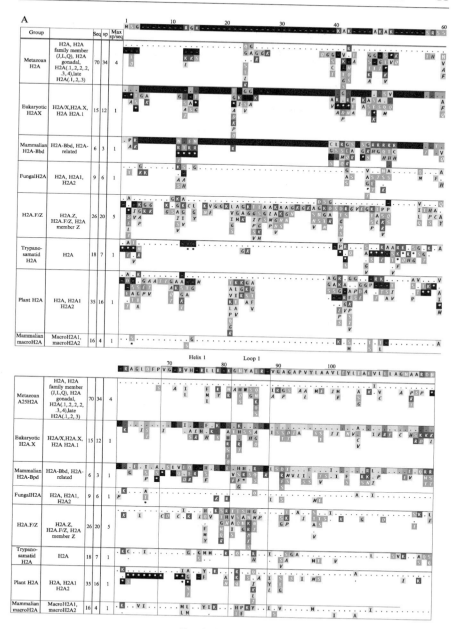

Fig. 1. (continued)

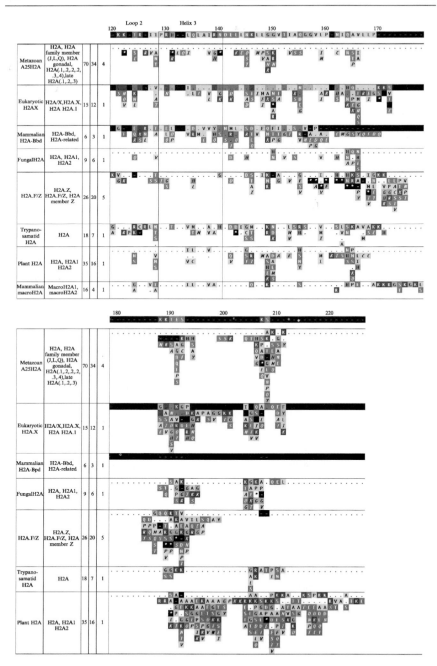

FIG. 1. *(continued)*

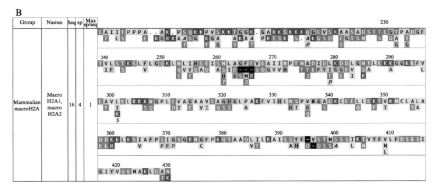

FIG. 1. Summary of H2A subclasses and variants. (A) A consensus sequence of all aligned H2A sequences is shown at the top. Dots in the sequences below indicate identity to the consensus. Groups are named on the basis of clustering patterns observed in neighbor-joining trees of aligned H2A sequences (not shown). *Names,* a selection of sequence descriptors found in the definition lines of the sequence records; *seq,* number of unique sequences in the group; *sp,* number of species in the group; *max sp/seq,* the greatest number of species having the same sequence in the group. For each group the first line is the consensus sequence for that group. Variations from the group consensus are indicated below it. Italic indicates a "singleton," i.e., the residue was found in only one sequence from one species in the group. An asterisk (*) indicates singleton identity or a gap. Background color key: white, identity to the anchored consensus; black, gap; orange, aromatic; yellow, aliphatic/hydrophobic; light green, glycine; green, hydrophilic; light blue, histidine; blue, basic; red, acidic. (B) C-terminal section of macroH2A. (See color insert.)

subclasses (Fig. 1A and B). H2A.X is found in species spanning the eukaryotic spectrum and features a conserved serine four residues from the carboxyl terminus (part of an SQ motif, positions 208 and 209 in Fig. 1A) that is phosphorylated in response to double-stranded DNA breaks, perhaps marking the site for repair (reviewed in Redon *et al.*[16]). Interestingly, the fungal H2A subclass clusters near the H2A.X subclass, and also features a conserved SQ motif at its C terminus. H2A.F/Z sequences constitute another pan-eukaryotic subclass and are necessary but not sufficient for H2A function in organisms tested. Characteristic H2A.F/Z residues in a C-terminal, H3-binding portion of the protein (positions 145–193 in Fig. 1A) have been suggested to impart a specific, although as yet unknown, function, as have the lysine residues in the amino-terminal portion (reviewed in Redon *et al.*[16]). Of these lysine

[16] C. Redon, D. Pilch, E. Rogakou, O. Sedelnikova, K. Newrock, and W. Bonner, *Curr. Opin. Genet. Dev.* **12,** 162 (2002).

residues, two (at positions 11 and 42 in Fig. 1A) appear to be specific to H2A.F/Z and not the major metazoan H2A. MacroH2A is a large bipartite histone divided into a recognizable H2A portion with many subclass-characteristic substitutions, and a long C-terminal extension found in no other histone subclass (residues 227–430 in Fig. 1B). MacroH2A has been found only in vertebrates and is concentrated in the inactive female X chromosome (reviewed in Brown[17]). H2A-Bbd is a highly

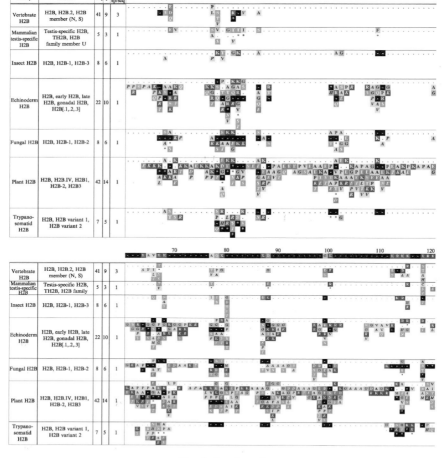

FIG. 2. *(continued)*

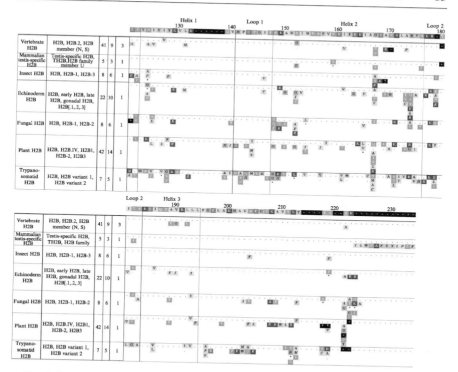

FIG. 2. Summary of H2B subclasses and variants. A consensus sequence of all aligned H2B sequences is shown at the top. Dots in the sequences below indicate identity to the consensus. Groups are named on the basis of clustering patterns observed in neighbor-joining trees of aligned H2B sequences (not shown). *Names,* a selection of sequence descriptors found in the definition lines of the sequence records; *seq,* number of unique sequences in the group; *sp,* number of species in the group; *max sp/seq,* the greatest number of species having the same sequence in the group. For each group the first line is the consensus sequence for that group. Variations from the group consensus are indicated below it. Italic indicates a "singleton," i.e., the residue was found in only one sequence from one species in the group. An asterisk (*) indicates singleton identity or a gap. Background color key: white, identity to the anchored consensus; black, gap; orange, aromatic; yellow, aliphatic/hydrophobic; light green, glycine; green, hydrophilic; light blue, histidine; blue, basic; red, acidic. (See color insert.)

divergent subclass, so far found only in mammals, which displays a complementary localization to macroH2A, that is, it is excluded from inactive chromosomes.[18]

[17] D. T. Brown, *Genome Biol.* **2,** Reviews 0006 (2001).
[18] B. P. Chadwick and H. F. Willard, *J. Cell Biol.* **152,** 375 (2001).

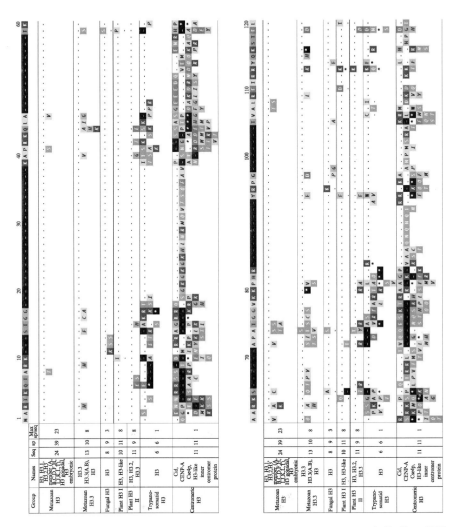

Fɪɢ. 3. Summary of H3 subclasses and variants. A consensus sequence of all aligned H3 sequences is shown at the top. Dots in the sequences below indicate identity to the consensus. Groups are named on the basis of clustering patterns observed in neighbor-joining trees of aligned H3 sequences (not shown). *Names,* a selection of sequence descriptors found in the definition lines of the sequence records; *seq,* number of unique sequences in the group; *sp,* number of species in the group; *max sp/seq,* the greatest number of species having the same sequence in the group. For each group the first line is the consensus sequence for that group.

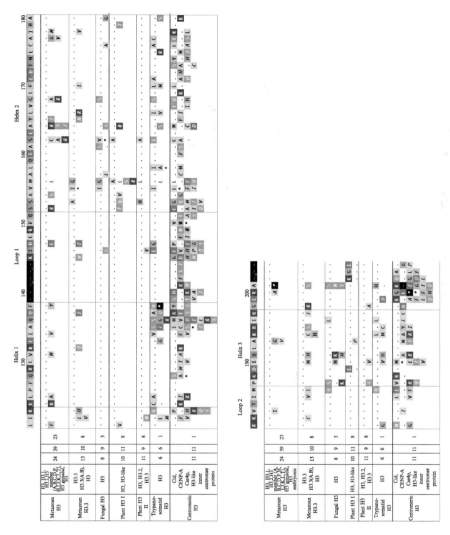

Variations from the group consensus are indicated below it. Italic indicates a "singleton," i.e., the residue was found in only one sequence from one species in the group. An asterisk (*) indicates singleton identity or a gap. Background color key: white, identity to the anchored consensus; black, gap; orange, aromatic; yellow, aliphatic/hydrophobic; light green, glycine; green, hydrophilic; light blue, histidine; blue, basic; red, acidic. (See color insert.)

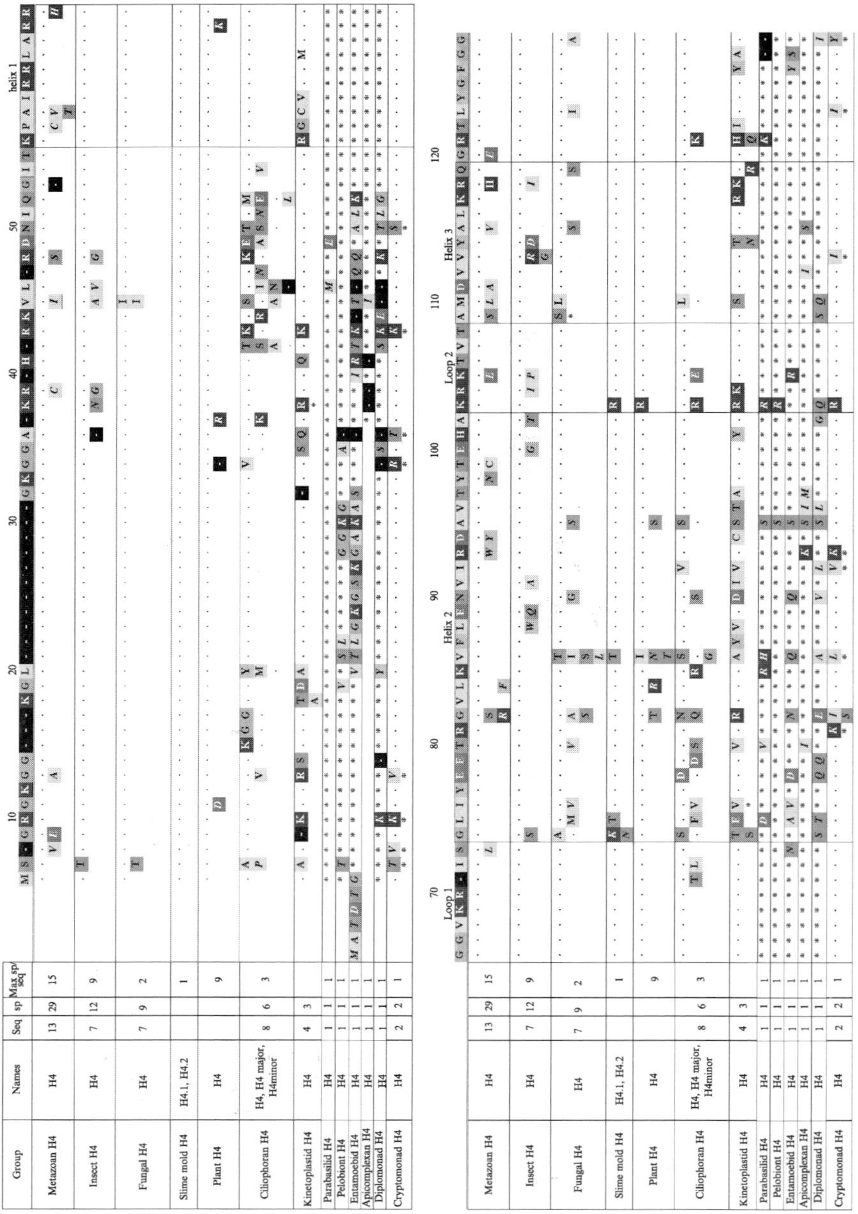

FIG. 4. *(continued)*

H2B

Functional subclasses of H2B sequences have not been positively identified, although at least one tissue-specific form has been identified in mammalian testis (Fig. 2). An echinoderm sperm variant featuring a repeating pentapeptide has also been described (reviewed in von Holt et al.[19]), indicating that the echinoderm group in Fig. 2 probably could be subdivided further. The N-terminal diversity seen within the plant subclass in Fig. 2 suggests that it, too, could be further subdivided.

H3

The H3 class notably contains two subclasses of replication-independent variants that are differentially localized within the cell. Histone H3.3 is an ostensibly homomorphous metazoan subclass that varies significantly from the predominant H3 in only four positions (positions 73, 153, 155, and 156 of Fig. 3). H3.3 can be deposited in nucleosomes of replicating DNA such as the major H3, but can also be deposited in nonreplicating DNA, preferentially in actively transcribed regions.[20] The replication independence of H3.3 may be mediated by any of the three H3.3-specific residues at positions 153–156.[21] Centromere-specific H3 is found in species ranging from yeast to human, and its deposition has been shown to be replication independent (reviewed in Smith[22]). It is thought to help specify centromere

[19] C. von Holt, W. N. Strickland, W. F. Brandt, and M. S. Strickland, FEBS Lett. **100**, 201 (1979).
[20] K. Ahmad and S. Henikoff, Proc. Natl. Acad. Sci. USA **99**(Suppl. 4), 16477 (2002).
[21] K. Ahmad and S. Henikoff, Mol. Cell. **9**, 1191 (2002).
[22] M. M. Smith, Curr. Opin. Cell Biol. **14**, 279 (2002).

Fig. 4. Summary of H4 subclasses and variants. A consensus sequence of all aligned H4 sequences is shown at the top. Dots in the sequences below indicate identity to the consensus. Groups are named on the basis of clustering patterns observed in neighbor-joining trees of aligned H4 sequences (not shown). *Names,* a selection of sequence descriptors found in the definition lines of the sequence records; *seq,* number of unique sequences in the group; *sp,* number of species in the group; *max sp/seq,* the greatest number of species having the same sequence in the group. For each group the first line is the consensus sequence for that group. Variations from the group consensus are indicated below it. Italic indicates a "singleton," i.e., the residue was found in only one sequence from one species in the group. An asterisk (*) indicates singleton identity or a gap. Background color key: white, identity to the anchored consensus; black, gap; orange, aromatic; yellow, aliphatic/hydrophobic; light green, glycine; green, hydrophilic; light blue, histidine; blue, basic; red, acidic. (See color insert.)

regional identity within the chromosome. Centromeric H3 displays some-what more subclass specificity (and considerably more diversity) within the histone fold than other H3 subclasses (Fig. 3), which may reflect a role in forming specialized nucleosomes.

H4

The H4 class is the most conserved of the four core histones. No functional, localization, or expression variants are known, and thus the clustering of its sequences falls entirely along phylogenetic lines (Fig. 4).

Complete alignments of all histone proteins can be found at http://genome.nhgri.nih.gov/histones/. Histones for the various species can be obtained by querying the Histone Sequence Database. The figures for this manuscript are available at http://www.ncbi.nlm.nih.gov/CBBresearch/Landsman/mie/.

Section II

Biochemistry of Histones, Nucleosomes, and Chromatin

[2] Reconstitution of Nucleosome Core Particles from Recombinant Histones and DNA

By Pamela N. Dyer, Raji S. Edayathumangalam, Cindy L. White, Yunhe Bao, Srinivas Chakravarthy, Uma M. Muthurajan, and Karolin Luger

Introduction

The ability to prepare nucleosome core particles (NCPs), or nucleosomal arrays, from recombinant histone proteins and defined-sequence DNA has become a requirement in many projects that address the role of histone modifications, histone variants, or histone mutations in nucleosome and chromatin structure. This approach offers many advantages, such as the ability to combine histone variants and tail deletion mutants, and the opportunity to study the effect of individual histone tail modifications on nucleosome structure and function.

We have previously described comprehensive protocols for the expression and purification of histones, for the refolding of the histone octamer, and for the reconstitution and purification of crystallization-grade mononucleosomes.[1] The previously published version has now been amended, and steps that can be omitted or simplified if high degrees of purity and homogeneity are not an issue are indicated. The cloning strategies for the construction of plasmids containing multiple repeats of defined DNA sequences, and the subsequent large-scale isolation of defined-sequence DNA for nucleosome reconstitution, are described in detail. We also describe adapted procedures to prepare nucleosomes with histones from other species, and for the refolding and reconstitution of (H2A–H2B) dimers and (H3–H4)$_2$ tetramers. Methods to reconstitute nucleosomes from different histone subcomplexes are also described. A flow chart for all procedures involved in the preparation of "synthetic nucleosomes" is given in Fig. 1. Procedures described here are indicated in gray in Fig. 1.

Cloning and Purification of Large Amounts of Defined-Sequence DNA

Cloning Strategy

A general procedure to construct a plasmid containing multiple repeats of a given DNA sequence, based on published strategies,[2,3] and to purify large amounts of defined-sequence DNA fragments is outlined below.

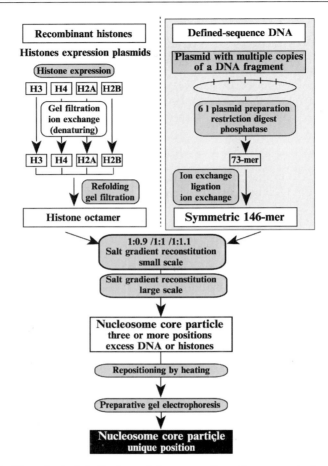

FIG. 1. Flow chart of methods used for preparation of components for nucleosomes. Procedures that are described in this chapter are shown in gray.

Figure 2 outlines the cloning strategy for fragments containing either the complete desired sequence (Fig. 2A), or one-half of a palindromic DNA fragment (Fig. 2B). Because of the recombination activities in most bacterial cells, long palindromic DNA fragments cannot be amplified, but must

[1] K. Luger, T. J. Rechsteiner, and T. J. Richmond, *Methods Enzymol.* **304,** 3 (1999).

[2] R. T. Simpson, F. Thoma, and J. M. Brubaker, *Cell* **42,** 799 (1985).

[3] T. J. Richmond, M. A. Searles, and R. T. Simpson, *J. Mol. Biol.* **199,** 161 (1988).

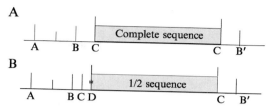

FIG. 2. Insert construction for the preparation of a defined-sequence DNA fragment. (A) Strategy for preparing an insert that encompasses the entire desired sequence (not suitable for palindromic sequences). (B) Strategy for designing inserts for ligation for palindromic (or partially palindromic) sequences. *, Site of large-scale ligation. Note that the two "halves" of the final product do not have to be identical if the restriction site for the final ligation step is chosen judiciously to prevent self-ligation (e.g., HinfI). A, Unique site; B and B', compatible cohesive ends; C, generates end(s) of actual fragments (large amounts needed); D, used for head–head ligation of two fragments; overhang can be chosen to allow or prohibit self-ligation.

be assembled by ligation of two halves. Figure 3 describes the strategy for duplication and outlines procedures for insert preparation. We use pUC-based vectors for these constructs.

In designing the cloning strategy for creating multiple DNA repeats, the DNA sequence of interest is flanked by restriction sites as shown in Fig. 2, where A is a unique site (e.g., KpnI), B and B' are sites for enzymes that are compatible, but nonidentical (e.g., BamHI and BglII), and C is a site for an enzyme that is used to excise the fragment from the plasmid (e.g., EcoRV). Here, blunt ends are desirable. If the final DNA fragment is to be generated by large-scale ligation of two shorter fragments (e.g., if palindromic 146-bp DNA fragments are the desired end-product), restriction enzyme D should generate overhangs suitable for high-efficiency ligation. We used EcoRI for a perfectly palindromic 146-bp DNA fragment,[4] and a HinfI site to generate 147-bp DNA fragments by ligation of two fragments.[5] Because large amounts of restriction enzymes cutting sites C and D will be used, economical considerations also come into play in the cloning strategy.

Digestion of the plasmid DNA with A and B creates a vector into which a fragment generated by A and B' can be ligated, destroying the restriction site at the B–B' junction (Fig. 3). Thus, with each cloning step, the number

[4] K. Luger, A. W. Maeder, R. K. Richmond, D. F. Sargent, and T. J. Richmond, Nature 389, 251 (1997).

[5] C. A. Davey, D. F. Sargent, K. Luger, A. W. Maeder, and T. J. Richmond, J. Mol. Biol. 319, 1097 (2002).

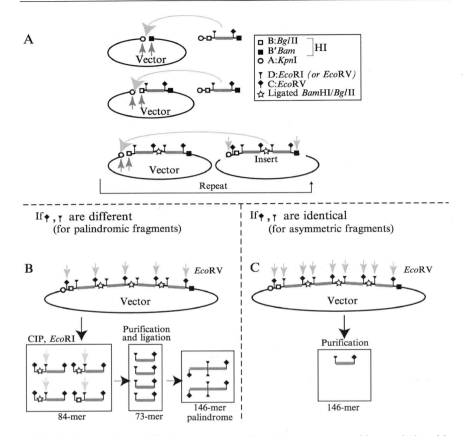

Fig. 3. Strategy for amplification and preparation of large amounts of inserts designed in Fig. 2. (A) Cloning and duplication strategy. Sites for restriction enzymes are indicated by symbols (see *inset* for legend). (B) Insert preparation from large-scale plasmid preparations (see text for details) for palindromic DNA fragments that undergo ligation. CIP, Incubation with calf intestine phosphatase. (C) Insert preparation for nonpalindromic DNA fragments that do not need to be self-ligated.

of inserts can be doubled. The individual steps for fragment insertion and amplification are described.

1. Synthesize and anneal pair(s) of suitable oligonucleotides (oligos). Follow standard cloning procedures to insert the fragment into a suitable high-copy plasmid via restriction sites A and B′.

2. Cut the plasmid containing the proper insert with restriction enzymes A and B (digest 1). Purify the vector DNA.

3. Cut the plasmid containing the insert with a second digest of restriction enzymes A and B' (digest 2). Purify the insert DNA away from the plasmid vector and keep the insert generated by the digest.

4. Ligate the insert DNA (created by digest 2) with the vector DNA (created by digest 1).

5. Repeat steps 2–4: Each repetition will duplicate the number of previously present insert copies. Depending on the length of the insert, about 16 to 24 inserts can be obtained easily. Use HB 101 cells or other host cells that are RecA minus for plasmid amplification. The following statistics give the experimental amplification efficiencies found by our laboratory for each doubling cycle: $1 \rightarrow 2$, ~100% efficiency; $2 \rightarrow 4$, ~70% efficiency; $4 \rightarrow 8$, ~60% efficiency; $8 \rightarrow 16$, ~40% efficiency.

6. Assay for total size of the insert by digestion with restriction enzymes A and B', and check for integrity of inserts by sequencing (early stages) and by cutting with C.

7. If efficiencies for duplication are low, try ligation of a 2-mer or 4-mer instead of duplication, to increase insert number.

Large-Scale Plasmid Purification

This method has been adapted from the original alkaline lysis protocol described earlier.[6] It has been optimized for high yields and purity of pUC-based plasmids, containing 24×146 bp (or 84-bp) inserts.

Equipment

Centrifuge
37° incubator/water bath
TSK-DEAE column
Orbital shaker
12 wide-bottom 4-L Fernbach flasks

Buffers and Reagents

Alkaline lysis solution I: 50 mM glucose, 25 mM Tris-HCl (pH 8.0), 10 mM EDTA (pH 8.0)

Alkaline lysis solution II: 0.2 N NaOH, 1% (w/v) sodium dodecyl sulfate (SDS)

Alkaline lysis solution III: 4 M potassium acetate, 2 N acetic acid

Ampicillin (100-mg/ml stock solution, sterile filtered)

Calf intestine alkaline phosphatase (CIAP; Roche Molecular Bio-chemicals, Indianapolis, IN)

[6] J. Sambrook and D. Russell, "Molecular Cloning: A Laboratory Manual." Cold Spring Harbor Laboratory Press, Cold Spring Harbor, NY, 2001.

CIA: Chloroform–isoamyl alcohol (24:1, v/v)
*Eco*RI (~100,000 U/ml)
*Eco*RV (~100,000 U/ml)
100% ethanol, ice cold
Isopropanol
Calbiochem Miracloth (EMD Biosciences, San Diego, CA)
4 *M* NaCl, autoclaved
3 *M* Sodium acetate (pH 5.2), autoclaved
PAGE [10% polyacrylamide, 0.2 × Tris–borate–EDTA (TBE)]
40% PEG 6000, autoclaved
Phenol, Tris–EDTA (TE) equilibrated
RNase A (DNase free[6])
TE 10/0.1: 10 m*M* Tris-HCl (pH 8.0), 0.1 m*M* Na-EDTA; autoclaved
TE 10/50: 10 m*M* Tris-HCl (pH 8.0), 50 m*M* Na-EDTA; autoclaved
T4 DNA ligase (200,000 U/ml)
Terrific broth (TB): 1.2% (w/v) Bacto Tryptone, 2.4% (w/v) yeast
extract, 0.4% (v/v) glycerol. Adjust autoclaved and cooled medium
to a final concentration of 17 m*M* KH_2PO_4 and 72 m*M* K_2HPO_4

Plasmid Purification

1. Inoculate each of four 5-ml precultures containing TB (or 2× TY; see Histone Expression and Purification, below) and ampicillin (100 μg/ml) with a colony from a freshly transformed plate. Shake for 3–4 h at 37°. Transfer all precultures to a 500-ml flask containing 100 ml of 2× TY and ampicillin (100 μg/ml), and incubate for 2–3 h at 37° until turbid. Do not grow to saturation. Transfer equal amounts of the preculture to 12 Fernbach flasks containing 500 ml of TB and ampicillin at 100 μg/ml. Incubate under vigorous shaking for 16–18 h at 37°. Harvest cells by centrifugation in 500-ml centrifuge bottles. Fresh weight yields ~125 g of cells. Cells should be processed immediately for optimal yields.

2. Resuspend cells from 6 liters of cell culture in a total of 360 ml of alkaline lysis solution I by passage through a 10-ml plastic pipette. Redistribute the cells equally back into the six centrifuge bottles. Add 120 ml of alkaline lysis solution II to each bottle. Mix by shaking vigorously at least 20 times, until the thick translucent suspension is completely free of any clumps of cells. Incubate on ice, and shake repeatedly for a total of 10 to 20 min. Break up large clumps that still remain after such treatment by passage through a 10-ml disposable plastic pipette.

3. Carefully pour 210 ml of ice-cold alkaline lysis solution III down the side of each bottle. Mix by inverting and swirling 10 times and incubate on ice for 20 min. This step is critical because plasmid DNA is renatured,

whereas chromosomal DNA precipitates. Viscosity is reduced dramatically during this step. Low yields, or large amounts of chromosomal DNA in the plasmid preparation, may result if mixing is done too slowly.

4. Centrifuge at 10,000g for 20 min at 4°. Warm the rotor to 20° by running empty at 8000g for 15 min. Pour the supernatant through Miracloth to remove remaining precipitate, and add 0.52 volume of isopropanol. Let stand at room temperature for 15 min.

5. Centrifuge at 10,000g for 30 min at 20° to collect the precipitate. Air dry for 30 min to 1 h. Using a clean spatula, distribute pellets between two 30-ml centrifuge tubes. Use 5 ml of TE 10/50 to rinse out centrifuge bottles, and adjust each tube to a final volume of 20 ml. Mix the DNA into a homogeneous solution, and then add 120 μl of RNase A (10 mg/ml) (an RNase A stock of 1.2 Kunitz units/μl should be diluted to 1:100 in relation to the final reaction, $\sim$0.01 Kunitz unit/μl reaction mix) and incubate at 37° overnight. The pellets should have dissolved completely. (Store at $-20°$ as necessary.)

6. If the suspension is viscous, dilute with TE 10/50 buffer to up to twice the volume. Extract each 20 ml of suspension with 10 ml of phenol. Centrifuge at 27,000g for 20 min at 20°. The DNA will be in the upper, aqueous phase, separated from the phenol phase by a thick white interphase. Repeat two more times or until the interface is clear. Extract the aqueous phase with 10 ml of CIA. Spin for 5 min (12,000g, 20°). Transfer the aqueous phase into a 50-ml centrifuge tube and adjust to a final volume of 30 ml with TE 10/50.

7. Precipitate plasmid DNA by adding one-fifth of the original volume of 4 M NaCl (to give 0.5 M NaCl) and two-fifths of 40% PEG 6000 [to give 10% (w/v) PEG 6000]. Mix at 37° for 5 min and incubate on ice for 30 min.

8. Centrifuge at 3000g in a swinging-bucket tabletop centrifuge for 20 min at 4°. Decant the supernatant, which contains RNA. Dissolve the pellets in a total of 15 ml of TE 10/0.1 (overnight at room temperature or for less time at 37°). Check both fractions by agarose gel electrophoresis. Fractionation should be complete, and there should be no traces of RNA visible in the plasmid fraction.

9. Extract two times with 10 ml of CIA to remove PEG. Ethanol precipitate DNA by addition of a 1/10 volume of 3 M sodium acetate (pH 5.2) and 2.5 volumes of 100% cold absolute ethanol. Pellet the DNA, dissolve in 10 ml of TE 10/0.1 by incubating for 1 to several hours at 37°, and determine the total concentration. Yields are usually between 150 and 200 mg.

Purification of Insert. Experimental details in this section depend on the restriction sites that were chosen in the design of the plasmid. Given the large amounts of DNA present, restriction digests can routinely be performed at plasmid concentrations of 1 mg/ml. Most restriction enzymes are more efficient under these conditions. Optimize reaction conditions before proceeding with large-scale digestions. Below we give conditions that were used for isolation of the palindromic 146-bp DNA fragment derived from human α-satellite DNA that is routinely used for crystallography.[4]

1. The insert is excised with EcoRV, at a concentration of 1 mg/ml plasmid, in sterile 50-ml centrifuge tubes. Use 30 units of EcoRV per nanomole of EcoRV site. Incubate at 37° for at least 16 h, and then check for completion by gel electrophoresis on 10% polyacrylamide gels (0.2× TBE). If the digest is not complete, add 50% more restriction enzyme and incubate for another 15 h. Check the digest as described above.

2. Separate the excised EcoRV fragment from the linearized plasmid by PEG precipitation. Add 0.192 volume of 4 M NaCl and 0.346 volume of 40% PEG 6000. Incubate on ice for 1 h and spin down the vector DNA at 27,000g and 4° for 20 min. Precipitate the EcoRV fragment contained in the supernatant by the addition of 2.5 volumes of 100% cold ethanol. Air dry the DNA briefly (~10 min) and dissolve in 5 ml of TE 10/0.1.

3. Determine the concentration. Check both precipitated PEG supernatant and PEG pellet on a 1% agarose gel and PAGE as described above (run series of 1:10 dilutions). There should be no cross-contamination between the two fractions. Yields should be close to 90% (i.e., if the fragments encompass 40% of the entire plasmid, ~40 mg of excised fragment should be obtained 100 mg of plasmid). Note: This procedure will not work for DNA fragments with sticky ends.

4. If the cloned fragment represents the entire sequence, either use as is (after phenol extraction and ethanol precipitation), or purify further by ion-exchange chromatography. If further cutting and ligation are required, proceed with step 5.

5. Dephosphorylate EcoRV fragment by combining EcoRV fragment (1 mg/ml) with calf intestine alkaline phosphatase (CIAP, 1 U/nmol of DNA end; Roche), using the conditions given by the manufacturer. Incubate at 37° for 24 h, and then add 50% of the original amount of CIAP and incubate for another 24 h at 37°. Complete phosphorylation is essential, because self-ligation of the blunt ends during subsequent steps needs to be avoided. If in doubt, perform a small-scale assay for blunt-end ligation. None should occur if dephosphorylation is complete.

6. Inactivate the CIAP by extracting the DNA solution two times with 50% of the original volume of phenol–CIA (1:1 mixture) and then ethanol precipitate by addition of a 1/10 volume of 3 M sodium acetate (pH 5.2)

and 2.5 volumes of cold ethanol. Spin down the precipitated DNA at 3000*g* (swinging bucket tabletop centrifuge), air dry the pellet briefly, and dissolve in 5 ml of TE 10/0.1.

7. To create cohesive ends for self-ligation, use *Eco*RI at 20–30 U/nmol of *Eco*RI site (substrate concentration, 1 mg/ml) and incubate at 37° for at least 15 h. Check completion of the digest by PAGE. Make sure the digestion is complete before proceeding with the next step.

8. FPLC purify the fragment by chromatography over a TSK-DEAE column (the sample can be loaded directly, or it can be ethanol precipitated to reduce the volume). Ethanol precipitate the FPLC fractions (no need to add salt), air dry the pellet briefly, and dissolve it in ~5 ml of TE 10/0.1 or 1× ligation buffer (see below). Yields are typically 85% of the starting amount.

9. Perform a small-scale ligation to test whether ligation can be driven to completion and to assess whether phosphorylation of *Eco*RV ends was complete. The latter should be visible in the formation of a ladder as a result of blunt-ended tail–tail ligation of the *Eco*RV fragments. Use ~0.5 U of ligase per microgram of fragment, at a substrate concentration of 1 mg/ml, under conditions as given by the manufacturer. Incubate at room temperature for at least 15 h, and check completion of ligation by PAGE. Add more ligase if necessary.

10. If necessary, purify ligated from unligated fragments by ion-exchange chromatography on a TSK-DEAE column (or another ion-exchange column of similarly high resolution). This separation depends strongly on the DNA sequence and must be optimized individually.

Histone Expression and Purification

These procedures, which utilize expression vectors for *Xenopus laevis* histones[7] have been described extensively.[1] We have since used this protocol to express and purify various H2A and H3 histone variants from different species (e.g., Suto *et al.*[8]), and histones from yeast (White *et al.*[9]; also see Wittmeyer *et al.*[10]), *Drosophila,* and mouse. All these histones have been subcloned in untagged form into the pET vector series (Novagen,

[7] K. Luger, T. J. Rechsteiner, A. J. Flaus, M. M. Waye, and T. J. Richmond, *J. Mol. Biol.* **272,** 301 (1997).

[8] R. K. Suto, M. J. Clarkson, D. J. Tremethick, and K. Luger, *Nat. Struct. Biol.* **7,** 1121 (2000).

[9] C. L. White, R. K. Suto, and K. Luger, *EMBO J.* **20,** 5207 (2001).

[10] J. Wittmeyer and T. Formosa, *Methods Enzymol.* **262,** 415 (1995).

Madison, WI). Histidine-tagged histones are also purified in the same way. In some cases, codon usage has been optimized for *Escherichia coli,* and the time after induction as well as the bacterial strain have been optimized for each case. In some cases, better results are obtained with BL21(DE3) strains that compensate for poor codon usage. All expressed proteins are invariably expressed in insoluble form and isolated from the insoluble fraction obtained after cell lysis (inclusion bodies).

Equipment

Dialysis tubing (6- to 8-kDa cutoff, 2.5- to 4-cm flat width)
Ion-exchange column, TSK SP-5 PW resin material
Lyophilizer
Orbital shaker
Peristaltic pump
Sephacryl S-200 high-resolution gel-filtration column (5×100 cm; Pharmacia, Uppsala, Sweden)
6 wide-bottom Fernbach flasks (4 L)
Tissumizer (Tekmar, Cincinnati, OH) or sonicator for cell lysis

Buffers and Reagents

Ampicillin (100-mg/ml stock solution, sterile filtered)
2-mercaptoethanol (2-ME)
BL21(DE3)pLysS, BL21(DE3)pLysS Codonplus or BL21(DE3) cells, competent
Isopropyl-β-D-thiogalactopyranoside (IPTG)
Centrifuge tubes, 50 ml
Chloramphenicol stock solution, 25 mg/ml in ethanol
Dimethyl sulfoxide (DMSO)
Glucose
Lysozyme
Liquid nitrogen
SDS–PAGE equipment: Standard equipment, 18% SDS gels
TYE agar plates: 1.0% (w/v) Bacto Tryptone, 0.5% (w/v) yeast extract, 0.8% (w/v) NaCl, 1.5% (w/v) agar, ampicillin (100 μg/ml), and chloramphenicol (25 μg/ml)
$2\times$ TY: 1.6% (w/v) Bacto Tryptone, 1.0% yeast extract, 0.5% NaCl, with antibiotics and 0.1% glucose
Unfolding buffer: 6 M guanidinium-HCl, 20 mM Tris-HCl (pH 7.5), 5 mM dithiothreitol (DTT)
Wash buffer: 50 mM Tris-HCl (pH 7.5), 100 mM NaCl, 1 mM benzamidine, 1 mM 2-ME

Histone Expression

1. Transfect BL21(DE3)pLysS cells with 0.1 to 1 μg of the pET-histone expression plasmid and plate on TYE agar plates with ampicillin (100 μg/ml) and chloramphenicol (25 μg/ml). Incubate at 37° overnight. For best and most reproducible results, a new transformation should be done each night for the protein that is expressed the next day. For some histones, BL21(DE3)pLysS Codonplus (RIL) or BL21(DE3) cells will give better results.

2. Expression conditions depend on the histone in question and should be optimized individually. For most histones, conditions given in Luger *et al.*[7] are adequate.

3. Inoculate each of four preculture tubes (4 ml of 2× TY with antibiotics and 0.1% glucose) with one colony from the culture plate. Incubate in a shaker at 37°.

4. When preculture tubes appear slightly turbid (2–3 h), add the contents of all four tubes to a flask containing 100 ml of 2× TY with appropriate antibiotics and glucose. Incubate in a shaker at 37°. For most reproducible results, do not let precultures grow to saturation.

5. When the 100-ml flask has reached an OD_{600} of ~0.4, distribute the contents evenly into six wide-bottom Fernbach flasks containing 1 liter each of 2× TY medium and appropriate antibiotics and glucose. Incubate in a shaker at 37° until the OD_{600} reaches about 0.4. Induce expression by addition of IPTG to a final concentration of 0.2–0.4 mM.

6. After 2 h, harvest the cells at room temperature and resuspend the cell pellets in a total of 35 ml of wash buffer. Flash freeze in liquid nitrogen and store at −20° in a 50-ml centrifuge tube.

Note. Cells expressing histone proteins (especially H4) are prone to lysis and should be centrifuged at room temperature. For the same reason, it is not recommended (or necessary) that the cell pellet be washed. Resuspend the cells well before freezing, as this will improve lysis on thawing. The cell suspension can be stored at −20 or −70°.

Inclusion Body Preparation

1. Lyse the cell suspension by thawing at 37°.

2. Pour the cell extracts into 250-ml centrifuge bottles. At this point, the cells should be viscous. If the cell suspension is still watery, then full lysis has not occurred. In this case, or if no pLysS plasmid has been present, add lysozyme to a concentration of 1 mg/ml and incubate on ice for 30 min. Repeated freeze–thaw cycles also facilitate lysis. Bring the total volume to 100 ml.

3. Blend the cell extracts with the Tissumizer to reduce viscosity. Blend until viscosity is reduced; avoid overheating of sample. A sonicator can also be used with similar results.

4. Spin at 4° for 20 min at 12,000g. Pour off the supernatant and resuspend the tight, solid pellet with 75 ml of wash buffer containing 1% Triton X-100. If the pellet is "spongy," sonicate/blend (Tissumizer) again. Spin for 20 min as described previously.

5. Repeat once as described above and once with wash buffer without Triton X-100. The drained pellet can be stored for a limited time at −20°.

Histone Purification

A two-step purification procedure yielding up to 1 g of highly pure histone protein from 6 liters of induced cells has been described previously.[1] The purification protocol involves gel filtration and HPLC/ion-exchange chromatography under denaturing conditions. If purity is not a major concern, one of the chromatography steps (usually the ion-exchange chromatography) can be omitted. The gel-filtration column can be scaled down accordingly if only small amounts of histones are purified. The purified proteins can be stored as lyophilisates for extended periods of time, to be used in refolding reactions as described subsequently.

Refolding of Histone Octamer

All possible combinations of recombinant *Xenopus laevis* full-length and globular domain histone proteins, as well as histone octamers from other species, or containing histone variants, can be refolded to functional histone octamers according to a previously described protocol.[1] The method works best for 6 to 15 mg of total protein; the limiting factor here is the size of the gel-filtration column. Much smaller samples can be prepared when using an analytical column. Some applications require the preparation of H2A–H2B dimers and $(H3–H4)_2$ tetramers. The same protocols can be used for refolding and purification of these histone subcomplexes.

Equipment

Dialysis tubing (6- to 8-kDa cutoff, 2.3-cm flat width)
HiLoad 16/60 Superdex 200 HR preparation-grade gel-filtration column (Pharmacia), equipped with UV detector and fraction collector
SDS-PAGE equipment: Standard equipment, 18% SDS gels

Concentration device: Devices suitable for up to 25-ml volumes [e.g., Centricon centrifugal filter devices; Amicon Bioseparations (Millipore, Bedford, MA)]

Buffers and Reagents

Purified and lyophilized histones (3- to 4-mg aliquots)

Unfolding buffer: 6 M guanidinium chloride, 20 mM Tris-HCl (pH 7.5), 5 mM DTT. Needs to be made fresh for good refolding efficiency

Refolding buffer: 2 M NaCl, 10 mM Tris-HCl (pH 7.5), 1 mM Na-EDTA, 5 mM 2-ME

Histone Octamer Refolding

1. Dissolve each histone aliquot to a concentration of approximately 2 mg/ml in unfolding buffer. Unfolding should be allowed to proceed for at least 30 min and for no more than 3 h. Determine the concentration of the unfolded histone proteins by measuring absorbance of the "undiluted" solution against unfolding buffer at 276 nm (remove any undissolved particulate matter by centrifugation, if necessary). Extinction coefficients can be obtained (see Table I for full-length *Xenopus* and yeast histones) or calculated (for histones from other species or histone variants) using the following Web site: http://ca.expasy.org/tools/protparam.html. *Note:* Using correct extinction coefficients is essential for good yields in refolding.

2. Mix histone proteins to exactly equimolar ratios and adjust to a total final protein concentration of 1 mg/ml, using unfolding buffer. Dialyze at 4° against at least three changes of 600 ml of refolding buffer (at least 6 h each; the second or third step should be overnight). Histone octamer should always be kept at 0–4° to avoid dissociation.

3. Remove any precipitated protein by centrifugation. Concentrate to a final volume of approximately 1 ml, using the concentration device. Histone octamers refolded with tailless histones often stick to the filter membrane of the concentration device and take a much longer time to concentrate. Make sure the octamer solution is mixed (pipette up and down) to avoid clogging filtration devices.

4. Load samples onto the gel-filtration column previously equilibrated with refolding buffer as described.[1] High molecular weight aggregates will elute after about 45 ml, histone octamer at 65 to 68 ml, (H3–H4)$_2$ tetramer at about 72 ml, and histone (H2A–H2B) dimer at 84 ml (Fig. 4).

5. Check the purity and stoichiometry of the fractions by 18% SDS–PAGE. Dilute sample by a factor of at least 2.5 before loading onto the gel to reduce distortion of the bands resulting from the high salt concentration.

TABLE I
MOLECULAR WEIGHTS AND MOLAR EXTINCTION COEFFICIENTS (ε) FOR FULL-LENGTH
Xenopus laevis AND *Saccharomyces cerevisiae* HISTONE PROTEINS

	Full-length *Xenopus* Histone		Full-length *S. cerevisiae* histone	
Histone	Molecular weight	ε (cm/M), 276 nm	Molecular weight	ε (cm/M), 276 nm
H2A	13,960	4050	13,858	4350
H2B	13,774	6070	14,106	7250
H3	15,273	4040	15,225	2900
H4	11,236	5400	11,237	5800

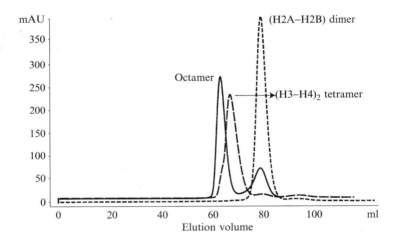

Fig. 4. Elution profile of histone subcomplexes from a Superdex S-200 gel-filtration column. See text for details. Histone octamer (solid line) elutes first, in accordance with its molecular weight (108,500). A small excess of H2A and H2B is apparent in the formation of a small dimer peak, which can be separated from the main peak by this method. In contrast, (H3–H4)$_2$ tetramer (MW, 53,000) elutes close to the octamer peak (dashed line). Note the small shoulder indicative of some octamer-like assemblies formed by (H3–H4)$_2$ tetramer. H2A–H2B dimer (MW, 27,000; dotted line) elutes last.

If octamer contains globular H3 histone, be aware that globular histone H3 comigrates with full-length H4, and only two bands will be seen on the gel.[7]

6. Pool fractions containing octamer and concentrate, using the concentration device, to 3–15 mg/ml. Determine the concentration of the octamer spectrophotometrically. Extinction coefficients can be

approximated by adding up those of individual histones (times two). Yields of pure histone octamer are usually between 50 and 75% of the input material (yields may be lower for octamer containing tailless histones).

7. Octamer can be stored on ice (for short-term storage), or at $-20°$ as a 50% (v/v) glycerol solution (for long-term storage). Octamer containing globular histones usually is stable only for a few months and can often form higher order aggregates on storage for longer periods of time. Octamer stored on ice can be used as such for nucleosome reconstitution. Concentrations of octamer in 50% glycerol are extremely inaccurate, and pipetting of accurate amounts is difficult. When stored in glycerol, dialyze octamer overnight at $4°$ against refolding buffer before use for nucleosome reconstitution and redetermine the concentration spectrophotometrically.

Reconstitution of Nucleosome Core Particles

In vitro (and *in vivo*) reconstitution of nucleosomes relies on the sequential binding of one (H3–H4)$_2$ tetramer and two H2A–H2B dimers onto the DNA. This may be achieved in two different ways. Salt gradient deposition[3,11] relies on the fact that (H3–H4)$_2$ tetramers bind at higher salt concentrations than do H2A–H2B dimers. Chaperone-assisted assembly makes use of specific histone–chaperone complexes that ensure the ordered addition of histone complexes onto the DNA.[10]

We have previously described a detailed method for the large-scale assembly of NCPs, using salt gradient deposition.[1] Here we describe three methods to reconstitute nucleosomes. Microscale reconstitutions (1 μg) are routinely done with radiolabeled DNA. Small-scale reconstitutions (25–100 μg) are used to carefully titrate histones and DNA to optimize yields for subsequent large-scale reconstitutions (0.5–4 mg).

We note that identical methods can be used with refolded and purified H2A–H2B dimers and (H3–H4)$_2$ tetramers instead of octamers, with essentially the same results (Fig. 5). However, special care must be taken to combine H2A–H2B dimer, (H3–H4)$_2$ tetramer, and DNA at a molar ratio of 2:1:1, because excess dimers and tetramers can lead to the formation of aggregate or of nonnative nucleosome species. These species are apparent by their relatively intense staining with ethidium bromide as compared with Coomassie Brilliant Blue, and by the inability to reposition to the energetically favored position by "heat shifting."

[11] J. O. Thomas and P. J. G. Butler, *J. Mol. Biol.* **116,** 769 (1977).

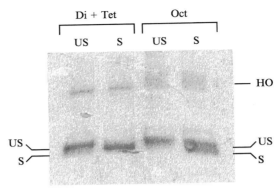

FIG. 5. Nucleosome core particles formed with DNA and octamer, or with (H2A–H2B) dimers and (H3–H4)$_2$ tetramer, are identical. High-resolution gel-shift assays (see text for details) demonstrate that the final nucleosome core particle product is independent of the type of histone subcomplexes used for assembly. US and S, samples before and after a 2-h incubation at 37°, respectively; HO, higher order aggregates, which are easily removed by preparative gel electrophoresis (Fig. 6). The gel was stained with Coomassie Brilliant Blue.

Equipment

Dialysis tubing (6- to 8-kDa cutoff, 1 or 2.3-cm flat width); dialysis membrane cut into a circle with a radius of 3 cm

Concentration device: Devices suitable for up to 25-ml volumes [e.g., Centricon centrifugal filter devices from Amicon Bioseparations (Millipore); Vivaspin devices from ISC Bioexpress (Kaysville, UT)]

Microdialysis devices that hold total volumes of 5–350 µl [e.g., dialysis buttons from Hampton Research (Laguna Hills, CA)]

Reconstitution apparatus with connected tubing, as introduced in Luger et al.[1]

Peristaltic pump with a double pump head, capable of maintaining a constant flow rate of 1–6 ml/min [e.g., Econo pump (Bio-Rad, Richmond, CA)]

Prep Cell apparatus: Model 491 Prep Cell (Bio-Rad) with a standard power supply, connected to a UV detector (e.g., Econo UV monitor; Bio-Rad), a fraction collector (e.g., model 2110 fraction collector; Bio-Rad), a chart recorder (e.g., model 1327 Econo recorder; Bio-Rad), and equipped with a peristaltic pump

Standard PAGE apparatus, 5% polyacrylamide gels (acrylamide–bisacrylamide 59:1, 0.2× TBE, 10 cm × 10 cm × 1.5 mm)

Buffers and Reagents

0.2× TBE (Prep Cell electrophoresis buffer) (2000 ml)

4 M KCl or 5 M NaCl stock solution

CCS (long-term storage buffer): 20 mM potassium cacodylate (pH 6.0), 1 mM EDTA

Purified octamer (3–15 mg/ml) or dimers and tetramers; DNA (3–6 mg/ml)

Reconstitution and storage buffers (make and prechill buffers at 4°):

    RB-high (reconstitution buffer): 2 M KCl, 10 mM Tris-HCl (pH 7.5), 1 mM EDTA, 1 mM DTT (400 ml)

    RB-low (reconstitution buffer): 0.25 M KCl, 10 mM Tris-HCl (pH 7.5), 1 mM EDTA, 1 mM DTT (2000 ml)

    TCS (short-term storage buffer, Prep Cell elution buffer): 20 mM Tris-HCl (pH 7.5), 1 mM EDTA, 1 mM DTT

Make sure all buffers are made fresh. During reconstitution of NCP with octamer containing globular histones, make the above-described buffers with 5 mM DTT (instead of 1 mM DTT).

Microscale reconstitution

Microscale reconstitution is a good method for reconstitution with radiolabeled DNA.[12,13] If performed at ambient temperatures, only one species will be observed by gel electrophoresis, and the "heat-shifting" step (see below) can be omitted. At 4°, different translational positions are observed, and heat shifting can be studied.

1. Mix 1 μg of radioactively labeled DNA in ~9 μl of 2 M NaCl (bringing the NaCl concentration to 2 M with 5 M NaCl), and then add the appropriate amount of octamers.

2. Incubate for 30 min at ambient temperature, then add an equal volume (10 μl) of 10 mM Tris-HCl, pH 7.6, and incubate for 1 h either at 4° or at room temperature.

3. The following additions of 10 mM Tris-HCl, pH 7.6, are each for 1 h: 5 μl (→0.8 M NaCl); 5 μl (→0.67 M NaCl); 70 μl (→0.2 M); and 100 μl (→0.1 M; optional).

4. Analyze by native PAGE as described below, and autoradiograph the gel.

---

[12] J. M. Gottesfeld, C. Melander, R. K. Suto, H. Raviol, K. Luger, and P. B. Dervan, *J. Mol. Biol.* **309**, 625 (2001).

[13] J. M. Gottesfeld and K. Luger, *Biochemistry* **40**, 10927 (2001).

Small-Scale Reconstitution of NCP

Small-scale reconstitution works well for amounts of NCP 25 and 500 μg. Multiple setups can be dialyzed in one vessel. The efficiency of reconstitution of octamers containing different histones (e.g., full-length histones, globular histones, and histone variants) into NCPs can vary, mainly because of inaccuracies in the concentration of histone octamer. If reconstitutions are performed with refolded H2A–H2B dimer and (H3–H4)$_2$ tetramers, titration of the three components is essential. An excess of DNA results in nonnative nucleosomal species that cannot subsequently be removed, whereas an excess of histones results in low yields. Hence, small-scale reconstitutions are recommended to establish the relative molar ratios of DNA to octamer [or of DNA, H2A–H2B dimer, and (H3–H4)$_2$ tetramer] before a large-scale experiment.

1. Titrations are performed by varying the molar ratio of DNA to histone complexes. A typical experiment using histone octamer includes three small-scale setups of 0.9:1.0, 1.0:1.0, and 1.1:1.0 DNA-to-octamer ratios in a volume of 100 μl or smaller. If H2A–H2B dimer and (H3–H4)$_2$ tetramers are used for reconstitution, the three components must be titrated accurately. Make sure to adjust to 2 M KCl or NaCl before adding histone octamer. The final DNA concentration is between 0.2 and 0.7 mg/ml (ideally, 0.7 mg/ml).

2. Use the reconstitution method and apparatus described below for large-scale reconstitution. The last dialysis step may be omitted if time is critical. Alternatively, stepwise dialysis against subsequent changes of more dilute buffers can be used. Dialyze at 4° (or at ambient temperatures; see above) against 300 ml each of TCS buffer containing 2, 0.85, 0.65, and 0.2 M KCl (at least 90 min per step).

3. Remove the contents from the dialysis buttons, incubate one-third each for 2 h at 37 and 55°, respectively (spin frequently to collect condensation), and leave one-third on ice.

4. Analyze the products by using the high-resolution gel-shift assay described earlier[1] and later in the chapter. Sample that has not been heat treated is run as a control. For subsequent large-scale reconstitutions, choose conditions that (1) contain only a small (if any) excess of free DNA and (1) are completely shifted to a single position.

Large-Scale Reconstitution

Large-scale reconstitution and purification of up to 4 mg of nucleosomes require about 5 days and involve the following steps.

1. Octamer and DNA are mixed at 2 M KCl. Make sure to adjust the salt concentration of the DNA to 2 M, using 4 M KCl. Always add histone octamer last. The mixture can incubate at 4° while the dialysis apparatus is being set up. Adjust to a final DNA concentration of 0.7 mg/ml with RB-high.

2. Set up the dialysis apparatus at 4° as described in Luger et al.,[1] but calibrate the peristaltic pump to a flow rate of 1.5 ml/min. This reduces the reconstitution time from 36 to 18 h. Transfer the sample to a dialysis bag and start dialysis against 400 ml of RB-high at 4° under constant stirring. Make sure the dialysis bag also spins vigorously to ensure mixing of the contents. The apparatus is set up in such a manner that the pump continuously replaces buffer from the dialysis vessel with RB-low.

3. After the gradient has finished, dialyze for at least 3 h against 400 ml of RB-low. If the samples are to be stored without any further purification, dialyze against an appropriate low-salt buffer (include DTT and 0.1 mM buffered cacodylic acid) and store at 4°. If samples will be further purified, dialyze against TCS buffer and store at 4° until the next step.

High-Resolution Gel Shift and Heat Shifting of NCPs

Reconstitution on longer DNA fragments usually results in a heterogeneous population of NCP with respect to the position of the DNA on the histone octamer. Surprisingly, this also holds true for DNA fragments with a limiting length of 146 bp, even if presumed "strong positioning sequences" are used. A simple heating step (37–55° for 20–180 min) results in a uniquely positioned NCP preparation for DNA 145 to 147 bp in length. Repositioning can be monitored by a high-resolution gel shift assay described below (e.g., Fig. 5). Incubation time and temperature necessary for repositioning depend on the sequence and the length of the DNA fragment, and must be checked individually for each combination of DNA fragment and histone octamer. For example, *Xenopus laevis* full-length histone octamer with the 146-bp fragment derived from the 5S RNA gene of *Lytechinus variegatus* is heated for 30 min at 37° for a complete shift, whereas other sequences might require as long as 2 h at 55°.

1. Prerun a 5% polyacrylamide gel (10 × 8 × 0.15 cm gel: 5% polyacrylamide; 59:1 acrylamide to bisacrylamide; 0.2 × TBE) for at least 1 h at 4° and 150 V.

2. Mix the buffer from the two chambers and redistribute. This significantly improves the resolution. Alternatively, use a gel apparatus in which the contents of the upper and lower chambers are recirculated continuously.

3. Shortly before loading the samples, the wells should be rinsed well with 0.2 × TBE.

4. Load 3–4 pmol of NCP [mixed in sucrose to a final concentration of 5% (v/v) sucrose for gel loading] in no more than 10 μl. Traces of bromphenol blue can be added for easier loading.

5. Run the gel at 150 V for a suitable length of time, or until bromphenol blue has reached the bottom of the gel.

6. Stain the gel first with ethidium bromide. Note that free DNA is stained significantly better by ethidium bromide than is DNA bound to the histone octamer. Subsequent staining with Coomassie Brilliant Blue sometimes gives better resolution on slightly overloaded gels because of the limited sensitivity of Coomassie Brilliant Blue compared with ethidium bromide; however, free DNA will not be evident in Coomassie Brilliant Blue staining.

Purification of NCP by Preparative Gel Electrophoresis

The method relies on the differential migration of free DNA, NCP, and high molecular weight aggregates on nondenaturing polyacrylamide gels. It gives rise to highly pure NCP preparations suitable for crystallization, is only marginally affected by covalent modification of the histones, and can be used for nucleosomes that are unstable at slightly elevated salt concentrations. The method works well for amounts between 200 μg and 2 mg. If larger amounts are purified, perform two runs with the same gel. For best results, the gel should be no more than 24 h old.

Given below are conditions that have been optimized for the purification of NCP containing 146 bp of DNA. Conditions for preparative gel electrophoresis should be optimized by analytical nondenaturing gel electrophoresis (see earlier), following the guidelines given in the instruction manual for the model 491 Prep Cell (Bio-Rad). In our hands, the correlation between analytical and preparative gels has been excellent, and the described protocol works well to separate nucleosomes from free DNA and aggregates. Note that the ratio between acrylamide and bisacrylamide, the length of the gel, and the elution speed can greatly alter the relative mobility and the separation of the components (Fig. 6). The choice of elution buffer and electrophoresis buffer may also influence the relative mobility of the different species. Improved resolution between different peaks is often a tradeoff with high dilution of the sample. For example, we are able to partially separate shifted from unshifted nucleosomes using longer gels (7.5 cm instead of 5 cm). Figure 6A shows fractions from preparative gel electrophoresis using a standard 5-cm gel, and Fig. 6B shows the same sample run on a 7.5-cm gel. Note the improved separation

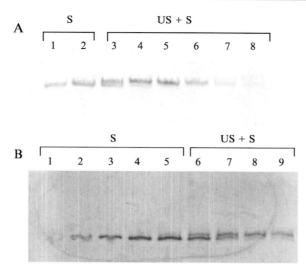

Fig. 6. Preparative gel electrophoresis is capable of separating different nucleosome species. Fractions from a 5-cm (A) or 7.5-cm (B) preparative gel are analyzed by high-resolution gel-shift assay. Note that the separation between shifted and unshifted (S and US, respectively) species is improved, although still incomplete, when a longer gel is used. This improved separation is not apparent when analyzing the OD_{260} chromatogram of the eluting material (not shown). Gels were stained with Coomassie Brilliant Blue.

between unshifted and shifted nucleosome core particles, which cannot be obtained by any other method. However, high dilution of NCP during purification using longer gels might result in a partial dissociation of DNA and octamer, and thus a 5-cm gel is the correct choice for most applications.

1. Prepare 20 ml of a 5% polyacrylamide gel mixture (same conditions as that for smaller 5% gels), and pour a cylindrical gel with an outer radius of 28 mm, an inner radius of 19 mm, and a height of 50 mm. Polymerize overnight at room temperature while recirculating water through the cooling core, and assemble the apparatus at 4° according to instructions given in the manual for the model 491 Prep Cell. Connect to the power supply, UV detector, fraction collector, and peristaltic pump. Use the circular dialysis membrane (see Materials). Prerun the gel under constant recirculation of the buffer for 90 min in 0.2 × TBE (2000 ml) at 4° and at a power of 10 W. Record a baseline at 260 nm, using TCS buffer (TCS buffer used during reconstitution can be reused for this purpose) as elution buffer.

2. Concentrate NCP (in TCS buffer) to a maximum of 600 μl for a 4-mg reconstitution. Mix with sucrose to a final concentration of 5% (v/v)

and load on the preparative gel, using a syringe with an attached piece of tubing. Electrophoresis is carried out at constant power of 10 W, and the complex is eluted at a flow rate of 1.0 ml/min with TCS buffer as the elution buffer and 0.2× TBE as the electrophoresis buffer. Record the elution at an OD of 260 nm, and collect fractions of appropriate size (0.7 to 2 ml). Free DNA will appear first, followed by NCP, and finally, higher molecular weight aggregates.

3. Analyze fractions on a 5% nondenaturing gel and pool peak fractions corresponding to NCP. Concentrate NCP immediately to at least 1 mg/ml.

4. Dialyze (23-mm dialysis bags; MWCO, 6000–8000 Da) semiconcentrated NCP against CCS buffer (400 ml for 3 h at 4° each time—change the buffer three times).

5. Finally, concentrate NCP to the desired final concentration (typically, 6–10 mg/ml for crystallization purposes) and store at 4°. NCP purified by this method can be stored up to several months.

Acknowledgment

We thank Joel Gottesfeld (Scripps Research Institute) for developing the protocol for microscale reconstitution.

[3] Preparation and Crystallization of Nucleosome Core Particle

By B. Leif Hanson, Chad Alexander, Joel M. Harp, and Gerard J. Bunick

Structural Biology and Chromatin Studies

The last half-decade has seen the development of experimentally determined atomic position models for the nucleosome core particle (NCP), based on palindromic DNA engineered from a human X chromosome alphoid satellite DNA repeat.[1] Modifications of that structure, such as NCP with variant histones and other DNA sequences, are being studied. One can anticipate that future diffraction-based studies of chromatin will range from further DNA and histone variants of the basic NCP structure to gene

[1] J. M. Harp, E. C. Uberbacher, A. Roberson, and G. J. Bunick, *Acta Crystallogr. D. Biol. Crystallogr.* **52**, 283 (1996).

modulation complexes and higher order structures. The primary value of molecular models is the heuristic support they provide to biological questions. For example, examining how and where the histone tails interact with DNA may lead to a better understanding of the role of specific histone variants. Equally important, some questions may be resolved only by experimentally determined structures. After more than 20 years and numerous studies, uncertainty remains concerning the position of H1 in the chromatosome. A crystallographically determined structure possibly will be required to elucidate the details of H1 binding in the chromatosome and higher order chromatin.

The role of sample preparation and purification cannot be overestimated in experimental structural biology studies, especially for the crystallization of macromolecules. Purity, homogeneity, conformational state, and sample yield are major considerations in the preparation and purification of biological materials for structural studies. This chapter delineates some of the techniques used to produce large quantities of crystallizable material for chromatin studies. The scale of the procedures can be adapted to suit the situation. The first protocol described is for the isolation of histones from eukaryotic sources. Subsequent topics include isolation and purification of defined sequence DNA from recombinant sources. Reconstitution of NCP from histone octamer and defined sequence DNA is detailed, followed by descriptions for crystallizing NCP. Other reviews of the production of crystallizable chromatin have been published[2,3] and the reader is urged to consult those articles as well. The techniques cited here, unless otherwise stated, have been developed or modified in our laboratory during the past 20 years of crystallographic study of the nucleosome.[4]

Materials and Instrumentation

Materials

Two white Leghorn hens from a poultry supply facility [Cherokee Poultry Unit, University of Tennessee (Knoxville, TN) School of Agriculture]

Escherichia coli DH5α cells containing pET plasmid (Stratagene, La Jolla, CA), Epicurian Coli SURE strain with pBluescript II KS(+) (Stratagene)

---

[2] K. Luger, T. J. Rechsteiner, and T. J. Richmond, *Methods Enzymol.* **304,** 3 (1999).

[3] L. M. Carruthers, C. Tse, K. P. Walker, III, and J. C. Hansen, *Methods Enzymol.* **304,** 19 (1999).

[4] J. M. Harp, B. L. Hanson, D. E. Timm, and G. J. Bunick, *Acta Crystallogr. D. Biol. Crystallogr.* **55,** 1329 (2000).

Ceramic hydroxyapatite (CHT) (Macro-Prep type II, 20-μm particle size; Bio-Rad, Hercules, CA), Superose 12 in a prepacked HR 10/30 column (Pharmacia, Uppsala, Sweden), Sephacryl S-400HR (Pharmacia)

Centriprep -10, -30 and Centricon -10, -30 microconcentrators (Amicon/millipore, Bedford, MA)

Scalpels

40-ml Oak Ridge tubes

15- and 50-ml Falcon tubes, 500-ml polycarbonate tubes, Eppendorf (Hamburg, Germany) Phase Lock Gel tubes (heavy)

Spectra/Por 1 dialysis tubing (6000–8000 molecular weight cutoff) (Spectrum Laboratories, Rancho Dominguez, CA)

Enzymes

Lysozyme (Sigma, St. Louis, MO)

Micrococcal nuclease (Worthington Biochemical, Lakewood, NJ)

RNase A (Sigma)

*Eco*RI, *Eco*RV, T4 DNA ligase (Stratagene)

Reagents

Benzamidine, 2-mercaptoethanol (2-ME), ethidium bromide, ethylenediaminetetraacetic acid (EDTA), ethylenebis (oxyethylenenitrilo) tetraacetic acid (EGTA)

Halothane, heparin, Nonidet P-40 detergent, phenylmethylsulfonyl fluoride (PMSF)

Spermidine, spermine, sucrose, phenol, chloroform, ampicillin, PEG 8000

Buffers

B1 (histone wash buffer, 4 liters): 15 mM Tris-HCl, 15 mM NaCl, 60 mM KCl, 0.5 mM spermidine, 0.15 mM spermine, 0.34 M sucrose, 2 mM EDTA, 0.5 mM EGTA, 1 mM benzamidine 15 mM 2-ME, 0.2 mM PMSF; pH 7.4 at 4°

B2 (histone lysis buffer, 4 liters): 15 mM Tris-HCl, 15 mM NaCl, 60 mM KCl, 0.5 mM spermidine, 0.15 mM spermine, 0.34 M sucrose, 2 mM EDTA, 0.5 mM EGTA, 0.5% Nonidet P-40, 1 mM benzamidine, 15 mM 2-ME, 0.2 mM PMSF; pH 7.4 at 4°

B3 (histone digestion buffer, 600 ml): 50 mM Tris-HCl, 25 mM KCl, 4 mM MgCl$_2$, 1 mM CaCl$_2$, 1 mM benzamidine, 0.2 mM PMSF; pH 8 at 4° (500 ml) and 37° (100 ml)

B4 (histone extraction buffer, 2 liters): 10 mM Tris-HCl, 0.25 mM EDTA, 1 mM benzamidine, 0.2 mM PMSF; pH 7.4 at 4°

B5 (histone second extraction buffer, 2 liters): 1 mM Tris-HCl, 0.1 mM EDTA, 0.2 mM PMSF; pH 7.4 at 4°

B6 (histone finish buffer, 2 liters): 10 mM Tris-HCl, 1 mM benzamidine, 0.2 mM PMSF; pH 7.4 at 4°

B7 (histone chromatography buffers for CHT): 10 mM sodium phosphate, pH 7.4. Elute with 0.7 M NaCl, 10 mM sodium phosphate, pH 7.4, followed by 3 M NaCl, 10 mM sodium phosphate, pH 7.4

B8 (histone octamer buffer): 2 M NaCl, 20 mM Tris-HCl, 0.1 mM EDTA, 0.1 mM PMSF; pH 8.0

P1 (cell lysis buffer 1): 50 mM sucrose, 25 mM Tris, 10 mM EDTA (pH 8), lyzozyme (1 mg/ml) and RNase A (10 μg/ml)

P2 (cell lysis buffer 2): 0.2% NaOH, 1% sodium dodecyl sulfate (SDS)

P3 (cell lysis equilibration buffer): 3 M potassium acetate plus 170 ml glacial acetic acid

TE buffer: 10 mM Tris, 1 mM EDTA, pH 8

P4 (plasmid precipitation buffer): 20% PEG 8000, 1.8 M NaCl

TBE buffer: 89 mM Tris-borate, 2 mM EDTA, pH 8.3

N1 (nucleosome reconstitution starting buffer): 2 M KCl, 20 mM Tris-HCl, 1 mM EDTA, 1 mM 2-ME, 0.04% NaN$_3$, 0.1 mM PMSF, 0.01% Nonidet P-40; pH 8.0 at 22°

N2 (nucleosome gradient buffer): 0.3 M KCl, 20 mM Tris-HCl, 1 mM EDTA, 1 mM 2-ME, 0.04% NaN$_3$, 0.1 mM PMSF, 0.01% Nonidet P-40; pH 8.0 at 29°

N3 (nucleosome final buffer): 50 mM KCl, 20 mM Tris, 1 mM EDTA, 1 mM 2-ME, 0.04% NaN$_3$, 0.1 mM PMSF, 0.01% Nonidet P-40; pH 8.0 at 29°

Crystallization buffer for NCP: 50 mM KCl, 10 mM potassium cacodylate, 0.1 mM EDTA, pH 6.0

Growth Medium

TB (terrific broth plus, 1 liter): 6 g of tryptone, 12 g of yeast, 1.5 g of NaCl, 16 ml of glycerol, diluted with water to 900 ml. Add 100 ml of 0.72 M K$_2$HPO$_4$, 0.17 M KH$_2$PO$_4$; pH 7. Add 50 mg of ampicillin

Instrumentation

Sorvall RC-5C centrifuge (Kendro Laboratory Products, Asheville, NC)

Heraeus Contifuge 17RS tabletop centrifuge (Kendro Laboratory Products)

Pharmacia fast protein liquid chromatograph (FPLC)

Bio-Rad electrophoresis apparatus

Model 3927 incubator (ThermoForma, Marietta, OH)
Tissue homogenizer (Tissuemizer; Tekmar, Cincinnati, OH)
Dounce homogenizer (Wheaton, Millville, NJ)

Histone Isolation and Purification from Chicken Erythrocytes

Histone octamer can be prepared by gentle salt extraction from soluble chromatin. Core histones are separated from DNA, using a linear NaCl gradient in hydroxyapatite chromatography. Stable octamer is then reassembled by dialysis against NaCl and purified by gel filtration. These are standard protocols for isolation of histones.[5-8] Chicken erythrocytes are nucleated, providing a readily available source of inactive chromatin. The main limitations for using chicken as a source for histones is access to live birds and permitted anesthetics for use during exsanguination. An alternative to in-house harvest of blood from chickens is to collect fresh material at a slaughterhouse for fowl. Balanced against collection of greater volumes of material is the limited access to clean working space in an abattoir. Commercially collected blood samples have been tested and shown to contain unacceptable levels of histone degradation, despite precautions taken at the source.

Preparation of Soluble Chromatin

Buffers are prepared the day before the preparation and the pH is adjusted at the 4° working temperature. All buffers are filtered with Millipore (Bedford, MA) filters of at least 0.45 μm and stored at 4°. B1 and B2 buffers may need to be gently heated to dissolve the sucrose. Nonidet P-40 is not added until after the sucrose has been dissolved and the buffers have been filtered. PMSF and 2-ME are added immediately before use. PMSF can be made up in a 100 mM stock solution in a 100-ml bottle of anhydrous isopropanol. The PMSF may precipitate at 4°, and thus may need warming to dissolve solid particles. Prepare two 500-ml plastic beakers containing 5000 units of heparin, 2 ml of B1 buffer, and 0.5 ml of 100 mM PMSF and store on ice.

To administer anesthesia, 2–3 ml of halothane is poured over gauze into a tall form beaker lined with a plastic bag. The chicken's head is placed in

[5] D. R. Hewish and L. A. Burgoyne, *Biochem. Biophys. Res. Commun.* **52**, 504 (1973).
[6] L. C. Lutter, *J. Mol. Biol.* **124**, 391 (1978).
[7] C. von Holt, W. F. Brandt, H. J. Greyling, G. G. Lindsey, J. D. Retief, J. D. Rodrigues, S. Schwager, and B. T. Sewell, *Methods Enzymol.* **170**, 431 (1989).
[8] J. Ausio, F. Dong, and K. E. van Holde, *J. Mol. Biol.* **206**, 451 (1989).

the beaker and the bag is partially closed. The bird will begin to breathe heavily and the eyes should close. Careful observation of the breathing is needed to avoid causing respiratory arrest and to maintain adequate anesthesia. When the bird fails to respond to stimuli, the feathers surrounding the neck are plucked to reveal the jugular vein beneath the skin. A 5- to 8-cm square window of skin is cut away and the vein is gently lifted away from the neck. The fascia attached to the vein are dissected away with a No. 11 scalpel blade. Once a 4-cm length is free, the vein is cut and the bird is held by the head and legs above the work area so that the blood can be collected into the beaker containing the B1–heparin–PMSF buffer on ice. The beaker should be swirled during the collection to mix the blood and heparin thoroughly. It is important for good yield to provide gentle cyclical chest compression to the bird during the bleeding. The average volume of blood per chicken is 50 ml, but occasionally 75–100 ml is obtained from one bird. Veterinarian-performed cardiac puncture has proved less successful than exsanguination from the jugular vein. A cervical dislocation is administered to the fowl after exsanguination.

The blood is filtered through sterile cheesecloth into 40-ml Oak Ridge tubes on ice and centrifuged at $5000g$ in a Heraeus 3746 rotor in a Heraeus Contifuge 17RS centrifuge for 10 min at 4°, removed, and placed on ice. The liquid layer and the buffy coat are removed by aspiration (it is better to lose some erythrocytes than to leave behind residual leukocytes). The remaining erythrocytes are suspended in B1 buffer and transferred to 500-ml polycarbonate bottles. These bottles should be dedicated solely for this usage. Aliquot the cells into six bottles and centrifuge at 5000 rpm for 10 min in a Sorvall RC-5C. Decant the supernatant and resuspend the cells in B1 buffer at 4° and spin. Repeat for a total of three washes.

The washed cells are suspended in B2 buffer, using the Tissuemizer at low speed to help lyse the cells and ensure resuspension. The suspension is centrifuged at 5000 rpm for 10 min. Repeat the suspension and washing three times, reducing the number of centrifuge bottles until only two are used in the last centrifugation. After completing the lysis washes, the nuclear pellet should be creamy white in color. Next, the nuclear pellet is suspended and washed in B3 buffer. The supernatant is discarded and the pellet is suspended in 20 ml of B3 buffer at 37°, for digestion with micrococcal nuclease. The DNA content is estimated from the measurement of A_{260} of a sample prepared by adding 10 μl of well-suspended nuclei to 990 μl of 0.1 N NaOH. The amount of nuclease used is based on the amount of DNA in the sample, assuming that 40 units of micrococcal nuclease is needed for each milligram of DNA. The material from two chickens typically requires 3000–4000 units of micrococcal nuclease. Micrococcal nuclease digestion proceeds at 37° for 30 min with gentle shaking in a water bath. After 30 min, the reaction is stopped by adding 0.4 ml of 0.5 M EDTA (pH 8.0).

The material is centrifuged and the nuclear pellet is transferred into the Dounce homogenizer after suspension in 15 ml of B4 buffer. The nuclear material is homogenized until the nuclear membranes are completely disrupted. The sample is transferred into dialysis tubing and dialyzed against B4 buffer overnight at 4° (with two changes of buffer). The following day the suspension is harvested from dialysis, placed in Oak Ridge tubes, and spun at 10,000g in a Heraeus 3746 rotor. The soluble chromatin is decanted and the quantity of extracted material is measured optically at 260 nm in a UV spectrophotometer. For this material, 1 AU is approximately equal to a soluble chromatin concentration of 10 mg/ml. The reserved pellet is suspended in B5 buffer and the dialysis step is repeated. The soluble chromatin material is pooled and the material is dialyzed into B6 buffer to remove the EDTA and prepare the sample for chromatographic extraction of histones.

Chromatography to Extract Histones and Purify Octamer

Histones are separated from DNA, using a column of ceramic hydroxyapatite in an FLPC system. A hydroxyapatite column (16 mm × 10 cm) is usually sufficient to purify 100 mg of soluble chromatin. The column is equilibrated with B7 buffer and then loaded with up to 100 mg of sample. The eluent from the column is monitored at 280 nm. After returning to baseline, an NaCl gradient is used to elute the histones from the column. Lysine-rich histones (H1 and H5) are eluted first with B7 buffer containing 0.7 M NaCl, followed by H2A–H2B dimer. After the dimer has eluted, the NaCl concentration in buffer B7 is increased to 3 M to elute the majority of the octamer. The elution profile is shown in Fig. 1. The dimer and octamer peaks should be pooled. The concentration of material can be determined spectrophotometrically, using a value of 0.43 AU for 1 mg/ml at 280 nm. A Centricon-10, Centriprep-10, or Amicon pressure cell concentrator can be used to concentrate the pooled histones.

The concentrated octamer should be reassembled overnight by dialysis against buffer B8. The reassembled material can be purified by size-exclusion chromatography (SEC), using a column packed with Superose-12HR resin. An elution profile from the octamer SEC run and an SDS–polyacrylamide gel of the purified product are shown in Fig. 2. The finished product is concentrated and stored at −20° until needed for reconstitution of nucleosomes or crystallization of histones. The stored octamer aliquots often exhibit a pronounced concentration gradient and therefore should be carefully but thoroughly mixed after being brought to room temperature.

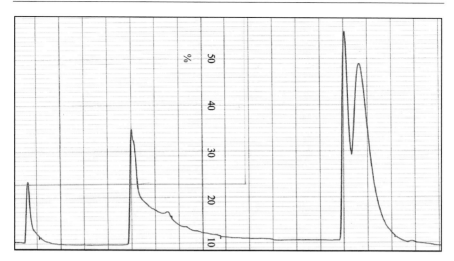

FIG. 1. A_{280} absorbance line tracing of elution from ceramic hydroxyapatite FPLC column. The left peak is the elution of H1 and H5, the middle peak is H2A–H2B dimer, and the right peak shows the elution of octamer (sharp peak is H2A–H2B dimer-depleted octamer and the broad peak is octamer). Fractions from the middle and right peaks are pooled before passage through the final purification column.

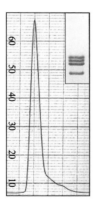

FIG. 2. A_{280} absorbance line tracing of elution from Superose 12 FPLC column, with SDS gel lane of peak material (*inset*). The peak contains purified octamer, and shows no depletion of H2A–H2B dimer or residual H1 and H5, which are present in the shoulder to the right of the main peak.

The sample should also be analyzed by SDS–polyacrylamide gel electrophoresis (PAGE) to check for proteolysis and assessment of stoichiometry.

Alternative Isolation Protocol

An alternative protocol for the isolation of histone octamer from chicken erythrocytes has been developed.[9] This procedure is less labor intensive (6 h of bench work) but requires access to an ultracentrifuge. The procedure can be implemented before the addition of B3 buffer in the previous protocol. Two parts 4 M NaCl is mixed with 1 part saturated ammonium sulfate at 4° and then swirled quickly into a suspension of 2×10^9 chick erythrocyte nuclei/ml at a 3:2 ratio of salt solution to nuclei. This forms a gel, which after being left on ice for ~1 h becomes cloudy with the precipitation of nonhistone nucleoproteins. No enzymatic lysis of the chromatin is performed. The sample is centrifuged at 100,000g for 18 h in an ultracentrifuge with a Beckman (Fullerton, CA) Ti45 rotor. The pellet is discarded and the supernatant contains only the core histones H2A, H2B, H3, and H4, plus the linker histones H1 and H5. The supernatant should have a UV spectrophotometer trace that peaks at a wavelength of 277 nm with no absorbency due to DNA at 260 nm.

The supernatant containing histones is concentrated to 10 mg/ml, using an Amicon pressurized cell concentrator with a YM10 10-kDa molecular mass cutoff membrane. An equal volume of cold saturated ammonium sulfate is then added dropwise while stirring on ice. The suspension is centrifuged at 100,000g for 1 h. The supernatant contains primarily the linker histones plus small quantities of histones H2A and H2B and is discarded. The pellet, containing only the core histones, with some excess of histones H3 and H4, is dissolved in 10 mM Tris-HCl (pH 7.4), 2 M KCl and the procedure is repeated for further purification of the histone core octamers.

Production of Recombinant Histones

The expression of histones in bacterial culture overcomes some problems with isolating material from higher organisms, related to animal welfare regulations and histone homogeneity. Recombinant material will not have N- or C-terminal tail modifications that can cause heterogeneity in isolated vertebrate material. However, unless grown by large-scale fermentation, the yields from recombinant histone production are most often less than can be derived from chicken erythrocyte preparations. Yields can

[9] S. J. Lambert, J. M. Nicholson, L. Chantalat, A. J. Reid, M. J. Donovan, and J. P. Baldwin, *Acta Crystallogr. D. Biol. Crystallogr.* **54**, 1048 (1999).

often be improved by optimizing the gene sequence for proper bacterial codon use. Systems for overexpression of soluble protein complexes, such as H3–H4 tetramer, have been reported.[10] Individual cloned histones are produced as inclusion bodies within the bacterial cell, and after production and purification are recombined to form histone octamer. A detailed discussion of the recombinant histone production process is described elsewhere.[2]

Defined Sequence DNA Production

Large-scale production of palindromic DNA is detailed in Palmer *et al.*[11] The primary feature of this method is the production of a defined half-palindrome DNA sequence in bacteria that can be ligated *in vitro* to produce a palindrome. Palindromic DNA on nucleosome core particles used for crystallization are less disordered and diffract to higher resolution. Such DNA sequences will serve as the basis for future structural studies of higher order chromatin assemblages.

Plasmid Construction

One-half of the desired DNA with a restriction site sequence at the palindromic center is ligated at the multiple cloning site located in the polylinker sequence of a pBluescript II KS(+) (Stratagene) plasmid. This provides adjacent *Eco*RI and *Eco*RV cutting sites. Cloning using T4 DNA ligase to ligate the fragment into prepared vector DNA and to transform the *E. coli* host cells for maintenance and production of DNA is done according to standard protocols.[12] The Epicurian Coli SURE strain (Stratagene) has been used for production because of its deficiency in homologous recombination. After the first copy of target DNA is cloned, adjacent restriction sites are used to increase the number of target DNA copies in the same plasmid. In pBluescript II we chose two restriction enzymes producing blunt ends, *Sma*I and *Hin*CII, one on either side of the DNA insert, and an additional enzyme site *Xba*I distal from the *Hin*CII site. After amplification, the plasmid containing the half-palindrome is separated into two process streams for further manipulation. One aliquot of plasmid has the half-palindrome sequence excised with *Sma*I and *Xba*I. The other aliquot is used to prepare vector for the *Sma*I–*Xba*I fragment by cutting with

[10] S. Tan, *Protein Expr. Purif.* **21,** 224 (2001).
[11] E. L. Palmer, A. Gewiess, J. M. Harp, M. H. York, and G. J. Bunick, *Anal. Biochem.* **231,** 109 (1995).
[12] J. Sambrook, E. F. Fritsch, and T. Maniatis, "Molecular Cloning: A Laboratory Manual," 2nd Ed. Cold Spring Harbor Laboratory Press, Cold Spring Harbor, NY, 1989.

*Hin*CII and *Xba*I, and removing the resulting small fragment. The *Sma*I–*Xba*I fragment is then ligated into the plasmid opened by *Hin*CII and *Xba*I, yielding two copies of the half-palindrome in direct tandem, separated by a 30-bp spacer. The ligation results in the destruction of the *Hin*CII site of the original plasmid construct, while a new copy is carried along with the *Sma*I–*Xba*I fragment. The sequence bounded by the *Sma*I–*Xba*I restriction fragment can be doubled by repeating this cloning procedure. The insert-doubling procedure is repeated with the two-copy plasmid to create a four-copy insert, each in direct tandem and separated by a spacer. This procedure is repeated three more times, generating plasmids with 8, 16, and 32 copies of the half-palindrome.

Production and Isolation of Half-Palindrome DNA Plasmid

Bacterial fermentation is done in 1 liter of TB containing ampicillin (50 μg/ml) in 2-liter baffle flasks. Plasmids containing multiple copies of the DNA sequence are used to transform Epicurian Coli SURE competent cells and an inoculum is prepared by incubating the cells overnight in TB containing ampicillin (50 μg/ml) at 37° in 15-ml sterile tubes with loose caps. Each 1-liter culture is inoculated with 2 ml of overnight culture and allowed to ferment with shaking until the cells grow out of log phase and plateau (usually overnight). The bacteria are then harvested by centrifugation at 4° in a Sorvall GS3 rotor at 5000 rpm for 45 min.

The bacterial pellet is then lysed by first suspending the pellet in 100 ml of sterile P1 at 4°. After supension, 100 ml of P2 is added. The bottle is gently inverted five times to mix the contents without shearing the chromosomal DNA and allowed to sit at room temperature for 10 min. After 10 min, 100 ml of P3 at 4° is added and mixed gently, and then stored at 4° for 30 min. The cell debris is removed by straining the lysate through several folds of cheesecloth, and the lysate is treated with 0.7 volume of isopropanol (to precipitate the plasmid DNA). The mixture is allowed to sit for 10 min at room temperature and is then centrifuged at 7000 rpm in a Sorvall GS3 rotor for 30 min at room temperature. Room temperature is necessary to prevent salt precipitation. The supernatant is discarded, and the bottle is placed in a vacuum desiccator until the pellet is dry.

The dried sample pellets are pooled into a single container by suspension in 50 ml of TE with ribonuclease A (20 μg). This solution should digest at room temperature for at least 30 min. After this digestion, about 33 ml of P4 is added dropwise while stirring. The P4 is added to the point at which the plasmid is visibly precipitating in "threadlike" strands. The sample is then stored at 4° overnight to complete the precipitation. The plasmid is then harvested from the precipitation by centrifugation for 1 h at 7000 rpm in a Sorvall GS3 rotor at 4°. The supernatant should be analyzed on a gel to assure that all the plasmid has precipitated.

The pellet is suspended in 20 ml of TE and transferred to a 50-ml sterile Falcon tube. To this is added 20 ml of phenol (equilibrated in 100 mM Tris-HCl, 0.1% hydroxyquiniline buffer, pH 7.8). The sample is vortexed to mix well. The Falcon tube is then spun in a clinical centrifuge for several minutes, and the upper aqueous layer is removed and retained. The organic phase is discarded. The aqueous phase is then mixed with 20 ml of phenol–chloroform (1:1 mixture equilibrated in 100 mM Tris-HCl, pH 7.8), and vortexed. The sample is centrifuged, the aqueous phase is separated and retained, and then mixed with 20 ml of chloroform, vortexed, and centrifuged. The aqueous phase is again separated and retained, using Eppendorf Phase-Lock Gel tubes (heavy), and the volume of the aqueous sample is determined. The plasmid is precipitated with a 1/10 volume of 3 M sodium acetate, pH 5.5, and 2 volumes of 100% ethanol. The plasmid is harvested by centrifugation at 7000 rpm in a Sorvall GS3 rotor at 4° for 60 min, the supernatant is discarded, and the pellet is dried in a vacuum desiccator. The dry pellet is dissolved in TE and transferred to a sterile 50-ml Falcon tube. This material is stored at −20° until isolation of the DNA fragments. The yield of plasmid can be determined spectrophotometrically and the quality of the plasmid isolation is established by electrophoresis in 1% agarose. A typical yield from 18 liters of bacterial culture is ~400 mg of plasmid.

Isolation of Half-Palindrome from Plasmid

The half-palindrome DNA is excised from the plasmid, using *Eco*RI followed by *Eco*RV restriction enzymes. This cleaves the half-palindrome fragment from the vector and spacer DNA. Because of the cost of restriction enzymes, all enzymatic stages are preceded by pilot reactions. The pilot reactions are used to calibrate the amount of enzyme for the digest. The completeness of the reaction is assayed before moving to the next step in the protocol. The plasmid is always cut first with *Eco*RI because of the inefficiency of *Eco*RV when cutting supercoiled plasmid. The protocol used with the restriction enzymes is always that of the manufacturer. After the initial *Eco*RI cut of the pooled plasmid sample has been assayed and deemed acceptable, NaCl is added to the material to meet the conditions needed for *Eco*RV activity. Typically, 25,000 units of *Eco*RI and 10,000 units of *Eco*RV are used for an 18-L bacterial culture preparation (~400 mg of plasmid). Pilot reactions of *Eco*RI and *Eco*RV are shown in Fig. 3.

The vector DNA is separated from the half-palindrome fragment by precipitating the vector with 20% PEG 8000–1.8 M NaCl, added dropwise.

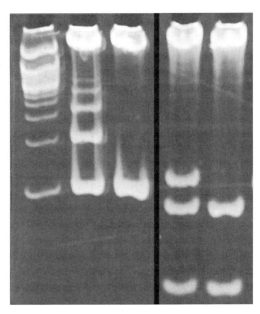

FIG. 3. Composite image showing ethidium bromide-stained DNA from pilot reactions used to assess amounts of *Eco*RI and *Eco*RV needed to isolate the half-palindrome from the plasmid. From left to right: DNA ladder standard; incomplete *Eco*RI digestion of plasmid; complete *Eco*RI digestion of plasmid; incomplete *Eco*RV digestion of *Eco*RI-digested plasmid; complete *Eco*RV digestion showing half-palindrome and spacer DNA.

A final concentration of 5% PEG 8000 is usually sufficient for precipitation. The material is stored at 4° overnight and then spun at 7000 rpm in a Sorvall GS3 rotor at 4° for 60 min. The supernatant is retained and then precipitated with 3 volumes of cold 100% ethanol and spun at 7000 rpm in a Sorvall GS3 rotor at 4° for 60 min. The half-palindrome DNA pellet is then dissolved in TE buffer and the residual PEG is removed by extracting twice with an equal volume of chloroform, using Eppendorf Phase-Lock Gel tubes (heavy). The half-palindrome DNA is again precipitated with a 1/10 volume of 3 M sodium acetate (pH 5.5) and 3 volumes of 100% ethanol. The sample is again centrifuged as previously described and the pellet is dried in a vacuum dessicator. The dried pellet is dissolved in 0.4 M NaCl, 20 mM Tris-HCl at pH 8.0, and the half-palindrome is purified by FPLC ion-exchange chromatography on a MonoQ HR16/10 column using a linear gradient from 0.4 to 1.0 M NaCl. The peak fractions are pooled and precipitated with a 1/10 volume of 3 M sodium acetate (pH 5.5) and 2.5 volumes of cold 100% ethanol and stored overnight at 4°. The sample is then centrifuged as previously described and the pellet is dried in the desiccator. The pellet is dissolved in H_2O, and the sample purity and concentration are established by PAGE and spectrophotometric absorbance.

Ligation and Purification of Palindrome

The ligation and purification of the half-palindrome fragments to make a uniform blunt-end palindrome requires ligation with T4 ligase, cutting with EcoRV, and purification of the full palindrome by ion-exchange chromatography. The purified half-palindrome is ligated into large multimers by T4 DNA ligase overnight at $17°$ in a reaction containing 0.2 unit of T4 DNA ligase per milligram of DNA fragment, 50 mM Tris-HCl (pH 7.5), 10 mM $MgCl_2$, 10 mM dithiothreitol, 1 mM adenosine triphosphate, bovine serum albumin (25 μg/ml), and a 5% final concentration of PEG 4000, which facilitates blunt-end ligations. As previously, a pilot reaction is performed to determine the lowest enzyme concentration needed for complete reaction of the half-palindrome into larger DNA fragments. After the ligation, the DNA is precipitated with a 1/10 volume of 3 M sodium acetate (pH 5.5) and 2.5 volumes of cold 100% ethanol, and incubated overnight at $4°$. After pelleting and drying, as described previously, the material is dissolved in TE and then supplemented with EcoRV digestion buffer (provided by the manufacturer). The sample is digested at $37°$ overnight with EcoRV. After precipitation and pelleting, the sample is dissolved in 0.4 M NaCl, 20 mM Tris-HCl at pH 8.0 and the sample is purified on a MonoQ 16/10 column as described in the previous section. The yield is determined spectrophotometrically by measuring the A_{260} (OD_{260} 1 = 50 μg/ml).

Nucleosome Reconstitution

Reconstitution of nucleosome core pare particles is initiated by combining purified nucleosomal DNA and purified histone octamer in approximately equimolar amounts in a high-salt solution. The ionic strength of the solution is then reduced to allow the DNA to bind to the octamer core. As with the DNA protocols, pilot reactions must be performed to calibrate the input ratio of DNA to octamer. The appropriate ratio of DNA to histone octamer is 1.13:1. Pilot samples should be varied around this value with a starting NaCl concentration of 1 M in a 1/10 concentration of TBE. Holding the pilot reaction material at $37°$, every 20 min add sufficient $0.1\times$ TBE so that in five steps the final NaCl concentration is 0.1 M. An aliquot of each pilot reaction is run on a 6% nondenaturing polyacrylamide gel in $0.1\times$ TBE, and migrating samples are visualized with ethidium bromide. The input ratio for the large-scale reaction is the point at which the free

DNA band in the stained gel is visually undetectable. Figure 4 shows the gel analysis of a typical reconstitution pilot reaction.

Once the input ratio of DNA to histone octamer has been established in the pilot reactions, large-scale reconstitution can proceed with the same starting materials. The appropriate mixtures are put into a dialysis bag made from 6000–8000 MWCO Spectra/Por 1 tubing, and equilibrated at 22° with 2 liters of N1 buffer in 1-liter changes. Once the starting sample has equilibrated, the reconstitution proceeds, using a modification of the slow salt gradient dialysis method[13] (Fig. 5). The gradient is started by pumping N2 buffer into an Erlenmeyer flask containing the dialysis bag of NCP material in the N1 buffer while stirring. A constant volume is maintained by pumping N2 buffer into the bottom of the dialysis vessel and allowing the excess to flow out through the side arm. The rate at which N2 buffer is added is adjusted to lower the salt concentration to 1 M KCl in 1 day, and to complete the dialysis over an additional 4 days at 29°. At concentrations greater than 1 M KCl the cloud point of the detergent is just above 22°, and at higher temperatures the solution will undergo a binary phase separation. As the sample approaches 0.5 M KCl, precipitate may begin forming in the dialysis bag. This is the result of an improper ratio of DNA to octamer. Even the most carefully calculated ratios will usually result in some precipitate. At the end of the salt gradient dialysis, the reconstituted material is dialyzed against N3 buffer and then harvested. The NCP is analyzed by 6% nondenaturing PAGE in 0.1× TBE and the yield is determined spectrophotometrically, using an extinction of 10 AU at 260 nm for NCP at 1 mg/ml. The sample is then concentrated using Centriprep-30 and Centricon-30 microconcentrators to 10 mg/ml.

The concentrated NCP is passed over a Sephacryl S-400 HR size-exclusion 10/30 column (Pharmacia) in 5-mg aliquots. This column is equilibrated with 50 mM KCl, 10 mM potassium cacodylate, 0.1 mM EDTA, pH 6.0. Peak fractions are concentrated with a Centricon-30 to 6–8 mg of nucleosomes per milliliter. The nucleosomes are then ready to be salted in with $MnCl_2$. The target $MnCl_2$ concentration is 70 mM in the NCP sample. The concentration of $MnCl_2$ for salting in the nucleosomes can be higher, 80–90 mM. An aliquot is added from a 1 M $MnCl_2$ stock sufficient to raise the sample concentration to the target value. The $MnCl_2$ is added all at once, causing the NCP to precipitate, and the sample is gently mixed by inversion until the NCP dissolves. If the precipitate persists, more $MnCl_2$ should be added with gentle mixing by inversion of the tube until the sample is fully dissolved.

[13] T. J. Richmond, M. A. Searles, and R. T. Simpson, *J. Mol. Biol.* **199,** 161 (1988).

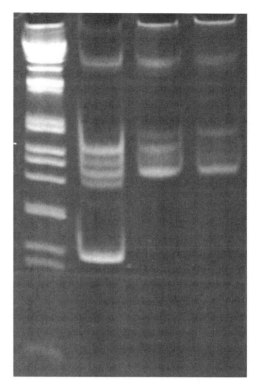

Fig. 4. Particle gel analysis from pilot NCP reconstitution reactions showing ethidium bromide-stained DNA. From left to right: 1-kb DNA ladder standard; reconstitution with excess DNA; reconstituted NCP with degenerate phased DNA; proper ratio of histone and DNA for well-phased NCP.

Crystallization of NCP

The NCP at 6–8 mg/ml and ~70 mM MnCl$_2$ concentration is now ready for crystallization. The final size-exclusion chromatographic run of the purified NCP has the added benefit of separating properly phased NCP from those with DNA misaligned on the histone core (a properly phased NCP migrates more rapidly than one with degenerate phases). If needed, a further polishing of the crystallizable material is possible using preparative gel electrophoresis.[14]

[14] J. M. Harp, E. L. Palmer, M. York, A. Gewiess, M. Davis, and G. J. Bunick, *Electrophoresis* **16,** 1861 (1995).

FIG. 5. NCP reconstitution setup within Thermo Forma incubator described in text. At any given time point, the concentration of $MnCl_2$ can be determined by the formula $C_a = C_{aIN} - (C_{aIN} - C_{a0})^{(-\nu/Vt)}$, where C_{aIN} is the molar concentration of $MnCl_2$ at the inlet, C_{a0} is the initial concentration of $MnCl_2$, ν is the volumetric flow rate in milliliters per minute, V is the volume of the reaction flask, and t is time in minutes.

Our preferred method for crystallization of nucleosome core particles is by dialysis. We used glass dialysis buttons placed in scintillation vials before the development of the diffusion-controlled apparatus for microgravity (DCAM) shown in Fig. 3 or an updated version, the counterdiffusion cell (CDC).[15] These devices possess two buffer chambers separated by a tube containing an agarose plug to prevent mixing of the buffers while allowing diffusion between the chambers (Fig. 6). A 45-μl crystallization button is molded into the end plate sealing one of the chambers. The

[15] D. C. Carter, B. Wright, T. Miller, J. Chapman, P. Twigg, K. Keeling, K. Moody, M. White, J. Click, J. R. Ruble, J. X. Ho, L. Adcock-Downey, G. Bunick, and J. Harp, *J. Crystal Growth* **196,** 602 (1999).

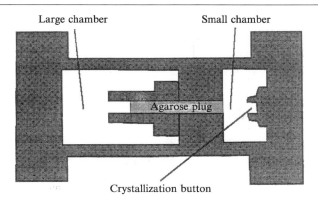

FIG. 6. Diagram of the DCAM in longitudinal section shows the two buffer chambers separated by an agarose plug. The NCP is placed within the crystallization button, covered with a dialysis membrane, and allowed to equilibrate with the buffer.

NCP preparation is placed in the crystallization button and covered with a Spectra/Por 1 dialysis membrane (6000–8000 molecular weight cutoff) and secured with an O-ring. The chamber containing the button is filled with buffer containing 50 mM KCl, 10 mM potassium cacodylate, 1 mM PMSF, and a concentration of $MnCl_2$ equivalent to or higher than that of the sample. The opposite chamber is filled with 50 mM KCl, 10 mM potassium cacodylate, 1 mM PMSF and a lower concentration of $MnCl_2$. Diffusion commences immediately on assembly of the DCAM; however, because crystallization of the NCP requires lowering of the $MnCl_2$ concentration rather than an increasing precipitant concentration, it is possible to employ a "time delay fuse" by preparing the agarose plug separating the two chambers with a high (200 mM) concentration of $MnCl_2$. With the concentration of $MnCl_2$ in the agarose plug much higher than that in either chamber, several days elapse before a Mn^{2+} concentration gradient begins to develop between the two chambers. The Mn^{2+} target equilibrium value for NCP crystallization is about 40 mM, but should be experimentally determined for each NCP preparation by construction of a phase diagram.

Crystallization usually takes 1 to 2 months at 22° in a Thermo Forma incubator. Crystals are harvested from crystallization buttons directly into stabilization buffer containing 30 mM KCl, 30 mM $MnCl_2$, 10 mM potassium cacodylate (pH 6.0), and 2% 2-methyl-2,4-pentanediol (MPD). The concentration of MPD is increased in steps by addition of buffer containing 30 mM KCl, 30 mM $MnCl_2$, 10 mM potassium cacodylate (pH 6.0), and 50% MPD. The amount of buffer containing 50% MPD added at each step is adjusted to provide a 2% increase in MPD concentration and a minimum of 8 h is allowed for equilibration of the crystals before the next addition of

MPD. The unit cell dimensions change as a function of MPD concentration. There is also a correlation between the MPD concentration and the quality of diffraction. The optimum concentration of MPD for the smallest unit cell parameters and best diffraction characteristics is about 22.5%, as adjusted by refractometry to verify the final value.

Acknowledgments

This research was sponsored by grants from the NIH (GM-29818), NASA (NAG8-1568), the Office of Biological and Environmental Research, the U.S. Department of Energy, and the Laboratory Directed Research and Development Program of Oak Ridge National Laboratory, managed by UT-Battelle, LLC, for the U.S. Department of Energy under Contract DE-AC05-00OR22725.

This chapter has been authored by a contractor of the U.S. Government under contract DE-AC05-00OR22725. Accordingly, the U.S. Government retains a nonexclusive royalty-free license to publish or reproduce the published form of this contribution, or allow others to do so, for U.S. Government purposes.

[4]   Creating Designer Histones by Native Chemical Ligation

By MICHAEL A. SHOGREN-KNAAK and CRAIG L. PETERSON

Introduction

Histones in all eukaryotic organisms are enzymatically altered to present a wide range of posttranslational modifications, including acetylation, phosphorylation, methylation, ubiquitination, and ribosylation. Moreover, these modifications play important roles in modulating nearly all DNA-associated processes, such as gene expression, epigenetic patterning, DNA repair, and replication. To probe how posttranslational modifications of histones influence both chromatin structure and the binding and function of chromatin-associated proteins, we describe a strategy that employs native chemical ligation[1,2] to generate designed histones. Specifically, we present the synthesis of wild-type *Xenopus laevis* histone H3 that can be modified at individual or multiple amino-terminal tail sites.[3]

[1] P. E. Dawson, T. W. Muir, I. Clark-Lewis, and S. B. Kent, *Science* **266**, 776 (1994).
[2] U. K. Blaschke, J. Silberstein, and T. W. Muir, *Methods Enzymol.* **328**, 478 (2000).
[3] M. A. Shogren-Knaak, C. J. Fry, and C. L. Peterson, *J. Biol. Chem.* **278**, 15744 (2003).

This strategy, in principle, can be used to study virtually any histone modification and offers a number of advantages over alternative methods. Unlike enzymatically modified histones or those purified from cellular lysates, ligation provides homogeneously modified histones. Furthermore, because these synthetic histones can be readily assembled *in vitro* into chromatin, they are more physiologically relevant than peptide substrates. Thus, it is possible to study the effects of histone modifications on chromatin structure and function, where recognition or interaction between histone domains, octamer subunits, nucleosomes, or nucleosomal DNA is important.

In this chapter we discuss the criteria used to select the H3 ligation site, and the protocol used for the synthesis and purification of modified histone H3 tail thioester peptides, the generation of amino-terminal cysteine-containing histone H3 core protein, and the ligation of these constituents (shown schematically in Fig. 1).

Histone H3 Ligation Strategy and Design

Native chemical ligation is a technique for generating full-length proteins from unprotected synthetic peptides and/or expressed protein fragments, and it is well suited for including amino acids that are not directly incorporated into proteins via the genetic code.[1,2,4] Native chemical ligation requires an N-terminal polypeptide fragment ending in a C-terminal thioester, and a C-terminal polypeptide fragment beginning with an N-terminal cysteine (Fig. 1). When mixed together, these fragments can form a reversible covalent association via *trans*-thioesterification of the C-terminal thioester by the N-terminal cysteine thiol. This reaction intermediate can then rearrange irreversibly via an S- to N-acyl shift, resulting in a canonical amide peptide bond at the ligation site (Fig. 2). Furthermore, this method is compatible with proteins containing other cysteines, because thioester products with these cysteines cannot rearrange to form an amide bond and thioesterification is reversible.

For histone H3, the majority of amino acid side chains that are posttranslationally modified reside in the amino-terminal tail. To incorporate modified residues of interest into a ligated histone, we chose solid-phase peptide synthesis to generate the N-terminal thioester-containing fragment and recombinant expression of the C-terminal cysteine-containing protein fragment. However, it is also possible to study C-terminal modifications (e.g., in histone H2B) by recombinantly expressing N-terminal protein

[4] M. Huse, M. N. Holford, J. Kuriyan, and W. W. Muir, *J. Am. Chem. Soc.* **122,** 8337 (2000).

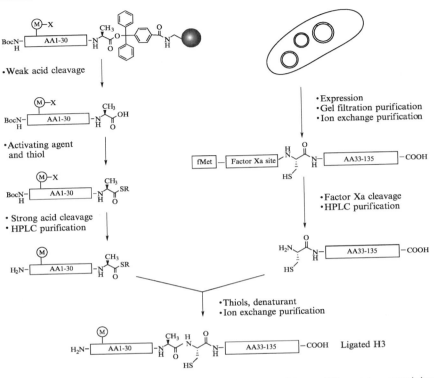

Fig. 1. Native chemical ligation strategy for generating histone H3 proteins containing specifically modified N-terminal residues. An N-terminal peptide fragment of histone H3 that contains specifically modified amino acid residues (in this example, a methylated lysine residue denoted by an encircled "M"), and a C-terminal thioester moiety (COSR), is produced by standard solid-phase peptide synthesis on an acid-hypersensitive support (*left*). A C-terminal protein fragment of histone H3 containing an N-terminal cysteine residue is generated by proteolytic trimming of recombinant protein (*right*). Reaction of these two fragments in the presence of thiol reagents produces native full-length histone H3 containing the modifications of interest. (See color insert.)

fragments that contain a C-terminal thioester and synthesizing C-terminal tail peptides containing an N-terminal cysteine.[2]

The bond between amino acid residues 31 and 32 was chosen as the histone H3 ligation site because it fulfills a number of criteria: first, a 31-amino acid peptide is synthetically tractable and allows many of the known H3 posttranslational histone modifications to be incorporated. Second, position 32 of histone H3 (threonine in wild-type *Xenopus* H3) is not highly

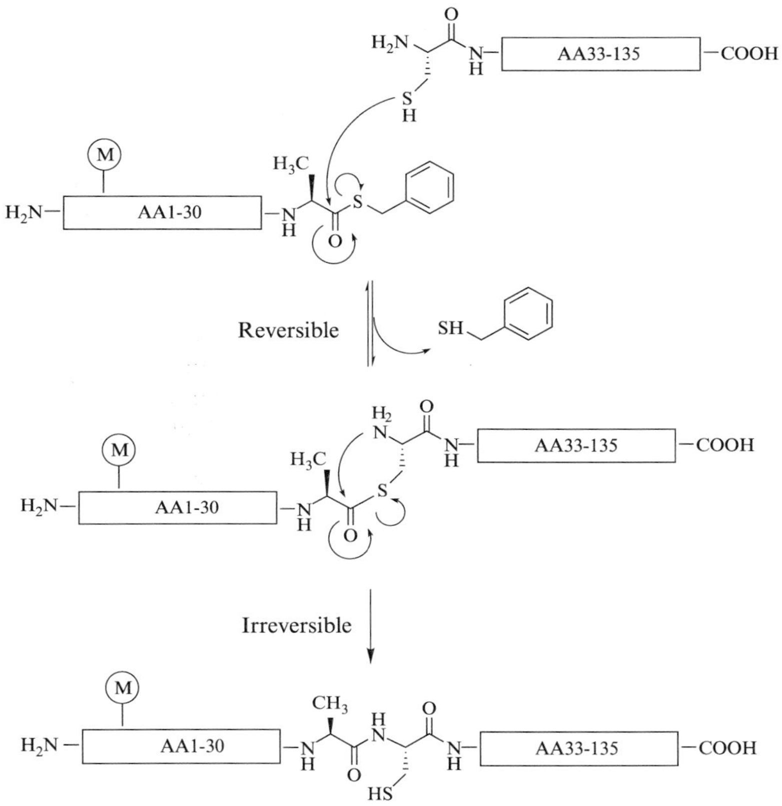

FIG. 2. Mechanism of histone H3 native chemical ligation. The C-terminal thioester of the N-terminal H3 peptide undergoes reversible *trans*-thioesterification with the cysteine thiol group of the C-terminal histone H3 fragment. This reaction intermediate rearranges irreversibly by an S- to N-acyl shift to generate native histone H3.

conserved among species,[5] suggesting it could be changed to cysteine without major structural or functional effects. Finally, positions 31 and 32 were expected to be compatible with ligation, as these residues are outside the structured core of the histone,[6] and position 31 (alanine) is not a proline, valine, or isoleucine residue, amino acids known to be detrimental to ligation.[2]

[5] S. Sullivan, D. W. Sink, K. L. Trout, I. Makalowska, P. M. Taylor, A. D. Baxevanis, and D. Landsman, *Nucleic Acids Res.* **30**, 341 (2002).
[6] K. Luger, A. W. Mader, R. K. Richmond, D. F. Sargent, and T. J. Richmond, *Nature* **389**, 251 (1997).

N-Terminal H3 Peptide Thioester Synthesis and Purification

Generation of peptides with a reactive C-terminal thioester group requires the selective thioesterification of the terminal peptide α-carboxylic acid group in the presence of other potentially reactive side-chain functional groups (Fig. 1, left side). To achieve this selectivity, peptides with a free α-carboxylic acid group, but protected side-chain groups, are generated by synthesizing the desired peptide on an acid-hypersensitive resin.[7] The synthesized peptides are then treated with a weak acid cleavage cocktail, which cleaves the peptide from the synthetic support to expose the free terminal α-carboxylic acid, while maintaining the side chain-protecting groups. The terminal α-carboxylic acid of the peptide is then thioesterified, using standard peptide-coupling reagents. Treatment of the peptide with a strong acid cleavage cocktail removes the side chain-protecting groups while maintaining the C-terminal thioester. Finally, high-pressure liquid chromatography (HPLC) is used to obtain the thioester ligation substrate in pure form.

Protected N-Terminal H3 Peptide Synthesis

Synthesis of the protected N-terminal H3 peptide from C to N terminus largely follows standard protocols for 9-fluorenylmethoxycarbonyl-(Fmoc)-based solid-phase peptide synthesis[8] and should be possible at many peptide synthesis facilities.

An acid-hypersensitive resin, NovaSyn-TGT resin (Calbiochem-Novabiochem; EMD Biosciences, San Diego, CA) preloaded with the N-α-Fmoc-protected residue 31 (alanine for wild-type *Xenopus* H3) is used as the synthetic support. This resin has a lower substitution (0.1–0.3 mmol/g) than many commonly used peptide resins, and thus reaction volumes may have to be adjusted accordingly to ensure proper resin solvation. However, in our experience this lower substitution resin provides better synthetic results than alternative higher substitution acid-hypersensitive peptide resins.

Standard side chain-protected Fmoc–amino acids are used to incorporate standard amino acids. Incorporation of acetylated lysine, phosphorylated serine or threonine, and mono-, di-, or tri-methylated lysine can be accomplished with N-α-Fmoc-N-ε-acetyl-L-lysine, N-α-Fmoc-O-benzyl-L-phosphoserine, N-α-Fmoc-O-benzyl-L-phosphothreonine (EMD Biosciences) N-α-Fmoc-N-ε-methyl-N-ε-Boc-L-lysine, N-α-Fmoc-N-ε-(methyl)$_2$-L-lysine·HCl, and N-α-Fmoc-N-ε-(methyl)$_3$-L-lysine chloride (Bachem Bioscience, King of

[7] S. Futaki, K. Sogawa, J. Maruyama, T. Asahara, M. Niwa, and H. Hoja, *Tetrahedron Lett.* **38,** 6237 (1997).

[8] G. B. Fields and R. L. Noble, *Int. J. Pept. Protein Res.* **35,** 161 (1990).

Prussia, PA), respectively. The N-terminal residue (alanine for *Xenopus* wild-type histone H3) should be incorporated as the *N*-α-butyloxycarbonyl (Boc)-protected residue (strong acid labile), because the amino terminus must be protected to prevent cross-reaction during thioesterification.

Fmoc deprotection is performed with standard bases, such as piperidine. Activation of the Fmoc–amino acid α-carboxylic acid can be accomplished with standard coupling reagents [e.g., 2-(1*H*-benzotriazole-1-yl)-1,1,3, 3-tetramethyluronium hexafluorophosphate (HBTU), 2-(1*H*-benzotr-iazole-1-yl)-1,1,3,3-tetramethyluronium tetrafluoroborate (TBTU), and benzotriazole-1-yl-oxy-tris-pyrrolidino-phosphonium hexafluorophosphate (PyBOP)]. However, the coupling additive *N*-hydroxybenzotriazo-le(HOBt) should not be added, as it may cause loss of the peptide from the solid support during synthesis.[9] Double coupling of all residues and/or *N*-α-acetyl capping of uncoupled α-amines is suggested to improve the purity and yield of the N-terminal H3 peptide.[8]

1. Using an automated solid-phase peptide synthesizer, synthesize the desired peptides from residues 2–30 (RTKQTARKSTGGKAPRKQLAT-KAARKSAP, for wild-type *Xenopus* H3) on Fmoc-Ala-NovaSyn-TGT resin (EMD Biosciences) at a scale from 50 to 100 μmol.

2. Couple residue 1 (alanine for wild-type *Xenopus* H3) as the *N*-α-Boc-protected amino acid.

3. Cleave and analyze a small amount of the resin (50 mg) by reversed-phase HPLC and mass spectrometry, using standard protocols, to determine the identity and purity of the synthetic product.

Cleavage of Side Chain-Protected N-Terminal H3 Peptide

Side chain-protected N-terminal H3 peptide is cleaved from the acid-hypersensitive trityl-TentaGel support with a mild acid solution of acetic acid.[10] Because of its length and the high density of hydrophobic, protected basic residues, the protected N-terminal H3 peptide is relatively difficult to remove from the synthetic support using standard organic solvents, and requires the addition of trifluoroethanol, a reagent believed to disrupt highly insoluble β strand-aggregated peptides. Because both acetic acid and tri-fluoroethanol are potentially reactive in the subsequent thioesterification reaction, their complete removal by trituration and washing of the protected peptide is necessary.

[9] K. Barlos, O. Chatzi, D. Gatos, and G. Stavropoulos, *Int. J. Pept. Protein Res.* **37**, 513 (1991).
[10] H. Benz, *Synthesis* **4**, 337 (1993).

1. Prepare a 1:2:7 solution of acetic acid–trifluoroethanol–dichloromethane and add 1 ml to 50-mg portions of the dry peptide resin. Rock gently for 3 h at 25°.

2. Filter the resin from the protected peptide solution using a glass wool plug. Wash the resin twice, each time with 250 μl of the 1:2:7 acetic acid–trifluoroethanol–dichloromethane solution, and pool all eluate.

3. Reduce the volume of the pooled eluate to approximately one-fifth the original volume under a gentle stream of nitrogen or compressed air. Add the cloudy solution dropwise to 3 ml of 3:1 ethyl acetate–hexane and vortex the resulting white precipitate.

4. Centrifuge for 5 min at 5000g to pellet the peptide and decant the supernatant.

5. Resuspend the pellet in 1 ml of hexane and transfer to a weighed Eppendorf tube. Wash the pellet by vigorously vortexing. Centrifuge at 14,000 rpm in a microcentrifuge and decant the hexane from the pellet. Repeat the wash two additional times.

6. Dry the peptide overnight in a SpeedVac (Thermo Savant, Holbrook, NY). Yields for the side chain-protected, free carboxy-terminal acid peptide are typically about 15 mg. The peptide can be stored at −20° in a dry environment.

Thioesterification of N-Terminal H3 Peptide

The C-terminal thioester is generated from the side chain-protected, free carboxy-terminal acid peptide by activation of the C-terminal α-carboxylic acid with the activating reagent HBTU and coupling to benzyl mercaptan. The crude reaction mixture is dried and subjected to strong acid cleavage to remove the amino acid side chain-protecting groups. Crude thioester peptide is removed from the cleavage reagents and thioesterification reagents by trituration of the peptide. The crude thioester peptide is further purified by preparative reversed-phase HPLC to provide peptide thioester competent for native chemical ligation.

1. For each free carboxy-terminal acid peptide aliquot, grind the peptide pellet into a fine powder to improve its solubility. Add 1 ml of dimethylformamide (DMF) and divide the reaction into two portions. Add an additional 750 μl of DMF to each tube, as well as 5.6 μl of N,N-diisopropylethylamine. Heat the solution to 45° with vigorous mixing for 30 min, or until the peptide is dissolved.

2. Lower the temperature to 35° and add 9.79 mg of HBTU. Mix by vortexing and let sit for 1 min. Add 3.03 μl of benzyl mercaptan. Let the reaction proceed for 20 h at 35° with vigorous mixing. Benzyl mercaptan has an extremely strong odor and should be handled carefully.

3. Remove the liquid components of the reaction *in vacuo* until dry. Wash the pellet with 1 ml of water–0.1% trifluoroacetic acid. Remove the water solution *in vacuo* until dry. Because side-chain deprotection of peptides by strong acid reagents is adversely affected by DMF, thorough drying is necessary.

4. Remove the side chain-protecting groups of the peptide by addition of standard strong acid cleavage reagents.[11] For peptides that do not contain either tryptophan or methionine, this can be accomplished by adding 1 ml of 95:2.5:2.5 trifluoroacetic acid–water–triisopropylsilane and allowing the reaction to proceed for 2 h at 25° with gentle rocking.

5. Reduce the volume of the cleavage mixture to roughly one-fifth the original volume under a gentle stream of nitrogen or compressed air. Add 1 ml of 2:1 ether–hexane to triturate the peptide as a white precipitate. Vortex vigorously. Centrifuge at 14,000 rpm in a microcentrifuge for 2 min and decant the supernatant. Repeat the addition of 2:1 ether–hexane, vortexing, centrifugation, and decanting an additional three times.

6. Allow the peptide pellet to air dry for 20 min, and then resuspend in 1 ml of water–0.1% trifluoroacetic acid. Transfer to a weighed Eppendorf tube and dry in a SpeedVac. Typical yields should be roughly 6 mg.

7. Resuspend the peptide in water–0.1% trifluoroacetic acid to a concentration of 10 mg/ml. Purify the crude peptide by reversed-phase high-pressure liquid chromatography, using a semipreparative reversed-phase C_{18} column (ZORBAX 300SB-C_{18}, 9.4 × 250 mm, 5-μm particle size, 300-Å pore size; Agilent Technologies, Palo Alto, CA). Use a mixed solvent system of water–0.1% trifluoroacetic acid and acetonitrile–0.1% trifluoroacetic acid and monitor at 228 nm. Typically up to 2 mg of crude peptide can be purified over a shallow gradient from 18 to 22% organic solvent over 20 min at 2 ml/min. A number of UV-active peaks will be present, but the desired peptide should be the second of two closely eluting peaks (Fig. 3A), where the first peak is an isomeric form of the second peptide peak. Modified peptides do not tend to elute at drastically different times.

8. Dry and weigh the purified N-terminal H3 thioester peptide. Confirm the identity and purity of the product by mass spectrometry. Typical yields by weight fall between 30 and 50% of the initial crude weight. Redissolve the thioester peptide in water–0.1% trifluoroacetic acid, divide into individual 1.3-mg aliquots, and dry in a SpeedVac.

---

[11] NovaBiochem, "NovaBiochem Catalog and Peptide Synthesis Handbook." CB Bioscience, San Diego, CA, 1999.

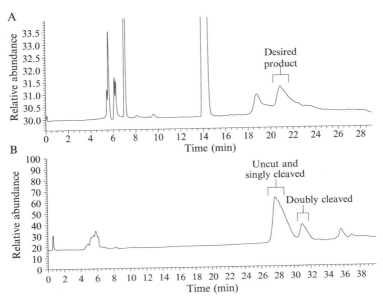

FIG. 3. High-pressure liquid chromatography purification of the N-terminal H3 thioester peptide and the C-terminal histone H3 fragment. (A) N-terminal H3 thioester peptide (20.75 min) is separated from other reaction products on a semipreparative C_{18} HPLC column. A representative trace of 0.5 mg of total protein is shown. Water–0.1% trifluoroacetic acid and acetonitrile–0.1% trifluoroacetic acid are used as the mobile phase, with protein eluted through a linear gradient of 18–22% organic solvent over 20 min at 2.0 ml/min and monitored at 228 nm. (B) Purification of histone H3 fragments after factor Xa protease treatment. Uncleaved and singly cleaved histone H3 digestion products (27.75 min) are separated from other reaction products on a semipreparative C_4 HPLC column. A representative trace of 2 mg of total protein is shown. Water–0.1% trifluoroacetic acid and acetonitrile–0.1% trifluoroacetic acid are used as the mobile phase, with protein eluted through a linear gradient of 37–48% organic solvent over 33 min at 2.5 ml/min and monitored at 235 nm.

C-Terminal Histone H3 Fragment Synthesis and Purification

Native chemical ligation requires a C-terminal peptide or protein fragment containing an N-terminal cysteine. Full-length *Xenopus* histone H3 can be expressed recombinantly in *Escherichia coli*.[12] Thus, it is possible to engineer the histone expression plasmid to contain an N-terminal cysteine residue at position 32. However, because the initiator formyl

[12] K. Luger, T. J. Rechsteiner, and T. J. Richmond, *Methods Enzymol.* **304,** 3 (1999).

methionine is not readily removed in *E. coli,* a protease site that can be cleaved C terminal to the recognition sequence (in this case, the factor Xa recognition site) is also engineered to allow production of a histone H3 fragment with an N-terminal cysteine.

Using this engineered histone H3 expression plasmid, the C-terminal histone H3 fragment is expressed in *E. coli,* inclusion bodies containing the expressed protein are isolated, and the C-terminal histone protein is purified by gel filtration and ion-exchange chromatography (Fig. 1, right side).[12] The protein is subjected to factor Xa proteolysis to remove both the N-terminal methionine and the factor Xa cleavage site and then purified away from overdigested histone by reversed-phase HPLC.

C-Terminal Histone H3 Fragment Plasmid Construction

DNA sequence corresponding to H3 amino acid residues 33 to 135 is amplified by polymerase chain reaction (PCR) from a wild-type *Xenopus* histone H3 expression plasmid, using an upstream primer containing an *Nde*I restriction site, a start codon, codons for a minimal factor Xa cleavage site (IEGR), and a cysteine codon for position 32 (5'-GCACTCGAGCCA-TATGATCGAAGGTCGTTGTGGCGGAGTCAAGAAACCTCACC-GTTAC-3') and a downstream primer containing a *Bgl*II restriction site (5'-AGCTCGCAATAGATCTAAGCCCTCTCGCCTCGGATTCT-3'). The resulting product is digested and ligated into an *Nde*I–*Bgl*II pET11c expression vector. The resulting plasmid is sequenced to confirm its identity and then transformed into a BL21(DE3) *E. coli* expression strain (Invitrogen, Carlsbad, CA).

C-Terminal Histone H3 Fragment Expression and Purification

Expression and purification of the C-terminal histone H3 fragment are performed largely as described by Luger and coworkers for recombinant wild-type *Xenopus* histone H3.[12] In our initial studies, we found yields and reproducibility of protein expression and purification, factor Xa cleavage, and the ligation steps were detrimentally affected by poor solubility of the expressed histone fragment. To address this issue, efforts were made to limit the exposure of the protein to conditions compatible with oxidation and to limit the frequency with which the histone was brought to dryness. In addition, we have found that mutation of Cys-111 to alanine appears to reduce aggregation and improve solubility.

Changes to the original histone purification protocol are as follows: after gel-filtration purification of the histone, fractions containing the histone are not dialyzed against a 2 m*M* 2-mercaptoethanol solution and dried, but are instead diluted into urea buffer [7 *M* urea, 20 m*M* sodium acetate, 1 mM dithiothreitol (DTT), 1 mM EDTA, pH 5.2] to a final sodium chloride concentration of 200 mM, and then directly subjected to ion-exchange purification, with a gradient from 200 m to 600 mM sodium chloride. After ion-exchange chromatography, fractions containing the purified histone protein are pooled and dialyzed against three changes of a 5 mM solution of DTT at 4°. A small aliquot of the slightly cloudy mixture is diluted into unfolding buffer (7 M guanidine-HCl, 20 mM Tris, 10 mM DTT, pH 7.5) and quantified at 276 nm (ε_{276} = 0.320 ml·mg^{-1}· cm^{-1}). The histone is then subjected to factor Xa cleavage.

Factor Xa Cleavage and Purification of C-Terminal Histone H3 Fragment

Factor Xa protease cleaves C terminal to a preferred recognition site of Ile-Glu/Asp-Gly-Arg. However, factor Xa can cleave other basic sites, and with high enzyme concentration and lengthy exposure the histone H3 C-terminal fragment can become completely degraded. Under limited exposure to factor Xa, cleavage occurs primarily at the introduced recognition site with a secondary C-terminal site that removes the last seven amino acids of the histone, cleaving between residues corresponding to Arg-128 and Arg-129 of the wild-type H3 sequence. Under proper conditions it is possible to achieve a cleavage product distribution of about 20% uncleaved, 55% singly cleaved, and 25% cleaved both N and C terminally. Proteins that do not remove the N-terminal recognition sequence are unreactive in the native ligation step (i.e., they do not present an N-terminal cysteine) and can readily be separated from the full-length ligated histone. However, proteins cleaved at both the N and C termini can be ligated and are difficult to resolve from nontruncated ligation products. Thus, it is necessary to remove the double-cleaved factor Xa product before ligation, using reversed-phase high-pressure liquid chromatography.

1. Perform a set a of pilot factor Xa cleavages to determine the amount of enzyme that gives the best cleavage results. In three Eppendorf tubes dilute the C-terminal histone H3 protein solution to a final concentration of 0.5 mg/ml in factor Xa buffer (20 mM Tris, 100 mM NaCl, 2 mM CaCl$_2$). Add factor Xa protease (New England BioLabs, Beverly, MA) to a final concentration of 4, 2, and 1 μg/ml and incubate with rocking at 25° for 30 min.

2. Quench the reactions with phenylmethylsulfonyl fluoride (PMSF) protease inhibitor at a final concentration of 1 mM. Incubate with rocking at 25° for 30 min.

3. Resolve 4, 2, and 1 μg of total protein for each of the reaction trials on a sodium dodecyl sulfate (SDS)–18% polyacrylamide gel. Visualize the

proteins with Coomassie blue stain. The uncleaved, singly cleaved, and N- and C-terminally cleaved histones have a mass of 12.6, 12.0, and 11.2 kDa, respectively.

4. Scale up the cleavage reaction described in step 1, using an amount of factor Xa that will give the greatest amount of singly cleaved histone product according to the pilot reactions. Quench and analyze the reaction products as described in the preceding steps.

5. Dialyze the reaction mixture against three changes of 0.1% trifluoroacetic acid in water. Flash freeze and lyophilize to dryness.

6. Dissolve the protein to a concentration of about 10 mg/ml in unfolding buffer. Separate the doubly cleaved histone protein from the other histone products by reversed-phase HPLC with a semipreparative reversed-phase C_4 column (Vydac 214TP C_4, 10 × 250 mm, 5-μm particle size, 300-Å pore size; W. R. Grace & Co., Columbia, MD), using a mixed solvent system of water–0.1% trifluoroacetic acid and acetonitrile–0.1% trifluoroacetic acid, and monitor at 228 nm (longer wavelengths can be used if the signal is saturating). Typically, up to 2 mg of crude total protein can be purified over a shallow gradient from 37 to 48% organic solvent over 33 min at 2.5 ml/min. The desired singly cleaved histone protein elutes as a broad peak, while the double-cleaved product elutes directly afterward (Fig. 3B). Note that the uncleaved histone protein coelutes with the singly cleaved histone protein and is carried into the ligation reaction.

7. Pool all the desired HPLC samples. To determine the protein concentration, take a small volume of the protein solution, dry it in a SpeedVac, and quantify at 276 nm ($\varepsilon_{276} = 0.320$ ml·mg^{-1}·cm^{-1}) in unfolding buffer. Divide the stock into individual aliquots corresponding to 1 mg of the singly cleaved product, flash freeze, and lyophilize.

Histone H3 Ligation

Ligation of polypeptides containing a C-terminal thioester and an N-terminal cysteine occurs spontaneously in aqueous solution. However, to drive the bimolecular reaction, an excess of peptide thioester over cysteine-containing protein is used. To promote ligation, guanidine is added to reduce secondary and tertiary structure. Benzyl mercaptan is added both to provide a reducing environment and to drive the reversibility of *trans*-thioesterification.[13] Benzene thiol is added to increase the rate of the reaction.[13] Both of these thiol additives have an extremely strong odor and should be handled carefully.

[13] P. E. Dawson, M. J. Churchill, M. R. Ghadiri, and S. B. H. Kent, *J. Am. Chem. Soc.* **119**, 4325 (1997).

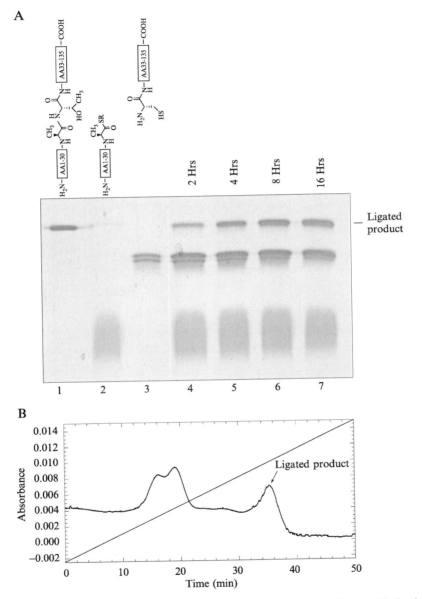

FIG. 4. Histone H3 ligation and purification. (A) The course of a histone H3 ligation reaction is analyzed on an SDS–18% polyacrylamide gel stained with Coomassie blue (lanes 4–7). For comparison, wild-type recombinant histone H3 (lane 1), synthetic N-terminal

Although almost all the singly cleaved C-terminal histone H3 fragment is converted into ligated product, some small fraction remains unreacted. Ion-exchange chromatography is used to remove the unreacted fraction as well as any uncleaved histone H3 and excess peptide. Because the C-terminal histone H3 fragment and the ligated product both have relatively poor solubility in the presence of the organic thiol additives, the reaction components are mixed with an acetonitrile–water solution before ion-exchange chromatography to aid solubility.

1. Perform a pilot ligation reaction by dissolving 1.00 mg of C-terminal histone H3 fragment (1×) and 1.3 mg of N-terminal H3 peptide (4× by weight) in 500 μl of ligation buffer (3 M guanidine-HCl, 100 mM potassium phosphate, pH 7.90). To the clear solution add 2.5 μl of both benzyl mercaptan and benzenethiol (0.5% each). Incubate the mixture at 25° with vigorous mixing.

2. Take 5-μl samples of reaction mixture (10 μg of total protein) over the course of 24 h. Add 20 μl of ligation buffer and then 75 μl of 25:75:0.1 acetonitrile–water–trifluoroethanol and desalt by dialysis. Separate 2 μg of protein on an SDS–18% polyacrylamide gel and visualize by Coomassie blue stain (Fig. 4A). Over the course of the reaction, the amount of ligated product should increase and then plateau. The amount of peptide thioester should decrease but not disappear.

3. Repeat step 1 with multiple 1-mg aliquots to that point in time at which the reaction does not proceed further as determined in step 2. Analyze the reaction on an SDS–18% polyacrylamide gel as described in step 2 to characterize the preparative reactions.

4. Prepare a stock of SAU-200 buffer (7 M urea, 20 mM sodium acetate, 200 mM sodium chloride, 1 mM DTT, 1 mM EDTA, pH 5.2), and SAU-600 buffer (7 M urea, 20 mM sodium acetate, 600 mM sodium chloride, 1 mM DTT, 1 mM EDTA, pH 5.2). Solubilize each reaction mixture by adding 2 ml of ligation buffer and then 22 ml of 25:75:0.1 acetonitrile–water–trifluoroethanol. Add 100 ml of SAU-200 buffer and load onto a Hi-Trap sulfopropyl HP ion-exchange column (Amersham Biosciences, Piscataway, NJ). Wash extensively with SAU-200. Elute the ligated histone with a linear gradient from 100% SAU-200 to 100% SAU-600 over 50 min (Fig. 4B).

---

histone H3 thioester peptide (lane 2), and undigested/singly cleaved C-terminal histone H3 proteins (lane 3) are included. (B) Ligation products are purified by ion-exchange chromatography. Shown is a representative trace from a 1-mg scale ligation reaction purified on a Hi-Trap sulfopropyl column. Over the course of 50 min, the sodium chloride concentration is increased linearly from 200 to 600 mM (7 M urea, 20 mM sodium acetate, 200–600 mM sodium chloride, 1 mM DTT, 1 mM EDTA, pH 5.2) and monitored at 280 nm. Full-length histone H3 elutes at 35 min.

5. Analyze fractions on an SDS–18% polyacrylamide gel and visualize by Coomassie blue stain. Pool the desired fractions. Dialyze against three changes of 5 mM DTT.

6. Quantify protein amounts by comparison with a wild-type histone H3 standard on an SDS–18% polyacrylamide gel. Characterize the identity of the histone H3 by mass spectrometry. Divide the samples into individual aliquots, flash freeze, and lyophilize.

Concluding Remarks

In our laboratory, we have generated histone octamers and nucleosomal arrays, using ligated wild-type and phosphorylated histone H3s, and have found that they display physical properties similar to those generated from recombinantly expressed histone H3.[3] In addition, as enzymatic substrates for remodeling and acetylation, these nucleosomal arrays display properties similar to those of wild-type, unligated arrays.[3] It should also be possible to incorporate ligated histones into other chromatin substrates, such as mononucleosomes.[12] In addition, with modification of the protocol described above, it should be possible to produce other histones with virtually any posttranslational modification and study them individually or in combination.

[5] Two-Dimensional Gel Analysis of Histones and Other H2AX-Related Methods

By Duane R. Pilch, Christophe Redon, Olga A. Sedelnikova, and William M. Bonner

Background

Polyacrylamide was introduced by Ornstein[1] and Davis[2] as a more reproducible medium than starch for electrophoretic separations. At the same time, they also introduced a discontinuous buffer system that greatly increases sample resolution by electrically forcing proteins in dilute samples to concentrate into zones only microns thick. However, histone

---

[1] L. Ornstein, *Ann. N. Y. Acad. Sci.* **121,** 321 (1964).
[2] B. J. Davis, *Ann. N. Y. Acad. Sci.* **121,** 404 (1964).

for histone analysis, it does remove a great many soluble and nonhistone chromatin proteins from the analysis.

3. For pellets <5 μl, add 25 μl of extraction solution to the tubes and loosen the pellets. For larger pellets, add 5 volumes of extraction solution or 1 volume of 2× extraction solution. Incubate the samples at 4° for at least 10 min; they may be stored for several days at 4°.

Extraction solution
H_2SO_4 (concentrated)	120 μl (0.22 M final concentration)
Glycerol	2 ml
2-Mercaptoethanolamine	0.1 g
Water	To 10 ml

4. Pellet the precipitates at 16,000g for 10 min. The supernatants containing the histones may be loaded onto AUT gels directly from these tubes without disturbing the pellets. Also, the supernatants may be transferred to 1.5-ml tubes and stored indefinitely at −20°.

Yeast Histone Extraction

Histones comprise a much smaller fraction of the total cellular protein in yeast than in mammals. Because yeast cells have tough cell walls, the strategy is to pulverize the yeast directly in acidic solutions, followed by concentration of the dilute histone extracts.

1. For small-format gel analysis, pellet ∼4 × 10^7 yeast cells in 1.5-ml screw-cap tubes.

2. Discard the supernatants and add 0.15 g of 0.5-mm glass beads and 90 μl of 0.2 M H_2SO_4 to the pellets.

3. Place the tubes in a TurboMix (Scientific Industries, Bohemia, NY; www.scientificindustries.com) tube holder attached to a Vortex-Genie for 5 min.

4. Centrifuge the beaten samples at 16,000g for 10 min and transfer the supernatants to fresh 1.5-ml tubes for TCA precipitation (described below).

5. *Alternative:* For larger scale preparations of up to ∼3.5 × 10^8 cells, use 0.45 g of 0.5-mm glass beads and 0.4 ml of 0.2 M H_2SO_4 in 1.5-ml screw-cap tubes. Use a Mini-Beadbeater-8 for 5 min at the highest setting instead of a TurboMix.

Sample Storage, Concentration, and Preparation for Gel Analysis

The histone-containing acid extracts obtained as described above can be stored indefinitely at −20°. Although acid extracts containing glycerol for density can be loaded directly onto AUT gels, they can also be processed for SDS gels by precipitating the histones and resuspending them in the appropriate loading solution. In addition, although the yeast histone acid extracts may be loaded on a small-format gel and yield detectable bands, precipitation is usually necessary to obtain enough material for ready visibility of minor components such as γ-H2A.

TCA Precipitation. To the acid supernatants, add 100% (w/v) trichloroacetic acid (TCA) to a final concentration of 20%. To prepare 100% (w/v) TCA, open a fresh bottle and add 40 ml of water for each 100 g of TCA. Pellet the precipitates at 16,000g for 10 min at 4°. Aspirate or drain off the supernatants. Wash the pellets with cold ethanol and let dry. The dried pellets may be stored indefinitely at $-20°$ and then dissolved in loading buffers for a variety of gels.

AUT Gels. Acid extracts may be loaded directly onto AUT gels, but if the samples loaded as acid extracts are found to smear during electrophoresis as determined by streaking in the 2D gel toward the top of the 1D gel, it may be useful to TCA precipitate the histones, dissolve the precipitate in AUT sample buffer, and clear the solution of any undissolved residue by centrifugation.

AUT sample buffer
Urea	4.8 g
Acetic acid	600 μl
Ammonia	40 μl of NH_4OH (28% NH_3)
Water	To 10 ml
Methylene blue	To give a light blue color

SDS Gels. Dried TCA precipitates can be dissolved in SDS sample buffer by heating to 95° for 5 min. If too much residual TCA is present, the acid pH may prevent the pellet from dissolving. This problem is usually signaled by the bromphenol blue in the SDS sample buffer turning yellow. In this case, add NH_4OH to raise the pH to neutral, signaled by the dye turning blue. TCA precipitates of acid extracts are more reliable for SDS gel analysis than are solutions of cells boiled in SDS sample buffers, because the viscosity of the DNA imparted to such samples often compromises histone resolution, leading to erroneous results.

Option for Yeast. It is also possible to disrupt the yeast directly in 20% TCA, letting the glass beads settle briefly before transferring the suspension to new tubes for processing as described above; however, this protocol results in samples in which the histones represent a smaller fraction of total protein. When these dried pellets are dissolved in either AUT sample buffer or SDS sample buffer, the sample should be cleared of undissolved residue by centrifugation.

Gel Preparation and Electrophoresis

AUT First-Dimension Gels

1. Prepare the gel shells for the first dimension. The gels need to be no more than two-thirds the thickness of the 2D shells, because the 1D gels will expand on preparation for the second dimension.

2. Prepare the 18% resolving gel solution:

Urea	48 g
Acrylamide	30 ml of a 60% (w/v) stock solution or 18 g of powder [*note:* 70 ml of water per 100 g of acrylamide makes 60% (w/v)]
Bisacrylamide	6 ml of a 2% (w/v) stock solution or 0.12 g of powder
Triton X-100	2 ml of a 25% (w/v) stock solution (26.75 g/100 ml)
Acetic acid	6 ml
Ammonia	400 μl of NH$_4$OH (28% NH$_3$)
TEMED	500 μl
Water	To 90 ml

The urea can be dissolved by placing the mixture in an uncapped tube in a microwave and heating in 5-s increments with mixing after each increment. If cloudy, clarify the solution through a 0.4-μm pore size filter. The solution can be stored at ambient temperature for about 1 week.

3. Add 1 ml of 0.004% riboflavin (kept at 4°) per 10 ml of resolving gel solution. Premade 1-mm Novex gel shells are useful for initial gel analyses, but optimum resolution may require shells with dimensions of about 18 × 16 × 0.075 cm (similar to the Hoefer SE600/SE400). Pour the solution into the shells to a line placed at twice the depth of the loading well below the top. Overlayer the gel solution with ~0.5 ml of water-saturated butanol and place the shells between two fluorescent light boxes. When the gels have polymerized (~20 min), rinse off the butanol with water.

4. Prepare the stacking gel solution (together with the resolving gel solution):

Urea	48 g
Acrylamide	8 ml of a 60% stock solution or 4.8 g of powder
Bisacrylamide	8 ml of a 2% solution or 0.16 g of powder
Acetic acid	6 ml
Ammonia	400 μl of NH$_4$OH (28% NH$_3$)
TEMED	500 μl
Water	To 90 ml

5. Add 1 ml of 0.004% riboflavin per 10 ml of stacking gel solution and pour into the gel shells to the top. Insert the sample combs and place the shells between two light boxes. The gels should polymerize in ~20 min.

6. Remove the samples from storage. Remove the combs from the shells. Liquid seeping into the wells from the stacking gel should be removed because it will interfere with sample loading. Load aliquots equivalent to 1–10 μg of histone (based on the cell numbers presented above) into the empty wells. Add extraction solution with a small amount of methylene blue to the empty wells. Cytochrome c may also be used as a visible protein marker; it migrates slightly slower than H4.

7. Overlayer each sample to the top of the well with reservoir buffer. Place the shells in the electrophoresis apparatus and add reservoir buffer to both chambers.

Reservoir buffer

Acetic acid	60 ml
Glycine	8 g
Water	To 1000 ml

8. Electrophorese at constant power to prevent overheating during a quick run, ~7 W per gel for the 1-mm Novex shell and 12 W for a 0.07-mm Hoefer SE600 shell. The time required for the methylene blue to reach the bottom is ~60 min for the Novex and 3 h for the Hoefer. However, the best resolution is obtained when electrophoresis is performed overnight at 7 mA for each Hoefer SE600 shell.

9. When the methylene blue dye reaches to within a few millimeters of the bottom, transfer the gels to a tray of staining solution for 2 h to overnight with gentle shaking. Then transfer the gels to destaining solution containing a little staining solution to prevent overdestaining.

Staining solution

Ethanol	1.6 liters
Acetic acid	200 ml
Water	To 4 liters
Coomassie blue	4 g
2-Mercaptoethylamine-HCl	A few pellets just before use

The 2-mercaptoethylamine-HCl helps prevent photooxidation of methionine residues in the histones. H2A2 often migrates as a doublet in the 2D AUC gel; the faster band is a photooxidized species formed when the 1D gel is viewed on a lightbox.

Destaining solution

Ethanol	800 ml
Acetic acid	200 ml
Water	To 4 liters
2-Mercaptoethylamine-HCl	A few pellets just before use

10. When sufficiently destained, the gels can be imaged with a digital camera on a light box or scanned directly into an image-processing program such as Paint Shop Pro or Photoshop (Jasc Software, Minneapolis, MN; Adobe, San Jose, CA). Gels can be stored indefinitely in heat-sealed plastic bags.

AUC Second-Dimension Gels

Histone species in the 1D AUT gels can be further analyzed on 2D AUC or SDS gels. In both 2D gel types, the core histone species are separated from a diagonal of protein species whose migrations are not retarded by Triton X-100 in the first dimension. However, the acetylated and phosphorylated histone species are well resolved from the unmodified parent species in AUC gels because their migration rates differ in both dimensions (Fig. 1),[11] whereas in SDS gels all modified forms migrate with the parent. The procedures for preparing AUC 2D gels are almost identical to those for the 1D AUT gels, but the small differences are essential. The AUT gels have 8 M urea and 0.5% Triton X-100 in the resolving gel but no acetyltrimethylammonium bromide (CTAB) in the upper reservoir buffer. The AUC gels have 1.5 M urea but no Triton X-100 in the resolving gel and 0.15% CTAB in the upper reservoir buffer. Also, the 2D gel shell needs to be at least 1.5 times the thickness of the 1D shell.

11. Incubate the whole gel or excised gel pieces to be analyzed on the second dimension for 2 h to overnight in at least 3 volumes of AUC gel soaking solution.

AUC gel soaking solution

Acetic acid	15 ml
Ammonia	0.75 ml
Water	Up to 250 ml
2-Mercaptoethylamine-HCl	2.5 g

12. Prepare the 2D gel shells. They should have a thickness at least 1.5 times that of the first dimension in order to accommodate the soaked and swollen 1D gel samples.

13. Prepare 18% 2D resolving gel solution:

Urea	9 g
Acrylamide	30 ml a 60% stock solution or 18 g of powder
Bisacrylamide	6 ml of a 2% solution or 0.12 g of powder
Acetic acid	6 ml
Ammonia	400 μl
TEMED	500 μl
Water	To 90 ml

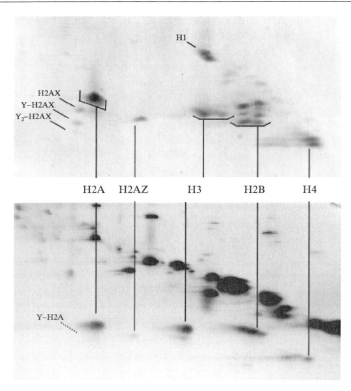

Fig. 1. Histone analysis on 2D gels. *Top:* Histones from mouse ES cells exposed to 200 Gy and analyzed on 14 × 16 cm gels. *Bottom:* Yeast histones analyzed on gels made in commercially available (Novex) 8 × 6 cm shells.

14. Add 1 ml of 0.004% riboflavin per 9 ml of resolving gel solution and pour it into the shells to a mark 3 cm below the top. Overlayer the gel solutions with ~0.5 ml of water-saturated butanol. Place the shells between two light boxes. When the gels are polymerized (~20 min), rinse out the butanol with water.

15. Prepare the 2D stacking gel solution:

Urea	9 g
Acrylamide	8 ml of a 60% stock solution or 4.8 g of powder
Bisacrylamide	8 ml of a 2% solution or 0.16 g of powder
Acetic acid	6 ml
Ammonia	400 μl
TEMED	500 μl
Water	To 90 ml

16. Add 1 ml of 0.004% riboflavin per 9 ml of stacking gel solution. Pour into the shells to a mark 1 cm from the top. Overlayer with ~0.5 ml of water-saturated butanol. Place the shells between two light boxes. When the gels are polymerized (~20 min), rinse out the butanol with water.

17. Prepare embedding solution from equal volumes of 2× gel prep solution and 2× agarose solution (melted in the microwave). Two separate stocks are used because agarose heated in acetic acid fails to gel. Keep the embedding solution warm, or make fresh as needed.

2× gel prep solution
Acetic acid	60 ml
Ammonia	3 ml
Water	Up to 500 ml

2× agarose solution (can be remelted as often as needed)
Agarose	2 g (any high melting point agarose)
Glycerol	~40 ml or 40 g
Water	To 100 ml

Embedding solution
2× gel prep solution	10 ml
2× agarose solution	10 ml (melted in microwave)

18. Cut out the desired pieces of the soaked 1D gels and place them in the 1-cm cavity above the stacking gels. Fill the remaining cavity of each gel with embedding solution, pushing the 1D gel pieces down until they touch the stacking gel. Let the embedding solution solidify.

19. Place the shells in the electrophoresis apparatus and add the reservoir buffers.

Upper reservoir buffer
Acetic acid	60 ml
Glycine	8 g
Water	To 1000 ml
CTAB	30 ml of 5% solution (microwave to dissolve)

Lower reservoir buffer
Acetic acid	60 ml
Glycine	8 g
Water	To 1000 ml

20. Perform electrophoresis, using the same settings as for the 1D gel (step 8).

21. The gels are finished when the Coomassie blue reaches the bottom. Open the shells and place the gels in staining solution. Placing all the gels in the same tray helps ensure even staining; corners of the gels may be cut to identify individual gels and their orientation. After staining for 2 h to overnight, transfer the gels to destaining solution.

Analysis and Storage of Gels

When the gels are destained, they can be placed on a light box and imaged with a handheld digital camera. A flashcard in the camera and a flashcard reader attached to the computer greatly facilitate image transfer. Gels may also be scanned directly into the computer. Using image manipulation software such as Photoshop or Paint Shop Pro, images may be converted to grayscale, cropped, reoriented, and optimized for contrast and brightness. Images are typically saved as BMP files for Word or Powerpoint for Windows, as TIF files for several quantitation programs such as ImageQuant, or as JPG files for Internet transmission.

Stained gels can be stored wet in heat-sealed plastic bags indefinitely or they may be dried onto paper or clear membrane. Dried gels should also be stored long term in heat-sealed bags because they may curl and crack from changes in ambient humidity.

Figure 1 presents the results of 2D gel analysis of histone proteins from mouse embryonic stem (ES) cells irradiated with 200 Gy (Fig. 1, top) and from unirradiated yeast (Fig. 1, bottom). The gel format was medium (Hoefer SE600) for the mammalian histones and small (ready-made Novex shells) for the yeast histones. H2AX forms (Fig. 1, top) and hence the percentage of γ-H2AX formation (γ-H2AX + γ^2-H2AX) in response to DNA double-strand breaks (DSBs) may be determined by a relative volume measurement (length × width × pixel value) in ImageQuant. For detailed studies of acetylated histone forms, see Pantazis and Bonner.[11]

Other Techniques

Labeling with Radioactive Precursors

Histones are easily labeled with ^{3}H-, ^{14}C-, or ^{35}S-labeled amino acids, keeping in mind that histone H1 and several H2A species lack sulfur-containing residues. Adequate incorporation can be obtained by overnight incubations of cultures proliferating in complete medium. Histone modifications can be labeled by short incubations with radioactive acetate (using cycloheximide to prevent label incorporation into amino acids) or phosphate. The gels are stained and dried onto clear membrane before exposure to film or imaging plates so that the X-ray film can be superimposed on the dried gel to assess the relative migration of mass and radioactivity.

Immunoblotting

The 1D or 2D gels, stained or unstained, may be electroblotted using SDS transfer buffer. Stained gels should be equilibrated for 1 h with equilibration solution. Histones, which have a positive net charge at pH values below 11, will migrate away from the membrane if SDS is absent from the solutions. SDS also releases the proteins from Coomassie blue-stained gels.

1. Soak stained gels in equilibration solution for 1 h at 20°.
   Equilibration solution

Tris base	0.15 g
SDS	1 ml of a 10% stock
Water	To 100 ml

2. Place unstained gels or equilibrated stained gels in the electroblotting apparatus. Fill with SDS transfer buffer and electrophorese for 1.25 h at 25 V (constant) or for 1 h at 100 mA (constant).
   Transfer buffer

Glycine	7.2 g
Tris base	1.5 g
SDS	1 ml of 10% stock
Water	To 800 ml
Methanol	200 ml

γ-H2AX Labeling with $^{32}PO_4$ or $^{33}PO_4$

In response to cellular DNA DSB formation, histone H2AX in mammals and its orthologs in other species rapidly becomes phosphorylated on conserved serine four residues from the C terminus. Cells are irradiated on ice, preventing γ-H2AX formation. With monolayer cultures, the cold medium is removed and a minimal amount (about 1 ml for a 10-cm dish) of warm phosphate-free medium containing ~1 mCi of $^{32}PO_4$ or $^{33}PO_4$ per milliliter of medium is added. With suspension cultures, the ice-cold cell pellets are irradiated and then resuspended in the warm medium. After a 10- to 20-min incubation at 37° in 5% CO_2, the cells are lysed with ice-cold lysis buffer and the nuclei are pelleted for salt washing and acid extraction.

UVA and Laser-Guided DNA Double Strand Break Formation

DNA DSBs may be induced in cells grown in bromodeoxyuridine (BrdU), incubated with Hoeschst dye 33258, and exposed to UVA light.[12] If the UVA light is in the form of a laser microbeam from a laser dissection or a laser scanning confocal microscope,[13] it is not necessary to grow the

[12] E. P. Rogakou, D. R. Pilch, A. H. Orr, V. S. Ivanova, and W. M. Bonner, *J. Biol. Chem.* **273**, 5858 (1998).

[13] E. P. Rogakou, C. Boon, C. Redon, and W. M. Bonner, *J. Cell Biol.* **146**, 905 (1999).

cells in BrdU.[14,15] H2AX is rapidly phosphorylated at these sites, followed by accumulation of other DNA repair proteins. Because these techniques are rapidly evolving, interested readers should consult the literature.

[14] J. S. Kim, T. B. Krasieva, V. LaMorte, A. M. Taylor, and K. Yokomori, *J. Biol. Chem.* **277,** 45149 (2002).
[15] J. Walter, T. Cremer, K. Miyagawa, and S. Tashiro, *J. Microsc.* **209,** 71 (2003).

[6] Assembly of Yeast Chromatin Using ISWI Complexes

By Jay C. Vary, Jr., Thomas G. Fazzio, and Toshio Tsukiyama

Introduction

Chromatin structure is known to play important roles in the regulation of cellular processes, such as transcription, recombination, replication, and repair. To more fully understand the role of chromatin and the enzymes that regulate its structure, it is often useful to study these enzymes in a minimal system, using purified components *in vitro*. The availability of suitable chromatin substrates is essential to these types of experiments, and a number of methods have been developed to reconstitute chromatin *in vitro*.

Nucleosomes have been reconstituted by salt dialysis since they were first described as repeating units of histones and DNA.[1,2] However, chromatin resulting from salt dialysis does not display the characteristic evenly spaced nucleosomes found *in vivo*. As a result, a number of methods using cell-free extracts from *Xenopus, Drosophila,* and human have been developed.[3–5] Although chromatin assembled using the endogenous factors in these extracts does exhibit normally spaced nucleosomes, the presence of other potential chromatin-binding proteins in the extract makes these substrates unsuitable for assays requiring more minimal systems.

A member of the *Drosophila* ISWI class of ATP-dependent nucleosome-remodeling factors, ACF, has been shown to both assemble and evenly space nucleosomes *in vitro* in the presence of the histone

[1] P. Oudet, M. Gross-Bellard, and P. Chambon, *Cell* **4,** 281 (1975).
[2] R. D. Kornberg, *Science* **184,** 868 (1974).
[3] T. Nelson, T. S. Hsieh, and D. Brutlag, *Proc. Natl. Acad. Sci. USA* **76,** 5510 (1979).
[4] G. C. Glikin, I. Ruberti, and A. Worcel, *Cell* **37,** 33 (1984).
[5] B. Stillman, *Cell* **45,** 555 (1986).

chaperone dNAP-1. Both the ACF and NAP-1 components utilized in this system can be purified from recombinant sources, thus producing a chromatin substrate likely to be free of contaminating proteins.[6] Three distinct complexes of this ISWI class are found in the yeast *Saccharomyces cerevisiae:* the Isw1a, Isw1b, and Isw2 complexes. Each of these complexes exhibits nucleosome-stimulated ATPase activity, and the ability to evenly space nucleosomes, *in vitro.*[7,8]

Here we describe the *in vitro* reconstitution of evenly spaced nucleosomal arrays, using a system consisting of only recombinant yeast histones, recombinant yeast Nap1, and a member of the yeast ISWI class of ATP-dependent chromatin-remodeling factors. This system produces a clean substrate, similar to the ACF method; however, the availability of histone mutants from *S. cerevisiae* allows reconstitution with a wide variety of mutant nucleosomes, all of which can be expressed recombinantly. In addition, these recombinant histones are devoid of posttranslational modifications, resulting in a homogeneous population, which may be uniformly modified if so desired. This system, therefore, provides a great degree of experimental flexibility in its application.

Reconstitution of Recombinant Histone Octamer

Both mononucleosomes and nucleosomal arrays can be reconstituted using nearly any desired DNA and histones purified from HeLa, *Drosophila,* or *Xenopus* extracts for use in biochemical assays. Purified native *S. cerevisiae* histones, however, have been difficult to obtain. Luger *et al.* have described a method for the expression and purification of recombinant histone subunits as well as their reconstitution into octamer and nucleosomes.[9] Here we describe a modified version of this protocol, which has been adapted for yeast histone subunits, and also for a smaller scale that is appropriate for many biochemical assays. For a comparison with other techniques used for reconstitution of octamer, please see [2] in this volume and [20] in volume 377.[9a]

[6] T. Ito, M. E. Levenstein, D. V. Fyodorov, A. K. Kutach, R. Kobayashi, and J. T. Kadonaga, *Genes Dev.* **13**, 1529 (1999).
[7] T. Tsukiyama, J. Palmer, C. C. Landel, J. Shiloach, and C. Wu, *Genes Dev.* **13**, 686 (1999).
[8] J. C. Vary, Jr., V. K. Gangaraju, J. Qin, C. C. Landel, C. Kooperberg, B. Bartholomew, and T. Tsukiyama, *Mol. Cell. Biol.* **23**, 80 (2003).
[9] K. Luger, T. J. Rechsteiner, and T. J. Richmond, *Methods Enzymol.* **304**, 3 (1999).
[9a] K. Luger, P. N. Dyer, R. S. Edayathumangalam, C. L. White, Y. Bac, S. Chakrabarthy, and U. M. Muthurajan, *Methods Enzymol.* **375**, [2], (2004) (this volume); B. Carins, J. Wittmeyer, and A. Saha, *Methods Enzymol.* **377**, [20] (2004).

Expression of Recombinant Histones and Purification

We have made the following modifications to the protocols for *Xenopus laevis* histone expression and yeast histone expression.[9,10] The scale has been reduced from 6 to 2 liters of culture and, unless otherwise noted, a corresponding reduction in solution volumes has been made.

1. Expression plasmids containing yeast histone subunits were kindly provided by T. Richmond and T. Rechsteiner (ETH Zurich, Switzerland). The H4 sequence has been extensively modified to more closely match the preferred codon usage of *Escherichia coli*. To account for other codon usage biases between *S. cerevisiae* and *E. coli*, all histones can be expressed in BL21-CodonPlus(DE3)-RIL cells (230245; Stratagene, La Jolla, CA).

2. After a small-scale expression test, two precultures are used to inoculate four 500-ml cultures, which are grown to an OD_{600} of 0.4.

3. Isopropyl-β-D-thiogalactopyranoside (IPTG) is added to 0.2 mM and induced at 37° for 3 h (H3 and H4) or for 4 h (H2A and H2B).

4. Cells are harvested at room temperature, resuspended in wash buffer [50 mM Tris-HCl (pH 7.6), 100 mM NaCl, 1 mM EDTA, 5 mM 2-mercaptoethanol (2-ME), and 0.2 mM phenylmethylsulfonylfluoride (PMSF)], and flash-frozen as described.

5. The pellet is thawed at 30° with constant gentle stirring followed by DNA shearing by sonication on a Branson Sonicator on ice with six 10- to 15-s pulses until the viscosity is reduced.

6. Inclusion bodies are recovered by centrifugation for 20 min at 14,000 rpm in a Beckman (Fullerton, CA) JA-17 rotor, and the pellet is resuspended/broken apart in TW buffer [wash buffer plus 1% (v/v) Triton X-100]. Additional sonication steps (four 10-s pulses) may be necessary to shear DNA if cells were not efficiently lysed in the prior step.

7. Centrifuge as described previously, and wash the pellet once more in TW buffer and twice in wash buffer before storing the pellet at −80°.

8. Pellets are soaked in dimethyl sulfoxide (DMSO) for 30 min at room temperature and then twice resuspended/gently rotated in unfolding buffer [7 M guanidinium-HCl, 20 mM Tris-HCl (pH 7.5), and 10 mM dithiothreitol (DTT)] for 1 h, followed by a room temperature 20-min spin at 14,000 rpm in a JA-17 rotor to remove cell debris. The pooled supernatants are dialyzed against three changes of freshly made urea dialysis buffer [7 M urea (from a deionized fresh 8 M stock), 10 mM Tris-HCl (pH 8.0), 100 mM NaCl, 1 mM EDTA, 5 mM 2-ME, and 0.2 mM

[10] M. E. Gelbart, T. Rechsteiner, T. J. Richmond, and T. Tsukiyama, *Mol. Cell. Biol.* **21**, 2098 (2001).

TABLE I
CONDITIONS USED FOR SALT GRADIENT ELUTION OF HISTONE PROTEINS

H2A, H2B		H3		H4	
Volume (ml)[a]	% B[b]	Volume (ml)[a]	% B[b]	Volume (ml)[a]	% B[b]
10	10	10	20	10	20
125	10–35	10	20–25	125	20–45
5	35–40	125	25–50	20	45–60
50	100	10	50–60	50	100
		50	100		

[a] The volume of buffer used at each salt concentration.
[b] The percentage of UBuffer–1 M NaCl, for example, 10% B is 100 mM NaCl. Ranges represent linear gradients of the corresponding volume.

PMSF] over the course of 5 h, and centrifuged for 20 min at 14,000 rpm in a JA-17 rotor to remove insoluble material.

9. Inject samples onto an HR10/10 column packed with 8 ml of Q Sepharose Fast Flow followed by an identical column packed with 8 ml of SP Sepharose Fast Flow (Amersham Biosciences, Uppsala, Sweden), both preequilibrated in UBuffer [7 M deionized urea, 10 mM Tris-HCl (pH 8.0), 1 mM EDTA, 1 mM DTT, and 0.2 mM PMSF] mixed to 10% (H2A or H2B) or 20% (H3 or H4) UBuffer–1 M NaCl (10% B or 20% B). After the (nonbinding) flow-through material has passed entirely through the system, remove the Q Sepharose column, which will have bound many nonhistone proteins and DNA.

10. Begin the salt gradient according to the histone subunit being purified (see Table I) and collect fractions.

11. Fractions are analyzed on 16% SDS PAGE gels, and peak fractions are pooled and dialyzed against three changes of H_2O, the second of which should be done overnight.

12. The dialyzed fractions are then lyophilized, followed by resuspension in H_2O and quantitated, using the extinction coefficients listed in Table II. Histone proteins can then be separated into aliquots appropriate for octamer reconstitution (~3 mg/tube) and lyophilized once more.

Reconstitution of Histone Octamer

Histone octamer can be reconstituted by salt dialysis as described previously for *X. laevis* or yeast.[10,11] What follows is a brief description.

[11] K. Luger, A. W. Mader, R. K. Richmond, D. F. Sargent, and T. J. Richmond, *Nature* **389**, 251 (1997).

TABLE II
EXTINCTION COEFFICIENTS AND MOLECULAR WEIGHTS FOR YEAST HISTONE PROTEINS

Yeast histone	Extinction coefficient[a] (276 nm)	Molecular weight (g/mol)
H2A	4350	13,858
H2B	7250	14,106
H3	2900	15,225
H4	5800	11,237

[a] From C. White and K. Luger.

1. Each histone aliquot is dissolved to 2 mg/ml in unfolding buffer for 2–3 h. Using a total of about 3 mg for each histone, the four histones are mixed in equimolar ratios and adjusted to 1 mg/ml with unfolding buffer.

2. Histones are then gradually refolded by dialysis against four changes of 2 liters of refolding buffer [2 M NaCl, 10 mM Tris-HCl (pH 7.5), 1 mM EDTA, 5 mM 2-ME, 2 mM PMSF] at 4°. The second and fourth dialysis steps should be performed overnight.

3. The refolded histones are then centrifuged for 20 min at 14,000 rpm in a JA-17 rotor at 4° to remove insoluble material, and then concentrated to 1 ml, using a Centricon-10 spin column (Millipore, Bedford, MA) and filtered with an Ultrafree-MC spin filter (Millipore).

4. The histone mixture is then injected into a Superdex-200 chromatographic column preequilibrated in refolding buffer and run at 1 ml/min. Fractions containing aggregate, octamer, tetramer, dimer, and monomers can be detected on Coomassie blue-stained 16% SDS PAGE gels. Pure peak fractions can then be pooled, further concentrated on a Centricon-10 column, and diluted to 2 M NaCl, 50% glycerol before storage at −20°. A 1-mg/ml solution of yeast octamer gives an OD_{276} absorbance of 0.45.

Reconstitution of Recombinant Nucleosomes

Nucleosomes can be reconstituted by salt dialysis on a DNA fragment using histone octamers from a variety of organisms. However, yeast histones do not form nucleosomes by this process as readily as those purified from other organisms. We therefore recommend using the histone chaperone Nap1, as described below, to assist in nucleosome reconstitution with recombinant yeast histones. We have, however, successfully reconstituted mononucleosomes using a nucleosome positioning sequence from the *Xenopus borealis* 5S RNA gene, and the protocol for this follows.

Reconstitution of Nucleosomes by Salt Dialysis

As the optimal octamer-to-DNA ratio may change depending on the sequence used, titration may be required to find the optimal ratio for each system. What follows is a sample reaction that can be used as a starting point for further refinements.

1. A reaction is assembled in a custom microdialysis chamber (see Fig. 1) in the following order: Hi buffer [10 mM Tris-HCl (pH 7.6), 1 mM EDTA, 2 M NaCl, 0.05% Nonidet P-40 (NP-40), 5 mM 2-ME] to 150 μl (final volume), 75 μg of bovine serum albumin (BSA), 1.5 μg of DNA, and finally 1.5 μg of octamer.

2. Add the dialysis chamber to a beaker with 600 ml of prechilled Hi buffer at 4° for 45 min, stirring rapidly to ensure adequate equilibration.

3. Using two pumps, each set to 3 ml/min, pump Lo buffer [10 mM Tris-HCl (pH 7.6), 1 mM EDTA, 50 mM NaCl, 0.05% NP-40, 5 mM 2-ME] into the beaker top (using one pump) while simultaneously removing liquid from the bottom (using the other pump) to prevent overflow and ensuring a linear gradient. Run for 20 h or until the Lo buffer runs out.

4. Recover chambers and blot away excess liquid with a Kim-Wipe (Kimberly-Clark, Roswell, GA). Pierce the dialysis membrane with a pipette tip and recover the dialysate. The reconstituted nucleosomes can be stored at 0° for 1 week.

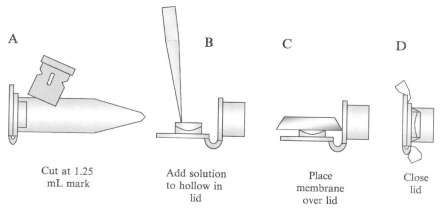

A	B	C	D
Cut at 1.25 mL mark	Add solution to hollow in lid	Place membrane over lid	Close lid

FIG. 1. Microdialysis chamber for nucleosome reconstitution. The lid of a siliconized 1.7-ml microcentrifuge tube is used to create a dialysis chamber for small volumes of liquid. (A) The tube is cut with a razor blade at the 1.25-ml mark. (B) The solution to be dialyzed is added into the hollow of the tube cap. (C) Dialysis membrane is placed over the hollow. (D) The lid is closed, forming a water-tight seal around the dialysis chamber. *Note:* It is recommended that each chamber be tested with dialysis membrane before adding the sample, as many will not close securely with the membrane in place.

Note. The efficiency of loading can be assessed by adding to the reaction a trace amount of ^{32}P-labeled DNA. A gel shift can be observed using a 5% polyacrylamide gel in 0.25× TBE by autoradiography. It is important not to use loading dyes, however, as they disrupt histone–DNA interactions.

Reconstitution of Nucleosomes by Nap1

Recombinant yeast Nap1 can be purified easily as described previously, permitting a highly defined assembly reaction.[12] The following reaction is relatively small in scale but may be scaled up as needed. As described above, the optimal octamer-to-DNA ratio needs to be titrated for efficient nucleosome formation.

1. ExB 5/50 buffer [10 mM HEPES-KOH (pH 7.6), 50 mM KCl, 5 mM MgCl$_2$, 0.5 mM EGTA, 10% glycerol, BSA (0.1 mg/ml)] is added to 30 μl (final volume), followed by 0.75 μg of Nap1 and 0.22 μg of octamer, and incubated on ice for 30 min.

2. DNA (0.25 μg) is then added, and the reaction is incubated at 30° for 4 h.

3. Efficiency can be assayed by gel shift as described above if necessary.

Note. The DNA utilized in this reconstitution may also be immobilized to streptavidin-coated magnetic beads by end-filling, using biotinylated dATP. This permits the use of a 600 mM NaCl wash following assembly to remove excess Nap1 and histone proteins from the substrate.[10]

Purification of Yeast ISWI Complexes

ISWI complexes can be purified to near homogeneity by the use of epitope tags on their subunits, and the ease of genetic manipulations in yeast permits the construction of these epitope-tagged proteins with relatively little effort. Several variations of yeast ISWI complex purifications have been published;[7,8,10,13] here we present our preferred methods for the purifications of each complex. The use of three tandem FLAG epitopes (3 × FLAG), or the tandem affinity purification (TAP) tag, is discussed in detail. In both instances, the epitope tag is fused to the C-terminal portion of the protein of interest, usually Isw1p or Isw2p. Although these constructs can be expressed from a plasmid, we find that integration of these constructs at their endogenous loci is preferable as it allows growth in nonselective rich media.

[12] T. Fujii-Nakata, Y. Ishimi, A. Okuda, and A. Kikuchi, *J. Biol. Chem.* **267,** 20980 (1992).
[13] S. R. Kassabov, N. M. Henry, M. Zofall, T. Tsukiyama, and B. Bartholomew, *Mol. Cell. Biol.* **22,** 7524 (2002).

Using equivalent amounts of extract, FLAG purification yields more complex than TAP purification; however, the purity is somewhat lower unless it is followed by ion-exchange chromatography. Purification of Isw1p results in a mixture of both Isw1a and Isw1b complexes. Although both are enzymatically active for nucleosome spacing, each can be purified separately, if desired, by using strains bearing *ioc2* (for Isw1a purification) or *ioc3* (for Isw1b) deletions, or by utilizing tags on the Ioc2p or Ioc3p subunits themselves. It should also be noted that purification of Isw1p–TAP results in little Isw1b complex recovery, presumably because the large size of the TAP tag destabilizes this complex. This bias is not observed with the FLAG tag, however.

Whole Cell Extract Preparation

Cultures are grown in rich medium to the desired density and harvested by centrifugation. We routinely use 2-liter cultures grown to near saturation or 12-liter cultures grown to midlog phase, although no significant differences in activity have been observed between yeast ISWI complexes purified at either density. The cell pellet is washed twice with H_2O (including 4 mM PMSF, 1 mM sodium metabisulfite, and 2 mM DTT) and once with buffer H-0.3 [25 mM HEPES-KOH (pH 7.6), 0.1 mM EDTA, 0.5 mM EGTA, 2 mM $MgCl_2$, 20% glycerol, 0.02% NP-40, 300 mM KCl, 1 mM PMSF, 2 μM pepstatin A, 0.6 μM leupeptin, chymostatin (2 μg/ml), 2 mM benzamidine, 0.5 mM sodium metabisulfite, and 1 mM DTT—the number following buffer H refers to the molar concentration of KCl]. The cell pellet can be flash-frozen in liquid nitrogen, if desired, and stored at $-80°$.

After harvest, a number of methods can be used to break the cell wall of the yeast. We prefer to use a microfluidizer (Microfluidics, Newton, MA) at 120 lb/in^2 ($\sim$800 kPa), because of its ease of use and reproducible results. The cell pellet is resuspended in an equal volume of buffer H-0.3 and is passed through the microfluidizer four times, resulting in 10–30% breakage of cells. Although additional passes result in higher levels of cell lysis, our total protein recovered diminishes, presumably because of denaturation of the proteins. If a microfluidizer is not available, the cells can be similarly processed with a French press or BeadBeater (BioSpec Products, Bartlesville, OK). Alternatively, the cell pellet may be frozen before resuspension and ground in any commercially available rotary coffee grinder, blender, mortar and pestle, or electric mortar grinder in the presence of excess dry ice, and resuspended in a volume of buffer H-0.3 equal to the volume of the original cell pellet. Please see [26] in volume 377[13a] for details on the use of a blender for cell lysis.

[13a] X. Shen, *Methods Enzymol.* **377,** [26] (2004).

The lysed mixture obtained by any of the above methods is then centrifuged in a Beckman SW40Ti rotor at 27,000 rpm for 90 min, and the supernatant (whole cell extract) is recovered.

FLAG Purification

FLAG-tagged ISWI complexes can be immunopurified efficiently with anti-FLAG M2 agarose beads (A1205; Sigma, St. Louis, MO) as described previously,[7] or with magnetic beads coupled to anti-FLAG M2 antibody [F3165 (Sigma) and 100.04 (Dynal, Oslo, Norway)]. Here we describe the precipitation using agarose beads, but magnetic beads can be substituted with minor modifications. Also see [26] in volume 377[13a] describing a similar method for FLAG purification of the INO80 complex.

1. Approximately 300 μl of agarose beads is preequilibrated in buffer H-0.3 and incubated with whole cell extract from the equivalent of 1 liter of saturated phase cells (or 6 liters of midlog phase cells) at 4° for 3 h to allow binding of the FLAG-tagged proteins.

2. The beads are washed twice in 10 ml of buffer H-0.3 for 5 min, pelleted, and transferred to a siliconized 1.7-ml microcentrifuge tube. The beads are washed six times in 1.2 ml of buffer H-0.3 for 5 min each, followed by three washes in buffer H-0.1 to reduce the salt concentration if ion-exchange chromatography is to follow.

3. Protein is then eluted by washing with 1 bed volume of 3 × FLAG peptide (1 mg/ml, F4799; Sigma) in buffer H-0.1 at 4° for 30 min. The beads are pelleted and the supernatant (containing the FLAG-tagged complex) is transferred to a siliconized microcentrifuge tube and quick-frozen in liquid nitrogen. This elution is repeated three more times, and each eluate is frozen as described previously.[7]

Ion-Exchange Chromatography

Although elution from anti-FLAG beads results in a relatively clean sample, further purification by ion-exchange chromatography removes many residual contaminants, including the FLAG peptide. In addition, a significant portion of Isw1p is recovered as a monomer after FLAG immunopurification. Chromatography allows removal of this enzymatically less active species, as well as the concentration of the active complexes.

Isw1 or Isw2 complexes in buffer H-0.1 are injected into HR5/2 columns (Amersham Biosciences) packed with a 0.2-ml bed volume of BioRex 170 (Bio-Rad, Hercules, CA) or Source 15Q (Amersham Biosciences) resin, respectively, which have been preequilibrated in Buffer H-0.2 (Isw1 complexes) or Buffer H-0.1 (Isw2 complexes). The column is washed with 40 column volumes before a linear salt gradient in buffer H is applied from 0.2 to 0.6 M KCl (Isw1 complexes) or 0.1 to 0.6 M KCl (Isw2 complex). Monomer Isw1p does not bind to the resins at 0.2 M KCl and is easily purified from the flow-through and early fractions of the wash, if desired. Isw2 complex and both Isw1 complexes elute between 0.25 and 0.4 M KCl.[7,8]

TAP Purification

Tandem affinity purification (TAP) is done essentially as described previously.[8,14] Briefly, the TAP tag consists of two IgG-binding domains and a calmodulin-binding peptide domain, which are separated by a site for the tobacco etch virus (TEV) protease. Tagged proteins are precipitated with IgG Sepharose 6 Fast Flow (Amersham Biosciences) and washed extensively. They are then eluted with TEV protease (Invitrogen, Carlsbad, CA), precipitated again with calmodulin-coupled agarose beads (Stratagene) in the presence of calcium, and washed extensively once more. The complexes are then eluted by washes with EGTA-containing buffer, which chelates the calcium, thereby releasing the bound complexes. We have found that Isw2 complexes elute easily from the calmodulin beads, whereas Isw1 complexes sometimes require additional final elution steps. Additional EGTA in the elution buffer may help sharpen the elution profile for Isw1 complexes.

Purification of Overexpressed Isw2–Itc1 Complex

For most of the biochemical assays utilized in our laboratory, Isw1- and Isw2-containing complexes purified as described previously in this chapter are sufficient. However, if one wishes to perform an experiment that requires large quantities of purified proteins, such as large-scale assembly of regularly spaced nucleosomal arrays or footprinting of ISWI complexes on chromatin, more concentrated preparations of protein complexes are desirable. Although the most common method for obtaining high concentrations of protein complexes is to express each subunit individually in bacteria, purify them separately, and reconstitute the complex *in vitro*, this method has the disadvantage of producing unmodified proteins that may not be properly folded.

In this section we describe a simple method of overproducing the two-subunit Isw2–Itc1 complex in yeast along with a two-step purification procedure that yields highly purified complex. In our case, we use a yeast strain

[14] G. Rigaut, A. Shevchenko, B. Rutz, M. Wilm, M. Mann, and B. Seraphin, *Nat. Biotechnol.* **17**, 1030 (1999).

in which the *ISW2* gene is tagged at its C terminus with two copies of the FLAG epitope, whereas the *ITC1* gene is C terminally tagged with three copies of the FLAG epitope. Both genes are placed under the control of separate copies of the *GAL1* promoter and integrated at the endogenous loci of the *ISW2* and *ITC1* genes. Thus, when grown in the presence of galactose, this strain produces large quantities of FLAG-tagged Isw2 and Itc1 proteins.

In addition to yielding significantly more Isw2 complex ($\sim$30- to 40-fold) than we obtain by TAP purification from cells producing wild-type levels, the overexpressed Isw2 complex is also more concentrated ($\sim$10-fold) than fractions from a typical TAP purification of Isw2 complex. Therefore, Isw2–Itc1 complex purified in this manner is suitable for a wide range of biochemical analyses.

1. An overnight culture is diluted 1000-fold into 6 liters of YEP plus 2% galactose and grown overnight until the optical density of the culture at 660 nm is 0.7 (midlog phase).

2. Whole cell extract is prepared with a microfluidizer as described previously in this chapter, except that buffer H-0.25 is used.

3. Approximately 1.5 ml of anti-FLAG M2 bead slurry (Sigma) is equilibrated in buffer H-0.25. The pelleted beads are then resuspended in approximately 13 ml of extract and the mixture is rotated at 4° for 3 h to allow binding of the FLAG-tagged proteins.

4. The beads are washed as described previously in this chapter for Isw1 and Isw2 complexes.

5. To elute FLAG-tagged Isw2 complex, the beads are resuspended in 1.4 ml of buffer H-0.1 containing 3 × FLAG peptide (1 mg/ml; Sigma) and rotated at 4° for 30 min. The elution is repeated three more times, with the final two performed at room temperature rather than 4°.

6. To concentrate the purified protein, as well as to remove FLAG peptide and residual contaminants, the four FLAG elutions are combined and run through an HR5/2 column with 0.2 ml of Source 15Q resin (Amersham Biosciences), using fast protein liquid chromatography (FPLC). After washing with three column volumes of buffer H-0.1, purified Isw2 complex is eluted on a 0.1–0.6 M KCl gradient over 10 column volumes, collecting 0.4-ml fractions. The fractions are transferred to siliconized microcentrifuge tubes and quick-frozen in liquid nitrogen.

The majority of the purified Isw2–Itc1 complex elutes at $\sim$0.25–0.35 M KCl (Fig. 2). We calculate the total yield from this procedure as approximately 20 μg of each subunit, amounts sufficient for many different biochemical assays.

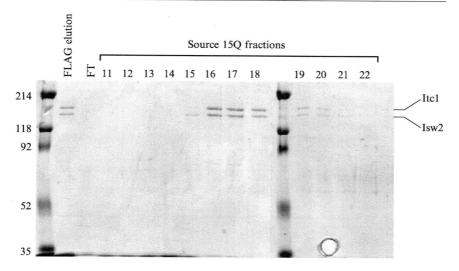

FIG. 2. Source 15Q purification of overexpressed Isw2–Itc1 complex. Isw2 and Itc1 were each overexpressed and purified from yeast by FLAG immunopurification. The eluates were loaded onto an HR5/2 column packed with 0.2 ml of Source 15Q resin (Amersham Biosciences), washed with three column volumes, and eluted in a linear salt gradient from 0.1 to 0.6 M KCl. The peak fractions (16–18) correspond to ~0.25–0.35 M KCl.

ATPase Assay for ISWI Complex Activity

To confirm the activity of ISWI complexes purified by the methods described in this chapter, we utilize an assay for the intrinsic ATPase activity of the Isw1p or Isw2p subunit, using radioactively labeled ATP. All three yeast ISWI complexes show minimal stimulation by DNA or histone octamer alone, yet show strong nucleosome-stimulated activity.[7,8]

This protocol has been slightly modified from the originally published protocol.[15] We have used both mononucleosomes and nucleosomal arrays with similar results. After incubation of ISWI complex with nucleosomes and ATP, the mixture is separated by thin-layer chromatography (TLC) to resolve ATP and ADP from free P_i, and the activity is determined by the ratio of P_i to ATP.

1. In a total volume of 5 μl, the following reagents are mixed in order: 25 ng of DNA or the equivalent of octamer or nucleosomes, or buffer alone; a premixed cocktail of 0.1 mM cold ATP, [γ-^{32}P]ATP at 1–2 mCi/ml, and 5 mM MgCl$_2$; and ISWI complex equivalent to Isw1p or Isw2p at 0.8 μg/ml. The reaction is allowed to proceed at 30° for 30 min.

[15] T. Tsukiyama and C. Wu, *Cell* **83**, 1011 (1995).

2. A 0.5-μl volume of each reaction is spotted onto a 20 × 10 cm cellulose PEI-F (Mallinckrodt Baker, Phillipsburg, NJ) TLC sheet, 2 cm from the bottom and at least 1.5 cm apart from its closest neighbor. A hair dryer or heat gun is then used to dry the spots to prevent diffusion. The sheet is then placed upright in a suitable container with enough 0.8 M LiCl, 0.8 M acetic acid to submerge the bottom ~1 cm of the sheet. When the solution front is just below the top of the sheet, it is removed and dried thoroughly with the hair dryer or heat gun as described above.

3. To quantitate the ratio of the faster moving free ^{32}P$_i$ to the nonhydrolyzed [γ-^{32}P]ATP, the sheet may be used to directly expose a phosphor screen. Alternatively, it may be used to expose a piece of X-ray film for ~30 min. The developed film can then be used as a guide to trace and cut out the regions of the sheet containing free ^{32}P$_i$ and [γ-^{32}P]ATP in each lane for scintillation counting. Either quantitation method may be used to calculate the percentage of ATP that was hydrolyzed by the ISWI complex.

Note. It is extremely important to use a source of [γ-^{32}P]ATP that has been purified to remove unincorporated ^{32}P$_i$, which will clearly interfere with the subsequent TLC analysis. We have found that Amersham Biosciences product AA0018 has been reliable, although other manufacturers may provide an equivalent product.

Assembly of Regularly Spaced Nucleosome Arrays

The addition of an ISWI complex and ATP to the Nap1-mediated nucleosome reconstitution protocol described earlier permits the assembly of regularly spaced nucleosome arrays. Histones are assembled onto DNA to create a nucleosomal array in the presence of Nap1, as described previously. In the presence of ISWI complex, the nucleosomal array will exhibit a characteristic spacing.[7,8] We describe a small-scale assay using this technique to confirm that all the reagents utilized are competent for assembly and spacing. By scaling up this nucleosomal spacing assay, large amounts of regularly spaced recombinant nucleosomal arrays can be generated for use in other types of biochemical assays, as described in the introduction to this chapter.

When Nap1 is used to assemble nucleosomes onto a long segment of DNA in the absence of an ISWI or other spacing activity, there is no characteristic spacing between adjacent nucleosomes. As a result, when a large population of these arrays of nucleosomes is partially digested with micrococcal nuclease (MNase), which preferentially cuts DNA between the nucleosomes, the resulting DNA fragments corresponding to dinucleosomes, trinucleosomes, and so on, will be a range of sizes. In contrast, MNase digestion of a regularly spaced nucleosome array results in similarly

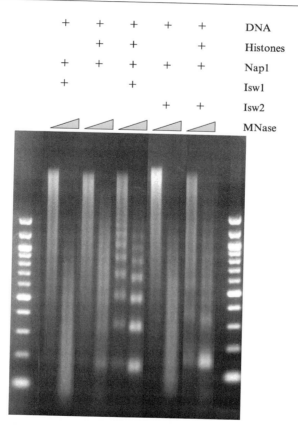

FIG. 3. Nucleosomal spacing assay for Isw1 and Isw2. DNA was incubated with histone octamer, Nap1, Isw1 complexes, and/or Isw2 complex, as indicated, in the presence of ATP for 30 min at 30°. Micrococcal nuclease (MNase) was used to digest the DNA for 3- and 15-min time periods. After proteolytic digestion, DNA samples and a 100-bp ladder were run on a 1.3% agarose gel and stained with ethidium bromide.

sized DNA fragments for each dinucleosome, trinucleosome, etc., species. When these digests are run on an agarose gel, a smear is observed for the irregularly spaced array (Fig. 3), whereas the regularly spaced array results in a characteristic "nucleosomal ladder." As described before, the histone octamer-to-DNA ratio given may be used as a starting point for titration of the optimal ratio required for each experimental system.

1. ExB 5/50 buffer is added to adjust reaction volume to 30 μl (final volume), followed by 0.75 μg of Nap1, 0.22 μg of octamer, and ISWI complex (4 ng of Isw1p or Isw2p equivalent). This is incubated on ice for 30 min.

2. To this is added 0.25 μg of λ DNA, 3 mM ATP, and an ATP regeneration system.[16] The reaction is incubated at 30° for 4 h.

3. CaCl$_2$ is added to 2 mM and 5–15 units of MNase is added to the samples.

4. At 3- and 15-min intervals, 14.5 μl of each reaction is removed and added to a new tube with 5 μl of stop solution [glycogen (4 mg/ml), 2% Sarkosyl, 80 mM EDTA]. After proteinase K digestion, the DNA is precipitated and run on a 1.3% agarose gel. The smaller bands can be obscured by running dyes such as bromphenol blue; we routinely use orange G, which migrates faster than the mononuclesome band.

Concluding Remarks

The purification and assembly protocols described in this chapter are the result of many refinements of previously existing techniques, and additional increases in efficiency will likely be achieved by careful titrations and modifications of these methods if larger quantities or specific activities are required. Once the quality and activity of the reagents described in this chapter have been assayed by the techniques mentioned, regularly spaced nucleosomal arrays using recombinant components can be assembled as desired by increasing the scale of the nucleosome spacing assay described earlier. The Isw1a and Isw1b complexes space nucleosomes ~175 bp apart, whereas the Isw2 complex yields an ~200-bp spacing, thus providing two options for regularly spaced arrays.

We have also found that these arrays can be assembled by the methods described, using immobilized DNA as a template.[8] As assembled nucleosomes are stable in 600 mM NaCl, this permits stringent washing to remove the assembly and spacing activities when no longer needed, so that this minimal system may be used for a number of sensitive applications. Given the large number of specified histone mutants and modifications, and DNA sequence variations that can be used in this system, a high degree of experimental flexibility can be achieved for subsequent studies.

Acknowledgments

We thank Marnie Gelbart for helpful advice and technical expertise in establishing the method of reconstitution of yeast histone octamers, and Drs. Karolin Luger and Tim Richmond for advice and for the yeast histone expression constructs.

[16] P. B. Becker, T. Tsukiyama, and C. Wu, *Methods Cell Biol.* **44,** 207 (1994).

[7] Reconstitution of Yeast Chromatin Using yNap1p

By SRIWAN WONGWISANSRI and PAUL J. LAYBOURN

Introduction

The eukaryotic genome is maintained in a DNA–protein complex known as chromatin. The fundamental unit is the nucleosome, in which the core histone octamer, (H3–H4–H2A–H2B)$_2$, has DNA wound around it in 1.65 turns.[1] Chromatin structure plays an important role in several biological processes, including transcription, through its function in organizing DNA and its ability to block access to the DNA. Transcription can occur in the chromatin environment through the help of accessory transcription coactivators that function to open chromatin structure, providing access to the transcription machinery.

The yeast, *Saccharomyces cerevisiae*, *PHO5* gene is used as a model to study transcription regulation in a chromatin context based on the detailed understanding of the *in vivo* chromatin structure and dynamics during the transition between activated and repressed conditions.[2,3] Yeast cells have proved to be a powerful experimental organism for investigating the role of chromatin structure in transcription regulation. The largest impediment to biochemical experiments was the lack of a chromatin reconstitution system. In late 1992, no one had developed a procedure for the purification of large quantities of yeast core histones or a method for reconstitution of yeast core histones into nucleosomal templates. In fact, the prevailing attitude was that native yeast core histones could not be purified without using methods such as acid extraction, and that purified yeast core histones could not be reconstituted into nucleosomes. Since then, although it presented many difficulties, we have succeeded in purifying yeast core histones in quantities (milligrams) sufficient for biochemical and biophysical analyses.[4] In addition, we have developed a chromatin reconstitution system wholly derived from yeast. Through the development of this system, we have been able to answer several important biological questions.

[1] K. Luger, A. W. Mader, R. K. Richmond, D. F. Sargent, and T. J. Richmond, *Nature* **389**, 251 (1997).

[2] M. Han and M. Grunstein, *Cell* **55**, 1137 (1988).

[3] M. Han, U. J. Kim, P. Kayne, and M. Grunstein, *EMBO J.* **7**, 2221 (1988).

[4] J. Pilon, A. Terrell, and P. J. Laybourn, *Protein Expr. Purif.* **10**, 132 (1997).

Purification of Native Yeast Core Histones

The procedure for purification of yeast core histones was built on the procedures for histone purification from calf thymus, chicken erythrocytes, and *Drosophila* embryos. The major steps include preparation of nuclei, solubilization of chromatin fragments by digestion with micrococcal nuclease digestion, fractionation of chromatin fragments from the bulk of the cellular components, and purification of the core histones from the genomic DNA by hydroxylapatite chromatography. There are two major differences between the yeast procedure and the original procedures. First, yeast cells have tough cell walls that must be removed before preparation of nuclei. Second, sucrose gradient fractionation was not effective for fractionation of the chromatin fragments and was replaced by gel-filtration chromatography. Another important difference is the addition of ribonuclease A (RNase A) to the micrococcal nuclease digestion. RNase A breaks up the abundant ribonuclear particles in yeast nuclei, thus preventing them from contaminating the chromatin peak of the gel-filtration column. Because of the high cost of using the commercially available Lyticase (Sigma, St. Louis, MO) in the large quantities required for spheroplasting hundreds of grams of yeast cells, we have opted to use recombinant endoglucanase.

β-1,3-Glucanase Production

This protocol is based on one sent to us by the Schekman Laboratory (University of California, Berkeley) in 1992.

Escherichia coli Transformation and Assay Yeast Cell Growth. On day 1, transform DH5α *Escherichia coli* cells with the glucanase expression vector pUV5-G1S (a gift from S.-H. Shen, National Research Council of Canada).[5] At the same time, or on day 2, start a culture of yeast cells for use as assay yeast cells. These cells need to be grown to an optical density at 600 nm (OD_{600}) of between 1 and 4 (a density of 1 to 2 is best). There is a clear decrease in how easily cells will spheroplast with rising culture density. After harvesting the assay cells by centrifugation, wash them with water and resuspend them in 50 mM Tris-HCl, pH 7.4, to about 20 OD_{600} units. Cells may be stored at 4° for up to 1 week. The cold shock buffer, 0.5 mM $MgSO_4$, can be made up ahead of time and chilled to 4°.

Cell Growth and Induction. On day 2, transfer two single colonies from the pUV5-G1S transformation into two 5-ml LB/AMP (Luria broth, 10 g of tryptone, 5 g of yeast extract, and 5 g of NaCl per liter containing ampicillin at 100 μg/ml) cultures and grow them overnight in a shaker at 37°. On

[5] S. H. Shen, P. Chretien, L. Bastien, and S. N. Slilaty, *J. Biol. Chem.* **266,** 1058 (1991).

day 3, inoculate a 100-ml culture of LB/AMP with one of the 5-ml cultures and grow for 2 h with shaking at 37°. Then, inoculate five 1-liter cultures of LB/AMP with 20 ml of the 100-ml culture and grow them in a shaker at 37° to an OD_{600} of 0.5 (about 4 h). At this point, add isopropyl-β-D-thiogalactopyranoside (IPTG) to 0.4 mM to induce expression. Continue shaking at 37° for 5 h. Harvest cells by centrifugation at 5000 rpm for 5 min, in a GS3 rotor or equivalent and freeze and store the bacterial pellets at −80°.

Osmotic or Cold Shocking. All the steps in osmotic or cold shocking are done at room temperature unless otherwise stated. Osmotic or cold shocking extracts periplasmic proteins from *E. coli* without lysing the cells. On day 4, thaw cells to room temperature and wash the cells by resuspending them in 1 liter of 25 mM Tris-HCl, pH 7.4, and centrifuging them out of the buffer as described previously. Resuspend the cells in 1/50th the original culture volume (100 ml) 25 mM Tris-HCl, pH 7.4. Measure the volume with a graduated cylinder and add EDTA to 2 mM. Add an equal volume of 40% sucrose, 25 mM Tris-HCl, pH 7.4, and mix the cells gently for **exactly** 20 min. Spin out the cells in a GSA rotor at 7500 rpm for 10 min. Pour off all the supernatant. Removal of as close to all of the sucrose buffer as possible is important. Place the cell pellets on ice and resuspend them in 1/50th the original volume (100 ml) of cold shock buffer (0.5 mM MgSO$_4$). Mix the cells gently for 20 min on ice. Spin out the cells in an SS34 rotor at 4°, 10,000 rpm for 10 min. The supernatant contains the glucanase activity.

Storage and Further Purification. We have found it most convenient to store the crude periplasmic proteins containing substantial gluconase activity by adding (NH$_4$)$_2$SO$_4$ to 90% saturation and storing at 4°. The Schekman Laboratory suggested an optional purification by CM Sepharose chromatography. For this purification, dialyze the supernatant against 50 mM sodium acetate, pH 5.0. Batch bind the crude extract to CM Sepharose resin preequilibrated in 50 mM sodium acetate, pH 5.0. Pour the resin into a short, wide column and wash the column with 2 to 3 column volumes of 50 mM sodium acetate, pH 5.0. Elute the glucanase activity with 150 mM NaCl, in 50 mM sodium acetate, pH 5.0.

Lyticase/Glucanase Assay. To determine how much glucanase to add to obtain complete spheroplasting, it is important to assay the activity of the preparation. Make fresh 2× assay buffer containing 100 mM Tris-HCl (pH 7.4) and 80 mM 2-mercaptoethanol. Centrifuge 1 ml of glucanase–(NH$_4$)$_2$SO$_4$ solution at 14,000 k rpm for 10 min at 4° in an Eppendorf tube, pour off the supernatant, and resuspend the protein pellet in 1 ml of 2× assay buffer. In each assay reaction, place 1 ml of 2× assay buffer, 0.92 ml of H$_2$O, and 0.08 ml of assay cells (to a final OD_{600} of 0.8 to 0.9). Then add 0 to 50 μl of Lyticase or glucanase. Incubate the reactions at 30° for 30 min while mixing (yeast cells settle quickly). Vortex the reaction and quickly measure the OD_{600}. One unit of glucanase activity is defined as a 10% decrease in OD_{600} in 30 min at 30°. The window of linearity of this assay is narrow, in the range of a 20 to 60% decrease. Measurements obtained with a change of less than 20% will tend to produce high unit-per-milliliter measurements. Measurements obtained with a change greater than 60% will tend to produce low unit-per-milliliter measurements.

Yeast Core Histone Purification

We have purified core histones from 100 g to 1 kg of yeast cells. The procedure does not work well for less than 100 g because of large losses during cell lysis. We have found it best to start with at least 250 g of yeast cells. Therefore, the protocol provided here is based on that amount of starting material. We have not found that use of protease-deficient yeast strains has made an appreciable difference in the quality or the yield of core histones obtained. However, we have found the use of yeast cells harvested in logarithmic growth phase to be critical, as cells become much more resistant to spheroplasting in late log and stationary growth phases.

Solutions. (*Note:* Components indicated with an asterisk are added just before use.) Prespheroplasting buffer (2.5 liters) contains 100 mM Tris-HCl, pH 7.9, and 60 mM 2-mercaptoethanol (2-ME) and is warmed to 37° to help thaw yeast cell pellets. Spheroplasting buffer (1.25 liters at 30°) contains 0.7 M sorbitol [2 M sorbitol stock, filtered through Whatman (Clifton, NJ) no. 1 paper, stored at 4°], 0.75% yeast extract, 1.5% peptone (made from a 5× yeast extract and peptone stock at 5 and 10%, respectively), 10 mM Tris-HCl (pH 7.4) (from a 2 M stock), and 10 mM 2-ME. YPD plus 1 M sorbitol (1.25 liters, 30°) consists of 1% yeast extract, 2% peptone, 2% dextrose, 1 M sorbitol, 10 mM Tris-HCl (pH 6.8), *1 mM phenylmethylsulfonyl fluoride (PMSF), *2 mM benzamidine, and *2 mM $Na_2S_2O_5$. Sorbitol buffer (2.5 liters, 4°) contains 1 M sorbitol, 10 mM Tris-HCl (pH 6.8), 1 mM EDTA, *1 mM PMSF, *2 mM benzamidine, and *2 mM $Na_2S_2O_5$. Lysis buffer (125 ml, 4°) is made up of 18% Ficoll 400, 20 mM KH_2PO_4, 0.25 mM EDTA (0.5 M, pH 8.0 stock), 0.25 mM EGTA (0.5 M, pH 8.0 stock), 0.5 mM spermidine (50 mM stock), 0.15 mM spermine (50 mM stock), and is adjusted to pH 6.8 with KOH. Reducing agents and protease inhibitors are then added to *3 mM dithiothreitol (DTT), *2 mM Benzamadine, *2 mM $Na_2S_2O_5$, *1 mM PMSF, *2 μM pepstatin A, *0.6 μM leupeptin, and *2 × 10^{-4}% chymostatin. Nuclear storage buffer (125 ml, 4°) contains 100 mM Tris acetate (pH 7.9), 50 mM potassium acetate, 20% glycerol, 2 mM EDTA, *3 mM

DTT, [*]2 mM Benzamadine, [*]2 mM $Na_2S_2O_5$, [*]1 mM PMSF, [*]2 μM pepstatin A, [*]0.6 μM leupeptin, and [*]2 × 10^{-4}% chymostatin. A separate batch of
nuclear storage buffer (3 liters, 4°) is made that contains only a subset of
reducing agent and protease inhibitors: [*]3 mM DTT, [*]2 mM Benzamadine,
[*]2 mM $Na_2S_2O_5$, and [*]1 mM PMSF. Hydroxylapatite chromatography
buffers (4°, filtered through 0.22-μm pore size filters) are buffer A (1.5
liters), containing 80 mM Na_2HPO_4, pH 6.8, and 10% glycerol, and buffer
B (1 liter) containing 80 mM Na_2HPO_4 (pH 6.8), 2.5 M NaCl, and 10%
glycerol. Both buffer A and buffer B contain [*]0.1 mM PMSF, [*]2 mM
$Na_2S_2O_5$, and [*]1 mM DTT.

   Preparation of Nuclei. Thaw 250 g (wet weight, ~ 2 × 10^{12} cells) of
yeast cells (stored at −80° as pellets) in 2500 ml (10 ml/g) of prespheroplasting buffer prewarmed to 37° (to help thaw yeast pellets). Save a 1-ml
sample as an untreated control while monitoring the OD change during
spheroplasting. Harvest the cells in a GS3 (or JA-10) rotor at 7000 rpm,
4° for 10 min. Resuspend the cells in 1250 ml of spheroplasting buffer
(5 ml/g) at 30°. Add 2 million units of recombinant β-1,3-glucanase and incubate at 30° while stirring. Spheroplasting is monitored by the change in
OD_{600} (dilute the cells to less than 1 OD_{600} in distilled H_2O, usually about
200- to 500-fold). The OD_{600} is measured every 15 to 20 min and should be
reduced to 30 to 40% of the initial OD_{600}. However, it is not advisable to
let the spheroplasting continue for more than 60 min. If the OD has not
changed enough in 30 min, add more endoglucanase. Spin out the cells at
7000 rpm, 4° for 10 min in a GS3 rotor. Resuspend the cells in YPD plus
1 M sorbitol (1.25 liters, 30°, 5 ml/g cells). The cells should be noticeably
more difficult to resuspend because of spheroplasting (they will tend to
clump). Spin out the cells at 7000 rpm, 4° for 10 min in a GS3 rotor. Resuspend the cells in 1 M sorbitol (1.25 liters, 4°, 5 ml/g cells). Spin out the cells
at 7000 rpm, 4° for 10 min in a JA-10 rotor. Repeat the 1 M sorbitol wash.
Resuspend the cells in at least 1 ml of lysis buffer (250 ml, 4°) per gram of
cells. Pass the total volume through a Yamato homogenizer three times
while it is turning at 1000 rpm at 4°. Then check the pH and make sure it
has not dropped. A low pH (<6.0) seems to promote clumping of organelles and this will drastically lower yield. If the pH has dropped, adjust
the pH to 7.0. Complete lysis and the lack of organelle clumping can be
checked microscopically. The organelles should be streaming between
lysed cells. Centrifuge the lysed cells at 6500 rpm, 4° for 10 min in a GSA
rotor. *Note:* If the volume is less than 100 ml, it works best to do these spins
in an SS-34 (or JA-20) rotor. Transfer the supernatants (everything that
will pour, including the looser components of the pellets) to new bottles
and recentrifuge at 6500 rpm. Transfer the supernatants as before to new
bottles and spin at 6500 rpm, 4° for 10 min in a GSA rotor. Pour off

the supernatants and recentrifuge as described previously. Combine the supernatants and spin at 5500 rpm, 4° for 10 min in a GSA rotor. Pour off the supernatants and recentrifuge as described previously two or three more times, depending on the amount of cell debris remaining in the supernatants. Under the microscope, the first two pellets will appear mostly as unlysed cells. The subsequent pellets should be mostly cell debris. The nuclei are visible in all the pellets as small round dots. After the five low-speed spins, the supernatant should be cloudy and essentially devoid of intact cells. Pellet the nuclei by spinning at 12,500 rpm, 4° for 30 min in a GSA rotor. The nuclear pellet is mostly white and the supernatant should be clear. Check the pellet and supernatant under a microscope. The pellet should be mostly nuclei with few cells. If the supernatant contains a large number of nuclei, recentrifuge it at 12,500 rpm. Good preparations will have a nuclear yield of 40–100 g; however, about 15–20 g is fine. If the yield is large (>70 g), split it into two aliquots. Resuspend the nuclei (pellets) at 0.5 ml/g cells (125 ml) in nuclear storage buffer. Flash-freeze the nuclei in liquid nitrogen and store at −80°.

Chromatin Digestion and Extraction of Nuclei. Thaw the nuclei and pellet at 12,500 rpm, 4° for 20 min in a GSA rotor. Resuspend the nuclei in 50 ml of nuclei storage buffer per aliquot. For a test micrococcal nuclease digestion, remove 1 ml of nuclei to an Eppendorf tube and warm to 37° for 10 min. Add 3 μl of 1 M $CaCl_2$ and 8 μl of micrococcal nuclease at 10 units/ml (Worthington, Lakewood, NJ) or 13.6 μl at 0.08 units/ml (Sigma), and incubate at 37°. Remove 100-μl aliquots at 0, 1, 2, 5, 10, 15, 20, and 30 min into tubes containing 50 μl of protease digestion buffer [20 mM EDTA (pH 8), 200 mM NaCl, 1% sodium dodecylsulfate (SDS), glycogen (0.25 mg/ml)] and 1 μl of RNase A (Worthington) at 10 mg/ml and incubate at 37° for 5 min. To each aliquot add 30 μl of distilled H_2O and 20 μl of 10% SDS, and mix. Then add 40 μl of 5 M NaCl, mix, and 200 μl of $CHCl_3$–isoamyl alcohol (24:1), mix, and centrifuge for 5 min. Transfer the upper, aqueous layer to new tubes and add 600 μl of ethanol, invert, and centrifuge for 15 min. Wash the pellets with 800 μl of 75% ethanol. Dry the pellets in a SpeedVac (Thermo Savant, Holbrook, NY) and resuspend the pellets in 12 μl of Tris–EDTA (TE). Load samples onto a 1.5% agarose gel (Tris–borate–EDTA, TBE): 3 μl of 0 through 30 min, and 10 μl of 1 through 10 min. Run the gel until the bromphenol blue has moved three-quarters of the way through the gel. Stain the gel with ethidium bromide or SYBR Gold (Molecular Probes, Eugene, OR) and photograph or scan.

Conduct a preparative micrococcal nuclease digestion of the rest of the nuclei by adding 1 M $CaCl_2$ to 3 mM and micrococcal nuclease at 200 units/ml (Worthington) or at 500 units/ml (Sigma) to 7 units/ml

(Worthington) or 0.08 units/ml (Sigma). Then add 50 μl of RNase A (10 mg/ml; Worthington). Incubate the digestion at 37° for the time determined in the test digestion to produce chromatin fragments corresponding primarily to 10-nucleosome lengths (1600 to 1700 bp). Stop the digestion by adding 0.5 M EDTA to 30 mM, and mix. Spin out the nuclei at 16,000 rpm, 4° for 15 min in an SS34 rotor. Save the pellets for further extraction. Second and even third extractions can be done by resuspending in 60 ml of 0.2 mM EDTA and centrifuging the nuclei as described previously. The chromatin is cleanest and at the highest concentration in the first extraction, however.

Sephacryl S-300 Gel-Filtration Chromatography: Run at 4°. Equilibrate the column (1 liter of Pharmacia Sephacryl S-300 resin in a XK 50/60 column) with nuclei storage buffer. Run the column at 5 ml/min, absorbance units full scale (AUFS) of 2 at 280 nm, and chart speed set at 0.15 cm/min. This column is run isocratically, but both intakes are put in the buffer and it is run at 50% B. Inject 50 ml of nuclear extract onto column and run at 5 ml/min. The chromatin will be found in the void volume (excluded peak). The chromatin peak in identified by running 100-μl samples of column fractions on a 1% agarose gel (stained with ethidium bromide or SYBR Gold) and on an 18% SDS–polyacrylamide gel (stained with Coomassie Brilliant Blue). The DNA is purified by adding 100 μl of protease digestion buffer [20 mM EDTA (pH 8), 200 mM NaCl, 1% SDS, glycogen (0.25 mg/ml)], and 5 μl of proteinase K (2.5 mg/ml) and incubating at 37° for 15 min. To this is added 200 μl of 0.6 M sodium acetate and 400 μl of CHCl$_3$. The mixture is vortexed and centrifuged for 5 min. The DNA is then ethanol precipitated from the aqueous phase, washed, and resuspended in 10 μl of agarose gel sample buffer. A judgment call must be made in deciding which fractions contain most of the chromatin but little of the higher molecular weight contaminating proteins. This is usually the first half of the excluded peak and perhaps one fraction past the fraction with the greatest absorbance. In the excluded peak, the large amount of DNA often causes these lanes of the protein gel to run somewhat blurrily. *Note:* Chromatin cannot be frozen for storage.

Hydroxylapatite Chromatography: Run at 4°. Check the chromatin concentration of the pooled S-300 peak by measuring the OD$_{260}$. This is best done by diluting the chromatin 1:20 (50 μl) to 1:50 (20 μl) fold into 1 ml of 1 M NaOH. Yield for the first extraction is typically approximately 20–40 mg of chromatin in about 250 ml.

Pour a hydroxylapatite (Bio-Rad Macro-Prep ceramic hydroxyapatite type I) column, using 1 ml of resin per 2 mg of chromatin. An approximately 30-ml column is usually required. Best results have been obtained with an XK 50/30 column with two adjustable adapters. A short, fat column is preferable. The resin should be loaded with chromatin to capacity. After the sample is loaded, if more than half the column is still white, too much resin has been used. *Note:* Removal of "fines," or small resin fragments, (defining) of the HT gel is important. Pack and equilibrate the column with 3 to 4 column volumes (CV) buffer at 0.3 M NaCl. Set the AUFS$_{280}$ at 0.5, the flow rate at 3.0 ml/min, and the chart speed at 0.1 cm/min.

Load the gel-filtration column chromatin peak pool onto the HT column at about 2 ml/min. It is often easiest to use a P1 peristaltic pump to load the sample. Wash the column at 3.0 ml/min with a 0.3 to 0.8 M NaCl gradient over 5 CV, then hold at 0.8 M NaCl for 2 to 4 column volumes. Continue to wash at 0.8 M NaCl until the absorbance has returned to baseline.

Elute the core histones by stepping to 2.5 M NaCl at 3.0 ml/min for 5 CV while collecting 4-ml fractions. The core histones should come off within the first 0.5 to 0.7 CV. The core histone peak fractions are identified by running 50 μl of each fraction on SDS–18% polyacrylamide gels. The peak is usually contained within four to five fractions, with a sharp leading edge containing the purest fractions and a somewhat broader trailing edge. Some contaminants will elute with the core histones and many of them will elute after the core histones. Peak fractions will have a concentration of 0.5 to 1 mg/ml as determined by A_{280} (extinction coefficient, 0.42 mg/ml Abs U^{-1}). If fractions are less than 95% pure, run them next to core histones of a known concentration to estimate their concentration. For an example of purified yeast core histones, see Pilon *et al.*[4]

Expression and Purification of Recombinant Yeast GST–Nap1p

Expression

In the late afternoon of day 1, transform plasmid (pGEX-NAP1) into BL21(DE3) *E. coli* cells. In the morning of day 2, pick six colonies into each of six 5-ml cultures of LB/AMP (12.5 μl of ampicillin at 40 mg/ml). Grow the cultures at 37° with shaking until turbid (~2 h). Pour 2 ml of each 5-ml culture into each of three 100-ml LB/AMP cultures in 250-ml flasks (250 μl of ampicillin at 40 mg/ml). Grow 100-ml cultures at 37° with shaking until turbid (~2 h). Pour 100-ml cultures into each of three 1-liter LB/AMP cultures in Fernbach flasks (2.5 ml of ampicillin at 40 mg/ml). Grow until the cultures reach an OD$_{600}$ of about 0.5 (~2 h). Add IPTG to 0.4 mM (0.4 ml of 1 M IPTG per liter) to induce expression. Grow the cells for another 3 h at 37°. Harvest the cells in a GS3 rotor (500-ml bottles) at 8000 rpm for 5 min at 4°. Pour off the supernatants (discard) and freeze the pellets at −80° overnight.

Cell Lysis

Resuspension buffer (50 ml) contains 20% sucrose (w/v), 100 mM Tris-HCl (pH 7.5), 0.5 mM EDTA (pH 8.0), 1 mM DTT, 1 mM PMSF, 1 mM Na$_2$S$_2$O$_5$, and 1 mM benzamidine.

Resuspend the cell pellets to 10 ml per liter resuspension buffer (30 ml total). Add lysozyme to 200 μg/ml (measure the volume; add, e.g., 8.8 mg/44 ml) and incubate on ice for 20 min. Add Brij 58 to 0.1% (e.g., add 0.44 ml of a 10% stock solution). Add KCl to 75 mM (e.g., add 1.1 ml of 3 M stock solution) and swirl gently to mix. Cell lysis should be obvious. Spin samples in a Beckman Ty70Ti rotor at 45,000 rpm for 90 min at 4°. Remove and save supernatants containing the GST–yNap1p, freeze in liquid N$_2$, and store at −80°.

Purification

Dialysis buffer (4 liters) consists of 25 mM Tris-HCl (pH 7.5), 10% glycerol, 50 mM NaCl, 2 mM DTT, 0.2 mM PMSF, 1 mM benzamidine, and 1 mM Na$_2$S$_2$O$_5$.

Thaw and dialyze the supernatant against 2 liters of dialysis buffer twice for 1.5 h each (save a 0.5-ml sample). Measure the volume of the dialyzed sample (usually 40 to 45 ml) and add 538 ml of 4 M (NH4)$_2$SO$_4$ per liter. Stir the sample in an ice–water bath for 20 min. Centrifuge at 15,000 rpm for 20 min at 4° in an SS-34 rotor and save the supernatant. Measure the volume of the supernatant (usually 60 to 65 ml), add (NH$_4$)$_2$SO$_4$ to 65% saturation (857 ml/liter, 4 M), and stir in an ice–water bath for 15 min. Centrifuge the samples at 15,000 rpm for 20 min at 4° in an SS-34 rotor. Discard the supernatants (save a 0.5-ml sample) and store the pellets at −80°.

Resuspend the pellets in 10 ml of PBS (140 mM NaCl, 2.7 mM KCl, 10 mM Na$_2$HPO$_4$, 1.8 mM KH$_2$PO$_4$, 1% Triton X-100, 1 mM DTT, 0.1 mM PMSF) at 4°. Dialyze the sample against 2 liters of PBS plus 1% Triton X-100 for 2 h twice at 4°. Save two 0.5-ml samples at −80°.

Equilibrate glutathione–Sepharose 4B resin (Pharmacia Biosciences). The procedure requires 2 ml of 50% slurry per 5 mg of GST–protein. We usually obtain 2.5 μg of GST–protein per milliliter of culture. We grow 3 liters, so we expect about 7.5 mg of GST–yNap1p. Therefore, we need 3 ml of a 50% slurry of the resin. To prepare the resin, remove 2.5 ml of resin from the stock bottle (75% slurry), spin down the resin at 500g in a clinical centrifuge (5 min), and remove the supernatant. Wash the resin with 18.8 ml of PBS (no Triton X-100) at 4° by inverting, spinning down the resin at 500g (5 min), removing the supernatant, and resuspending the resin in 1.88 ml of PBS to produce a 50% slurry.

Test the affinity purification of GST–yNap1p with glutathione resin by adding 0.5 ml of dialyzed extract to 150 μl of resin (50% slurry) and incubating the mixture on a rotator at 4° for 60 min. Pellet the resin at 500g (5 min) and remove the supernatant (save as a sample). Wash the resin with 750 μl of PBS (no Triton X-100, 10 bed volumes) by centrifuging the suspension at 500g (5 min) and discard the wash (repeat the wash two more times). Elute with 75 μl of elution buffer [10 mM glutathione, 50 mM Tris-HCl (pH 8.0), 1 bed volume] while mixing on a rotator at room temperature for 10 min. Centrifuge the resin at 500g for 5 min. Repeat the elution two more times and pool the eluates.

Run samples (first dialysis, first precipitation supernatant, second precipitation supernatant, second salt pellet resuspended in PBS, glutathione–Sepharose 4B input dialyzed versus PBS–Triton X-100, supernatant from incubation with resin, glutathione resin after elution, and eluate pool), 1 and 5 μl of each, on SDS–8% polyacrylamide gels and stain the gels with Coomassie Brilliant Blue.

Affinity purify the rest of the GST–yNap1p with glutathione resin by adding 3 ml of 50% slurry in PBS (1.5 ml into each of two 15-ml tubes) to the rest of the dialyzed sample. Incubate the resin on a rotator at 4° for 60 min. Pellet the resin at 500g for 5 min, remove the supernatant, and save. Wash the resin (resuspend) with 15 ml of PBS (no Triton X-100, 10 bed volumes; 7.5 ml/tube). Centrifuge the resin at 500g for 5 min and discard the wash. Repeat the wash two more times. Elute the GST–yNap1p with 1.5 ml of elution buffer [10 mM glutathione, 50 mM Tris-HCl (pH 8.0), 1 bed volume; 0.75 ml/tube] while mixing on a rotator at room temperature for 10 min. Centrifuge the resin at 500g for 5 min. Repeat the elution two more times, and pool the eluates.

Further purify the GST–yNap1p to separate it from nuclease activity by MonoQ column chromatography (run the column at 0.5 ml/min, AUFS = 2.0, fraction size of 1 ml). Dialyze the eluant against 4 liters of column buffer containing 150 mM NaCl, 10 mM Tris-HCl (pH 8.0), 1 mM EDTA, and 0.5 mM PMSF. Column buffers contain 20 mM Tris-HCl (pH 7.5), 5% glycerol, 1 mM EDTA, 0.5 mM PMSF, 1 mM DTT, 1 mM Na$_2$S$_2$O$_5$, and either no NaCl (buffer A) or 1 M NaCl (buffer B). Load sample onto a MonoQ column (1 ml) equilibrated with column buffer at 100 mM NaCl (10% B). Wash the column with 5 column volumes of 0.1 M NaCl (5 ml of 10% B). Elute the GST–yNap1p with a 0.1 to 0.8 M NaCl linear gradient over 50 column volumes (50 ml of 10 to 80% B). Determine the peak fractions by SDS–8% PAGE followed by Coommassie Brilliant Blue staining. Dialyze the pooled peak fraction into 10 mM HEPES-KOH (pH 7.6), 10 mM KCl, 1.5 mM MgCl$_2$, 10% glycerol, 10 mM β-glycerophosphate, 1 mM DTT, and 0.1 mM PMSF.

Reconstitution of Chromatin with Yeast Core Histones and yNap1p

Purified yeast core histones are diluted with 10 mM HEPES-KOH (pH 7.6), 10 mM KCl, 1.5 mM MgCl$_2$, 10% glycerol, 10 mM β-glycerophosphate, 1 mM DTT, and 0.1 mM PMSF to 350 μg/ml. Yeast core histone and recombinant GST–yNap1p (at 1 mg/ml) are combined at a ratio of 4:1 (w/w) and diluted 2-fold with 50 mM HEPES (pH 7.6), 0.1 mM EDTA (pH 8.0), and 10% glycerol. The core histones and GST–yNap1p are incubated together on ice for 30 min. The appropriate amount of core histones and GST–yNap1p are then combined with DNA in 10 mM HEPES (pH 7.6), 50 mM KCl, 5% glycerol, and 1% PEG 8000. Reconstitution is allowed to proceed at 30° for 4 h and reconstituted chromatin templates can be used immediately or stored at 4° (*Note:* never freeze) for up to 1 week.

Topological Assay for Degree of Chromatin Reconstitution

Topological assays are the best way to measure the degree of assembly of chromatin onto plasmid templates. Topoisomerase I enzyme relieves torsional stress in DNA through single-strand breakage and rejoining reactions and requires no energy input (ATP). We pretreat the closed circular plasmid DNA in order to start with relaxed DNA in order to know that any supercoils in the DNA are the result of nucleosome formation. Each nucleosome formed constrains one negative supercoil in the DNA wrapping the core histone octamer and forms a compensating positive supercoil in the DNA linking the nucleosomes. The positive supercoils in the nonnucleosomal DNA are subject to removal by the topoisomerase I. On extraction and purification, the DNA retains the negative supercoils constrained by the nucleosomes. The degree of supercoiling is assayed by electrophoretic mobility on native agarose gels.

One-Dimensional Topological Analysis

To analyze the degree of reconstitution, 6 μg of plasmid DNA (for 10 assembly assay reactions) is relaxed with 3.5 U of topoisomerase I [MBI Fermentas (Hanover, MD), or recombinant *Drosophila* topoisomerase I purified according to Hsieh *et al.*[6]] in 50 mM Tris-HCl (pH 7.5), 50 mM KCl, 10 mM MgCl$_2$, 1 mM DTT, 0.5 mM EDTA, and BSA (30 μg/ml) for 1 h at 30° in a 100-μl reaction volume. After the initial relaxation, an additional 20 U of topoisomerase I is added and the DNA is combined with the core histone–GST–yNap1p complex at various ratios. Each reconstitution reaction contains 500 ng of relaxed DNA and the appropriate amount

[6] T. S. Hsieh, S. D. Brown, P. Huang, and J. Fostel, *Nucleic Acids Res.* **20,** 6177 (1992).

octamer:yNap1p complex in 10 mM HEPES (pH 7.6), 50 mM KCl, 5.5 mM MgCl$_2$, 5% glycerol, 1% PEG 8000, and 140 ng of BSA in a reaction volume of 19 μl. Reconstitution is allowed to proceed for 4 h at 30°. The reaction is stopped by the addition of 100 μl of STOP [20 mM EDTA, 0.1% SDS, 200 mM NaCl, and glycogen (0.25 mg/ml)] and 12.5 μg of proteinase K, and incubated at 37° for 20 min. The DNA is extracted with phenol–chloroform–isoamyl alcohol (25:24:1), followed by ethanol precipitation. The samples are split in half and run on 1% TBE–agarose gels with or without chloroquine and then stained with ethidium bromide or SYBR Gold. The optimal chloroquine concentration in the second gel must be titrated for the specific plasmid DNA used, but is usually in the 2- to 4-μg/ml range. By comparing the sample run in the absence and presence of chloroquine, we obtain a clear measure of the degree of assembly.

Two-Dimensional Topological Assay

To quantitate the number of nucleosomes formed on the plasmid DNA, we use two-dimensional topological analysis based on a procedure described by Peck and Wang[7] as modified by Shimamura *et al.*[8] The sample preparation is identical to that for one-dimensional topological analysis. Load 750 ng of DNA purified from chromatin reconstituted in the presence of topoisomerase I on a 20 × 20 cm 1% TBE (1×) gel. Load the samples in the far left lanes with two empty lanes in between them. Typically, we can load two experimental samples, one topological marker and one fully relaxed DNA sample, on a single gel (for examples, see Pilon *et al.*[4] and Laybourn and Kadonaga[9]). The first sample is loaded into lane 2 and the gel is run for 20 min at 160 V, and then the next sample is loaded in lane 5 and the gel is run for 20 min at 160 V. Finally, the markers (supercoiled and relaxed, respectively) are loaded at the same time into lanes 8 and 10. The gel is electrophoresed in the absence of chloroquine for 16 h at 58 V. The gel is then equilibrated in chloroquine (the optimal concentration differs for each plasmid, but generally ranges from 2 to 4 μg) for 6 to 8 h in 1× TBE at room temperature while protected from light. The gel is turned 90° (this often requires some trimming) such that the left side of the gel is now at the top, and electrophoresed for 16 h at 67 V. Topological markers can be produced by mixing DNA extracted from partial assembly reactions (intermediate histone:DNA ratios) in the presence of topoisomerase I. Alternatively, markers can be made by mixing DNA combined with increasing amounts of ethidium bromide in the presence of topoisomerase I, followed

[7] L. J. Peck and J. C. Wang, *Proc. Natl. Acad. Sci. USA* **80,** 6206 (1983).
[8] A. Shimamura, D. Tremethick, and A. Worcel, *Mol. Cell. Biol.* **8,** 4257 (1988).
[9] P. J. Laybourn and J. T. Kadonaga, *Science* **254,** 238 (1991).

by extraction with butanol and purification. After electrophoresis, the DNA is visualized by ethidium bromide or SYBR Gold staining.

Micrococcal Nuclease Digestion Analysis of Quality of Reconstituted Chromatin

To assess the quality of the chromatin formed, as measured by evenness of nucleosome spacing, we use micrococcal nuclease digestion. The DNA fragments produced are purified by extraction and precipitation and resolved by native agarose gel electrophoresis. Micrococcal nuclease makes double-stranded cleavages preferentially in the DNA linking nucleosomes. A limited digestion of the chromatin DNA produces the characteristic "ladder" of DNA fragments corresponding to mono-, di-, tri-, tetra-(etc.) nucleosomes. Generally, we consider a seven (or greater)-rung ladder to be acceptable. The fragments produced from assembly with yeast core histones and yNap1p are multiples of 160 to 170 bp. Fragments of less than 100 bp indicate incomplete assembly.

Assemble chromatin on 2.1 μg of supercoiled DNA (no topoisomerase I or MgCl$_2$) in 210 μl. To the assembly reaction add 290 μl 10 mM HEPES (pH 7.6), 50 mM KCl, 1% PEG 8000, 5% glycerol, and 5 mM CaCl$_2$ (freshly made). Warm the sample to 37° for 5 min and then add 15 μl of a 1:20 dilution of micrococcal nuclease stock [2 μl of 500-U/μl micrococcal nuclease stock (Worthington) plus 38 μl of 10 mM HEPES (pH 7.6), 50 mM KCl, 1% PEG 8000, 5% glycerol, and 5 mM CaCl$_2$] and incubate the sample at 37°. Remove 125-μl aliquots at 1, 2, 4, and 8 min into tubes containing 25 μl of 10 mM Tris-HCl (pH 8.0) and 0.5 M EDTA. Add 200 μl of chromatin STOP (see above) and then add 15 μl of proteinase K (2.5 mg/ml) and incubate at 37° for 20 min. Add 300 μl of phenol–CHCl$_3$–isoamyl alcohol, vortex, and spin for 5 min. Transfer 350 μl of the upper aqueous layer to new tubes containing 23.3 μl of 4 M ammonium acetate and mix. Add 990 μl of 100% ethanol, invert, and centrifuge for 15 min. Remove the supernatant with a drawn pipette and dry the pellets for 5 min on a 60° heating block or until the ethanol is gone. Resuspend the pellets in 15 μl of distilled H$_2$O, heat at 60° for 2 min, vortex, spin down, and then add 3 μl of 6× loading buffer. Run the samples on a 1.2% agarose–TBE gel with 100-bp markers until the bromphenol blue dye has run 8/10s the length of the gel. The gel is then stained with ethidium bromide or SYBR Gold, destained, and the image is digitally scanned. For examples, see Pilon et al.[4] and Terrell et al.[10]

[10] A. R. Terrell, S. Wongwisansri, J. L. Pilon, and P. J. Laybourn, J. Biol. Chem. **277**, 31038 (2002).

Sucrose Gradient Purification of Assembled Chromatin Templates

We have found it useful at times to purify our chromatin templates away from the GST–yNap1p. The most reliable method we have found is sucrose gradient sedimentation. We have found sizing column chromatography to work, as well. However, sucrose gradient sedimentation has the added benefit of fractionating fully assembled chromatin from partially assembled DNA templates. We have found that trying to remove GST–yNap1p with glutathione–agarose beads results in disassembly of the chromatin.

The only drawback to purification of chromatin templates is a slight (2- to 3-fold) dilution. The effect of the dilution can be minimized by assembling a larger quantity of chromatin at a higher concentration. For example, we typically assemble 50 μg of DNA in a 0.6- to 1-ml volume.

We prepare two solutions containing 5 and 30% sucrose, respectively, and 10 mM Tris-HCl (pH 7.8), 0.1 mM EDTA, 1 mM DTT, 0.1 mM PMSF, and 0.1 mM benzamidine. We form 13-ml 5 to 30% linear sucrose gradients in Beckman SW41 ultracentrifuge tubes. The assembled chromatin is layered onto the top of a gradient and the gradient is centrifuged at 40,000 rpm for 4 h at 4° in a Beckman SW41 rotor. We collect 1-ml fractions from the bottom of the gradient. The peak is determined by DNA–agarose gel (1%) and by SDS–PAGE (18%) (for an example, see Pilon et al.[4]). The concentration of the chromatin can be estimated by comparison with a known standard sample. Concentration determination by UV spectroscopy is generally unsatisfactory.

General Comments

Assembly of yeast chromatin, using purified core histones and yNap1p, works efficiently and produces good-quality chromatin in the absence of cofactors or additional assembly factors. Early on, we attempted to form chromatin with yeast core histones by salt dialysis methods, but met with no success. For undetermined reasons, this approach results in insoluble aggregates at core histone-to-DNA ratios above 0.4 (w/w). We have shown that chromatin formed with purified yeast core histones and yeast Nap1p recapitulates the chromatin structure, including nucleosome positioning, on the *PHO5* promoter.[10] We have also used these chromatin templates to investigate the mechanism of transcriptional activation on this promoter. We have found that yNap1p assembles nucleosomes as efficiently on linear fragments as on closed circular plasmids. Finally, we have found that yNap1p assembles chromatin with recombinant yeast core histones or purified native yeast core histones with equivalent efficiency and quality. In fact, we have found that yNap1p works well with a broad range of core histones, recombinant or native, from yeast, *Drosophila,* and *Xenopus.*

Acknowledgments

We thank Tomoko Fujii-Nakata for the GST–yNap1p expression plasmid (pTN2) and Shi-Hsiang Shen for β-glucanase expression plasmid (pUV5-G1S). We also thank Stephanie Abernathy for critical reading of the manuscript. Yeast chromatin work in our laboratory has been supported in part by a Research Grant from the National Science Foundation (MCB-9505644) and by a Junior Faculty Research Award from the American Cancer Society awarded to P.J.L.

[8] DNA Synthesis-Dependent and -Independent Chromatin Assembly Pathways in *Xenopus* Egg Extracts

By DOMINIQUE RAY-GALLET and GENEVIÈVE ALMOUZNI

Introduction

In vivo, de novo nucleosome assembly occurs mainly during DNA replication after passage of the replication fork. Parental histones are distributed between daughter strands in a random fashion and a complement of newly synthesized histones must be incorporated to fully duplicate the original chromatin structure. This *de novo* nucleosome assembly pathway has been studied in powerful biochemical assays that reproduce the coupling of the assembly reaction with DNA synthesis.[1] A key factor specifically involved in this nucleosome assembly coupled to DNA synthesis is the histone chaperone CAF-1 (chromatin assembly factor-1), which was isolated from nuclear extracts derived from cultured human cells. Its purification was based on its capacity to assemble histone octamers selectively onto newly synthesized DNA, using a plasmid containing the simian virus 40 (SV-40) origin, which can replicate *in vitro* in the presence of T antigen.[2] The CAF-1 complex consists of three evolutionary conserved subunits, p150, p60, and p48, which are associated with newly synthesized histones H3 and H4 to facilitate their deposition on DNA during replication.[3] Moreover, CAF-1 is also implicated in nucleosome assembly coupled to

[1] P. D. Kaufman and G. Almouzni, *in* "Chromatin Structure and Gene Expression" (S. C. R. Elgin and J. L. Workman, eds.), p. 24. Oxford University Press, New York, 2000.
[2] S. Smith and B. Stillman, *Cell* **58**, 15 (1989).
[3] A. Verreault, P. D. Kaufman, R. Kobayashi, and B. Stillman, *Cell* **87**, 95 (1996).

nucleotide excision repair (NER),[4] a process associated with DNA synthesis to ensure the maintenance of nucleosomal organization after DNA repair. In both cases, the proliferating cell nuclear antigen (PCNA) is required for CAF-1 to be recruited.[5,6] Thus, a CAF-1–PCNA chromatin assembly pathway can be delineated that is tightly associated with DNA synthesis.[7,8]

Attention has been drawn to other nucleosome assembly pathways that occur independently of DNA synthesis. Theoretically, such pathways could be used to restore chromatin structures after disruptive events that are not associated with DNA synthesis. These chromatin assembly pathways may be involved in the deposition of histone variants, for example, the histone H3 variants, CENP-A located at centromeric regions,[9] and H3.3 synthesized in quiescent G_1 and G_2 cells.[10] For both of these variants, several lines of evidence support deposition outside of the S phase.[11–13] However, the mechanism and the potential chaperone(s) involved in the deposition of these histone variants are not known. Among the histone chaperones known to date, we focused on HIRA,[14] the homolog of the two *Saccharomyces cerevisiae* proteins Hir1p and Hir2p, that are cell cycle-regulated transcriptional repressors of histone gene expression.[15,16] Using *Xenopus* egg extracts depleted of HIRA, we found that HIRA is critical *in vitro* for a nucleosome assembly pathway independent of DNA synthesis.[17]

Thus, two chromatin assembly pathways, dependent on DNA synthesis or independent of it, can be monitored in an *in vitro* system. To gain insight into their specific features, we determined conditions under which the two pathways can be dissociated. The strategy was to abolish either the

[4] P. H. Gaillard, E. M. Martini, P. D. Kaufman, B. Stillman, E. Moustacchi, and G. Almouzni, *Cell* **86,** 887 (1996).

[5] K. Shibahara and B. Stillman, *Cell* **96,** 575 (1999).

[6] J. G. Moggs, P. Grandi, J. P. Quivy, Z. O. Jonsson, U. Huebscher, P. B. Becker, and G. Almouzni, *Mol. Cell. Biol.* **20,** 1206 (2000).

[7] P. Ridgway and G. Almouzni, *J. Cell Sci.* **113,** 2647 (2000).

[8] J. A. Mello and G. Almouzni, *Curr. Opin. Genet. Dev.* **11,** 136 (2001).

[9] D. K. Palmer, K. O'Day, H. L. Trong, H. Charbonneau, and R. L. Margolis, *Proc. Natl. Acad. Sci. USA* **88,** 3734 (1991).

[10] R. S. Wu, S. Tsai, and W. M. Bonner, *Cell* **31,** 367 (1982).

[11] R. D. Shelby, K. Monier, and K. F. Sullivan, *J. Cell Biol.* **151,** 1113 (2000).

[12] B. Sullivan and G. Karpen, *J. Cell Biol.* **154,** 683 (2001).

[13] K. Ahmad and S. Henikoff, *Mol. Cell* **9,** 1191 (2002).

[14] S. Lorain, J. P. Quivy, F. Monier-Gavelle, C. Scamps, Y. Lecluse, G. Almouzni, and M. Lipinski, *Mol. Cell. Biol.* **18,** 5546 (1998).

[15] V. Lamour, Y. Lecluse, C. Desmaze, M. Spector, M. Bodescot, A. Aurias, M. A. Osley, and M. Lipinski, *Hum. Mol. Genet.* **4,** 791 (1995).

[16] P. W. Sherwood, S. V. Tsang, and M. A. Osley, *Mol. Cell. Biol.* **13,** 28 (1993).

[17] D. Ray-Gallet, J. P. Quivy, C. Scamps, E. M. Martini, M. Lipinski, and G. Almouzni, *Mol. Cell* **9,** 1091 (2002).

nucleosome assembly associated with DNA synthesis or the nucleosome assembly independent of DNA synthesis by eliminating, respectively, either CAF-1 or HIRA activity. *Xenopus* egg extracts obtained after high-speed centrifugation, high-speed egg extracts (HSEs), represent a convenient system to study both chromatin assembly pathways.[18] Large quantities of material can be generated rapidly and the nucleosome assembly mechanisms active in these extracts are also observed in other cell-free systems, including extracts derived from human cells. This conservation of functions is attested by the observation that assembly reactions can be complemented with extracts or proteins derived from different species.[19,20] Therefore, data gathered in this system can be relevant in a wide range of eukaryotic organisms. We describe here how, in the *Xenopus* egg extract, it is possible to distinguish between these different chromatin assembly pathways and to analyze their properties.

High-Speed Egg Extract Preparation and Immunodepletion

For *in vitro* nucleosome assembly assay and NER, interphasic high-speed egg extracts (HSEs) are prepared. We present here briefly the protocol that we use currently. For details concerning the preparation of various extracts derived from *Xenopus* eggs, several publications provide further information.[21,22] *Xenopus laevis* female frogs are induced to lay eggs with human chorionic gonadotropin hormone (hCG) and eggs are collected in 0.1 *M* NaCl. Incubation in fresh 2% cysteine solution (in 0.1 *M* NaCl with pH adjusted to pH 7.8) for 3 to 4 min allows removal of the jelly coat. After extensive washes in large volumes of 0.1 *M* NaCl, the eggs are rinsed in extraction buffer [10 m*M* KOH-HEPES (pH 7.8), 70 m*M* KCl, 5% sucrose, 0.5 m*M* dithiothreitol (DTT), and protease inhibitors] and transferred to chilled centrifugation tubes. Excess buffer is removed to avoid dilution of the extract and tubes are subjected to low-speed crushing at 10,000g for 30 min at 4°. After centrifugation three layers are visible. The middle ooplasmic layer is collected by inserting a glass Pasteur pipette through the upper yellow lipid layer. This material is then clarified by ultracentrifugation at 150,000g for 1 h at 4° (in ultraclear tubes; Beckman, Fullerton,

[18] G. Almouzni and M. Mechali, *EMBO J.* **7**, 665 (1988).
[19] R. T. Kamakaka, M. Bulger, P. D. Kaufman, B. Stillman, and J. T. Kadonaga, *Mol. Cell. Biol.* **16**, 810 (1996).
[20] J. P. Quivy, P. Grandi, and G. Almouzni, *EMBO J.* **20**, 2015 (2001).
[21] A. W. Murray, *in* "Methods in Cell Biology" (B. K. Kay and H. B. Peng, eds.), p. 581. Academic Press, San Diego, CA, 1991.
[22] G. Almouzni, *in* "Chromatin: A Practical Approach," p. 195. Oxford University Press, New York, 1998.

CA). A syringe with a needle is used to pierce the tube and recover the clear ooplasmic fraction, aliquots of which are frozen in liquid nitrogen and stored at $-80°$. In general, we obtain a yield of about 3 to 5 ml of extract from three injected frogs. For each new HSE preparation, the following parameters are measured in order to ensure a reproducible quality of the extract: (1) the protein concentration, by the Bradford method,[23] usually about 40 mg/ml; (2) the ionic strength, using a conductivity meter (MeterLab CDM210; Radiometer Analytical, Lyon, France), which gives an equivalent of 60 to 70 mM salt; and (3) the chromatin assembly activity, by performing nucleosome assembly followed by supercoiling assay (described below) or micrococcal nuclease digestion (not described here).[24,25] An aliquot of 10 μl of a standard HSE can fully assemble 150 ng of a 3.2-kb plasmid in 3 h.

We use an immunodepletion strategy to remove either HIRA or p150 (the large subunit of the CAF-1 complex) from HSE (see Fig. 1A). We thus generate p150-depleted HSE, which is competent only for DNA synthesis-independent chromatin assembly, and HIRA-depleted HSE, which is competent only for DNA synthesis-dependent chromatin assembly. To immunodeplete the chosen protein from HSE, we select antibodies that are both highly specific and efficient in immunoprecipitation. We obtain specific rabbit polyclonal anti-HIRA and anti-p150 antisera by immunization with recombinant *Xenopus* HIRA and p150 proteins, respectively.[17,20] For immunodepletion, antibodies are first bound to a solid-phase matrix, which is then mixed with the extract to deplete the protein of interest. In a typical immunodepletion protocol, equal volumes of protein A–Sepharose slurry (CL-4B; Amersham Biosciences, Piscataway, NJ) and serum (anti-p150 or anti-HIRA) are mixed and incubated on a rotating wheel at $4°$ for 2 to 3 h. The bound IgG/protein A–Sepharose beads are collected by centrifugation at 5000g for 1 min, and a 10-fold volume of phosphate-buffered saline (PBS) is used to wash the beads by inverting the tube several times. This washing step is repeated four times; the PBS is then eliminated carefully to leave the bound IgG/protein A–Sepharose beads in a minimal volume in order to avoid dilution of the extract to be depleted. Equal volumes (usually 200 μl) of HSE and IgG/protein A–Sepharose are mixed and incubated on a rotating wheel at $4°$ for 1 to 2 h. After centrifugation as described above, the supernatant corresponding to the depleted HSE is collected, frozen in liquid nitrogen, and stored at $-80°$ in

[23] M. M. Bradford, *Anal. Biochem.* **72,** 248 (1976).
[24] P. H. Gaillard, D. Roche, and G. Almouzni, *in* "Methods in Molecular Biology" (P. B. Becker, ed.), p. 231. Humana Press, Totowa, NJ, 1999.
[25] J. G. Moggs and G. Almouzni, *Methods Enzymol.* **304,** 333 (1999).

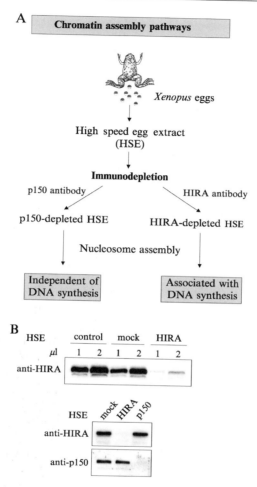

FIG. 1. Two distinct chromatin assembly pathways: one associated with DNA synthesis and CAF-1 mediated, and another uncoupled from DNA synthesis and HIRA dependent. (A) Immunodepletion strategy used to monitor these two distinct chromatin assembly processes individually. *Xenopus* high-speed egg extracts (HSEs), supporting both pathways, were depleted either of the p150 subunit of CAF-1 or of HIRA by using antibodies against *Xenopus* p150 or HIRA, respectively. p150-depleted HSE supports nucleosome assembly independently of DNA synthesis whereas HIRA-depleted HSE supports only synthesis-dependent assembly. (B) Analysis of immunodepleted HSE. *Top:* Two amounts (1 and 2 μl) of control HSE, mock HSE, or HIRA-depleted HSE were analyzed by Western blotting using anti-*Xenopus* HIRA antibody. *Bottom:* 1-μl samples of mock, HIRA, or p150-immunode-pleted HSE were analyzed by Western blotting using anti-*Xenopus* HIRA or p150 antibodies.

aliquots to avoid multiple freeze–thawing cycles that could affect the HSE activity. As controls, the preimmune serum of the corresponding antibody bound to protein A–Sepharose or protein A–Sepharose alone is used for a mock HSE depletion. The efficiency of the depletion is monitored by Western blotting. Briefly, the HSEs are subjected to sodium dodecyl sulfate (SDS)–polyacrylamide gel electrophoresis (aliquots of 0.5 to 2 μl), electrotransferred on a nitrocellulose membrane, and then incubated with a primary antibody in PBS containing 0.1% Tween 20 and 5% nonfat milk. After incubation with horseradish peroxidase-conjugated secondary antibody (Jackson ImmunoResearch Laboratories, West Grove, PA), the immune complexes are revealed by a chemiluminescence reaction, using a SuperSignal substrate detection kit (Pierce Biotechnology, Rockford, IL) and visualized by exposure to X-ray film. As illustrated in Fig. 1B (top), the HIRA depletion is analyzed by comparing the amount of HIRA protein in control HSEs and in the mock and HIRA-depleted extracts by Western blotting. For a semiquantitative estimation, two aliquots (1 and 2 μl) of the different HSEs are run in parallel so that we can ensure at the detection step that the signal obtained after exposure to X-rays is proportional to the amount of protein initially loaded. Note that we usually detect slightly less HIRA in mock-depleted extract than in the control HSE because of dilution of the extract during the depletion process. This is supported by Ponceau staining of the membrane, which shows a general decrease in the total amount of protein when compared with control HSE (not shown). In the HIRA-depleted extract, faint and proportional bands for HIRA are observed with 1 and 2 μl, showing that the depletion is successful. This semiquantitative Western blot allows us to estimate that the efficiency of the HIRA depletion reaches almost 90%. If the depletion is incomplete, the procedure is modified by increasing the amount of antibody/protein A–Sepharose relative to the amount of HSE. This has been found to be more effective than performing two rounds of depletion or increasing the incubation time of the HSE with the antibody/protein A–Sepharose. These latter methods generally lead to a decrease in the nucleosome activity of the extract. In Fig. 1B (bottom) the depletion of both HIRA and p150 from HSEs is analyzed by Western blotting with anti-HIRA and anti-p150 antibodies. The amounts of both proteins are largely decreased in their respective depleted extracts. In contrast, HIRA is present at a similar level in mock and p150-depleted HSEs and, conversely, the amount of p150 is the same in mock and HIRA-depleted extracts. This shows that the depletion is specific. The use of antibodies directed against other proteins present in the extract also helps to assess whether the depletion process affects the amount of other proteins in the HSE. In particular, when concerned with nucleosome formation, it is important to examine whether the amounts of other known histone chaperones such as nucleoplasmin, N1/N2, and Asf1, and also of histones, are affected by the HIRA/CAF-1 depletions.[17]

In Vitro Nucleosome Assembly Assays

The two chromatin assembly pathways can be monitored by *in vitro* nucleosome assembly assays using a 3.2-kb circular plasmid pBS (Stratagene, La Jolla, CA) that is UV irradiated or not UV irradiated, and the size of which allows a simple resolution of topoisomers by gel electrophoresis (see Fig. 2A). To monitor the nucleosome assembly pathway independent of DNA synthesis, we use this pBS plasmid with no lesion induced (named p 0). To monitor the CAF-1-dependent assembly pathway associated with DNA synthesis, we use an assay analyzing concomitly NER and nucleosome assembly.[4] The pBS plasmid is damaged by UV-C (named p UV), which creates two major DNA lesions [cyclobutane pyrimidine dimers and (6–4) photoproducts] that are both repaired by NER pathway. Here, the pBS plasmid DNA is prepared with a Qiagen (Chatsworth, CA) plasmid purification kit. The UV-damaged plasmid is created by using a germicidal lamp (wavelength, 254 nm). The height of the lamp is adjusted to obtain the desired UV fluence rate measured with a Latarjet dosimeter. A dose of 100 J/m^2 induces one pyrimidine dimer photoproduct in 1000 bp [about 0.75 cyclobutane pyrimidine dimer and 0.25 (6–4) photoproduct].[26] The dose usually used in our assays is 500 J/m^2, resulting in intense signals for the repair synthesis compared with nondamaged DNA.

The assembly reaction is performed as follows. In a final volume of 25 μl, 10 μl of HSE (depleted or mock depleted) is added to 150 ng of p 0 or p UV in a reaction buffer containing 5 mM MgCl$_2$, 40 mM HEPES-KOH (pH 7.8), 0.5 mM DTT, 4 mM ATP, 40 mM phosphocreatine, 2.5 μg of creatine phosphokinase and 5 μCi of [α-^{32}P]dCTP in order to monitor the repair synthesis. The reaction is incubated for 3 h at 23° and stopped by adding 25 μl of stop mix (30 mM EDTA, 0.7% SDS). The nucleosome assembly reaction is then monitored by a supercoiling assay allowing analysis of the topological forms of the plasmid after protein extraction by agarose gel electrophoresis. The circular plasmid, when incubated in HSEs, undergoes conformational changes corresponding to the progressive deposition of nucleosomes while the topoisomerase activity allows the absorption of the constraints. After deproteinization, topoisomers with an increasing number of negative supercoils reflect the extent of assembly. The stopped nucleosome assembly reaction is completed to

[26] R. D. Wood, M. Biggerstaff, and M. K. K. Shivji, *Methods* **7**, 163 (1995).

100 μl with H$_2$O, and incubated with 10 μg of DNase-free RNase A for 45 min at 37° and with 20 μg of proteinase K for 1 h at 37°. After protein extraction with phenol–chloroform–isoamyl alcohol (25:24:1, v/v), the DNA in the aqueous phase is precipitated by adding 100 μl of 5 M ammonium acetate and 250 μl of ethanol in the presence of glycogen (2 μg) as carrier and subjected to centrifugation at 20,000g for 30 min at 4°. The pellet is washed with 70% ethanol, dried, and then resuspended in 16 μl of TE [10 mM Tris-HCl (pH 7.5), 1 mM EDTA (pH 8)] and 4 μl of 5× loading buffer (0.42% bromphenol blue, 50% glycerol). Generally, half (10 μl) is loaded on a 1% agarose gel in TAE 1×buffer (40 mM

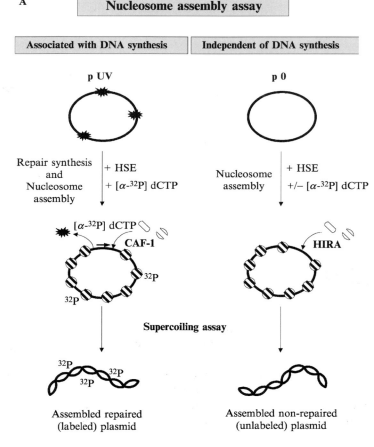

FIG. 2. (*continued*)

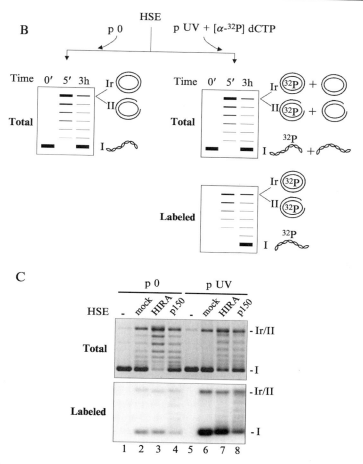

FIG. 2. DNA synthesis-dependent or -independent nucleosome assembly assays. (A) To examine chromatin assembly associated with DNA synthesis (mediated by CAF-1), the NER-coupled nucleosome assembly pathway was used. UV-irradiated plasmid (p UV) was incubated with HSE in the presence of $[\alpha\text{-}^{32}\text{p}]\text{dCTP}$ and the nucleosome assembly on repaired (labeled) DNA was analyzed by supercoiling assay and autoradiography. To examine nucleosome assembly independent of DNA synthesis (HIRA dependent), nonirradiated plasmid (p 0) was incubated with HSE. The nucleosome assembly of the nonrepaired (unlabeled) DNA was analyzed by supercoiling assay and visualized by ethidium bromide staining. (B) Schematic representation of supercoiling gel analysis. After nucleosome assembly, the plasmid DNA is deproteinized and analyzed by agarose gel electrophoresis. For a nucleosome assembly assay with HSE and nonirradiated plasmid (p 0), the DNA is visualized by ethidium bromide staining (total DNA). For a nucleosome assembly assay with HSE and UV-irradiated plasmid (p UV) in the presence of $[\alpha\text{-}^{32}\text{p}]\text{dCTP}$, the repaired DNA is visualized by autoradiography or phosphoimaging (labeled DNA). Migration positions of

Tris-acetate, 1 mM EDTA) and electrophoresis performed at 1.5 V/cm for 20 h at 4°. After migration, the gel is stained with ethidium bromide to visualize total DNA (repaired and nonrepaired molecules) and then dried and autoradiographed or analyzed by phosphorimager to visualize only the labeled, (repaired) DNA (for more details about the nucleosome assembly protocol, see Gaillard et al.[24]). A schematic representation of supercoiling gels resulting from these nucleosome assembly assays is presented in Fig. 2B. For p 0, the gel is stained with ethidium bromide to visualize total DNA. Input plasmid (time 0) is highly supercoiled (fast migrating form I). After 5 min, the supercoiled plasmid becomes relaxed (slow migrating form Ir) as a consequence of the action of topoisomerases present in the HSE, and then progressive supercoiling appears, corresponding to progressive nucleosome assembly. In general, maximal assembly is achieved after 3 h of incubation, leading to the accumulation of form I corresponding to the position of fully supercoiled DNA. Note that both the relaxed DNA form (Ir) and the nicked DNA form (II) migrate at the same position on these gels. For p UV, staining of the gel with ethidium bromide allows one to visualize total DNA, the migration profile of which is comparable to that observed with p 0. It is important to bear in mind that in the case of p UV, total DNA corresponds to a mixed population of molecules containing repaired (labeled) and nonrepaired DNA, and in these extracts only a fraction corresponding to 10–15% of the damaged plasmid molecules is repaired.[27] By autoradiography or PhosphorImager analysis of the gel, only repaired (labeled) plasmid molecules are revealed. After 5 min of reaction, faint labeled bands are detected corresponding to first repair events with incorporation of [α-^{32}P]dCTP, and after 3 h there is an accumulation of fast-migrating supercoiled form (I) resulting from both completed repair and nucleosome assembly of the p UV plasmid.

Nucleosome assembly assays are performed by combining each depleted extract (mock, HIRA, and p150) with either nonirradiated plasmid (p 0) or UV-C-irradiated plasmid (p UV) in the presence of [α-^{32}P]dCTP

[27] P. H. Gaillard, J. G. Moggs, D. M. Roche, J. P. Quivy, P. B. Becker, R. D. Wood, and G. Almouzni, *EMBO J.* **16,** 6281 (1997).

DNA plasmid form I (supercoiled), form II (nicked circular), and form Ir (relaxed, closed circular) are indicated. (C) Analysis of the nucleosome assembly activities in HIRA- and p150-depleted HSEs. Mock, HIRA-depleted, or p150-depleted HSEs were used in nucleosome assembly assays that were carried out over 3 h in the presence of [α-^{32}P] dCTP with 150 ng of either nonirradiated plasmid (p 0) or UV-C-treated plasmid (p UV). Nucleosome assembly was analyzed by supercoiling assay as described above. Input DNA run in parallel (lanes 1 and 5) and migration positions of DNA plasmid forms are indicated.

to monitor the repaired labeled DNA. After 3 h of incubation, the topological form of the plasmid DNA, obtained in each nucleosome assembly reaction, is analyzed by supercoiling assay (Fig. 2C). The p150-depleted HSE, like mock-depleted HSE, supports efficient nucleosome assembly on unirradiated DNA (p 0); indeed, the plasmid molecules become almost completely supercoiled (form I) (Fig. 2C, total DNA, lanes 2 and 4). With HIRA-depleted HSE, only a few supercoiled DNA molecules are observed and DNA is present mostly as the relaxed form (Ir), indicating that this synthesis-independent nucleosome assembly reaction is impaired (Fig. 2C, total DNA, lane 3). Note that by autoradiography, faint supercoiled labeled bands are observed with p 0 in each depleted HSE (Fig. 2C, labeled DNA, lanes 2–4). This corresponds to a low level of background incorporation on the nonirradiated p 0, essentially due to the presence of residual nicks in the plasmid preparation, which can promote PCNA–CAF-1-dependent assembly.[25] Therefore, HIRA-depleted HSE fails to support nucleosome assembly on nondamaged DNA as opposed to p150-depleted HSE, which remains efficient in this nucleosome assembly pathway. In contrast, HIRA-depleted HSE, as well as mock-depleted extract, support efficient assembly on UV-irradiated plasmid (p UV), labeled DNA molecules being mainly supercoiled (Fig. 2C, labeled DNA, lanes 6 and 7). With p150-depleted HSE, a low level of supercoiled repaired DNA and a higher level of relaxed repaired DNA are observed (Fig. 2C, labeled DNA, lane 8), showing that this extract is unable to assemble nucleosomes coupled to NER. The amount of supercoiled plasmid visualized by ethidium bromide (Fig. 2C, total DNA, lane 8) is nevertheless almost unchanged as compared with the reaction with p 0 (Fig. 2C, total DNA, lane 4) and corresponds to nonrepaired molecules. We conclude that p150-depleted HSE fails to support nucleosome assembly coupled to DNA repair whereas HIRA-depleted HSE retains this activity.

In conclusion, with HIRA- and p150-depleted HSEs, we effectively gain access to extracts that are competent, respectively, only for the chromatin assembly pathway associated with DNA synthesis (coupled to NER) or only for the assembly pathway independent of DNA synthesis.

Control Experiments for Nucleosome Assembly with Depleted HSE Analyzed by Supercoiling Assays

To draw valid conclusions in these nucleosome assembly assays with depleted extracts, important control experiments are necessary. One is to ensure that the depleted extract retains its topoisomerase activity which is essential to follow assembly of nucleosomes. This can be easily verified by analyzing the topological form of the plasmid at an early time point in

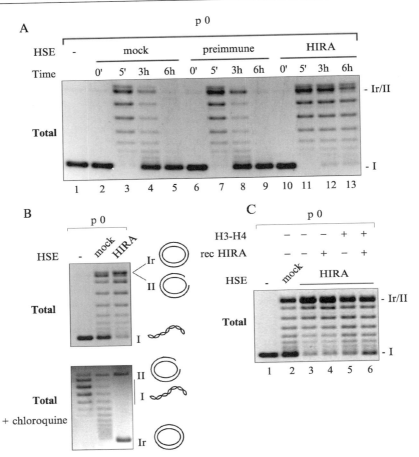

FIG. 3. Control experiments involving nucleosome assembly with depleted HSE. (A) Time course analysis of nucleosome assembly with mock, preimmune, or HIRA-depleted HSE. Nucleosome assembly assays were carried out with nonirradiated plasmid p 0 (150 ng) and either mock, preimmune, or HIRA-depleted HSE. The nucleosome assembly reactions were processed for supercoiling assay after 0 min, 5 min, 3 h, or 6 h of incubation. After electrophoresis on agarose gels, DNA was visualized by ethidium bromide (total DNA). Input p 0 DNA run in parallel (lane 1) and the migration positions of the different plasmid forms are indicated as in Fig. 2B. (B) Nucleosome assembly assays, with nonirradiated plasmid p 0 (150 ng) and with mock or HIRA-depleted HSE, were analyzed after 3 h of incubation by supercoiling assay. Agarose gel electrophoresis was performed in the presence or in the absence of chloroquine (10 μg/ml) and DNA was visualized by ethidium bromide (total DNA). The input DNA run in parallel and the migration positions of the different plasmid forms are indicated in Fig. 2B. (C) Significant rescue of nucleosome assembly activity of HIRA-depleted HSE by the addition of both HIRA and (H3–H4)$_2$ tetramers. Recombinant *Xenopus* HIRA (rec HIRA) or (H3–H4)$_2$ tetramers or both were added to the nucleosome assembly reaction mix with HIRA-depleted HSE and p 0 (150 ng). A nucleosome assembly

the chromatin assembly reaction. As shown in Fig. 3A, we have observed that, after 5 min of incubation, the supercoiled plasmid becomes relaxed similarly in mock, preimmune, or HIRA-depleted HSEs, indicating that the depletions do not affect the topoisomerase activity (lanes 3, 7, and 11). It is also important to test longer chromatin assembly incubation times to determine whether the defect in the assembly reaction observed at 3 h remains after a longer incubation time and that the decrease in assembly is not just the consequence of a slower activity of the depleted extract. For HIRA-depleted extract, we showed no increase in supercoiled plasmid even after 6 h of incubation (Fig. 3A, compare lanes 12 and 13). Another control experiment is to verify that the plasmid migrating as a Ir/II form is, in fact, a relaxed circular plasmid (form Ir), corresponding to nonassembled DNA, and not a nicked circular plasmid (form II) that could be generated by the presence of nuclease activity gained during the depletion process. One way to address this issue is to perform electrophoresis in a chloroquine-containing agarose gel (10 μg/ml). Chloroquine adds positive supercoils to DNA, and therefore negatively supercoiled plasmids become relaxed and relaxed plasmids migrate as a supercoiled plasmid, whereas nicked circular plasmids still migrate at the same position as in a nonchloroquine gel. By this approach, we show that after 3 h of nucleosome assembly reaction in the presence of HIRA-depleted HSE, the plasmid DNA recovered is truly in the relaxed form (Ir) and therefore corresponds to nonassembled DNA and not to nicked DNA (Fig. 3B). Finally, it is also important to be able to rescue the defect of the depleted HSE: first, by adding back a small amount of control HSE to be sure that the depletion did not introduce contaminating inhibitory activity, and second, by adding back the protein that was depleted. However, the complementation can be complicated if the serum depletes not only the protein against which it was derived but also associated proteins that could be partners for the nucleosome assembly. For HIRA-depleted HSE, the addition of recombinant HIRA (purified from *Escherichia coli,* using the Impact T7 purification system; New England BioLabs, Beverly, MA) is not sufficient to rescue the supercoiling activity (Fig. 3C, lane 4). However, we have succeeded in recovering nucleosome assembly by adding back both recombinant *Xenopus* HIRA and (H3–H4)$_2$ tetramers (purified from chicken erythrocytes[28])

---

[28] R. H. Simon and G. Felsenfeld, *Nucleic Acids Res.* **6,** 689 (1979).

---

assay with mock-depleted HSE was performed in parallel as a positive control for assembly. The reactions were stopped after 3 h and analyzed by supercoiling assay, and the DNA was visualized after agarose gel electrophoresis by ethidium bromide staining. DNA input run in parallel and the migration positions of the different plasmid forms are indicated as in Fig. 2B.

(Fig. 3C, lane 6). These complementation studies are performed in siliconized tubes to circumvent problems related to the sticking feature of histones. Moreover, it is important to preincubate HIRA and (H3–H4)$_2$ for at least 10 min on ice and to add this mix in chromatin assembly after 5 min of reaction at 23°. The addition after 5 min is crucial to allow the topoisomerases to relax the plasmid; otherwise, the histones could bind directly to the supercoiled DNA and prevent the nucleosome assembly process. The complementation that we observe suggests that HIRA complexed with (H3–H4)$_2$ is necessary to initiate nucleosome assembly. However, the rescue being only partial, we propose that either the recombinant HIRA lacks some posttranslational modifications or some HIRA-codepleted partners are missing to restore a completely efficient nucleosome assembly.

Conclusions and Perspectives

We describe here an immunodepletion strategy for *Xenopus* HSE in order to dissociate the DNA synthesis-dependent and -independent chromatin assembly pathways that operate efficiently in these extracts. By depleting p150 (CAF-1 large subunit) we can obtain an HSE competent only for nucleosome assembly independent of DNA synthesis, and by depleting HIRA we can generate a system that is efficient only for DNA synthesis-dependent nucleosome assembly pathway that we analyze by NER. We previously described a specific suppression of synthesis-independent nucleosome assembly in *Drosophila* preblastoderm embryo extracts.[25] In that case, this suppression was achieved by increasing the salt in the nucleosome assembly assay, but the biochemical basis of this defect was not fully understood. It is possible that salt can affect the behavior of a critical component in this assembly pathway. With specific immunodepleted HSE, we gain access to an *in vitro* system allowing the separate investigation of these two chromatin assembly pathways, dependent on or independent of DNA synthesis.

We believe that chromatin assembly operates by an "assembly line" mechanism in which histones are successively transferred between dedicated chaperones before their deposition onto DNA to prevent unwanted interactions of histones due to their charges. It may be possible that the two nucleosome assembly pathways delineated here differ entirely in the make-up of this assembly line or that some components are common between them. An immunodepletion strategy, similar to that described here for CAF-1 and HIRA, can be used for other known histone chaperones in order to determine whether they have a function assigned to one or the other pathway.

Acknowledgments

We are grateful to Catherine Green for critical reading of the manuscript. This work was supported by la Ligue Nationale contre le Cancer (Equipe Labellisée la Ligue), Euratom (FIGH-CT-1999-00010 and FIGH-CT-2002-00207), the Commissariat à l'Energie Atomique (LRC no. 26), and RTN (HPRN-CT-200-00078 and HPRN-CT-2002-00238).

[9] Preparation of Defined Mononucleosomes, Dinucleosomes, and Nucleosome Arrays *In Vitro* and Analysis of Transcription Factor Binding

By LISA ANN CIRILLO and KENNETH S. ZARET

Introduction

Eukaryotic genes are transcriptionally controlled by regulatory factors bound to promoters and enhancers. Because the eukaryotic genome is assembled into chromatin, it follows that we must understand how regulatory factors function in that context. The repeating structural unit of chromatin is the nucleosome, with 146 bp of DNA wrapped nearly twice around an octamer of the four core histones H2A, H2B, H3, and H4. Long arrays of nucleosomes can fold into more compact structures, which in turn are stabilized by the binding of the linker histone H1.[1–3] This chapter describes ways to reconstitute nucleosomes and nucleosome arrays with defined DNA sequences, in order to investigate transcription factor interactions and histone modifications in those contexts.

Accessibility of DNA-binding proteins to nucleosomal DNA is contingent on the rotational and translational position of the DNA with respect to the histone octamer. DNA sequence determines the curvature, flexibility, and kinking ability of DNA, which in turn determine overall nucleosome stability and positioning. Although some naturally occurring gene regulatory regions harbor strong rotational and translational nucleosome-positioning sequences,[4–6] nucleosome positioning most often appears to

[1] C. Tse and J. C. Hansen, *Biochemistry* **36,** 11381 (1997).

[2] L. M. Carruthers, J. Bednar, C. L. Woodcock, and J. C. Hansen, *Biochemistry* **37,** 14776 (1998).

[3] L. M. Carruthers and J. C. Hansen, *J. Biol. Chem.* **275,** 37285 (2000).

[4] H. Richard-Foy and G. L. Hager, *EMBO J.* **6,** 2321 (1987).

[5] E. Verdin, P. Paras, Jr., and C. Van Lint, *EMBO J.* **12,** 3249 (1993).

[6] R. T. Simpson, F. Thoma, and J. M. Brubaker, *Cell* **42,** 799 (1985).

be weak or random *in vitro*. The rotational position of a DNA-binding site with respect to the core particle surface can be a determining factor in whether or not a given transcription factor will bind to nucleosomal DNA, because it may not be able to bind at a site facing toward the nucleosome core.[7] Wrapping of DNA around the histone octamer results in kinking and distortion of DNA, with such perturbations being more predominant for nucleosomal DNA located at the dyad axis of the core particle. For this reason, some, but not all, transcription factors bind better to their sites if they are located at the edge of the nucleosome particle[8]; therefore, the translational position of the binding site with respect to the nucleosome dyad axis will be an important determinant in transcription factor binding. In addition, electrostatic interactions between the core histones and the DNA phosphate backbone strongly influence transcription factor binding to nucleosomal DNA. Modification of the lysine residues within the amino-terminal tails of the core histones can have positive or negative effects on transcription factor binding to nucleosomal DNA, depending on the modification.[9] A final influence on transcription factor binding to nucleosomal DNA is the presence of prebound DNA-binding factors and nucleosome-remodeling complexes. Transcription factor binding can perturb nucleosome structure through the alteration of nucleosome position,[10] destabilization or stabilization of underlying nucleosomes,[11–14] or linker histone displacement.[15,16] ATP-dependent nucleosome-remodeling complexes can perform these activities[17,18] except that they are blocked by linker histones[19] and, because they do not bind specific target sequences, they can act on virtually any nucleosomal template.

The ability to reconstitute nucleosomes *in vitro* with defined nucleotide sequences has allowed a detailed analysis of nucleosome interactions

[7] Q. Li and Ö. Wrange, *Mol. Cell. Biol.* **15,** 4375 (1995).
[8] M. Vettese-Dadey, P. Walter, H. Chen, L.-J. Juan, and J. L. Workman, *Mol. Cell. Biol.* **14,** 970 (1994).
[9] B. D. Strahl and C. D. Allis, *Nature* **403,** 41 (2000).
[10] E. Y. Shim, C. Woodcock, and K. S. Zaret, *Genes Dev.* **12,** 5 (1998).
[11] J. L. Workman and R. E. Kingston, *Science* **258,** 1780 (1992).
[12] L. A. Cirillo and K. S. Zaret, *Mol. Cell* **4,** 961 (1999).
[13] T. K. Archer, P. Lefebvre, R. G. Wolford, and G. L. Hager, *Science* **255,** 1573 (1992).
[14] M. Truss, J. Bartsch, A. Schelbert, and M. Beato, *EMBO J.* **14,** 1737 (1995).
[15] M. Kermekchiev, J. L. Workman, and C. Pikaard, *Mol. Cell. Biol.* **17,** 5833 (1997).
[16] L. A. Cirillo, C. E. McPherson, P. Bossard, K. Stevens, S. Cherian, E.-Y. Shim, E. A. Clark, S. K. Burley, and K. S. Zaret, *EMBO J.* **17,** 244 (1998).
[17] R. E. Kingston and G. J. Narlikar, *Genes Dev.* **13,** 2339 (1999).
[18] P. B. Becker and W. Horz, *Annu. Rev. Biochem.* **71,** 247 (2002).
[19] P. J. Horn, L. M. Carruthers, C. Logie, D. A. Hill, M. J. Solomon, P. A. Wade, A. N. Imbalzano, J. C. Hansen, and C. L. Peterson, *Nat. Struct. Biol.* **9,** 263 (2002).

with transcription factors and remodeling complexes.[17,18,20,21] Most early studies of transcription factor binding to nucleosomal templates were conducted on *in vitro*-assembled mono- and dinucleosome particles. More recent studies have examined the binding of transcription factors to *in vitro*-assembled nucleosome arrays, permitting the interaction of transcription factors with higher order chromatin structure to be examined.

In the first section of this chapter, protocols for the preparation of mono-and dinucleosome templates are described, followed by protocols to examine transcription factor binding and its effects on nucleosome structure and positioning, as well as facilitated factor binding, on these templates. In the second section, protocols for the preparation of extended and compacted nucleosome arrays are described, followed by protocols to examine transcription factor binding to these templates and the consequent effects on local chromatin structure. Each of the chromatin templates described below is assembled from purified components, that is, DNA fragments and core histones. Assembly of chromatin templates using extracts prepared from *Drosophila* embryos and *Xenopus* oocytes,[22,23] as well as purified chromatin assembly factors,[24] has also been described. However, chromatin templates assembled by these methods are often contaminated by the retention of substoichiometric amounts of chromatin assembly proteins and their corresponding remodeling activities, which can have an unwanted influence on transcription factor binding.

Assembly of Mono- and Dinucleosome Templates by Salt Gradient Dialysis

This protocol details the assembly of mono- and dinucleosome particles, using purified core histones and DNA fragments by the salt–urea gradient dialysis method.[25] At 2 M NaCl, a salt concentration at which core histone octamers are dissociated from DNA, a mono- or dinucleosome-sized DNA template is mixed with purified core histones. The salt concentration is then gradually decreased by dialysis, resulting in the reconstitution of histone octamers onto the DNA template. (H3–H4)$_2$ tetramers bind the DNA when the salt concentration is decreased to 1 M,

[20] J. L. Workman and R. E. Kingston, *Annu. Rev. Biochem.* **67**, 545 (1998).
[21] Q. Li, U. Bjork, and O. Wrange, *Methods Enzymol.* **304**, 313 (1999).
[22] E. Bonte and P. B. Becker, *Methods Mol. Biol.* **119**, 187 (1999).
[23] D. J. Tremethick, *Methods Enzymol.* **304**, 50 (1999).
[24] T. Ito, M. Bulger, M. J. Pazin, R. Kobayashi, and J. T. Kadonaga, *Cell* **90**, 145 (1997).
[25] D. Y. Lee, J. J. Hayes, D. Pruss, and A. P. Wolffe, *Cell* **72**, 73 (1993).

followed by association of $(H2A-H2B)_2$ tetramers at 0.6 M. The protocol is in three parts: synthesis and purification of DNA templates, nucleosome assembly, and purification of assembled nucleosome templates that are labeled at either one end or uniformly. The assembly reaction requires highly purified, carefully quantitated DNA templates. We synthesize our DNA fragments by polymerase chain reaction (PCR), gel purify the fragments, and then quantitate the purified DNA templates by two different methods. Finally, because the nucleosome assembly and purification are carried out at 4°, all necessary equipment, tubes, and solutions should be equilibrated at 4° before use.

Solutions

   10× Tris–borate–EDTA (TBE): 0.89 M Tris-HCl, 0.74 M boric acid, 8.9 mM EDTA
   6× bromphenol blue gel loading buffer: 30% glycerol, 0.09% bromphenol blue
   6× gel loading buffer: 30% glycerol, 0.09% bromphenol blue, 0.09% xylene cyanol

Protocols

Synthesis and Purification of End-Labeled and Internally Labeled DNA Templates. To assemble mononucleosomes, we typically use DNA fragments from 150 to 180 bp in length, whereas we have used 428-bp fragments to assemble dinucleosome templates. Our laboratory uses the transcriptional enhancer of the serum albumin gene as an experimental system to study chromatin structure and function, and these DNA fragments correspond to regulatory sequences to which positioned nucleosomes have been mapped *in vivo*.[26] In general, histone octamers can be reconstituted onto any double-stranded DNA; an excellent discussion detailing the choice of DNA fragments for nucleosome reconstitution can be found in Li *et al.*[21] Each assembly reaction requires 20 μg of DNA. To obtain this amount of DNA, we normally perform ≈8 PCRs and pool the products.

To make end-labeled nucleosomes for electromobility shift, DNase footprinting, and exonuclease III (ExoIII) digestion analysis, the 5′ end of one of the two PCR primers is first kinased with $[\gamma\text{-}^{32}P]$ATP. An 8.5-μl volume of primer 1 (25 pmol/μl), 2.5 μl of 10× T4 kinase buffer, 1.25 μl of 0.1 M dithiothreitol (DTT) (1.25 mM), 0.5 μl of $[\gamma\text{-}^{32}P]$ATP (150 mCi/ml; PerkinElmer Life Sciences, Boston, MA), 11.25 μl of distilled H_2O, and 1.0 μl of T4 kinase (New England BioLabs, Beverly, MA) are

[26] C. E. McPherson, E.-Y. Shim, D. S. Friedman, and K. S. Zaret, *Cell* **75**, 387 (1993).

combined in a microcentrifuge tube for a total volume of 25 μl. The kinase reaction is incubated at 37° for 45 min, followed by incubation at 68° for 10 min to inactivate the T4 kinase. The end-labeled primer is then used in the following PCR: 10.0 μl of 10× *Taq* polymerase buffer, 2 to 4 m*M* MgCl$_2$, 7.0 μl of 5 m*M* dNTPs (0.35 m*M* final), 1.0 μl of primer 2 (25 pmol/μl), 2.94 μl of kinased primer 1, 0.5 μl of bovine serum albumin (BSA, 10 mg/ml; 5 μg final), 0.5 μl of template DNA (10 ng/μl), 0.3 μl of *Taq* polymerase (0.2 U, Amplitaq; Applied Biosystems, Foster City, CA), and water to 100 μl. Conditions for a single PCR are given; a cocktail containing 8.5 times the indicated amounts is made and dispensed into a total of eight PCR tubes.

To make uniformly labeled nucleosome templates for micrococcal nuclease (MNase) digestion analysis of nucleosome positioning, PCRs are performed with unlabeled primers in the presence of [α-^{32}P]dATP and dCTP as follows: 10.0 μl of 10× *Taq* polymerase buffer, 2 to 4 m*M* MgCl$_2$, 7.0 μl of 5 m*M* dNTPs (0.35 m*M*), 1.0 μl of primer 1 (25 pmol/μl), 1.0 μl of primer 2 (25 pmol/μl), 0.5 μl BSA at 10 mg/ml (5 μg), 0.5 μl of template DNA (10 ng/μl), 0.5 μl of [α-^{32}P]dCTP (3000 Ci/mmol; PerkinElmer Life Sciences), 0.5 μl of [α-^{32}P]dATP (3000 Ci/mmol; PerkinElmer Life Sciences), 0.3 μl of *Taq* polymerase, and water to 100 μl. Once again, these conditions are for a single PCR, which is normally scaled up 8.5-fold and dispensed into eight PCR tubes. The reactions are then subjected to PCR under preoptimized conditions for the specific template and primers.

After PCR, the reactions are combined and loaded into a single 14-cm-wide well on a 6% native polyacrylamide gel in 1× TBE and electrophoresed for 2 h at 150 V. Loading dye containing only bromphenol blue is used when purifying probes for nucleosome assembly in order to prevent contamination of the DNA templates with xylene cyanol. The gel is placed onto a fluor-coated thin-layer chromatography (TLC) plate (Ambion, Austin, TX) and DNA is visualized with a hand-held UV lamp (shadow casting). The band corresponding to the DNA fragment is excised from the gel and cut in half. Each half is placed into 32-mm-wide dialysis tubing (6000–8000 MW cutoff) containing 0.5× TBE. The DNA is then electroeluted from the gel pieces into the dialysis tubing at 50 V overnight. DNA recovered from the dialysis tubing is combined, ethanol precipitated, resuspended in 100 μl of 0.1× TE, and quantitated by determining the OD$_{260}$. To confirm the DNA concentration, 25 and 50 ng of DNA is run on a 6% mini-native polyacrylamide gel and compared with known amounts of DNA markers run alongside. The specific activity is determined by counting 2 μl of the labeled DNA in a scintillation counter. This specific activity will be used to determine the final nucleosome concentration after nucleosome assembly and purification. We typically obtain specific activities of 10^6 cpm/μg for the 180-bp mononucleosome template and 10^5 cpm/μg for a 428-bp dinucleosome template.

Nucleosome Assembly. The mono- and dinucleosome assembly reactions are set up in low adhesion (USA Scientific, Ocala, FL) microcentrifuge tubes to prevent the histone proteins from adhering to the walls of the tubes. Twenty micrograms of DNA is used per assembly. The core histones used in our assemblies are purified from either sheep or rat liver nuclei by salt extraction as previously described[27,28] and quantitated by a Bradford assay, using BSA as a standard. The ratio of histones to DNA is normally 0.8 to 1 μg of histone per microgram of DNA; the optimal ratio may vary for different histone preparations and different DNA fragments. Selecting the appropriate histone-to-DNA ratio is important. Oversaturation of the DNA with improperly assembled histone protein will be enhanced at high histone-to-DNA concentrations; conversely, a low histone-to-DNA ratio will leave excess DNA unreconstituted. When setting up a nucleosome assembly with a new template or a new preparation of core histones, it is advisable to determine the optimal histone-to-DNA ratio experimentally, assessing the extent of reconstitution on nucleoprotein gels as described below.

In a typical assembly reaction, 20 μg of labeled, gel-purified DNA fragment, 10 mM HEPES (pH 7.5), 2 M NaCl (Sigma, St. Louis, MO), 1 mM EDTA, 10 mM 2-mercaptoethanol, 16–20 μg of core histones, and water up to 200 μl is combined in a microcentrifuge tube, transferred into 18-mm-wide dialysis tubing (3500 MW cutoff), and dialyzed against 500 ml of 2 M NaCl, 10 mM HEPES (pH 7.4), 1 mM EDTA, 10 mM 2-mercaptoethanol, and 5 M urea at 4° overnight. The following day, the assembly is dialyzed against successive changes of the same buffer with 5 M urea and 1.2, 1.0, 0.8, and 0.6 M NaCl at 4° for at least 90 min each, followed by dialysis in two changes of the same buffer containing 0.6 M NaCl but without urea at 4° for 2 h each, followed by a final dialysis against 10 mM HEPES (pH 7.4), 10 mM NaCl, 0.1 mM EDTA, and 1 mM 2-mercaptoethanol at 4° overnight. Each dialysis step is carried out in 500 ml of dialysis buffer. The dialysis buffers, minus the 2-mercaptoethanol, should be prepared beforehand and allowed to equilibrate to 4°. 2-Mercaptoethanol should be added to each buffer just before use. After the final dialysis step, the assembly reaction (normally a little more than 300 μl) is transferred to a low-adhesion microcentrifuge tube and stored at 4°.

Purification of Nucleosome Templates. The mononucleosome assembly will contain a mixture (usually 50:50 for a good assembly) of

[27] J.-K. Liu, Y. Bergman, and K. S. Zaret, *Genes Dev.* **2,** 528 (1988).
[28] J. L. Workman, I. C. A. Taylor, and R. E. Kingston, *Methods Cell Biol.* **35,** 419 (1991).

mononucleosomes and free DNA, whereas the dinucleosome assembly will contain a mixture of dinucleosomes, mononucleosomes, and free DNA. The mono- and dinucleosomes are purified from their respective assembly reactions on different 5-ml glycerol gradients in 50 mM HEPES (pH 7.4), 1 mM EDTA, and BSA (0.03 mg/ml). For purification of mononucleosome templates, a 5–30% glycerol gradient is used. For purification of dinucleosome templates, a narrow 16–18% glycerol gradient is used. The gradients are formed in 5-ml Beckman (Fullerton, CA) ultraclear ultracentrifuge tubes, using a gradient maker (Hoefer, San Francisco, CA). The nucleosome assembly is loaded onto the top of the gradient, after setting aside 5 μl of the nucleosome assembly for later analysis (see below). The gradients are centrifuged at 35,000 rpm in a Beckman SW50.1 rotor for 18 h at 4°, without a brake. Fractions are isolated by puncturing the ultracentrifuge tube near the bottom with a 18-gauge needle and then collecting fractions of four drops (~150 μl) each into numbered tubes.

For mononucleosome assemblies, 5 μl of each gradient fraction and 1 μl of the unfractionated assembly are run on a 4% native polyacrylamide-0.5× TBE gel until the bromphenol blue has migrated 75% of the way down the gel. Loading dye is not added to the gradient fractions or to the unfractionated assembly, as it might disrupt the assembled nucleosomes. The gradient fractions contain enough glycerol to sink to the bottom of the well on loading, and glycerol to 5% is added to the unfractionated assembly before loading. Five microliters of 1× loading dye containing bromphenol blue and xylene cyanol is run in an empty lane at one side of the gel, to assess migration. The gel is dried and subjected to autoradiography, in order to determine the fractions containing mononucleosomes and free DNA. The mononucleosomes migrate more slowly through the gel than the corresponding free DNA. If the gradient fractions are carefully isolated, there should be a distinct separation between the gradient fractions containing mononucleosomes and those containing free DNA. Fractions containing mononucleosomes are pooled and dialyzed two times against 10 mM HEPES (pH 7.4), 10 mM NaCl, 0.1 mM EDTA, and 1 mM 2-mercaptoethanol at 4° for 2 h each. Immediately after dialysis, the mononucleosomes are concentrated in a Centricon 30 (Amicon: Millipore, Bedford, MA) by centrifugation at 5000 rpm in a Sorvall SS34 rotor (Kendro Laboratory Products, Asheville, NC) at 4° to a final volume of 100 to 150 μl. The concentrated nucleosome assembly is transferred to a low-adhesion microcentrifuge tube and stored at 4°. We have found that nucleosome assemblies are stable for more than 1 year when stored at 4°. If desired, the free DNA fractions can also be pooled, phenol extracted to eliminate glycerol and residual histone proteins, ethanol precipitated, and resuspended in the buffer used to dialyze the pooled mononucleosomes.

To determine the concentration of DNA and nucleosomes, 2 μl of the concentrated mononucleosomes (and free DNA) is counted in a scintillation counter. The DNA concentration is then calculated by comparing the number of counts with the specific activity of the labeled DNA fragment added to the nucleosome assembly. Dilute nucleosome samples can dissociate,[29] so it is important to maintain a concentration of 10 μg/ml or higher. Integrity of the purified mononucleosomes can be ascertained by comparing equal amounts of purified mononucleosomes and free DNA on a 4% native polyacrylamide–0.5× TBE gel.

For dinucleosome assemblies, 3 μl of each gradient fraction and 1 μl of the unfractionated assembly are run on a 0.7% agarose–0.5× TBE, 20 × 25 cm gel at 4° until the bromphenol blue run in a side lane is 5 to 10 cm from the bottom. The gel is dried and subjected to autoradiography in order to determine the fractions containing mononucleosomes, dinucleosomes, and free DNA, each of which should run in distinct positions on the gel. Fractions containing dinucleosomes (and, if desired, mononucleosomes and free DNA) are pooled and treated as described for the mononucleosome assembly above. Integrity of the purified dinucleosomes is ascertained by comparing equal amounts of purified dinucleosomes, mononucleosomes, and free DNA on a 0.7% agarose–0.5× TBE gel.

Gel Shift Analysis of Protein Binding to Mono- and Dinucleosomes

One of the simplest methods of determining whether a transcription factor will bind to its sites in nucleosomal DNA is the gel shift assay. The protein is incubated with end-labeled mono- or dinucleosome particles containing one or more binding sites for the transcription factor, and the resulting complex is run on a nondenaturing polyacrylamide gel. Protein-bound nucleosome particles will migrate more slowly through the gel than the unbound nucleosome particles. The specificity of binding can be determined by performing the same assay with nucleosomal templates that contain (a) mutation(s) in the transcription factor-binding site or by competition with a binding site oligonucleotide (oligo) containing (a) similar mutation(s). It is best to work out the binding conditions for the particular transcription factor on the corresponding free DNA templates or binding site oligos before conducting the assay with the nucleosome templates. Important considerations include protein concentration, salt concentration in the binding buffer, incubation temperature for the binding reaction, and gel running conditions (percent acrylamide, gel running buffer, and temperature).

[29] J. S. Godde and A. P. Wolffe, *J. Biol. Chem.* **270**, 27399 (1995).

Solutions

10× binding buffer: 100 mM Tris(pH 7.5), 10 mM MgCl$_2$, 10% Ficoll

10× Tris–borate–EDTA (TBE): 0.89 M Tris-HCl, 0.74 M boric acid, 8.9 mM EDTA

6× gel loading buffer: 30% glycerol, 0.09% bromphenol blue, 0.09% xylene cyanol

Protocol

In a typical gel shift reaction performed with mononucleosomes, the following are aliquoted into a low-adhesion 1.5-ml microcentrifuge tube: 0.42 pmol of end-labeled mononucleosome template (for a nucleosome concentration of 4 nM) or 0.42 pmol of the corresponding free DNA, 2 μl of 10× binding buffer, 50 to 100 mM KCl, 1 mM DTT, BSA (3 mg/ml), 0.5% glycerol, 5 to 50 nM purified, recombinant protein in protein dilution buffer, and water up to 20 μl. Optimal concentrations of KCl and protein will need to be determined. Because transcription factors generally display a lower affinity for their binding sites on nucleosomal as compared with free DNA, 5 to 50 times more protein may be required to gel shift the mononucleosome than the corresponding free DNA templates. The reactions are then electrophoresed through a 4% native polyacrylamide–0.5× TBE gel until the bromphenol blue has run 75% of the way down the gel. The reactions are loaded with the gel running at 50 V, to prevent dilution of the protein–DNA complexes. No loading dye is added to the reactions, which contain enough Ficoll and glycerol to fall to the bottom of the well, because loading dye might disrupt the protein–DNA interactions. Therefore, 10 μl of 1× loading dye is run in a lane at the side of the gel to assess migration. The gel is dried and exposed to film. The gel shift reaction for the dinucleosome templates is carried out in a similar manner, with the exception that the gel is run at 4° at 60 V for 18 h or until the xylene cyanol dye front is located 2 cm from the bottom of the gel. This is necessary because of the longer size of the templates.

Analysis of Transcription Factor Binding, Using
   DNase I Footprinting

A second method of determining protein occupancy on mono- and dinucleosome templates is DNase I footprinting. In this assay, end-labeled mono- or dinucleosomes are bound by protein, followed by digestion with DNase. Levels of DNase are used so that only one cleavage occurs in each DNA molecule, generating a ladder of bands when the purified DNA is run on a denaturing polyacrylamide gel. Protein occupancy protects the corresponding DNA from digestion with DNase and is observed as a region devoid of bands, or "footprint," when compared with the DNase digestion pattern for unbound nucleosomal DNA. The underlying core histones inhibit DNase digestion of nucleosomal DNA, which sometimes makes it difficult or impossible to visualize protections due to protein binding. However, protein binding may enhance the sensitivity of bound DNA to DNase digestion, resulting in the generation of hypersensitive sites.[16] Generation of these hypersensitive sites is only partially inhibited by nucleosomal histones, and is not necessarily reflective of release of DNA from the nucleosomes.

In addition to protein occupancy, DNase footprinting can be also be used to determine the effects of protein binding on the rotational positioning of bound nucleosome cores. The periodicity of DNA wrapped around the histone octamer is such that at every 10 base pairs, the minor groove of the DNA is exposed on the core particle surface. Because DNase cleaves within the minor groove of the DNA, nucleosomes that are rotationally positioned on a DNA fragment will generate a 10-base pair ladder of hypersensitive sites when subjected to DNase digestion. However, in many instances, gene regulatory regions do not possess intrinsic nucleosome-positioning properties. When the corresponding DNA fragments are assembled into mono- or dinucleosomes, DNase digestion does not produce a 10-base pair ladder of hypersensitive sites, although such positioning may be induced by factor binding.[10,12]

Solutions

10× DNase binding buffer: 100 mM Tris(pH 7.5), 10 mM MgCl$_2$, 10% Ficoll

DNase dilution buffer: 44 mM MgCl$_2$, 2.2 mM CaCl$_2$

Stop buffer: 350 mM NaCl, 30 mM EDTA, 0.1% sodium dodecyl sulfate (SDS), tRNA (30 μg/ml)

Formamide loading buffer: 24.5 ml of deionized formamide, 0.5 ml of 0.5 M EDTA, 50 mg of bromphenol blue, 50 mg of xylene cyanol in 25 ml total

10× Tris–borate–EDTA (TBE): 0.89 M Tris-HCl, 0.74 M boric acid, 8.9 mM EDTA

Protocol

The following are aliquoted into low-adhesion 1.5-ml microcentrifuge tubes: 0.42 pmol of end-labeled mono- or dinucleosome template (for a nucleosome concentration of 4 nM) or 0.42 pmol of the corresponding free DNA, 2 μl of 10× DNase binding buffer, 50 to 100 mM KCl, 1 mM DTT,

BSA (3 mg/ml), 0.5% glycerol, a range of 1 to 200 nM purified transcription factor in protein dilution buffer or an equal amount of the buffer in which the protein is diluted as a control for DNase digestion, and water up to 20 μl. Optimal concentrations of KCl and protein must be determined. In the case of HNF3, 1 to 20 nM purified protein is required to footprint the free DNA template, 25 to 100 nM protein is required to footprint mononucleosomes, and 20 to 40 nM protein is required to footprint dinucleosomes. We normally use 5 nM HNF3 protein to footprint the albumin enhancer containing free DNA templates. Fifty and 100 nM HNF3 give good protection on the corresponding mononucleosome templates, whereas 20 nM HNF3 saturates its binding sites on the corresponding dinucleosomes. The template is incubated with the protein for at least 30 min at room temperature in order to allow the protein to bind. DNase (Worthington, Lakewood, NJ) is added and the reaction is allowed to incubate for an additional 1 min at room temperature, followed by the immediate addition of an equal reaction volume of DNase stop buffer. The amount of DNase must be titrated; typically, we use 1 ng on free DNA versus 10 ng on nucleosomal DNA, and the amount of DNase may need to be increased when transcription factors are added to nucleosomes. TE is added to bring the reaction to a total volume of 100 μl, and then the reactions are extracted once with 25:24:1 phenol–chloroform–isoamyl alcohol and once with 24:1 chloroform–isoamyl alcohol, followed by precipitation with 2.5 volumes of ethanol at $-20°$ for 1.5 h. Precipitation at $-20°$ is preferred to precipitation on dry ice, to minimize the amount of salt in the final pellet and make the pellet easier to resuspend. After centrifugation of the precipitated DNA, the DNA pellet is dried in a SpeedVac (Thermo Savant, Holbrook, NY) and resuspended in 6 μl of loading buffer. This buffer contains formamide to aid in the denaturation of the DNA. Because prolonged storage in formamide-containing buffer will subject the DNA to nonspecific nicking, it is best to store the DNA in ethanol at $-20°$ until it is ready to be analyzed by gel electrophoresis. The reactions are loaded onto a 6% polyacrylamide, 6 M urea, 1× TBE sequencing gel that has been prerun for at least 45 min. The gel should be poured, using a rabbit tooth comb, to generate the wells for loading. Before loading, the reactions should be heated to 90° for 5 min and then placed immediately on ice in order to denature the DNA. The gel is run at 1800 V; the length of running time will depend on the distance of the protein-binding site(s) from the labeled DNA end. A G-cleavage sequencing ladder corresponding to the nucleosomal DNA should be run beside the footprinting reactions.[30] After electrophoresis, the gel is transferred to Whatman (Clifton, NJ) 3-mm filter paper, dried on a gel dryer,

---

[30] A. Maxam and W. Gilbert, *Proc. Natl. Acad. Sci. USA* **74,** 560 (1977).

and then exposed to a phosphoimage screen and/or X-ray film with an intensifying screen.

ExoIII Analysis of Nucleosome Boundaries

Knowledge of the translational positioning, in addition to the rotational positioning, of the assembled nucleosome particles is essential for proper interpretation of the results obtained from the gel shift and DNase footprinting experiments described above. Translational positioning of the DNA with respect to the histone octamer will determine where, and if, the binding site for a particular transcription factor is located on the nucleosome particle. This will in turn, have an important influence on transcription factor access and affinity for its binding sites in nucleosomal DNA. Translational positioning of mononucleosome particles can be determined by exonuclease III (ExoIII) digestion, or with a micrococcal nuclease (MNase) assay, which is described in the next section. ExoIII digests each strand of DNA in the 3′-to-5′ direction and is impeded by histones or bound transcription factors. Therefore, when carried out on end-labeled nucleosome templates, distinct stops in ExoIII digestion corresponding to the nucleosome boundary opposite the labeled DNA end are observed. To accurately determine translational positioning, ExoIII analysis should be carried out on separate nucleosome preparations, each labeled at different 5′ ends. Transcription factor binding to mononucleosome particles may alter nucleosome boundaries by changing the translational position of nucleosomes, as assessed by ExoIII digestion.[16]

Solutions

10× DNase binding buffer: 100 mM Tris (pH 7.5), 10 mM MgCl$_2$, 10% Ficoll

ExoIII dilution buffer: 66 mM Tris-HCl (pH 8.0), 0.66 mM MgCl$_2$, 50% glycerol

Stop buffer: 350 mM NaCl, 30 mM EDTA, 10% SDS, tRNA (30 μg/ml)

Formamide loading buffer: 24.5 ml of deionized formamide, 0.5 ml of 0.5 M EDTA, 50 mg of bromphenol blue, 50 mg of xylene cyanol in 25 ml total

10× Tris–borate–EDTA (TBE): 0.89 M Tris HCl, 0.74 M boric acid, 8.9 mM EDTA

Protocol

The following are aliquoted into low-adhesion 1.5-ml microcentrifuge tubes: 0.42 pmol of end-labeled mononucleosome template (for a nucleosome concentration of 4 nM) or 0.42 pmol of the corresponding free

DNA, 2 μl of 10× DNase binding buffer, 50 to 100 mM KCl, 1 mM DTT, BSA (3 mg/ml), 0.5% glycerol, 1 to 200 nM purified, recombinant protein in protein dilution buffer or an equal amount of the buffer in which the protein is diluted as a control for ExoIII digestion, and water up to 20 μl. In experiments in which the effects of transcription factor binding on nucleosome position are to be examined, a range of protein concentrations that provide a good DNase footprint with the particular transcription factor on nucleosomal DNA is used. The template is incubated with the protein for 30 min to 2 h at room temperature in order to allow the protein to bind. ExoIII (Worthington) is then added and the reaction is allowed to incubate for an additional 1 min at room temperature, followed by the immediate addition of an equal reaction volume of stop buffer. The amount of ExoIII needs to be titrated for each template and DNA-binding protein; we typically use 0.5 U for free DNA and 5 U for nucleosomal DNA. The reactions are then treated exactly as indicated for the DNase digestion reactions in the previous section.

Mapping of MNase Cleavages on Nucleosome Particles by Restriction Endonucleases

This protocol, which involves successive micrococcal nuclease (MNase) and restriction endonuclease digestion of uniformly labeled nucleosome templates, provides positions for nucleosome cores residing within mononucleosome or dinucleosome templates.

Solutions

10× binding buffer: 100 mM Tris (pH 7.5), 10 mM MgCl$_2$, 10% Ficoll

MNase dilution buffer: 5 mM Tris-HCl (pH 7.5), 25 μM CaCl$_2$

6× gel loading dye: 30% glycerol, 0.09% bromphenol blue, 0.09% xylene cyanol

10× Tris–borate–EDTA (TBE): 0.89 M Tris-HCl, 0.74 M boric acid, 8.9 mM EDTA

Protocol

The following are aliquoted into low-adhesion 1.5-ml microcentrifuge tubes: 0.42 pmol of uniformly labeled nucleosome template (for a nucleosome concentration of 4 nM), 2 μl of 10× DNase binding buffer, 50 to 100 mM KCl, 1 mM DTT, BSA (3 mg/ml), 0.5% glycerol, 1 to 200 nM purified, recombinant protein in protein dilution buffer or an equal amount of the buffer in which the protein is diluted as a control for MNase digestion, and water up to 20 μl. In experiments in which the effects of transcription factor binding on nucleosome position are to be examined, a range of protein concentrations that provide a good DNase footprint with the particular transcription factor on nucleosomal DNA is used. The template is incubated with the protein for at least 30 min at room temperature in order to allow the protein to bind. Then MNase (Worthington) is added and the reaction is allowed to incubate for an additional 5 min at room temperature, followed by the immediate addition of 50 mM EGTA to terminate the digestion. Double-stranded DNA cuts by MNase are impeded by nucleosomal histones. Therefore, MNase will initially digest the linker DNA between nucleosome particles, before extending into the nucleosomal DNA. In this assay, limiting amounts of MNase are used such that all the linker DNA between the nucleosome particles is digested, leaving intact the DNA wrapped around each nucleosome particle. The concentration of MNase used must therefore be titrated to obtain the optimal concentration of MNase that will produce mononucleosome-sized particles. More MNase may be required to digest factor-bound templates than unbound templates. TE is added to bring the reaction to a total volume of 100 μl, and then the reactions are extracted once with 25:24:1 phenol–chloroform–isoamyl alcohol and once with 24:1 chloroform–isoamyl alcohol. A one-tenth volume of 3 M sodium acetate, pH 7, is added, followed by precipitation with 2.5 volumes of ethanol at $-20°$. The precipitation is carried out overnight, because no carrier DNA is used. The DNA pellet is dried in a SpeedVac and resuspended in 40 μl of TE.

Two microliters of BPB loading dye is added to 10 μl of the reaction products, and then the reaction products are run on a nondenaturing 8% polyacrylamide gel in 1× TBE until the BPB has just run off the gel. The gel is then dried and exposed to film. The purpose of this gel is to determine which MNase-cleaved templates are suitable for restriction enzyme digestion analysis; that is, those for which the predominant MNase digestion product is mononucleosomal. The remaining 30 μl corresponding to these samples is then subjected to digestion with a restriction enzyme the recognition site of which falls within the nucleosome whose position is being determined. A 30-μl volume of reaction product is digested with a 10-fold excess of restriction enzyme in 10× enzyme buffer in a total of 100 μl for 4 h at the recommended temperature for the particular restriction enzyme. After addition of 1 μg of tRNA carrier and a one-tenth volume of 3 M sodium acetate, pH 7, the digestion products are precipitated in 2.5 volumes of ethanol at $-20°$ overnight. Use of more than 1 μg of carrier tRNA makes resuspension of the final DNA pellet difficult. The DNA pellet is dried in a SpeedVac and resuspended in 5 μl of TE. After the addition of 1 μl of 6× BPB loading dye, the reactions are run on a sequencing gel-sized nondenaturing 10% polyacrylamide gel until the BPB dye is 2 cm from the bottom of the gel (24 to 36 h). A long gel is used in order to be able to separate the small, similar-sized bands that will result from the restriction enzyme digestion. An end-labeled DNA sizing ladder is run next to the reactions. The gel is then dried, exposed to a phospho-image screen and/or X-ray film with intensifying screens, and nucleosome position is determined from the sizes of the fragments that are generated. Unbound nucleosome templates, or nucleosome templates bound to a transcription factor that is incapable of nucleosome positioning, will contain randomly positioned cores, and therefore generate near-randomly sized fragments.

Assembly of Nucleosome Arrays

Gene regulatory regions occur in long arrays of nucleosomes *in vivo;* therefore, transcription factor binding to chromatin *in vivo* actually takes place in the context of a nucleosome array, not mono- or dinucleosome particles. Because of this, a growing number of researchers have begun to utilize nucleosome arrays as an *in vitro* chromatin system to examine transcription factor binding to nucleosomal DNA. Extended and linker histone-compacted nucleosome arrays exhibit folding dynamics similar to cellular chromatin,[2,31] permitting the interaction of transcription factors with higher order chromatin structure to be examined. The nucleosome array templates were initially developed in the laboratory of Simpson *et al.*,[6] and Hansen *et al.* have further refined the array assembly process and extensively characterized the biophysical properties of the assembled arrays.[2,31–33] Nucleosome arrays are assembled from DNA templates containing 10 or more tandem repeats of the sea urchin 5S rDNA sequence. Each of the 5S sequences assembles into a translationally positioned nucleosome after reconstitution with purified core histones, so that the resulting nucleosome arrays exhibit the conformational dynamics of nuclear chromatin.[2,31,32] Owen-Hughes and Workman[34] demonstrated that it was possible to insert a mononucleosome-sized DNA sequence containing the DNA-binding site for the transcription factor Gal4 between the fifth and sixth 5S rDNA sequences and use the resulting DNA templates

[31] J. C. Hansen, J. Ausio, V. H. Stanik, and K. E. van Holde, *Biochemistry* **28,** 9129 (1989).

[32] P. M. Schwarz and J. C. Hansen, *J. Biol. Chem.* **269,** 16284 (1994).

[33] P. M. Schwarz, A. Felthauser, T. M. Fletcher, and J. C. Hansen, *Biochemistry* **35,** 4009 (1996).

[34] T. Owen-Hughes and J. L. Workman, *EMBO J.* **15,** 4702 (1996).

to assemble stable nucleosome array templates capable of being bound by Gal4-AH. This has since been repeated for di- and trinucleosome-sized DNA fragments corresponding to the HIV and albumin enhancer DNA sequences, respectively,[35,36] and well as for mononucleosome-sized fragments containing concatemerized binding sites for various transcription factors.[37,38] It is outside the scope of this chapter to detail all the intricacies of nucleosome array assembly and analysis; an excellent discussion of nucleosome arrays as a model chromatin system is found in Carruthers et al.[39] A protocol detailing the assembly of nucleosome arrays from end-labeled DNA fragments by salt step dilution is described below. The end-labeled nucleosome arrays are used in gel shift, DNase hypersensitivity, and enzyme accessibility assays. DNase footprinting is performed on internally labeled nucleosome arrays; the assembly of nucleosome arrays from internally labeled DNA fragments is described together with the protocol for DNase footprinting.

Solutions

50× Tris–acetate–EDTA (TAE): 2 M Tris-acetate, 0.05 M EDTA
10× Tris–borate–EDTA (TBE): 0.89 M Tris base, 0.74 M boric acid, 8.9 mM EDTA
6× bromphenol blue loading dye: 30% glycerol, 0.09% bromphenol blue

Protocols

Preparation of DNA for End-Labeled Nucleosome Arrays. The DNA fragment that we use to assemble nucleosome arrays contains 10 repeats of the 5S rDNA sequence with albumin enhancer/promoter sequences cloned between the fifth and sixth repeats; however, as mentioned above, nucleosome arrays containing one or more inserted nucleosomes containing a variety of DNA sequences have been assembled and assayed. This entire array fragment was cloned as a 2700-bp *Mlu*I–*Pvu*II fragment in pBR322.[36] Two micrograms of array fragment is required for each array assembly reaction. Digestion, labeling, and gel purification of the

[35] D. J. Steger and J. L. Workman, *EMBO J.* **16,** 2463 (1997).
[36] L. Cirillo, F. R. Lin, I. Cuesta, M. Jarnik, and K. Zaret, *Mol. Cell* **9,** 279 (2002).
[37] S. Malik, A. E. Wallberg, Y. K. Kang, and R. G. Roeder, *Mol. Cell. Biol.* **22,** 5626 (2002).
[38] A. E. Wallberg, K. Pedersen, U. Lendahl, and R. G. Roeder, *Mol. Cell. Biol.* **22,** 7812 (2002).
[39] L. M. Carruthers, C. Tse, K. P. Walker, 3rd, and J. C. Hansen, *Methods Enzymol.* **304,** 19 (1999).

array fragment result in the loss of approximately two-thrids of the starting DNA, so we normally start with 75 μg of DNA to obtain ~20 to 30 μg of end-labeled array fragment, for use in multiple assembly reactions.

Seventy-five micrograms of the nucleosome array plasmid is digested with 30 units each of *Mlu*I, *Pvu*II, and *Hha*I in 10× NEB buffer 3 supplemented with BSA (1 mg/ml) in a total volume of 75 μl at 37° for no greater than 6 h. Digestion in excess of 6 h results in excessive nicking of the DNA fragment. *Mlu*I and *Pvu*II release the 2-kb array fragment while *Hha*I digests the vector DNA into multiple fragments. The array fragment is then end labeled by the addition of 0.25 mM dATP, dTTP and dCTP, 6 μl of [α-^{32}P]dGTP (3000 Ci/mmol; PerkinElmer Life Sciences) and 10 units of (exo-)Klenow (New England BioLabs) in a total volume of 90 μl followed by incubation at 37° for 45 min. *Pvu*II and *Hha*I are blunt cutters, so only the *Mlu*I ends, which contain a 5' overhang, are filled in and labeled. Twenty microliters of 6× bromphenol blue-containing loading dye (xylene cyanol runs at the same position as the array fragment in the gel) is added, and the entire 110 μl is run in a single 5-cm well on a 1% agarose 14 × 12 cm gel in 1× TBE until the bromphenol blue is three-quarters of the way down the gel. The gel is placed onto a fluor-coated TLC plate (Ambion) and DNA is visualized with a hand-held UV lamp. The ~2-kb band corresponding to the end-labeled DNA fragment is excised from the gel and cut into eight equal pieces. DNA from each piece is purified with a Qiagen (Valencia, CA) gel extraction kit. The final eluates are combined and the DNA is ethanol precipitated, resuspended in 25 μl of TE, and quantitated by determining the OD$_{260}$. To confirm the DNA concentration, 25 and 50 ng of DNA are run on a 1% agarose 5 × 7 cm gel and compared with quantitated DNA of approximately the same size, run alongside.

Nucleosome Array Assembly. Depending on the histone-to-DNA input ratio, the reconstituted arrays will be subsaturated, saturated, or supersaturated with core histones. Subsaturated arrays contain one or more octamer-free gaps, whereas supersaturated arrays are nonspecifically complexed with additional core histones. Only saturated arrays, in which all possible DNA repeats are occupied by core histones, undergo appropriate salt and linker histone-dependent folding and compaction.[39] We therefore normally assemble the nucleosome arrays every time using several different ratios of DNA to histone, in order to ensure that one assembly reaction contains an optimal saturation of the array DNA with histone octamers. We also perform a mock assembly, to which no histones are added. The assembly is set up in a 10-μl volume, so the labeled DNA fragment needs to be at a final concentration of 1 mg/ml or higher. To 1.5-ml low-adhesion microcentrifuge tubes, add, in this order: distilled H_2O to 10 μl; 2 M NaCl; 1 μg of BSA; 2 μg of labeled DNA fragment; and 0, 2, 2.4, or 2.6 μg of core histones (mock assembly, 1:1, 1:1.2, and 1:1.4 DNA-to-histone ratio, respectively). This reaction is incubated at 37° for 15 min. Then, at 15-min intervals, 3.3, 6.7, 5, 3.6, 4.7, 6.7, 10, 30, and 20 μl of 50 mM HEPES-NaOH (pH 7.5), 1 mM EDTA, 5 mM DTT, and 0.5 mM phenylmethylsulfonyl fluoride (PMSF) are added, with incubation at 30°. Finally, 100 μl of 10 mM Tris-HCl (pH 7.5), 1 mM EDTA, 1% Nonidet P-40 (NP-40), 20% glycerol, BSA (100 μg/ml), 5 mM DTT, and 0.5 mM PMSF are added, followed by incubation at 30° for 15 min and at room temperature for 5 min. The assemblies are then stored at 4°.

Array assembly is verified by comparative gel shift analysis of the assembled and mock-assembled arrays. Five microliters of each of the array assemblies and 5 μl of the mock-assembled DNA are run on a 1% agarose 5 × 7 cm gel in 1× TAE at 50 V for 1.5 h. Because they are more compacted, the nucleosome arrays migrate more quickly through the gel than the corresponding free DNA.[36] The degree of array saturation is determined by *Eco*RI digestion analysis.[39] This assay takes advantage of the fact that each of the 5S rDNA repeats is flanked on either side by an *Eco*RI digestion site; when the 5S repeats are assembled into a positioned nucleosome, these flanking sites are accessible to *Eco*RI digestion. A 3-μl volume (2 nM) of each of the array assemblies and 3 μl of the mock-assembled DNA are digested with *Eco* RI at 2 U/μl in 20 mM HEPES (pH 7.5), 50 mM KCl, 2% glycerol, 5 mM DTT, and BSA (100 μg/ml) for 2 h at 37°, and the digestion products are resolved on a 1% agarose 0.5× TBE gel. *Eco*RI digestion of the mock-assembled arrays results in the release of ten 208-bp fragments. In contrast, the predominant product of *Eco*RI digestion of the assembled arrays is the corresponding mononucleosome particle, which runs at a higher position in an agarose gel. A small amount of 208-bp fragment, corresponding to nonassembled 5S repeats, is also observed; arrays are considered saturated when the ratio of nucleosomal to free DNA is 9:1.

Gel Shift Analysis of Protein Binding to Nucleosome Arrays

As for mono- and dinucleosome particles, one of the simplest methods to determine whether a transcription factor will bind to its sites on a nucleosome array is the gel shift assay. Once again, it is best if the binding conditions for the gel shift assay are optimized for the particular transcription factor on free DNA templates or binding site oligos before conducting the assay with the nucleosome array templates.

Solutions

10× Tris–borate–EDTA (TBE): 0.89 M Tris-HCl, 0.74 M boric acid, 8.9 mM EDTA

6× gel loading buffer: 30% glycerol, 0.09% bromphenol blue, 0.09% xylene cyanol

Protocol

In a typical reaction, 200 fmol (2 nM final) nucleosome arrays are incubated with 0 to 50 nM protein in 10 mM HEPES (pH 7.5), 50 mM KCl, 5 mM DTT, BSA (0.25 mg/ml), 5% glycerol in a total volume of 20 μl at room temperature for 30 min. The reactions are then electrophoresed through a 1% agarose–0.5× TBE 20 × 35 gel until the xylene cyanol is located 10 cm from the bottom of the gel. No loading dye is included in the gel shift reactions, so 1× loading dye is run in a well at the side of the gel to determine migration. The gel is fixed in 10% methanol, 10% acetic acid for 30 min, dried, and exposed to a phosphoimage screen and/or to X-ray film with intensifying screens.

DNase Footprinting of Nucleosome Arrays Containing a Unique
    Internal Label

*Preparation of DNA for Nucleosome Arrays Containing a Unique,
Internally Labeled Site*

DNase footprinting of the nucleosome arrays is carried out on nucleosome arrays containing a unique internal label. The nucleosome array is internally labeled at a unique restriction site that lies immediately up- or downstream of the region to be footprinted. This procedure, illustrated in Fig. 1A, consists of digestion of the array plasmid with restriction enzyme, dephosphorylation and 5′ end labeling of the restriction site, recircularization of the plasmid, analysis of the ligation products, and gel purification of the internally labeled array fragment.

The DNA fragment used to assemble the nucleosome arrays shown in Fig. 1A and B contains a unique *Xba*I digestion site located downstream of an inserted nucleosome containing the transcription factor-binding sites of interest. Once again, DNA is lost over the course of this procedure, more than in the preparation of the end-labeled array fragments. DNA (100 μg) is digested with 100 units of *Xba*I in 1× NEB buffer 2 supplemented with BSA (1 mg/ml) in a total of 100 μl for 2 h at 37°. The digestion reaction is incubated for 15 min at 75° to inactivate the *Xba*I. The digestion site is

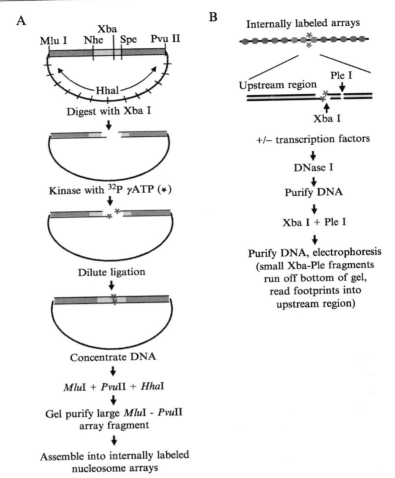

FIG. 1. DNase I footprinting of nucleosome arrays. (A) Strategy for preparation of DNA for nucleosome arrays containing a unique, internally labeled site. (B) Flow chart for DNase footprinting of internally labeled nucleosome arrays.

dephosphorylated by the addition of 5 units of shrimp alkaline phosphatase (Promega, Madison, WI) and incubation for 1 h at 37°. Shrimp alkaline phosphatase is used instead of calf intestinal phosphatase (CIP) because it is easily inactivated by incubation at 75° for 15 min, eliminating the need for additional purification steps and the accompanying loss of DNA. A one-tenth volume of 3 M sodium acetate, pH 7.0, and 2.5 volumes of ethanol are added and the DNA is precipitated for at least 1.5 h at −20°. The precipitated DNA is pelleted by centrifugation in a microcentrifuge, dried, and resuspended in 100 μl of 1× NEB T4 kinase buffer. The DNA is then quantitated by determining the OD_{260} of a 1-μl aliquot.

The digested, dephosphorylated DNA is end labeled at the internal *Xba*I site, using a 5-fold molar excess of [γ-^{32}P]ATP. For each 25 μg of DNA in 1× NEB T4 kinase buffer, 3 μl of [γ-^{32}P]ATP (15 Ci/ml; Perkin-Elmer Life Sciences) and 20 units of T4 kinase (New England BioLabs) are added. The final reaction volume will be dependent on the amount of DNA recovered from the preceding steps. The reaction is incubated for 45 min at 37°. The enzyme is inactivated by incubation at 68° for 10 min. The kinased DNA is then recircularized by dilute ligation at a DNA concentration no greater than 1 μg/ml. Therefore, each 15 μg of DNA is ligated in a total volume of 15 ml containing 50 mM Tris-HCl (pH 7.5), 10 mM MgCl$_2$, 10 mM DTT, 1 mM ATP, BSA (25 μg/ml), and 4000 units (8 Weiss units/ml) of T4 ligase (New England BioLabs) at 16° for greater than 12 h. After ligation, each 15-ml ligation reaction is extracted three or four times with butanol to reduce the reaction volume to 1.5 ml. Each ligation reaction contains a large amount of salt, which must be removed in order to carry out the subsequent digestion and gel purification steps. Therefore, the DNA from each ligation reaction is desalted with a Wizard clean-up system (Promega). The maximum input DNA and volume for each column is 5 μg/500 μl, so three columns are used per original 15-ml ligation reaction. A 500-μl volume of the ligation reaction is mixed with 1 ml of purification matrix and applied to one of the purification columns supplied with the kit. The column is washed with 2 ml of 80% isopropanol and placed into a fresh 1.5-ml microcentrifuge tube, and the DNA is eluted from the column by the addition of 50 μl of TE that has been prewarmed to 68° and centrifugation at top speed in a microcentrifuge for 1 min. The elution step is repeated once in order to maximize DNA recovery. The eluted DNA is pooled and ethanol precipitated after the addition of a one-tenth volume of 3 M sodium acetate, pH 7.0, and 2.5 volumes of ethanol at −20° for greater than 1.5 h. The DNA is pelleted by centrifugation in a microcentrifuge and resuspended in a total volume of 75 μl of TE, and the concentration is determined by determining the OD_{260}. The ligation is checked by running 200 ng of the ligated plasmid next to 200 ng of the uncut plasmid and 200 ng of the linearized plasmid on a 0.7% agarose–1× TAE gel. The ligated plasmid should run above the linearized plasmid but at or below the position of the nicked circular DNA in the uncut plasmid lane. It is normal to obtain a small percentage of oligomerized plasmid (runs above the nicked circular uncut DNA) and unligated plasmid; these will be eliminated in the subsequent digestion and gel purification steps.

The internally labeled array fragment is then isolated from the vector DNA by digestion with the same enzymes used to liberate the complete nucleosome array; in this case *Mlu*I, *Pvu*II, and *Hha*I (see Preparation of DNA for End-Labeled Nucleosome Arrays, above). The desired array fragment is purified, quantitated, and used to assemble nucleosome arrays as described above for the end-labeled array fragment.

DNase Footprinting

A diagram illustrating the strategy used to footprint the internally labeled arrays is shown in Fig. 1B. The footprinting reactions are carried out in a manner similar to that described above for the DNase footprinting of mono- and dinucleosomes. After purification and ethanol precipitation, the products of DNase digestion are digested with the restriction enzyme used to create the labeled site and a second restriction enzyme that cuts immediately up- or downstream to eliminate labeled products in the undesired direction. In the case of the nucleosome array shown in Fig. 1B, the restriction enzyme *Ple*I, which recognizes a site just downstream of the *Xba*I digestion site in the array, is used. Digestion with *Ple*I results in downstream DNase-cleaved fragments not exceeding 40 bp, corresponding to *Xba*I–*Ple*I, which run off the bottom of the gel. This allows most of the gel to display the footprinted enhancer sequences upstream of the *Xba*I site.

Solutions

10× DNase binding buffer: 100 mM Tris (pH 7.5), 10 mM MgCl$_2$, 10% Ficoll

DNase dilution buffer: 44 mM MgCl$_2$, 2.2 mM CaCl$_2$

Stop buffer: 350 mM NaCl, 30 mM EDTA, 0.1% SDS, tRNA (30 μg/ml)

Formamide loading buffer: 24.5 ml of deionized formamide, 0.5 ml of 0.5 M EDTA, 50 mg of bromphenol blue, 50 mg xylene cyanol in 25 ml total

10× Tris–borate–EDTA (TBE): 0.89 M Tris-HCl, 0.74 M boric acid, 8.9 mM EDTA

Protocol

The following are aliquoted into a low-adhesion 1.5-ml microcentrifuge tube: 200 fmol (2 nM final) of internally labeled nucleosome array or corresponding free DNA, 2 μl of 10× DNase binding buffer, 50 to 100 mM KCl, 1 mM DTT, BSA (3 mg/ml), 0.5% glycerol, 20 to 60 nM purified, recombinant protein in protein dilution buffer or an equal amount of the buffer in which the protein is diluted as a control for DNase digestion, and water up to 20 μl. Optimal concentrations of KCl and protein must be determined. In the case of HNF3, 20 to 40 nM protein is required to foot-print albumin enhancer-containing nucleosome arrays. The template is incubated with the protein for at least 30 min at room temperature in order to allow the protein to bind. DNase is then added and the reaction is allowed to incubate for an additional 1 min at room temperature, followed by the immediate addition of an equal reaction volume of DNase stop buffer. The amount of DNase must be titrated; usually, 10 to 20 ng of DNase is used. TE is added to bring the reaction to a total volume of 100 μl, and then the reactions are extracted once with 25:24:1 phenol–chloroform–isoamyl alcohol and once with 24:1 chloroform–isoamyl alcohol followed by precipitation with 2.5 volumes of ethanol at $-20°$ for 1.5 h. After centrifugation of the precipitated DNA, the DNA pellet is dried in a SpeedVac and resuspended in 18 μl of 1$\times$ restriction enzyme buffer corresponding to the restriction enzymes to be used. Ten units of each restriction enzyme is added and the digestion is allowed to proceed at 37° for 2 h. TE is added to 90 μl, followed by the addition of one-tenth volume of 3 M sodium acetate, pH 7.0, and 2.5 volumes of ethanol and precipitation at $-20°$ for 1.5 h. The precipitated DNA is centrifuged, dried, and resuspended in formamide loading buffer. The reactions are loaded onto a 6% polyacrylamide–6 M urea–1$\times$ TBE sequencing gel at 1800 V; the length of running time will depend on the distance of the protein-binding site(s) from the labeled DNA end. A sequencing ladder corresponding to the nucleosomal DNA should be run beside the footprinting reactions. After electrophoresis, the gel is transferred to Whatman 3-mm filter paper, dried on a gel dryer, and then exposed to a phosphoimage screen and/or to X-ray film with intensifying screens.

Hypersensitive Site DNase Analysis of Transcription Factor Binding to Nucleosome Arrays

Digestion of the end-labeled nucleosome arrays with limiting amounts of DNase results in the preferential digestion of linker DNA, generating a ladder of bands corresponding to each of the nucleosomes in the array. When a transcription factor binds to its sites within the nucleosome array, this binding will sometimes perturb chromatin structure; these perturbations are manifested as regions of hypersensitivity on either side of the bound nucleosome. For example, Owen-Hughes and Workman have shown that binding of the transcription factor Gal4 to its sites within a nucleosome array generates hypersensitive sites that flank the nucleosome to which it is bound (see Fig. 4 in Owen-Hughes and Workman[34]). This low-resolution DNase assay is conceptually similar to looking for regulatory regions in genomic DNA by determining the location of hypersensitive sites generated by DNase digestion of nuclei.

Solutions

10× Tris–borate–EDTA (TBE): 0.89 M Tris-HCl, 0.74 M boric acid, 8.9 mM EDTA

6× gel loading buffer: 30% glycerol, 0.09% bromphenol blue, 0.09% xylene cyanol

Protocol

In the typical assay, 200 fmol (2 nM) of end-labeled nucleosome arrays is incubated with 20–40 nM purified transcription factor in 10 mM HEPES (pH 7.5), 50 mM KCl, 5 mM DTT, BSA (0.25 mg/ml), and 5% glycerol in a total of 20 μl at room temperature for 1 h. The reactions are then digested for 1 min at room temperature with 25–100 ng of DNase. The optimal amount of DNA-binding protein and DNase needs to be determined by titration. DNase digestion is stopped by the addition of 20 μl of 20 mM Tris-HCl (pH 7.5), 50 mM EDTA, 1% SDS, tRNA (0.25 mg/ml), and proteinase K (0.25 mg/ml; Roche Molecular Biochemical, Indianapolis, IN). The reactions are incubated at 50° for 1 h and DNA is precipitated by the addition of a one-tenth volume of 3 M sodium acetate, pH 7.0, and 2.5 volumes of ethanol and incubation at −20° for more than 2 h. After centrifugation in a microcentrifuge, the pelleted DNA is dried and resuspended in 10 μl of TE and 2 μl of 6× loading dye containing bromphenol blue and xylene cyanol. The reactions are then electrophoresed through a 1% agarose–0.5× TBE 20 × 35 gel until the bromphenol blue has run off of the gel. The gel is fixed in 10% methanol–10% acetic acid for 30 min, dried, and exposed to a phosphoimage screen and/or to X-ray film with intensifying screens.

Generating Histone H1-Compacted Nucleosome Arrays and Assessing Transcription Factor-Mediated Chromatin Opening

In vivo footprinting and chromatin immunoprecipitation experiments have begun to reveal the order of protein-binding events that occur at specific promoters and enhancers during transcription activation *in vivo*.[40,41] These studies have led to the concept of initial or pioneer transcription factors, regulatory proteins that potentiate or poise the promoter or enhancer of a gene for activation by initiating the chromatin-remodeling events required for the binding of additional factors and preinitiation complex (PIC) assembly. It was shown that HNF3 and GATA-4, the initial transcription factors bound to the albumin enhancer *in vivo*, are able to

[40] M. P. Cosma, T. Tanaka, and K. Nasmyth, *Cell* **97,** 299 (1999).
[41] E. Soutoglou and I. Talianidis, *Science* **295,** 1901 (2002).

bind to and perturb or "open" linker histone-compacted nucleosome arrays.[36] By definition the initial proteins bound to a promoter or enhancer, pioneer factors must be capable of accessing their sites in silent compacted chromatin. We use two different assays to assess chromatin opening by DNA-binding factors: a DNase assay and a restriction enzyme accessibility assay. Both assays take advantage of the fact that nucleosome arrays compacted with the linker histone are particularly resistant to nuclease digestion. It is presumed that other pioneer transcription factors possess the same chromatin-opening capabilities as HNF3 and GATA-4.

Linker histone-compacted nucleosome arrays are assembled by combining equimolar amounts of extended nucleosome arrays and linker histone. Before the compacted nucleosome arrays are used in the chromatin-opening assays for the first time, electron microscopic analysis of the arrays should be performed, if possible, to determine proper assembly of the linker histone onto the arrays. Proper assembly of the linker histone onto the arrays should also be assessed by a chromatosome assay. In this assay, the H1-compacted arrays are digested with limiting amounts of MNase. While MNase digestion of the extended nucleosome arrays results in a terminal digestion product of 145 bp, or the amount of DNA wrapped around each nucleosome particle, appropriate binding of linker histone to the nucleosome array protects ~15 bp of linker DNA extending from each nucleosome particle from MNase digestion, resulting in a 165-bp digestion intermediate or "chromatosome." The products of MNase digestion are run on a native polyacrylamide gel, electroblotted to a nylon filter, hybridized to oligo probes corresponding to the 5S and insert-specific regions, followed by detection by phosphoimage and/or autoradiography. A protocol for the chromatosome assay, followed by protocols for the DNase and restriction enzyme accessibility chromatin-opening assays, is described below.

Solutions for Chromatosome Assay

10× Tris–borate–EDTA (TBE): 0.89 M Tris-HCl, 0.74 M boric acid, 8.9 mM EDTA

6× BPB gel loading buffer: 30% glycerol, 0.09% bromphenol blue

20× SSC: 3 M NaCl, 0.3 M sodium citrate, pH 7.0

100× Denhardt's: 2% BSA, 2% Ficoll, 2% polyvinylpyrrolidone (PVP) in 3× SSC

Protocol for Chromatosome Assay

To assemble linker histone-compacted nucleosome array templates, an equimolar concentration of linker histone to nucleosomes is incubated with 200 fmol (2 nM final) of unlabeled nucleosome arrays (26 nM nucleosomes of 13-mer array) in 10 mM HEPES (pH 7.5), 50 mM KCl, 5 mM DTT, BSA (0.25 mg/ml), and 5% glycerol in a total volume of 20 μl at room temperature for 1 h. We have used both commercially available linker histone (Roche Molecular Biochemicals) and linker histone purified from pig liver by salt extraction[42] in our assays. The linker histone-compacted arrays, and extended nucleosome arrays as a control, are then subjected to digestion with MNase in 30 mM CaCl$_2$. The concentration of MNase used must be titrated to obtain the optimal concentration of MNase that will produce mononucleosome-sized particles. We typically use 0.5 to 2 units of MNase to digest extended nucleosome arrays and 15 to 30 units to digest H1-compacted arrays. MNase digestion is stopped by the addition of 20 μl of 20 mM Tris-HCl (pH 7.5), 50 mM EDTA, 0.1% SDS, and proteinase K (0.05 mg/ml) (Roche Molecular Biochemicals). The reactions are incubated at 37° for 1 h, and DNA is precipitated by the addition of one-tenth volume of 3 M sodium acetate, pH 7.0, and 2.5 volumes of ethanol and incubation at $-20°$ for more than 2 h. After centrifugation in a microcentrifuge, the pelleted DNA is dried and resuspended in 20 μl of TE and 2 μl of 6× loading dye containing bromphenol blue. MNase digestion products (10 μl) are separated on two 8% native 1× TBE polyacrylamide gels, each of which is blotted to nylon membrane (Duralon UV; Stratagene, La Jolla, CA) by electrotransfer in 0.5× TBE at 80 mA at 4° for 16 h. The filters are placed DNA side up onto a piece of Whatman filter paper wet with 0.4 N NaOH in order to denature the transferred DNA, and then rinsed twice in 2× SSC and allowed to dry. The DNA is then fixed to the filters by UV cross-linking. The duplicate filters are prehybridized in 50% deionized formamide (International Biotechnologies, New Haven, CT), 5× SSC, 2× Denhardts, 10 mM NaHPO$_4$, 0.25% SDS, and denatured salmon sperm DNA (0.1 mg/ml) at 42° for 1 h and then hybridized to a nick-translated probe (10^6 counts/ml) containing the 5S rDNA 208 repeat or DNA insert [i.e., the inserted nucleosome(s) containing the transcription factor-binding sites] in 50% formamide, 5× SSC, 2× Denhardt's, 10 mM NaHPO$_4$, 0.25% SDS, 10% dextran sulfate, and denatured salmon sperm DNA (0.1 mg/ml) at 42° for 16 h. The filters are washed in three changes of 0.1× SSC, 0.1% SDS at 65° (30 min for each wash), air dried, and exposed to a phosphorimage screen and/or X-ray film with intensifying screens. A 165-bp digestion intermediate or "chromatosome," indicative of appropriate assembly of the linker histone onto the 5S and albumin enhancer-containing nucleosomes within the arrays, should be observed when the MNase-digested H1-compacted arrays are probed with either the 5S or DNA insert probes, respectively.

[42] G. E. Croston, L. M. Lira, and J. T. Kadonaga, *Protein Expr. Purif.* **2**, 162 (1991).

Solutions for DNase Assay

   10× Tris–borate–EDTA (TBE): 0.89 *M* Tris-HCl, 0.74 *M* boric acid,
   8.9 m*M* EDTA
   6× gel loading buffer: 30% glycerol, 0.09% bromphenol blue, 0.09%
   xylene cyanol

Protocol for DNase Assay

   An equimolar concentration of linker histone is assembled onto
200 fmol (2 n*M* final) of end-labeled nucleosome array templates in
10 m*M* HEPES (pH 7.5), 50 m*M* KCl, 5 m*M* DTT, BSA (0.25 mg/ml),
and 5% glycerol in a total of 20 µl as described above for the chromatosome assay. Twenty to 40 n*M* transcription factor is added, followed by incubation at room temperature for 1 to 2 h. The reactions are then digested
with DNase and processed as described above for the hypersensitive site
DNase analysis. The linker histone-compacted nucleosome arrays are resistant to digestion with DNase, while incubation with transcription factors
that are capable of chromatin opening, such as HNF3 and GATA-4, generates a region of hypersensitivity in the vicinity of their binding sites within
the nucleosome array and extending one or two nucleosomes beyond (see
Fig. 4A and B in Cirillo *et al.*[36]). The ability to bind to a nucleosome array
does not necessarily portend chromatin-opening capability, as demonstrated by the inability of Gal4-AH to open linker histone-compacted
nucleosome arrays harboring Gal4-binding sites (Fig. 4D in Cirillo *et al.*[36]).

Solutions for Restriction Enzyme Accessibility Assay

   10× Tris–borate–EDTA (TBE): 0.89 *M* Tris-HCl, 0.74 *M* boric acid,
   8.9 m*M* EDTA
   6× gel loading buffer: 30% glycerol, 0.09% bromphenol blue, 0.09%
   xylene cyanol

Protocol for Restriction Enzyme Accessibility Assay

   In this assay, the transcription factor-bound, linker histone-compacted
nucleosome arrays are digested with a restriction enzyme that digests
the nucleosome array just downstream of the inserted nucleosome(s) containing the transcription factor-binding sites. Because the restriction
enzyme digestion site thereby falls within linker DNA normally protected
by the linker histone, the linker histone-compacted nucleosome arrays are
resistant to restriction enzyme digestion. However, when bound by a transcription factor capable of chromatin opening, the restriction site is recognized and cleaved by the restriction enzyme, as observed for HNF3 and

GATA-4-bound, linker histone-compacted, albumin enhancer-containing nucleosome arrays (see Fig. 4E in Cirillo *et al.*[36]). Linker histone-compacted nucleosome arrays are incubated with protein under conditions similar to those described for the DNase chromatin-opening assay. $MgCl_2$ is then added to the reaction to a final concentration of 5 mM, followed by the addition of restriction enzyme and incubation at 37° for 1 h. As for the DNase assay, a range of restriction enzyme concentrations is used. Digestion is stopped by the addition of 20 μl of 20 mM Tris-HCl (pH 7.5), 50 mM EDTA, 1% SDS, tRNA (0.25 mg/ml), and proteinase K (0.25 mg/ml). The reactions are incubated at 50° for 1 h, and DNA is precipitated by the addition of a one-tenth volume of 3 M sodium acetate, pH 7.0, and 2.5 volumes of ethanol and incubation at −20° for greater than 1.5 h. After centrifugation, the pelleted DNA is dried and resuspended in 10 μl of TE and 2 μl of 6× loading dye containing bromphenol blue and xylene cyanol. The reactions are then electrophoresed through a 1% agarose–0.5× TBE 14 × 12 cm gel at 80 V until the bromphenol blue has run three-quarters of the way down the gel. The gel is fixed in 10% methanol, 10% acetic acid for 30 min, dried, and exposed to a phosphoimage screen and/or X-ray film with intensifying screens.

Acknowledgments

We thank Isabel Cuesta for development of the internally labeled array template strategy. This research was supported by an NIH postdoctoral fellowship (DK09529-01) to L.A.C., by an NIH grant (GM47903) to K.S.Z., and by a core grant (CA06297) to F.C.C.C.

[10] Purification of Native, Defined Chromatin Segments

By Robert T. Simpson, Charles E. Ducker,
John D. Diller, and Chun Ruan

Introduction

Understanding the biochemistry of an enzyme usually requires knowledge of its composition, structure, and function. These areas of information are usually acquired sequentially. In the case of attempts to understand regulation of genetic activity, the three cardinal knowledge areas are the same, but the order of acquisition of data seems to be reversed. Genetic and physiological information tells us something about the function of a given gene, particularly in organisms in which disruption of the gene can be accomplished easily. Our rather crude means of looking at structure, mostly involving chemical or enzymatic attack on the DNA and mapping accessibility, can give us inklings of structural information.[1] Yet, in most cases, we lack definitive information about the composition of a genetic locus as chromatin, beyond the likely presence of histones. Genetic information exists, in eukaryotes, in the context of chromatin. Increasingly, it is realized that the interactions of DNA with both histones and nonhistone proteins, as well as protein–protein interactions that do not involve direct binding to the nucleic acid, are critical for function of the DNA in transcription, recombination, and replication.

We describe here an approach to learning the composition, structure, and function of unique genes packaged as chromatin in different functional states. The approach has been developed in the context of a multicopy, episomal *Saccharomyces cerevisiae* DNA element, but should be generalizable to other species. The approach relies on the use of protein–nucleic acid affinity with the *Escherichia coli lac* repressor and operator to achieve a purification of >10^4-fold in a single step. The key elements in the procedure are as follows:

> A multicopy plasmid (although we are now attempting purifications of minichromosomes presented two or three copies per cell)
> Knowledge of the chromatin structure of the minichromosome
> Design and manufacture of a multicloning site and insertion of the *lac* operator, both in nucleosome-free regions

A current limitation of the procedure is the need to increase ionic strength to 0.2–0.3 *M* NaCl for elution of the minichromosome from the affinity matrix, making further purification of molecules with interactions that are unstable at this ionic strength questionable. The time of exposure to the higher ionic strength is brief, typically <30 min. We have an approach in progress that should eliminate ionic strengths >10–50 m*M* NaCl.

In this chapter, we first describe preparation of the reagents for purification, the *lac* repressor, and the backbone vector for the yeast minichromosome. We then present a protocol for purification of the minichromosome, using affinity methods and preparative ultracentrifugation. Last, we briefly provide leading references to analytical methods appropriate for characterization of the composition, structure, and function of chromatin containing unique genes.

---

[1] R. T. Simpson, *Methods* **15**, 283 (1998).

Reagent Preparation

lac *Repressor*

At least anecdotally, the *lac* repressor is a notably unstable protein. In addition, the protein is relatively difficult to assay, usually requiring an electrophoretic separation of radiolabeled operator DNA bound to the repressor from free DNA and quantification by autoradiography. About 30 years ago, Muller-Hill and Kania isolated a mutant *E. coli* strain that was constitutive for β-galactosidase expression.[2] The strain has a deletion of the intergenic *I-Z* region leading to production of a fusion protein. The repressor–galactosidase fusion from strain BMH 72-19-1 retains both *lac* operator-binding and galactosidase activities. Highly stable β-galactosidase confers some of its solidity to the repressor protein. Using the polymerase chain reaction, the fusion protein DNA sequences were amplified from genomic DNA of this strain, adding restriction sites for *Nde*I and *Xho*I at the amino- and carboxyl-terminal regions of the open reading frame, respectively. After cutting with the appropriate restriction enzymes, the DNA fragment, purified by agarose gel electrophoresis, was ligated into a similarly cut IMPACT pTYB2 (New England BioLabs, Beverly, MA) plasmid vector. The IMPACT vectors contain a self-cleaving intein protein-coding sequence fused to the carboxyl-terminal 52 amino acids of the *Bacillus circulans* WL-12 chitin-binding domain. The inserted DNA makes a fusion protein that has LacI, LacZ, intein, and a chitin-binding domain. The RNA for this protein is expressed in a Studier T7 RNA polymerase system induced by isopropylthio-β-D-galactopyranoside (IPTG).[3] Transfection into *E. coli* BL21(DE3) is effected by any of several standard methods. Ampicillin (100 μg/ml) selection defines transformants and restriction enzyme mapping of plasmid minipreparations confirms success in making the fusion protein expression bacterial strain. Alternatively, it is available from the authors' laboratory on request as strain TLIZ.

Bacteria containing the cloned composite DNA are grown in rich medium (2 × YT) with ampicillin for selection to an A_{600} of 0.4 at 37°. Typically, we grow 300 ml inoculated with an appropriate aliquot of a log-phase overnight culture. For both the bacterial and the yeast cultures, we think it is important to have cultures in log phase as opposed to stationary phase. The culture is cooled for 15 min on ice and transferred to a

---

[2] D. Levens and P. M. Howley, *Mol. Cell. Biol.* **5**, 2307 (1985).
[3] F. W. Studier, A. H. Rosenberg, J. J. Dunn, and J. W. Dubendorff, *Methods Enzymol.* **185**, 60 (1990).

shaker (300 rpm) incubator at 15° with addition of IPTG to a final concentration of 40 μM. The culture is incubated overnight (16 h) under these conditions.

Bacteria are harvested by centrifugation at 5000 rpm in a Sorvall GS-3 rotor (Kendro Laboratory Products, Asheville, NC) for 10 min at 4°, for example, and resuspended in 10% of the culture volume in chitin column buffer (CCB).

CCB: 20 mM HEPES (pH 8.0), 1 mM EDTA, 0.1% Tween 20, 500 mM NaCl

The suspended bacteria are lysed by sonication. Routinely, we use a large horn on a Branson Ultrasonics (Danbury, CT) sonifier 450 with a 50-ml disposable plastic centrifuge tube cut off at about the 35-ml mark to allow the tip of the horn to be placed within 1 mm of the tube bottom. Sonication is performed on ice with, typically, six 10-s bursts at 50% duty cycle and a power setting of 4, spaced by 30-s rest periods. Output of sonifiers varies with model, geometry, and age, so optimization of conditions for a particular setup is mandatory. Fortunately, assay of galactosidase is fast and simple, so the establishment of conditions is easy. Galactosidase assays are done with the following reagents.

ONPG: o-Nitrophenyl-galactopyranoside, 4 mg/ml in 10 mM PIPES (pH 6.5)

2× galactosidase buffer: 120 mM Na$_2$HPO$_4$, 80 mM NaH$_2$PO$_4$, 20 mM KCl, 2 mM MgSO$_4$, 10 mM 2-mercaptoethanol

1 M Na$_2$CO$_3$

1% sodium deoxycholate in water

Toluene

For whole cells (only), add toluene and deoxycholate at 1% of culture volume and shake for 5 min. Centrifuge in an Eppendorf microcentrifuge for a few seconds and use supernatant for enzyme measurements. For all other samples, mix an aliquot with water to a final volume of 0.5 ml. Add 0.5 ml of 2× galactosidase buffer and 0.2 ml of ONPG. Hold at room temperature for an appropriate time (judging yellow color by eye) and stop the reaction with 0.5 ml of Na$_2$CO$_3$. Read the optical density of the product at 420 nm. Absolute measurements of activity are not necessary; the goal is relative activities during sonication, adsorption of the fusion protein to column matrix, and so on.

Chitin beads are obtained from New England BioLabs as a 50% suspension in 20% ethanol. Typically, 2 ml (1-ml bed volume) of the suspension is loaded into a 1 cm diameter × 10 cm height column (737-1011; Bio-Rad, Hercules, CA) operating under gravity flow. The column is washed with two 5-ml aliquots of CCB at 4°. The sonicated bacterial culture is clarified by centrifugation at 18,000 rpm in the Sorvall SS-34 rotor at 4° for 30 min. Supernatant from the centrifugation is applied to the chitin bead column which is then washed successively with 10 ml of CCB and two 5-ml volumes of minichromosome binding buffer (MBB50):

MBB(number): 20 mM HEPES, 0.5 mM EDTA, 0.1% Tween 20 (pH 8.0)

where (number) refers to the NaCl concentration (mM). In published work, we have used MBB150.[4] More recently, we have lowered the ionic strength of the binding buffer to further expand minichromosome chromatin and hopefully facilitate interactions of repressor with the operator. This has neither hindered nor favored the purification of SALT10, our most studied minichromosome. We have routinely used a *lacI-Z* column within a few days of its preparation. About 50% of the β-galactosidase activity is bound to the column and sodium dodecyl sulfate (SDS) gel electrophoresis of the bound material shows a major band of protein with a molecular weight about that of the intein–chitin-binding domain portion of the fusion protein, suggesting that some intein self-cleavage has occurred during the protein preparation (our unpublished observations).

pALT

The starting point for our minichromosome backbone vector is the 1453-bp TRP1/ARS1 plasmid described by Zakian and Scott.[5] This circularized *Eco*RI fragment contains the gene for *N*-phosphoribosylanthranilate isomerase (*TRP1*), enabling selection of transformants on Trp⁻ medium, and an autonomously replicating sequence (*ARS1*), a yeast replication origin, which allows its maintenance as an autosome. Typically the minichromosome is present at 10–80 copies, on average, per cell. The chromatin structure of the minichromosome has been established.[6] Four imprecisely located nucleosomes are present on the *TRP1* coding sequences. Two apparently nucleosome-free regions, ~100 bp in length each, flank the open reading frame. The remainder of the circular minichromosome is packaged as three precisely positioned nucleosomes. Knowledge of the chromatin structure of the parent plasmid allows selection of a nucleosome-free region (the replication origin, 3′ of the *TRP1* gene) into which to insert the *lac* operator for use in protein–nucleic acid affinity purification. Stillman's studies of the DNA sequence elements necessary for replication origin function allowed this insertion into a region that does

[4] C. E. Ducker and R. T. Simpson, *EMBO J.* **19,** 400 (2000).
[5] V. A. Zakian and J. F. Scott, *Mol. Cell. Biol.* **2,** 221 (1982).
[6] F. Thoma, L. W. Bergman, and R. T. Simpson, *J. Mol. Biol.* **177,** 715 (1984).

not alter origin efficiency or reduce copy number of the minichromosome.[7] The 21-bp operator sequence (AATTGTGAGCGGATAACAATT) replaces sequences 780–800 base pairs from the *Eco*RI site, between ARS elements B2 and B3.

Construction of a multicloning site enhances the flexibility of the backbone vector. The unique 44-bp polylinker, called DL, is inserted at an engineered *Bam*HI site, located downstream of *TRP1*, 215 bp from the *Bgl*II site in *ARS1*. The modified backbone *Eco*RI fragment is ligated into the *Eco*RI site of a derivative of pUC19 lacking most of its multicloning site, to produce pDTL. This plasmid, pDTL, is the vector we use currently for most of our DNA manipulations in bacteria before introduction of episomal sequences into *S. cerevisiae*.

Yeast Transformation, Culture, and Extract Preparation

After cloning the desired gene into pALT or the pDTL multicloning site in *E. coli,* and confirming that the desired DNA sequences are present in the episome, bacterial plasmid sequences are removed by digestion with *Eco*RI (or at another pair of sites if the cloned insert contains an *Eco*RI site). Yeast DNA is then circularized by ligation. The plasmid is then introduced into an appropriate *S. cerevisiae* strain. We routinely utilize YPH499 (**a**-cell) or YPH500 (α-cell) for our studies of mating type-specific gene expression or repression, although any other appropriate strain is suitable, depending on the experimental protocol and information desired. Transformation is carried out using any common protocol ranging from spheroplast isolation with lithium-mediated transfection to electroporation. Using selection for presence of the *TRP1* gene offers an economic advantage, because only tryptophan, of all the amino acids, is destroyed on acid hydrolysis of proteins. Thus, Casamino Acids (BD Biosciences, Franklin Lakes, NJ), an acid hydrolysate of casein, lacks only tryptophan. Supplementation of minimal synthetic medium with Casamino Acids is less costly than preparation or purchase of synthetic media from commercial suppliers. The synthetic medium we routinely employ for minichromosome-containing yeast strains is the following:

Trp$^-$ medium (per liter)
    6.7 g Yeast nitrogen base without amino acids
    5 g Casamino acids
    20 g Dextrose
    0.02 g Uracil
    0.02 g Adenine

[7] Y. Marahrens and B. Stillman, *Science* **255**, 817 (1992).

After selection of appropriate transformed strains, cultures are grown in the above-described selective medium and doubling times are determined for future experiments. Because the minichromosomes often contain yeast genes that are regulated as solo copies in the genome but are present in multiple copies in our experiments, we think it critical to assess regulation of the amplified gene in the cells from which we propose to isolate chromatin for structural and compositional studies.

Figure 1 illustrates the desired results showing that a multicopy plasmid gene can be regulated in identical fashion to the solo genomic copy. In this case, it is the **a**-cell-specific *STE6* gene and a portion thereof present in the SALT10 minichromosome.[4] Expression of both was evaluated in **a**- and α-cell types and in an α-cell strain mutant for the corepressor *TUP1*. The multicopy minichromosome is present at about 60–70 molecules per cell, on average, although the absolute number per cell is a function of plasmid partitioning during cytokinesis. The episomal gene is fully repressed in α-cells, in spite of being present in high copy number. In the strain with a multicopy plasmid in **a**-cells, messenger RNA is ~60 times the level of the solo genomic copy, suggesting full expression in **a**-cells in addition to full repression in α-cells. For both episomal and genomic genes, a *tup1* mutation derepresses expression by about 40–50% in α-cells.

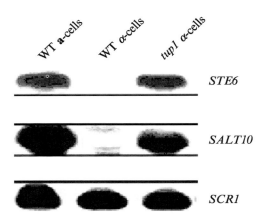

Fig. 1. Correct regulation of a multicopy minichromosome-borne gene. The **a**-cell-specific gene *STE6* was cloned into pALT and a portion of the central region of the open reading frame was deleted, leaving the flanking regions and a 10-nucleosome length of DNA from the α2 operator to the 3' end of the coding sequence and creating a plasmid called SALT10. The minichromosome was present at ~70 copies per cell. A northern blot of total RNA from wild-type **a**- and α-cells is shown together with a deletion strain lacking Tup1p. The blot was probed with labeled oligonucleotides specific for the genomic *STE6* gene, the minichromosome *SALT10* sequences or a loading control, *SCR1*.

We routinely grow our yeast cultures for several days with daily dilutions to ensure log phase the next day. A simple EXCEL program that allows calculation of these parameters enables culture planning if one knows the doubling time for the yeast. The copy number of some plasmids using *TRP1* selection drops as the culture goes into stationary phase.

Yeast are grown in selective media at 30° with shaking at 300 rpm to an A_{600} of 1. The volumes given here are for a 6-liter culture, but can be scaled down to 1 liter or up to 50 g of cell paste from fermenter cultures. Cell paste from large-scale cultures is conveniently stored as a frozen thin sheet in a Zip-Loc plastic bag at −80°. Vigorous application of a hammer to the frozen paste creates a rapidly dispersed sample for suspension in the next buffer. For smaller scale cultures, cells are harvested by centrifugation at 5000 rpm for 5 min at 20° in the Sorvall GS-3 rotor. The pellet is resuspended in 20 ml of sorbitol buffer supplemented with 1 mM phenylmethanesulfonyl fluoride (PMSF; a 200× stock solution is 0.2 M in methanol or dimethyl sulfoxide) and 10 mM 2-mercaptoethanol. PMSF must be added before the sulfhydryl reagent.

Sorbitol buffer: 1.4 M sorbitol, 40 mM HEPES (pH 7.5), 0.5 mM MgCl$_2$

Incubate at 30° for 15 min and centrifuge in a tared tube, using the HB-6 (or HB-4) rotor at 5000 rpm and 20° for 5 min. Weigh to determine cell mass. Suspend in 4 volumes (ml/g) of sorbitol buffer containing 1 mM PMSF. Add Zymolyase 100T (Seikagaku, Tokyo, Japan) to 0.5 mg/ml (stock solution, stored frozen, is 10 mg/ml in sorbitol buffer). Incubate at 30° with occasional shaking until spheroplast formation is complete, generally about 30 min.

To judge spheroplast formation, take a 2-μl aliquot and place it on a microscope slide with a coverslip. Focus the microscope on the yeast cells. Remove the slide from the stage and firmly press on the coverslip and try to slide it laterally. Replace on the stage and view the cells. If adequately spheroplasted, few if any intact yeast will have survived the shearing force. Add 20 ml of cold sorbitol buffer with 1 mM PMSF and harvest the cells by centrifugation, using the HB-6 rotor at 3500 rpm and 4° for 5 min. All steps from this point on are carried out on ice and in a cold room. Resuspend the spheroplasts (gently, using a 25-ml disposable pipette) in 25 ml of sorbitol buffer plus 1 mM PMSF and centrifuge as described above. Repeat this washing. Spheroplasts are sticky and somewhat difficult to resuspend, unlike intact yeast. The supernatants after centrifugation should be only slightly turbid—more turbidity indicates cell lysis from overly vigorous pipetting in resuspension.

Resuspend the pellet in 20 ml of MBB50 plus protease inhibitors. The following are routinely used; each is at 200× final concentration.

Pepstatin	0.4 mg/ml in methanol
Leupeptin	0.4 mg/ml in water
Aprotinin	0.6 mg/ml in water
PMSF	0.2 M in methanol

Transfer to a 40-ml Teflon–glass Potter-Elvehjem homogenizer (Thomas Scientific, Swedesboro, NJ). Incubate on ice for 15 min. The homogenizer pestle is mounted in a 3/8-in. electric drill (Sears). Homogenize eight full strokes at maximum drill speed. Incubate on ice for 2–4 h with occasional mixing. Clarify the homogenate by centrifugation at 18,000 rpm for 20 min at 4° in a Sorvall SS-34 or equivalent rotor. Discard the pellet.

Minichromosome Purification

Remove the chitin beads with absorbed fusion protein from the column. This is conveniently accomplished by retrograde flow through the column bottom, using a syringe with ~5 ml of MBB50. Mix the clarified homogenate supernatant with the column matrix and hold at 0–4° for 1 h. Repack the column. Wash the column extensively with MBB50, taking care to rinse thoroughly any part of the physical column that contacted the extract—at least 10,000 times more RNA is present than minichromosome DNA, as is apparent in Fig. 2. In our hands, this is a convenient stage to pause the preparation overnight. Prepare 10 ml of elution buffer, MBB300 with 1 mM IPTG. Add 5 ml to the column, run about 1.5 ml into the column, and stop the flow. Hold for 20 min and then complete the elution with the remainder of the 5 ml. Immediately dilute the eluate with an equal volume of water. Concentrate with a 15-ml Ultrafree Biomax 30,000 MWCO (Millipore, Bedford, MA) to a volume of ~400 μl. Typically, this requires 60–90 min in a Sorvall RC-3B centrifuge at 4° and 2500 rpm, using the swinging bucket rotor.

Prepare linear 15–40% sucrose gradients for the SW-41 rotor (Beckman), using 6 ml of each sucrose concentration in 10 mM HEPES (pH 8.0), 50 mM NaCl, 0.2 mM EDTA. Chill at 4° for 1–2 h. Apply the sample and centrifuge at 4° and 30,000 rpm for 10–14 h or 41,000 rpm for 5–8 h. Empty the gradient tubes, collecting 0.5-ml fractions, and determine the position of the minichromosome peak by agarose gel electrophoresis after DNA purification with proteinase K, phenol–chloroform extraction, and ethanol precipitation (Fig. 3). Fractions containing the minichromosome are pooled and concentrated as needed for further analysis. Dialysis of small volumes is usually carried out with Slide-A-Lyzer minidialysis units (Pierce Biotechnology, Rockford, IL).

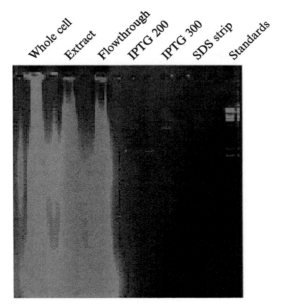

FIG. 2. Affinity purification of SALT10 minichromosome on a *lacI-Z*–intein–chitin-binding domain protein immobilized on chitin beads. Aliquots (0.5% for the left three lanes and 1% for the right three experimental lanes) were deproteinized (but not RNase treated) before electrophoresis. The gel was stained with ethidium bromide and visualized on a Typhoon scanner. Lanes labeled IPTG 200 and IPTG 300 were eluted from the column with 1 mM IPTG in MBB200 and MBB300, sequentially.

Analysis of Minichromosomes

Composition of regulated gene chromatin can be addressed by several methods. If the protein of interest is known (e.g., Tup1p for repressed **a**-cell-specific genes in α-cells), quantitative western blots with specific antibody and a standard series of recombinant protein can define the level of protein present for the regulated gene. For example, we determined that repressed **a**-cell-specific genes contained 2–2.4 Tup1p molecules per nucleosome.[4] This finding contributes significantly to a model for structure of these repressed domains (R. T. Simpson, unpublished data). The minichromosomes, after further purification by sucrose gradient ultracentrifugation, contain a highly limited repertoire of proteins (Fig. 4). In addition to histones (not shown in Fig. 4), the major proteins present in a SALT10 minichromosome have mobilities corresponding to the molecular weights

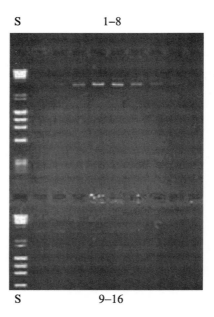

Fig. 3. Sucrose gradient purification of SALT10 minichromosome. The affinity-purified minichromosome was run on a linear 15–40% sucrose gradient centrifuged for 13 h at 30,000 rpm in an SW-41 rotor at 4°. Fractions of 0.5 ml were collected, beginning at the bottom of the tube. After deproteinization, 1% aliquots were run on this agarose gel and stained with ethidium bromide. The top half of the gel contains fractions 1–8 and the bottom half contains fractions 9–16, as indicated. Fractions 17–25 were not analyzed. S is a λ–HindIII plus ϕX174RF–HaeIII digest as standards.

of Ssn6p and Tup1p. The relative amounts of these two (putative) proteins are appropriate for a tetramer of Tup1p associated with a single molecule of Ssn6p,[8] assuming similar staining intensities with SYPRO Orange (Molecular Probes, Eugene, OR). Parallel analyses are underway to determine the content of the Sir protein corepressors at the silent mating type loci.

A third method of compositional analysis uses a proteomic approach. As one example in which the idea of minichromosome rescue of relevant DNA-interacting proteins has been successful, we recount identification of a protein involved in the function of the recombination enhancer (RE). This <1000-bp DNA sequence located ~29 kbp from the left end of yeast chromosome III is responsible for directionality in mating type interconversion in S. cerevisiae.[9] Comparison of the proteins present in

[8] U. S. Varanasi, M. Klis, P. B. Mikesell, and R. J. Trumbly, Mol. Cell. Biol. 16, 6706 (1996).
[9] J. E. Haber, Annu. Rev. Genet. 32, 561 (1998).

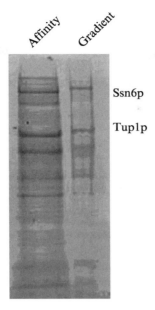

Fig. 4. Proteins present in purified SALT10 minichromosome. Total proteins present in the affinity-purified or affinity plus sucrose gradient-purified minichromosome were analyzed by SDS–PAGE. Histones ran off the gel. A Typhoon image of a gel stained with Sypro Orange is shown. The bands labeled Ssn6p and Tup1p are presumptive identifications. Their molecular masses are consistent with these identifications.

minichromosomes that consisted of the ALT backbone alone or of the ALT backbone containing the RE revealed several additional proteins in the RE minichromosome. One of these, on analysis by trypsin digestion and mass spectrometry, was identified as yKu80p. Further analysis of the possible role of this protein in mating type interconversion revealed that it was associated with RE DNA sequences, as analyzed by chromatin immunoprecipitation experiments, and that its deletion significantly altered directionality of switching in **a**-cells (C. Ruan and R. T. Simpson, unpublished observations).

A direct and appropriate analysis of the structure of a minichromosome containing a unique yeast gene parallels the functional analysis described above and in Fig. 1. Micrococcal nuclease and/or DNase I cutting site mapping should be compared for the multicopy minichromosome and the solo, nuclear genomic copy of the gene. Analysis can be at low resolution, by indirect end label mapping, or at base pair-level resolution using primer extension methodology. If the minichromosome is to serve as a valid model for the genomic situation, chromatin structure of the amplified, isolated gene and the genomic, native gene should be identical.

More direct structural analysis of the regulated gene chromatin in different functional states is possible with isolation of minichromosomes. For example, we have found distinctive chromatin architectures for active and repressed a-cell specific genes and for the silent mating type loci (C. L. Woodcock and R. T. Simpson, unpublished observations), using electron microscopy of minichromosomes. The possibility of three-dimensional reconstructions of either total minichromosomes or the structured domains by negative staining, or possibly by cryoelectron microscopy, so successful with the yeast SWI/SNF chromatin remodeling complex,[10] is enticing.

Functional analysis leading to mechanistic insight concerning gene regulation has been more difficult, even with isolated minichromosomes. One group has shown that minichromosomes (isolated by a different method) retain RNA polymerase II on an active gene and can generate run-on transcripts.[11] The goal of using isolated minichromosome transcription templates with an *in vitro* transcription system remains elusive.

Acknowledgment

Work in the authors' laboratory is supported by NIH Grant GM52311.

[10] C. L. Smith, R. Horowitz-Scherer, J. F. Flanagan, C. L. Woodcock, and C. L. Peterson, *Nat. Struct. Biol.* **10,** 141 (2003).
[11] C. H. Shen, B. P. Leblanc, J. A. Alfieri, and D. J. Clark. *Mol. Cell. Biol.* **21,** 534 (2001).

[11] Purification of Defined Chromosomal Domains

By Joachim Griesenbeck, Hinrich Boeger,
J. Seth Strattan, and Roger D. Kornberg

Introduction

A mechanistic understanding of transcriptional activation ultimately depends on its reconstitution *in vitro*. Genetic and biochemical evidence has shown that the activation of transcription involves the alteration of chromosomal proteins and chromatin structure. The alterations include covalent modifications of histones, exposure of promoter DNA, and changes in protein composition. Mechanistic studies to date have been limited to the use of artificial chromatin, reconstituted *in vitro*. It is unclear how these templates relate to the chromatin of transcriptionally active or repressed genes *in vivo*. The isolation of chromatin assembled *in vivo* may bridge the gap between natural and artificial chromosomal material. Here we describe a method for the isolation of genes from *Saccharomyces cerevisiae* in the form of chromatin in different transcriptional states. The method is an extension of the work of Gartenberg and co-workers, who employed site-specific recombination *in vivo* and differential centrifugation to separate selected chromosomal regions from bulk chromatin.[1] We have used affinity chromatography to purify the selected chromosomal regions to near homogeneity.

Genetic Manipulations

The isolation of a defined chromatin domain requires its excision from the chromosomal locus and can be facilitated by the inclusion of a DNA sequence to serve as a binding site for affinity purification (Fig. 1). The requirement for excision may be met by flanking the genomic region of interest with recognition sites (RS elements; Fig. 1) for the R-recombinase of *Zygosaccharomyces rouxii*. The recombinase excises the region between the two RS sites, releasing it in circular form. For affinity purification, we have incorporated a cluster of LexA-binding sites (L; Fig. 1) adjacent to an RS site, such that it is included in the excised chromatin domain. The cluster serves as a target sequence for a recombinant adapter protein, which contains the entire coding sequence for LexA from *Escherichia coli* at its N terminus, followed by a simian virus 40 (SV-40) nuclear localization signal, a linker region bearing the Garnier–Robson helix 6 from rat plectin,[2] and a C-terminal TAP tag.[3]

Standard molecular cloning techniques[4] and published procedures for genomic manipulation of yeast[5] are employed as follows.

[1] A. Ansari, T. H. Cheng, and M. R. Gartenberg, *Methods* **17**, 104 (1999).
[2] F. A. Steinbock and G. Wiche, *Biol. Chem.* **380**, 151 (1999).
[3] G. Rigaut, A. Shevchenko, B. Rutz, M. Wilm, M. Mann, and B. Seraphin, *Nat. Biotechnol.* **17**, 1030 (1998).
[4] J. Sambrook, E. F. Fritsch, and T. Maniatis, "Molecular Cloning: A Laboratory Manual." Cold Spring Harbor Laboratory Press, Cold Spring Harbor, NY, 1989.
[5] R. Rothstein, *Methods Enzymol.* **194**, 281 (1991).

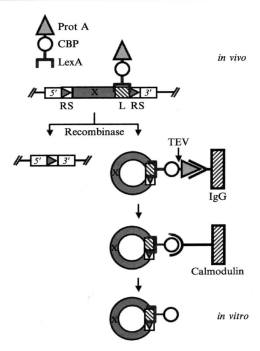

FIG. 1. Flow chart for the purification of a chromosomal locus. A chromosomal locus (X) is genetically modified by flanking it with recognition elements for an inducible recombinase (RS) and by introducing a cluster of LexA binding sites for a coexpressed recombinant adapter molecule (L). The adapter molecule is depicted with a bracket representing the LexA moiety (LexA). The circle and triangle represent two different affinity tags (CBP and Prot A). Induction of R-recombinase releases the chromosomal locus as a circle. Adapter-bound chromatin circles are isolated in a two-step affinity purification, using IgG- and calmodulin-coated chromatographic supports. The arrowhead indicates an internal cleavage site for TEV protease within the adapter molecule. 5' and 3' mark flanking DNA sequences up- and downstream of the genomic locus.

1. A chromosomal region including 500 bp of additional DNA both upstream and downstream of the sequence of interest is amplified by polymerase chain reaction (PCR) from yeast genomic DNA and cloned into a standard cloning vector (pX; Fig. 2A).

2. The sequence of interest is inserted into plasmid pM49.2, fusing it to a cluster of LexA-binding sites and flanking it with RS elements. Reinsertion into plasmid pX yields the modified locus (Fig. 2A) used for gene replacement by homologous recombination.

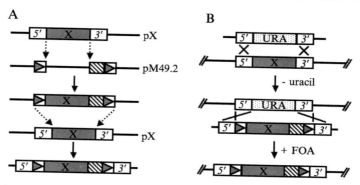

FIG. 2. Construction of a genetically modified locus in yeast. (A) Cloning strategy to modify a locus of interest. pX and pM49.2, cloning vectors. (B) Incorporation of the modified locus in the genome. A two-step replacement strategy using *URA3* as a counterselectable marker is employed.

3. The chromosomal wild-type copy of the gene is exchanged with the modified locus by a two-step gene replacement approach, based on homologous recombination, employing positive and negative selection for the *URA3* marker gene (Fig. 2B).

4. Yeast strains are transformed with two plasmids: plasmid pB3 contains the coding sequence for the R-recombinase under the control of the inducible *GAL1* promoter; plasmid pJSS3.1 expresses the recombinant adapter protein under the control of a constitutive glyceraldehyde phosphate dehydrogenase (GPD) promoter.

Plasmids and Yeast Strains

Plasmid pM49.2 is a derivative of pABX22, a kind gift of M. Gartenberg (the University of Medicine and Dentistry of New Jersey, Piscataway), and has been modified by addition of a LexA-binding cluster juxtaposed to an RS element. The LexA-binding cluster is a concatemer of three copies of the LexA operator from the *ColEI* gene of *E. coli* with the sequence GCTGTATATAAAACCAGTGGTTATATGTACAGTA. The creation of single restriction sites flanking the RS and LexA-RS elements in pM49.2 simplifies cloning strategies for the modification of many genomic loci.

Plasmid pB2 was obtained by flanking the *LEU2* gene in pRS415-RecR[1] with RS elements. Yeast cells transformed with this plasmid and grown in the absence of leucine eliminate clones that express the recombinase prematurely (i.e., before induction with galactose). Plasmid pB3 was derived by digesting plasmid pB2 with *Bae*I and religating, thereby deleting a potential LexA-binding site.

To generate plasmid pJSS3.1, full-length LexA, plectin spacer, and TAP tag were amplified by three separate PCRs, using *E. coli* genomic DNA, rat genomic DNA, and plasmid pBS1479[6] as templates, respectively. Primers were designed to introduce novel restriction sites in-frame for subsequent ligation into the complete construct. The final three-part ligation product was subcloned into the *E. coli*/yeast centromeric shuttle vector p416-GPD[7] (Funk, ATCC 87360).

Yeast strains used must be deficient in *URA3* and *LEU2* expression to allow genomic manipulation and to stably harbor pB3 and pJSS3, containing *LEU2* and *URA3* markers, respectively.

Cell Growth and Cell Disruption

Yeast strains transformed with plasmids pB3 and pJSS3 are grown at 30° in Hartwell's synthetic medium lacking leucine and uracil, containing 2% (w/v) D-glucose (SCD) or 2% (w/v) D-raffinose (SCR) as a carbon source. Starter cultures are routinely grown in SCD, because glucose repression of the *GAL1* promoter prevents premature excision of chromatin circles from the endogenous locus due to leaky recombinase expression. Induction of the recombinase by the addition of galactose releases about 75% of the chromatin from the endogenous locus.

Because the intactness of genomic DNA is a prerequisite for its removal in the subsequent differential centrifugation steps (see below), cells are lysed in liquid nitrogen, which limits fragmentation of chromosomes. Efficiency of lysis is between 60 and 70% in the following procedure.

1. Cells from a 200-ml culture grown overnight in SCD are collected by centrifugation (5 min, 2200*g* at room temperature), washed twice with water, and used to inoculate 12 liters of SCR in a fermentor. Typical starting densities for the 12-liter cultures are between 3.5×10^6 and 5×10^6 cells/ml. Cells are grown overnight, stirred at 500 rpm with mild aeration.

2. Recombination is induced at cell densities between 0.5×10^8 and 1×10^8 cells/ml, by adding 600 ml of 40% (w/v) D-galactose. Cells are grown for an additional 1 h and 30 min.

[6] O. Puig, F. Caspary, G. Rigaut, B. Rutz, E. Bouveret, E. Bragado-Nilsson, M. Wilm, and B. Seraphin, *Methods* **24**, 218 (2001).
[7] D. Mumberg and M. Funk, *Gene* **156**, 119 (1995).

3. Cells are harvested by centrifugation (10 min, 9000g at 4°), yielding a wet weight between 30 and 40 g. After washing twice with water, cells are pelleted in sealed 20-ml syringes by centrifugation (5 min, 3000g at 4°). Syringes are unsealed by cutting with a razor blade, supernatants are decanted, and cells are extruded into liquid nitrogen.[8] Cells can be stored at −80°.

4. Frozen cells are ground in a liquid nitrogen–dry ice mixture in a 1-liter stainless steel container, using a commercial blender (Waring, 7011HS). The blender is run for 10 min, alternating between low and high blending speed at 30-s intervals. Evaporated nitrogen is replaced while grinding at low blending speed. Occasional tapping with a spatula against the outside of the container prevents ground cells from sticking in a layer to the inside wall of the blender. The fine powder of ground yeast can be stored at −80°.

Extraction and Differential Centrifugation

A differential centrifugation protocol[1] allows separation of the circular chromatin domains released by site specific recombination from bulk genomic DNA. Cell debris and bulk genomic DNA are removed in an initial low-speed spin, leaving chromatin circles in the supernatant, which can be sedimented in a second spin at higher centrifugal force. Centrifugal forces and centrifugation times must be adjusted according to the chromatin domain under study, and should be determined in pilot experiments. About half of the starting circles can be recovered in the second pellet, in the case of circles containing 2.2 kb of DNA. For smaller circles, with a size of 0.75 kb, a recovery of 65% is obtained.

The following steps are performed at 4°.

1. Ground cells are extracted by the addition of buffer A (5 ml/g of cells), containing protease inhibitors, and stirring for 15 min.

2. The crude extract is clarified by centrifugation at 72,700g for 1 h.

3. Chromatin circles present in the supernatant are sedimented by centrifugation at 371,000g for 2 h (for circles bearing 2.2 kb of DNA), or 3 h (for circles bearing 0.75 kb of DNA) and resuspended in 1/10 of the original volume of buffer B, containing protease inhibitors. The protein concentration of the resuspended circles is between 5 and 10 mg/ml. The resuspended chromatin circles can be stored at −80°.

---

[8] M. C. Schultz, S. Y. Choe, and R. H. Reeder, *Proc. Natl. Acad. Sci. USA* **88**, 1004 (1991).

Reagents

Buffer A: 0.2 *M* potassium acetate, 2 m*M* EDTA, 10% (v/v) glycerol, 125 µ*M* spermidine, 50 µ*M* spermine, 5 m*M* 2-mercaptoethanol, and 25 m*M* HEPES-KOH (pH 7.4)

Buffer B: 0.1 *M* potassium acetate, 1 m*M* EDTA, 10% (v/v) glycerol, 125 µ*M* spermidine, 50 µ*M* spermine, 5 m*M* 2-mercaptoethanol, and 25 m*M* HEPES-KOH (pH 7.4)

Protease inhibitors (100× concentrate): benzamidine (33 mg/ml), PMSF (17 mg/ml), pepstatin A (137 µg/ml), leupeptin (28.4 µg/ml) in ethanol

Affinity Purification

Differential centrifugation enriches not only for chromatin circles, but also for ribosomes, minichromosomes (plasmids pB3, pJSS3.1, and 2 µ circle), and high molecular weight protein complexes. The resuspended pellet is a crude mixture of proteins and nucleic acids. The presence of nucleases and ATP-dependent chromatin-remodeling activities at this stage of the purification makes the material unsuitable for use in biochemical assays.

The recombinant LexA adapter molecule expressed in the yeast cells permits tandem affinity purification (TAP)[6] for the isolation of adapter-bound chromatin domains. Most contaminating ribosomal material is removed in a first affinity step, involving interaction between the protein A component (Prot A; Fig. 1) of the adapter and IgG–agarose beads. Because the bound material is released from the beads with recombinant His-tagged tobacco etch virus (TEV) protease, the protease itself contaminates the affinity eluate. If the stringency of washes in the subsequent calmodulin-binding peptide (CBP; Fig. 1)/calmodulin–agarose affinity purification does not suffice to remove all traces of the protease, the His-tagged enzyme can be eliminated by an additional incubation with Talon beads.

We have found that stringent wash conditions, such as 0.4 *M* potassium acetate or 0.1% Nonidet P-40 (NP-40), do not adversely affect the structural or topological integrity of the isolated chromatin domains (see later). The efficiency of the tandem affinity purification is dependent on the size of the chromatin circles: 22% of the input can be recovered for 2.2-kb circles, whereas 45% recovery is obtained for 0.75-kb circles.

Affinity chromatography is performed at 4°, under gravity flow, with insulin added at 0.1 mg/ml to all buffers except buffer C, as follows.

1. IgG–agarose beads (1/20 of the applied volume) are equilibrated with 20 bead volumes of buffer B in 20-ml Econo-Pac plastic columns (Bio-Rad, Hercules, CA), and incubated overnight in 2 bead volumes of the same buffer.

2. The resuspended pellet from differential centrifugation is diluted with an equal volume of buffer B containing protease inhibitors, resulting in protein concentrations between 2.5 and 5 mg/ml. The mixture is centrifuged at 21,200g for 5 min and the clear supernatant is applied to the IgG–agarose beads. The top and bottom of the columns are sealed and the contents are mixed by slow rotation for 2 h.

3. The flow-through is collected and the beads are washed with 40 bead volumes of buffer B without glycerol, containing 0.4 M potassium acetate, and subsequently with 40 bead volumes of buffer B without glycerol, containing 50 mM potassium acetate and 0.1% NP-40.

4. Chromatin circles are released by proteolytic cleavage for 1 h with recombinant TEV protease (80 ng/ml) in 1.2 bead volumes of buffer C, with occasional mixing. To wash residual chromatin circles off the column, 4 bead volumes of buffer B is added sequentially. For freezing, the glycerol concentration is adjusted to 10% and the eluates are stored at $-80°$.

5. Calmodulin beads (1/36 of the applied volume) are equilibrated with 20 bead volumes of buffer D in 20-ml Econo-Pac plastic columns and incubated overnight in 2 bead volumes of the same buffer.

6. The magnesium and calcium ion concentrations in the IgG–agarose eluates are adjusted by addition of 1 M magnesium acetate (2 μl/ml) and 1 M calcium chloride (3 μl/ml). The mixture is centrifuged at 21,200g for 5 min and the clear supernatant is applied to the calmodulin beads. The top and bottom of the columns are sealed and the contents are mixed by slow rotation for 1 h.

7. The flow-through is collected and the beads are washed with 80 bead volumes of buffer D containing 0.3 M potassium acetate and 0.1% NP-40.

8. Chromatin circles are eluted with buffer E, in six fractions of 2 bead volumes each. Chromatin circles are recovered in the first three fractions. For freezing, the glycerol concentration is adjusted to 10% and the eluates are stored at $-80°$.

Reagents

IgG–agarose beads (Sigma, St. Louis, MO)
Calmodulin affinity resin (Stratagene, La Jolla, CA)
Talon metal affinity resin (BD Biosciences Clontech, Palo Alto, CA)
TEV protease (GIBCO-BRL, Gaithersburg, MD)
Buffer C: 0.2 M potassium acetate, 1 mM EDTA, 5 mM 2-mercaptoethanol, and 25 mM HEPES-KOH (pH 7.4)

Buffer D: 0.1 M potassium acetate, 1 mM imidazole, 2 mM calcium chloride, 1 mM magnesium acetate, 125 μM spermidine, 50 μM spermine, 10 mM 2-mercaptoethanol, and 25 mM HEPES-KOH (pH 7.4)

Buffer E: 0.2 M potassium acetate, 2 mM EGTA, 1 mM EDTA, 10 mM 2-mercaptoethanol, and 25 mM HEPES-KOH (pH 7.4)

Comments

The combination of differential centrifugation and tandem affinity purification permits the separation of a specific chromosomal locus from virtually all nucleic acids and most of the proteins present in the cell. After purification of *PHO5* chromatin as described above, *PHO5* DNA is present in 20- to 30-fold excess over all other chromosomal DNA. The only nucleic acid detectable in the final eluate by agarose gel electrophoresis and ethidium bromide staining is *PHO5* circle DNA. The purified material is free from chromatin-remodeling activities, nucleases, proteases, and topoisomerases. Using this approach, we have found that different transcriptional states of *PHO5* chromatin are stable *in vitro,* conserving important structural and topological features of the *PHO5* locus *in vivo.*[9]

Even with the procedures described above, the goal of chromatin purification has not been fully achieved. Purification of a single-copy gene entails its separation from a 10,000-fold excess of genomic DNA and a 1 million-fold excess of contaminating proteins. Our procedures yield 0.1 to 0.3 pmol of *PHO5* DNA from 10^{12} cells (corresponding to 30–40 g wet weight). This quantity is at the lower end of sensitivity of mass spectrometry. The total amount of contaminating proteins carried through the purification is still 10-fold above the amount of histone associated with the isolated DNA circle. Additional steps for enrichment of isolated circles are under investigation.

[9] H. Boeger, J. Griesenbeck, J. S. Strattan, and R. D. Kornberg, *Mol. Cell* **11,** 1587 (2003).

[12] Probing Core Histone Tail–DNA Interactions in a Model Dinucleosome System

By CHUNYANG ZHENG and JEFFREY J. HAYES

Introduction

The core histone tail domains influence DNA accessibility at multiple levels of chromatin structure. The wrapping of DNA within the nucleosome is stabilized by the tail domains[1] and the binding of *trans*-acting factors to nucleosomal DNA is stimulated by acetylation or removal of these domains.[2–5] Oligonucleosome arrays are folded and compacted into "30-nm fibers" that are further folded and coiled into higher order structures even in the interphase nucleus.[6,7] The folding of nucleosome arrays requires the core histone tail domains and is modulated by acetylation of these domains.[8–11] Given the primary role of histone acetylation in gene activation and other nuclear processes utilizing chromatin, it is clear that the core histone tail domains are key regulators of chromatin structure and function.

Although the detailed arrangement of nucleosomes in chromatin fibers remains unknown, elements both within the core histone fold domains and the tail domains have been found to influence the formation of the higher order chromatin structures. For example, oligonucleosome arrays containing H4 SIN mutations (H4R45C or H4R45H) are impaired in their ability to undergo salt-induced condensation, whereas long-range interarray interactions required for the formation of higher chromatin structures appear unaffected.[12] On the other hand, model nucleosome arrays containing the

[1] J. Ausio, F. Dong, and K. E. van Holde, *J. Mol. Biol.* **206**, 451 (1989).
[2] D. Y. Lee, J. J. Hayes, D. Pruss, and A. P. Wolffe, *Cell* **72**, 73 (1993).
[3] K. J. Polach, P. T. Lowary, and J. Widom, *J. Mol. Biol.* **298**, 211 (2000).
[4] M. Vettese-Dadey, P. Walter, H. Chen, L. J. Juan, and J. L. Workman, *Mol. Cell. Biol.* **14**, 970 (1994).
[5] M. Vettese-Dadey, P. A. Grant, T. R. Hebbes, C. Crane-Robinson, C. D. Allis, and J. L. Workman, *EMBO J.* **15**, 2508 (1996).
[6] K. E. van Holde, "Chromatin." Springer-Verlag, New York, 1989.
[7] A. S. Belmont and K. Bruce, *J. Cell Biol.* **127**, 287 (1994).
[8] M. Garcia-Ramirez, F. Dong, and J. Ausio, *J. Biol. Chem.* **267**, 19587 (1992).
[9] M. Garcia-Ramirez, C. Rocchini, and J. Ausio, *J. Biol. Chem.* **270**, 17923 (1995).
[10] J. Allan, N. Harborne, D. C. Rau, and H. Gould, *J. Cell Biol.* **93**, 285 (1982).
[11] C. Tse, T. Sera, A. P. Wolffe, and J. C. Hansen, *Mol. Cell. Biol.* **18**, 4629 (1998).
[12] P. J. Horn, K. A. Crowley, L. M. Carruthers, J. C. Hansen, and C. L. Peterson, *Nat. Struct. Biol.* **9**, 167 (2002).

H2A variant H2A.Z exhibit enhanced folding behavior, but the intermolecular self-association was significantly impaired compared with arrays containing the major histones.[13] Moreover, it has been known for some time that arrays lacking the core histone tail domains are unable to form maximally folded conformations or exhibit self-association.[8,10,14–16] Interestingly, results suggest that the H4 N-terminal tail plays a pivotal role in compaction of the array.[17] While electrostatic interactions between the histone tails and DNA clearly contribute to tail-directed folding of the chromatin fiber, the special role of the H4 tail may be due to H4 interactions with an H2A–H2B dimer within adjacent nucleosomes.[17,18] Nonetheless, interactions of the core histone tail domains remain poorly characterized.

Previously, our laboratory developed a site-specific cross-linking method to detect histone tail–DNA interactions within mononucleosomes.[19] Now we present a technique to study internucleosome protein–DNA interactions in a model dinucleosome system.[20] Model dinucleosomes are generated in which one nucleosome typically contains a radiolabeled DNA template and native histones, whereas the other contains a single histone site specifically modified with a UV-activatable cross-linking agent. Each of the nucleosomes is reconstituted independently and then ligated together, and cross-linking is induced by UV irradiation under various conditions. After cross-linking, the two DNA templates are separated so that internucleosomal histone–DNA cross-links can be distinguished from intranucleosomal cross-links (Fig. 1). In addition, we describe a novel method for the preparation of H3-H4 tetramers.

Construction of Histone Cysteine-Substitution Mutant Coding Sequences

To attach the UV-activated cross-linker APB (4-azidophenacyl bromide) to histones, cysteine substitution mutations are introduced into N-terminal region of histones by polymerase chain reaction (PCR). All

[13] J. Y. Fan, F. Gordon, K. Luger, J. C. Hansen, and D. J. Tremethick, *Nat. Struct. Biol.* **9,** 172 (2002).

[14] S. C. Moore and J. Ausio, *Biochem. Biophys. Res. Commun.* **230,** 136 (1997).

[15] C. Tse and J. C. Hansen, *Biochemistry* **36,** 11381 (1997).

[16] P. M. Schwarz, A. Felthauser, T. M. Fletcher, and J. C. Hansen, *Biochemistry* **35,** 4009 (1996).

[17] B. Dorigo, T. Schalch, K. Bystricky, and T. J. Richmond, *J. Mol. Biol.* **327,** 85 (2003).

[18] K. Luger, A. W. Mader, R. K. Richmond, D. F. Sargent, and T. J. Richmond, *Nature* **389,** 251 (1997).

[19] K. M. Lee, D. R. Chafin, and J. J. Hayes, *Methods Enzymol.* **304,** 231 (1999).

[20] C. Zheng and J. J. Hayes, *J. Biol. Chem.* **278,** 24217 (2003).

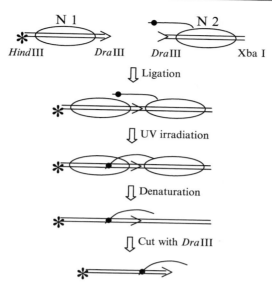

Fig. 1. Experimental strategy. DNA templates and predicted major nucleosome positions (ovals) are shown; the asymmetric *Dra*III site is indicated by the arrowheads. A typical configuration of radiolabel (asterisk) and cross-linker in a tail domain (line and filled circle) is shown. After *Dra*III digestion, only internucleosomal interactions will be detected as protein covalently cross-linked to radiolabeled DNA.

constructions were based on sequences of *Xenopus* histone as described by Perry and co-workers[21] and cloned into pET3 expression plasmids (Novagen, Madison, WI). To minimize the effect of the substitution on tail interactions, we have chosen to replace glycine or threonine residues with cysteine. In the case of H2B, we have added an alanine and cysteine to the N terminus of H2B (to generate the sequence NH_2-GCPEP...) because other single-substitution constructs express only poorly. In addition, *Xenopus* H3 contains a single cysteine at position 110, which may be a potential location for APB modification, so we have replaced this cysteine with alanine to create H3C110A. PCR-generated coding sequence for H2A-G2C, H2B-2C, H3-T6CC110A, or H4-G6C is ligated into pET3a or pET3d plasmids and transformed into *Escherichia coli* DH5α strains by standard methods. The positive transformations are screened by restriction enzyme digestion of plasmids and confirmed by sequencing. Each construct is transformed into the *E. coli* BL21(DE3) strain and glycerol stocks are made and stored at −70° until used for protein expression.

[21] M. Perry, G. H. Thomsen, and R. G. Roeder, *J. Mol. Biol.* **185,** 479 (1985).

Preparation of Recombinant Histones: A New Method for Fast Purification of Recombinant H3/H4 Tetramers

The recombinant dimers with or without cysteine substitutions are prepared as described previously.[19] Briefly, the bacteria cultures containing the H2A or H2B expression plasmids (typically 1 liter) are induced with 0.4 mM isopropyl-β-D-thiogalactopyranoside (IPTG) in midlog phase for 2.5 h and then the cells pelleted. The pellet is suspended in 10 ml of buffer containing lysozyme (0.2 mg/ml) and 0.2% Triton X-100 and incubated for 30 min at room temperature. The mixture is then diluted 2-fold with Tris–EDTA (TE) containing 2 M NaCl and sonicated for 2 min (Branson sonifier 250, output 5, duty cycle 50%; Branson Ultrasonics, Danbury, CT) while keeping the Oak Ridge centrifuge tube on ice. The sonification is repeated two more times (6 min total). Insoluble material is pelleted by centrifugation at 20,000g for 30 min and the amount of each protein in the supernatants is determined by SDS–PAGE. Supernatants containing equimolar amounts of the expressed H2A and H2B are combined, and the resulting dimers are purified by affinity chromatography with BioRex resin as described[19] (see below).

Unlike the preparation of H2A–H2B dimers, preparation of (H3–H4)$_2$ tetramers is more problematic because of the propensity of these proteins to aggregate during the refolding process. In addition, current widely used methods for preparation of recombinant tetramer are complicated and, in some cases, require HPLC.[18] We have found these methods to be time-consuming and costly, especially when several different histone tetramers are to be prepared. Thus we have developed a simple, fast, and low-cost method to isolate (H3–H4)$_2$ tetramer.

The evening before the day the H3 and H4 proteins are to be expressed, inoculate two 1-liter volumes of LB, each containing ampicillin (100 μg/ml), with the bacteria bearing the expression plasmids for H3 and H4 (or corresponding cysteine mutants of H3 and H4) from frozen stocks or fresh plates. The two flasks containing LB are kept in a 37° incubator overnight without shaking. In our experience, the growth of cells will slow when the scattering at 600 nm reaches an optical density (OD) of 0.4. Start the shaker the next morning and within 2 h the density of cells will reach an $A_{600 \text{ nm}}$ of 0.6. (Alternatively, in our hands it takes more than 9 h for cells in 1 liter of LB to reach an OD of 0.6 at 600 nm by constant shaking. The overnight incubation can also be applied to H2A and H2B expression. This procedure does not seem to affect the histone expression level nor the quality of expressed products.) Once the midlog density has been reached, a 100-μl sample of cells is spun down and frozen to be used as an expression control (Fig. 2A, lanes 2 and 3). IPTG is added to 0.4 mM (0.4 ml of a 1 M

stock) and incubation is continued for 2.5 h. After induction, another 100 μl of cells is spun down and, along with the before-induction control, resuspended in SDS loading buffer and applied to an SDS–polyacrylamide gel to check the expression level of histones (Fig. 2A, lanes 4 and 5).

The induced bacterial cells are harvested by centrifugation at 5000g for 15 min. Each pellet is then resuspended in 100 ml of P1 buffer [50 mM Tris, 10 mM EDTA, RNase (100 mg/liter), pH 8.0]. Suspensions containing approximately stoichiometric quantities of expressed H3 and H4 are then combined and distributed to eight 50-ml Falcon tubes for cell lysis. Under our experimental conditions, the expression levels of both proteins are typically similar, so equal amounts of suspensions are combined. Specifically, in each tube mix 12.5 ml of H3 bacterial suspension, 12.5 ml of H4 bacterial suspension, 100 μl of 0.1 M phenylmethylsulfonyl fluoride (PMSF), 100 μl of 1 M dithiothreitol (DTT), and 200 μl of 2-mercaptoethanol. Then 25 ml of buffer P2′ (0.2 M NaOH and 0.4% Triton-100) is added and mixed thoroughly by gentle swirling. The mixture is incubated at room temperature for 1 h, inverting several times during the incubation.

The lysate is put into several dialysis bags (typically, eight bags, 2.4 cm in width by ~35 cm in length) with a molecular mass cutoff of 8000 (Spectrum Laboratories, Rancho Dominguez, CA). The total volume of lysate is 400 ml, and it is dialyzed against 4 liters of 10 mM Tris (pH 8.0), 1 mM EDTA (TE) containing 2 M NaCl, at 4° for at least 4 h. The mixture is then dialyzed against another 4 liters of TE–2 M NaCl overnight. The next morning, the mix is again dialyzed against a third batch of 4 liters of TE–2 M NaCl for at least 4 h. During this process, the total volume of lysate is reduced to approximately 250–300 ml. The lysate is then centrifuged for 30 min at 15,000g, 4° (11,500 rpm for the SS-34 rotor of a Sorvall RC-5B centrifuge; Thermo Savant, Holbrook, NY), and the pellet is discarded. The soluble protein remaining in the supernatant is shown in lane 6 of Fig. 2A. Because of the estimated amount of final product (2–3 mg) and the amount of the (H3–H4)$_2$ tetramer in the supernatant loaded in the gel (13.5 μl from 270 ml), we estimate that the bands correspond to 0.1–0.15 μg of protein, just at the limit of detection by Coomassie blue staining. At this stage, the DNA concentration in the sample is about 12 μg/ml.

For every 10 ml of supernatant (in 2 M NaCl–TE), mix in 30 ml of TE, 100 μl of 1 M DTT, and 1 ml of BioRex 70 resin (50–100 mesh, 50% slurry in TE buffer) in 50-ml Falcon tubes. The tubes should be continually mixed (preferably by gentle rotation) for 4 h at 4°. Pour the mixture into 25-ml Bio-Rad (Hercules, CA) chromatography columns with the total bed volume of beads no greater than 5 ml in each column. Typically, we require

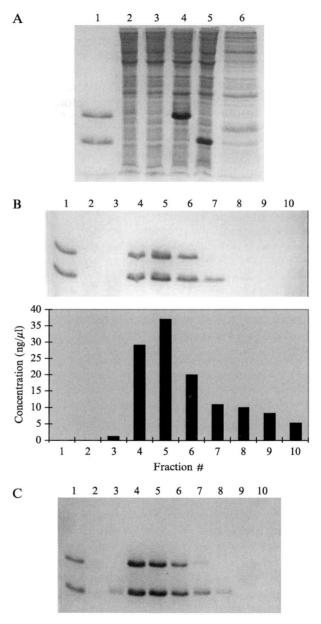

FIG. 2. *(continued)*

three or four columns, with a total bed volume of about 15 ml. Wash the beads thoroughly with TE containing 0.6 M NaCl. The volume of washing buffer should be 100-fold the bed volume. Almost all the remaining bacterial proteins are eluted in the wash step, as well as most of the bacterial nucleic acid. The tetramer is eluted by the careful application of TE containing 2.0 M NaCl and 1.0-ml fractions are collected in microcentrifuge tubes, and immediately frozen on dry ice.

Remove 10 μl from each fraction and analyze on a standard SDS–polyacrylamide gel to determine the tetramer elution profile (Fig. 2B). In this step, the tetramer already appears pure; there is no contaminated bacterial protein evident after Comassie blue staining of the gel, and no sign of degradation. However, we find DNA present in the fractions containing the tetramer. For example, the samples shown in lanes 2–10 of Fig. 2B contain DNA at 0, 1, 29, 37, 20, 11, 10, 8, and 5 ng/μl, respectively. The elution peak of DNA correlates well with the tetramer, suggesting that the DNA is bound to the protein. In this gel, the total estimated tetramer present within fractions 3, 4, and 5 combined is about 1 mg, with about 86 μg of contaminating DNA. In the presence of this amount of DNA, if there is no salt present, the tetramers will aggregate with the DNA and form precipitates. Depending on whether the presence of the DNA will affect the subsequent usage of the recombinant tetramer, these fractions may be used directly.

If needed, the contaminating DNA may be removed by additional BioRex and hydroxyapatite chromatography. Combine the peak fractions containing the eluted tetramer (the total volume should be about 12 ml, including fractions 3–5 in Fig. 2B and similar fractions from other columns) in an Oak Ridge (or similar capped) centrifuge tube and briefly sonicate the sample (Branson sonifier 250, output control 5, duty cycle 50%) for 2 min on ice. The samples are then mixed with 12 ml of a 50% slurry of

---

FIG. 2. SDS–polyacrylamide gels depicting the purification of recombinant tetramer. (A) *Xenopus* H3 and H4 are expressed in *E. coli* BL21(DE3) strains. Each lane except lane 1 contains cells from 100 μl of bacterial culture loaded directly onto the gel. Lane 1, 1.8 μg of native H3–H4 purified from chicken erythrocyte nuclei; lanes 2 and 3, uninduced bacteria bearing H3 and H4 expression plasmids, respectively; lanes 4 and 5, induced cells containing H3 or H4 expression plasmids, respectively; lane 6 contains supernatant from centrifuged cell lysate from 100 μl of cell culture. (B) *Top:* Elution profile of (H3–H4)$_2$ tetramer from the BioRex 70 column. Shown are fractions eluted with 2 M NaCl–TE from a 4-ml column. Lane 1, 1.8 μg of purified chicken erythrocyte core histones; lanes 2–10, 10-μl fractions of 1-ml fractions 1–9, respectively. *Bottom:* Concentration of DNA (ng/μl) found in the fractions shown in the gel above. (C) Elution profile of (H3–H4)$_2$ tetramer from the second BioRex 70 column after sonication. Shown are fractions eluted with TE–2 M NaCl from the 3-ml column. Lane 1, 1.8 μg of purified chicken core histones; lanes 2–8, successive 1-ml fractions.

BioRex resin (40% of the total bead volume used originally), and diluted with TE until the total volume is four times that of the eluted tetramer, so the NaCl concentration is 0.5 M. The DTT should be adjusted to 1 mM and the mixture rotated for 4 h at 4°. The material is transferred to a column and washed, and tetramer is eluted as before. Figure 2C shows the protein present within 1-ml fractions eluted from a 3-ml bead volume column (half the total production). Depending on how well the sonication shears the DNA and on the efficiency of the washing step, the DNA concentration in the final fractions is usually below 2 μg/ml, sometimes below the limit of detection by fluorography on a Hoefer (San Francisco, CA) DyNA Quant 200 fluorometer. This is the case for the samples shown in Fig. 2C, so fractions 4–6 are combined and stored at −20° for later use. If residual DNA is still detectable, the fractions of tetramer can be mixed with 50 μl of properly prepared hydroxyapatite resin per milliliters of tetramer (Bio-Rad laboratories, 50% slurry) for 30 min at 4°, and the resin removed by centrifugation. This process is repeated three times to eliminate any traces of contaminating nucleic acids. Alternatively, HCl precipitation can be used to remove the contaminating DNA.

The purified (H3–H4)$_2$ tetramers can be stored at −70° for several years and used for nucleosome reconstitutions. Nucleosomes formed by salt gradient dialysis with our recombinant histones and native histones purified from chicken erythrocytes are indistinguishable. By this protocol, we have successfully prepared recombinant (H3–H4)$_2$, (H3C110AT6C–H4)$_2$, (H3C110A–H4G6C)$_2$, and other tetramers with different cysteine substitutions.

The rationale underlying our purification procedure lies in the fact that recombinant H3 and H4 are typically present in insoluble form in traditionally prepared bacterial lysates, even after extensive sonication. Instead, we use a mild alkali lysis followed by dialysis in high-salt solution to obtain a large amount of soluble protein within the lysate. The basis for solubilization of the proteins is unclear but may be due to the denaturing effect of basic pH on the proteins, nucleic acids, or both. The plasmid and genomic DNAs remain intact under these conditions, so we surmise that, on gradual neutralization of the alkali some portion of the solubilized H3 or H4 folds together correctly as tetramer in 2 M NaCl, perhaps facilitated by loose binding to the large excess of DNA, avoiding self-aggregation. This idea is supported by the relative masses of protein and DNA in the soluble lysate, which contains approximately 2–3 mg of H3 and H4 and 3–3.6 mg of DNA. A simple calculation shows that under these conditions the DNA would be saturated by tetramers. If this is the case, we predict that adding more DNA during lysis will increase the final tetramer production. Furthermore, increasing the volume of lysis may reduce aggregation and improve the yield. We estimate the total amount of H3 and H4 in the combined 2 liters of LB is 200–400 mg and the final yield from our protocol is approximately 2–3 mg.

APB Modification of Histone Mutants with Cysteine Substitutions

H2A–H2B dimers and (H3–H4)$_2$ tetramers are prepared with the reducing agents DTT and 2-mercaptoethanol present until washing and elution from the BioRex column, and then the samples are quick-frozen and stored at −70°. Thus cysteines within the mutant proteins are already reduced when the stocks are quick frozen and APB modification is carried out directly on samples from the thawed stocks essentially as previously described.[19] An APB stock (0.03 M) in methanol is stored at −70° in the dark. On the basis of the protein concentration, a 5:1 molecular ratio of APB to cysteine is used in the conjugation reaction. The reaction proceeds at room temperature in the dark for 1 h and then is quickly frozen at −70° and stored for reconstitution. The APB modification efficiency is checked by the subsequent modification with fluorescein-5-maleimide (Molecular Probes, Eugene, OR). This dye reacts with the free thiol group in the protein, emitting green fluorescence after excitation. If the thiol group is reacted with APB, then it will not be modified with the fluorescein reagent. By comparing the incorporation of the fluorescent tag within the cysteine-containing histone before and after APB modification, we find that typically more than 95% of available cysteines are modified by APB (Fig. 3A). The reactivity of APB-modified protein can be checked by UV irradiation of 10 μl of the modified protein sample, which is then loaded onto a SDS–polyacrylamide gel. We find that in all cases tested so far irradiation of APB-modified dimers and tetramers results in a new band on the gel, corresponding to a covalent cross-link between proteins within the complex (Fig. 3B). Interestingly, this intracomplex cross-linking is not observed when the dimers or tetramers are reconstituted into nucleosomes (not shown).

Construction of Dinucleosome DNA Template

To allow the protein composition of each nucleosome to be different, each nucleosome in the dinucleosome is prepared separately and then both are ligated together. DNA templates for nucleosomes 1 and 2 were derived from plasmid Xbs-1[22] containing a *Xenopus borealis* somatic 5S rRNA

---

[22] R. C. Peterson, J. L. Doering, and D. D. Brown, *Cell* **20,** 131 (1980).

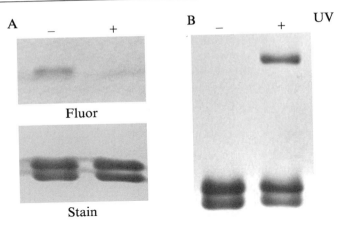

FIG. 3. APB modification of histone cysteine mutant H2A–G2C. (A) Unmodified and APB-modified H2A–G2C were incubated with fluorescein-5-maleimide and products were separated by SDS–PAGE. The gel was imaged by UV fluorography (*top*) and Coomassie staining (*bottom*). Left lane, before APB modification; right lane, after modification. (B) Comassie-stained SDS–polyacrylamide gel shows intradimer cross-linking by H2A–G2C–APB. Left lane, before UV irradiation; right lane, same amount of protein after UV irradiation.

gene. To ensure proper orientation on ligation, we placed an asymmetric recognition sequence for restriction enzyme *Dra*III at appropriate ends of the DNA templates. To prevent ligation at the other end of the fragments, we dephosphorylated the 5′ end with calf intestinal phophatase according to standard procedures. However, a considerable amount of self-ligation involving the dephosphorylated end still occurred in the high concentrations of DNA and ligase used in the preparation. Thus, we adapted the procedure as follows: after digestion with restriction enzymes, the four-nucleotide 5′ overhang was dephosphorylated as described previously and incubated with Klenow polymerase to fill two deoxy nucleotides, then a dideoxy nucleotide. Note that this procedure leaves a 1-nucleotide overhang at the treated end lacking both a 5′-phosphate and a 3′-hydroxyl for ligation.

To prepare the DNA fragments, PCR is used to add *Hind*III and *Dra*III sites to the ends of the nucleosome 1 DNA template and *Dra*III and *Xba*I sites to the nucleosome 2 template (Fig. 1). The resulting nucleosome 1 is 205 bp, and nucleosome 2 is 209 bp. Ligation of the two templates at the asymmetric *Dra*III site forms a tandem repeat of 5S sequences with a repeat of 210 bp. The two templates are individually cloned into pBlue-script plasmids pBSXN1 and pBSXN2. Unlabeled (cold) DNA fragments from each are then prepared in large quantity. pBSXN1 or pBSXN2 plasmid (10 mg) is cut with 3000 units of *Hind*III or *Xba*I, respectively, overnight at 37°. The linear plasmid is precipitated once and dephosphory-lated at 37° for 4 h in the presence of 1250 units of calf intestinal phosphat-ase (New England BioLabs, Beverly, MA). SDS (0.1% final concentration) is added to stop the reaction, and proteins are removed by two successive phenol extractions and finally by ethanol precipitation. The fill-in reaction is carried out at 37° for 20 min with 600 units of Klenow in the presence of dATP, dGTP, dTTP, and ddCTP (100 μM each) for pBSXN1 or dCTP, dTTP, dGTP, and ddATP (100 μM each) for pBSNX2. After ethanol pre-cipitation, the plasmid is cut with 4800 units of *Dra*III overnight at 37° to release the template DNA fragment. The fragment is isolated on prepara-tive 1.2% agarose gels and recovered by electroelution. After phenol ex-traction and repeated ethanol precipitation, the pellet is dissolved in 10 mM Tris (pH 8.0), checked for concentration, and stored frozen at −20° until needed. A 0.15- to 0.2-mg quantity of final product can be obtained from 10 mg of original plasmid.

Radioactively labeled nucleosome template 1 or 2 is prepared by digesting 10 mg of either parent plasmid with *Hind*III or *Xba*I, respect-ively. The free DNA ends are dephosphorylated as described above and the DNA is stored frozen. The DNA fragment is labeled on the day before nucleosome reconstitution. For each labeling, 100 μg of either digested plasmid is incubated with 25 units of Klenow in the presence of 500 μCi of [α-^{32}P]dATP and in the presence of dGTP, dTTP, and ddCTP (100 μM each) for pBSXN1 (or [α-^{32}P]dCTP, dTTP, dGTP, and ddATP for pBSXN2) for 15 min at room temperature, and then 100 μM cold dATP (dCTP for pBSXN2) is added for an additional 10 min. The labeled template is isolated on a 6% acrylamide gel by standard techniques and stored in TE at −20°. The problem of self-ligation of nucleosome templates is completely circumvented by this procedure, and the fragments are ligated at the *Dra*III site with the proper orientation (Fig. 4A).

Preparation of Model Dinucleosomes

Nucleosomes 1 and 2 are prepared separately and then ligated together, typically one reconstituted on a cold template and the second reconstituted on a radioactively end-labeled template prepared as described above. Nucleosome reconstitution is carried out by a standard salt dialysis pro-cedure,[23] in the dark if an APB-modified protein is included in the

[23] J. J. Hayes and K.-M. Lee, *Methods* **12**, 2 (1997).

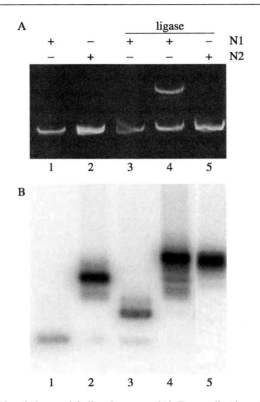

FIG. 4. Assembly of the model dinucleosome. (A) Proper ligation of template DNAs. Ligation reactions (total volume of 10 μl in T4 ligase buffer) were incubated at room temperature for 1 h and then loaded onto a 6% acrylamide gel. Samples shown in lanes 1 and 2 contained no ligase, while those in lanes 3, 4, and 5 contained 400 units of T4 DNA ligase. Samples shown in lanes 1 and 3 contained 0.1 μg of nucleosome 1 template (N1); lanes 2 and 5, 0.2 μg of nucleosome template 2 (N2); lane 4, 0.1 μg of N1 and 0.2 μg of N2. (B) Mononucleosomes were efficiently ligated into dinucleosomes. In this experiment, nucleosome 1 was radioactively labeled and samples were run in a 0.7% agarose nucleoprotein gel. Lane 1, mononucleosome DNA; lane 2, gradient-purified mononucleosome 1; lane 3, ligated dinucleosome DNA; lanes 4 and 5, nucleoprotein products from the ligation reaction before and after gradient purification of the dinucleosome, respectively.

reconstitution. For each cold reconstitution, 6.6 μg of unlabeled DNA template is mixed with 15 μg of H2A–H2B, 15 μg of H3–H4 (or their corresponding APB-modified proteins), and 40 μg of linearized Bluescript plasmid with blunt ends (as competitor). For reconstitution containing radiolabeled templates, the labeled fragment from 100 μg of end-polished plasmid (about 3 μg) is mixed with the same amount of protein and DNA competitor as cold reconstitution. Thus after reconstitution, the cold nucleosome is present in 2- to 3-fold molar excess compared with the radiolabeled nucleosome (see below). Because of slight variations in different preparations of DNA fragments and APB-modified histone mutants, the ideal histone-to-DNA ratio is empirically adjusted for each combination. Reconstituted mononucleosomes are loaded onto 10-ml 5–20% sucrose gradients in the presence of 10 mM Tris (pH 8.0), 1 mM EDTA at 4° (41 Ti rotor, 34,000 rpm, 18 h; Beckman, Fullerton, CA). Fractions containing pure mononucleosomes are identified by analysis in 0.7% agarose nucleoprotein gels. The nucleosome fractions are saved for the ligation reaction.

We have tried various ligating conditions, such as different temperatures and salt concentrations, and found that a critical factor is nucleosome concentration. Under our experimental conditions, each mononucleosome needs to be at least 10 nM for efficient ligation. A Microcon YM-50 filter concentrator (Millipore, Bedford, MA) is used to concentrate nucleosomes isolated from the sucrose gradients 10-fold. Fractions (500 μl) containing nucleosomes are pipetted into the sample reservoir inside the filter vial and centrifuged at 3000g for 15 min at room temperature. The flow-through is discarded, and the sample reservoir is placed upside down in a new vial. The filter and vial are spun for 3 min at 1000g to transfer the concentrate to the vial. The final volume should be about 50 μl. The ligation reactions typically contain 50 μl (5 or 2.5 pmol of unlabeled or labeled nucleosomes, respectively) each of nucleosomes 1 and 2, 20 μg of bovine serum albumin (BSA), and 3200 units of T4 ligase (New England BioLabs) and are kept at room temperature for 30 min in a total volume of 160 μl, followed by the addition of 20 μl of 10 mM ATP and 20 μl of 20 mM Mg^{2+} for an additional 60 min. The ligation product is purified by sucrose gradient centrifugation as described above except for 15 h at 4°. Figure 4B shows the preparation of dinucleosomes from mononucleosmes.

UV-Induced Cross-Linking and Identification of Cross-Linking Bands

Because in some instances we find that the presence of sucrose dampens the UV-induced cross-linking reaction, the buffer containing the gradient purified dinucleosomes is exchanged as follows: The gradient fractions (500 μl) containing purified dinucleosomes are placed in the concentrator filter (see previously), and the buffer solution is filtered out by centrifugation until ~10 μl is left. The concentrator is refilled with 500 μl of 10 mM Tris and centrifuged again. This process is repeated three times, and the dinucleosomes are finally resuspended in 500 μl of 10 mM Tris and stored at 4°. To carry out the cross-linking reaction, aliquots of the dinucleosome

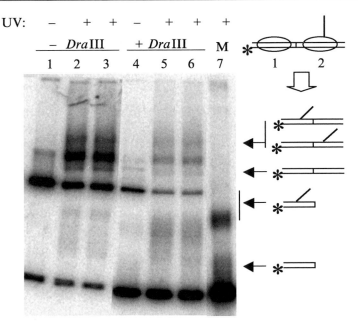

FIG. 5. Internucleosome cross-linking of the H2A tail. Dinucleosomes were prepared with a radioactive label at the *Hind*III end of nucleosome 1 and H2AG2C–APB in nucleosome 2 and then irradiated with UV light. Cross-links were analyzed by SDS–PAGE as described in text. Samples shown in lanes 1 and 4 are unirradiated controls; samples in lanes 2, 3, 5, and 6 were irradiated. Samples in lanes 2 and 5 were irradiated in TE whereas those in lanes 3 and 6 were irradiated in TE containing 50 and 100 mM NaCl, respectively. Samples in lanes 4–6 were digested with *Dra*III after UV irradiation. Lane 7 contains nucleosome 1 assembled with H2A–G2C–APB and irradiated as a mononucleosome cross-linking control. The scheme (*top right* of the gel) indicates the location of the radiolabel on the DNA (asterisk), nucleosomes 1 and 2 (ovals, with numbers below), and the location of the cross-linker-modified histone tail domain (vertical line from oval) within the model dinucleosome. The species represented by the bands on the gel are also indicated, with the location of the cross-linked protein shown as a slanted bar.

can be removed, adjusted to the appropriate salt concentration, and irradiated. For example, we typically remove 32 μl of dinucleosomes (~0.1 pmol) and rapidly mix it with 4 μl of BSA (1 μg/μl) and 4 μl of concentrated NaCl solution. The mix is placed into the bottom of a Falcon 5-ml polystyrene tube, which is itself inside a 15-ml Pyrex no. 9820 glass tube. The sample is irradiated with 365-nm UV light generated by a VMR LM-20E light box for 1 min. The polystyrene and glass tube exclude wavelengths of <290 nm.[24] The sample is then mixed with 5 μl of 1 M

DTT, 2 μl of 10% SDS, and 2 μl of calf thymus DNA (1.5 μg/μl) and incubated at 37° for 10 min, followed by adding 85 μl of doubly distilled H$_2$O and standard ethanol precipitation. The pellet is dissolved in 30 μl of 10 mM Tris, pH 8.0, and then 13 μl of the sample is digested with 60 units of *Dra*III, in a total volume of 30 μl at 37° for 3 h. Another 13-μl portion is diluted to 30 μl with doubly distilled H$_2$O and mock digested at 37° for the same time. The samples are then mixed with SDS and loaded onto a 6% polyacrylamide gel containing 1× standard SDS electrophoresis buffer and run at 100 V for 14 h before gel drying and phosphorimager analysis (a typical result is shown in Fig. 5).

---

[24] P. S. Pendergrast, Y. Chen, Y. W. Ebright, and R. H. Ebright, *Proc. Natl. Acad. Sci. USA* **89**, 10287 (1992).

[13] Site-Directed Histone–DNA Contact Mapping for Analysis of Nucleosome Dynamics

By Stefan R. Kassabov and Blaine Bartholomew

Introduction

Chromatin structure is dynamically modulated in the cell in order to facilitate or impede the access to DNA of proteins involved in essential functions such as transcription, replication, and DNA repair.[1,2] This process, termed chromatin remodeling, is carried out by a large group of enzymes that utilize energy derived from ATP hydrolysis to disrupt histone–DNA interactions in the nucleosome.[3,4] Despite intense investigation, the mechanisms by which these enzymes alter the chromatin structure and the exact nature of the remodeled nucleosomal state have remained elusive. A major obstacle to understanding this process has been the inability of the most commonly employed enzymatic assays to distinguish between changes in histone–DNA interactions caused by translational repositioning (sliding) of the octamer along DNA and those arising from alterations in the core nucleosome structure. Although other methods such as assessing the level

[1] J. T. Kadonaga, *Cell* **92**, 307 (1998).
[2] G. J. Narlikar, H. Y. Fan, and R. E. Kingston, *Cell* **108**, 475 (2002).
[3] K. Havas, I. Whitehouse, and T. Owen-Hughes, *Cell Mol. Life Sci.* **58**, 673 (2001).
[4] P. B. Becker and W. Horz, *Annu. Rev. Biochem.* **71**, 247 (2002).

of supercoiling of circular chromatin templates have produced clear evidence that some remodeling activities can cause alterations in the core nucleosome structure apart from sliding,[5,6] the nature of these alterations remains unclear because of the multiple possible alternative interpretations of the data and the fact that such assays cannot provide information about particular histone–DNA interactions in the nucleosome. Further complicating the task of characterizing the structural changes in remodeled nucleosomes has been the difficulty in assigning their accurate translational positions. Most traditional methods for mapping nucleosome positions rely on digestion with processive nucleases such as micrococcal nuclease and exonuclease III for detection of the outer border of the histone–DNA interface. Because remodeling complexes have been shown to weaken or even eliminate the histone–DNA contacts in this region, allowing such nucleases to penetrate with little or no impediment into the remodeled nucleosome core,[6] these methods are ineffective for mapping translational positions of remodeled nucleosomes.

Building on methods developed by other groups for site-directed mapping of histone–DNA interactions with single base pair resolution,[7,8] we have devised an experimental approach that allows simultaneous determination of the precise translational position(s) as well as direct structural assessment of the remodeled nucleosome at the level of specific histone–DNA contacts in the regions of the highly conserved histone fold domains.[9] This approach involves selective attachment of photoreactive aryl azide groups to unique cysteines engineered at sites near main interaction points of the nucleosomal DNA with the globular domains of histones H2B, H2A, and H4, respectively, 54, 39, and 2 bp away from the dyad axis of the nucleosome. Determination of the precise points of interaction of these residues with DNA before and after remodeling following UV-induced covalent cross-linking provides an accurate assessment of the distance/path of DNA on the histone octamer surface between six structural reference points (two copies of each of the three residues). This photochemical approach offers several additional advantages over methods used previously for probing the remodeled nucleosomal state. First, there is no requirement for special reaction conditions (as with nucleases) or addition of chemicals blocking the remodeling activity (as in hydroxy radical footprinting), so

[5] J. R. Guyon, G. J. Narlikar, S. Sif, and R. E. Kingston, *Mol. Cell. Biol.* **19**, 2088 (1999).
[6] Y. Lorch, M. Zhang, and R. D. Kornberg, *Mol. Cell* **7**, 89 (2001).
[7] A. Flaus, K. Luger, S. Tan, and T. J. Richmond, *Proc. Natl. Acad. Sci. USA* **93**, 1370 (1996).
[8] K. M. Lee and J. J. Hayes, *Proc. Natl. Acad. Sci. USA* **94**, 8959 (1997).
[9] S. R. Kassabov, N. L. Henry, M. Zofall, T. Tsukiyama, and B. Bartholomew, *Mol. Cell. Biol.* **22**, 7524 (2002).

mapping can be carried out simultaneously with remodeling. Second, in contrast to enzymatically based assays, the cross-linking reaction occurs on the time scale of milliseconds, making this method perfectly suited for rapid kinetic analysis of the remodeling mechanism. Third, the direct readout of this technique allows detection and differentiation of distinct structural intermediates and products in the remodeling process, which can reveal additional details of the remodeling mechanism. Finally, this method allows more unambiguous assignment of low occupancy and closely overlapping nucleosomal translational positions than any other previously used assay. The H4 S47C derivative used in the original hydroxy radical cleavage method, developed by Flaus *et al.*, cross-links/cleaves both DNA strands per modified residue, resulting in two cuts 8 bp apart on each strand per dyad.[7] When there are multiple dyad positions separated by similar or smaller distances, this could lead to overlapping cuts from different dyads and consequently to difficulties in assigning individual positions. In contrast, the H2B A53C and H2A S47C derivatives cross-link only one DNA strand per modified residue, producing a single strong cut, respectively, 54 bp 5' and 39 bp 3' away from the dyad, eliminating any ambiguity in assigning individual dyad positions. An additional advantage of the photochemical method is the ability to quantitatively estimate the overall level of contact between the modified cysteine and DNA through determination of the cross-linking efficiency on sodium dodecyl sulfate (SDS)–polyacrylamide gels. In the analysis of remodeled nucleosomes, this enables differentiation between loss of distinct cleavage sites due to translational position randomization, as opposed to loss of contact between the modified cysteine and DNA due to structural alterations or dissociation.

In this chapter, we describe this technical approach and present examples of its application for analysis of the altered translational positioning and structural conformation of nucleosomes remodeled by the SWI/SNF complex.

Identifying Histone Residues for Cysteine Substitutions Suitable for Protein–DNA Cross-Linking

To map the histone–DNA interface in the remodeled nucleosome from multiple specific positions, we had to identify additional histone fold domain residues similar to H4S47 for attachment of photoreactive groups. The choice of residues for cysteine replacement was guided by the published high-resolution crystal structure of the nucleosome. Target residues met the following criteria: they were not conserved, were located in close proximity to nucleosomal DNA, and were in less structurally important loop regions to avoid perturbing the nucleosomal structure. Cysteine replacements were carried out by site-directed mutagenesis of the pET-based *Xenopus laevis* histone expression vectors obtained from T. Richmond (ETH Zurich, Switzerland), using a Quick Change site-directed mutagenesis kit from Stratagene (La Jolla, CA) and following the manufacturer's instructions. The suitability of the selected residues for site-directed mapping was then tested functionally by overexpressing the mutagenized histones, assembling them into functional octamers with a complement of wild-type histones, and assessing their efficiency of DNA cross-linking after reconstitution on an end-labeled DNA and modification with the photoreactive reagent azidophenacyl bromide (APB) as described below. Using this approach, we were able to identify two substitutions—H2A A45C and H2B S53C—that were able to cross-link DNA with even higher efficiency than the H4 S47C derivative. Both H2A A45 and H2B S53, like H4 S47, are located at the beginning of the α2 helix of the histone fold, which places them in close proximity to the minor groove of the DNA on the two opposite sides of the L1L2 DNA-binding site of the H2A–H2B dimer (Fig. 1).

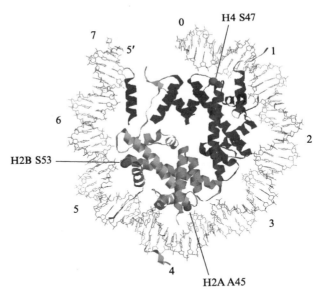

Fig. 1. Positions of cysteine replacements for attachment of photoreactive groups. Locations of the specific modification sites in the nucleosome at histone H4 residue 47, histone H2B residue 53, and histone H2A residue 45. The targeted residues are space filled. One-half of the nucleosome structure is viewed from the top of the particle. Note that the H2B and H2A modification sites are proximal to only one of the DNA strands. Reprinted from S. R. Kassabov, B. Zhang, J. Persinger, and B. Bartholomew, *Mol. Cell* **11**, 391 (2003), with permission from Elsevier.

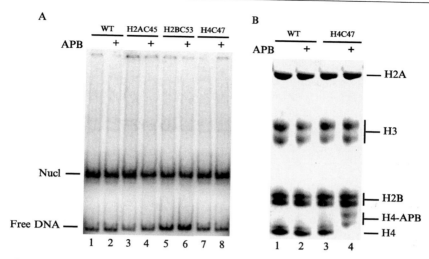

FIG. 2. Selective APB modification does not perturb the nucleosome structure. (A) Gel shift analysis of APB-modified derivatives. Nucleosomes assembled with octamers containing either wild-type or H2A C45, H2B C53, or H4 C47 engineered histone as indicated were modified with APB as described and 3-μl reaction aliquots were loaded on a 4% native polyacrylamide gel for comparison with unmodified samples. (B) Analysis of the selectivity and efficiency of APB modification. Nucleosomes assembled with wild-type or H4 C47 histone were modified with APB as described, the nucleosomal DNA was chemically digested with formic acid and diphenyl amine (DPA), and the histones were acetone precipitated, resuspended in TAU gel loading buffer containing 8 M urea, and loaded on a 15% Triton–acid–urea gel (60:1, acrylamide-to-bisacrylamide ratio), which was run for 20 h at 125 V and stained with Coomassie blue. (Histones H3 and H2B appear as doublets on the gel because they are cleaved during the formic acid–DPA cleavage step.)

Importantly, both H2A A45 and H2B S53 are located closer to one DNA strand than the other, which is the basis of their strand-specific cross-linking pattern (discussed below). High-resolution gel shift analysis and DNase footprinting were used to verify that neither the cysteine replacement nor the modification with APB had affected the nucleosome core structure (Fig. 2A and our unpublished data).

Histone Overexpression and Purification and Octamer Refolding

We have utilized the procedure published by Luger and colleagues, with some minor modifications.[10] Six-liter cell cultures were grown and the histones were expressed as described. The preparative TSK SP-5PW

[10] K. Luger, T. J. Rechsteiner, and T. J. Richmond, *Methods Enzymol.* **304,** 3 (1999).

HPLC column used for the second histone purification step in the original procedure was replaced with a Source 15S FPLC column (1.5 × 10 cm; Pharmacia, Uppsala, Sweden). Octamers were refolded and purified exactly as described except that a slightly higher ratio of H2A–H2B to H3–H4 (1.2:1) was used in the octamer refolding step to minimize the presence of free (H3–H4)$_2$ tetramer, which is difficult to separate from the octamer. The refolded octamer can be stored in refolding buffer [2 M NaCl, 10 mM Tris-HCl (pH 7.5), 1 mM Na-EDTA, 5 mM 2-mercaptoethanol] adjusted to 50% glycerol at a concentration of 5–10 mg/ml at −20° for several months.

Nucleosome Reconstitutions

End-labeled probes for reconstitution can be generated by a number of standard protocols. We are currently using a polymerase chain reaction (PCR)-based method, which offers good flexibility for generating different length probes without the need to rely on the availability of restriction sites. After designing and obtaining the appropriate primer set for the PCR amplification of the desired DNA sequence, one of the oligonucleotides is 5′ labeled with T4 polynucleotide kinase (PNK; New England BioLabs, Beverly, MA). Twenty picomoles of the oligonucleotide is incubated with 20 units of PNK and 100 μCi of [γ-^{32}P] ATP (6000 Ci/mmol) in 1× PNK reaction buffer (supplied by the manufacturer) in a total of 20 μl for 1 h at 37°. The kinase is heat inactivated for 20 min at 65° and the entire sample is used directly in the PCR amplification reaction. Amplification is carried out over 25 cycles, using an Ampli Taq Gold PCR kit from Stratagene with 100–200 ng of plasmid DNA template and a 2-fold excess of the unlabeled primer over the labeled primer. MgCl$_2$ concentrations and annealing/extension temperatures are optimized individually for each primer/target set. The amplified DNA probe is purified with a PCR purification kit from Qiagen (Valencia, CA) that removes the unincorporated primer. The DNA probes can generally be used for reconstitution without further purification but occasionally a reduction in the cross-linking efficiency has been observed with PCR-generated probes, presumably because of the presence of unidentified contaminants. In such cases, the probes are further purified by batch binding to DEAE-Sephadex resin (20 μl of 50% slurry per 100-μl PCR) in Tris–EDTA (TE) plus 10 mM NaCl binding buffer [10 mM Tris-HCl (pH 8.0), 1 mM Na-EDTA, 10 mM NaCl]. The resin is washed three times with 200 μl of the same binding buffer and once with TE plus 300 mM NaCl elution buffer and the probe is eluted with 200 μl of TE plus 1 M NaCl buffer. The resin is washed once with 100 μl of the elution buffer and the eluates are combined and concentrated to 10–20 μl, using a Microcon-30 filter concentration device (Amicon; Millipore, Bedford, MA).

Mononucleosomes are assembled by the rapid salt dilution method.[11] Starting reactions are 10 μl in volume and contain 2.0 M NaCl, 10 μg of salmon sperm DNA (sheared to 100- to 700-base pair length by sonication), 100–200 fmol of end-labeled DNA probe, and 10–15 μg of recombinant octamer (optimal octamer-to-DNA ratio depends on the affinity of the particular DNA sequence used for the histone octamer and is empirically determined). Octamer should be added last to avoid aggregation with DNA at low salt. The NaCl concentration in the reactions is reduced stepwise to 1.2, 0.79, and 0.3 M by dilution with 25 mM Tris-HCl (pH 7.5) with a 10-min incubation at 37° at each step (final volume is 67 μl). Nucleosome reconstitution efficiency (typically 70–95%, depending on the probe) can be analyzed by taking a small (3-μl) aliquot, adding glycerol to 10%, and loading it on a 4% native polyacrylamide gel (acrylamide–bisacrylamide, 39:1) at 200 V in 0.5× Tris–borate–EDTA (TBE) at 4° (Fig. 2A). Assembled nucleosomes can be stored, in principle, at least for several weeks at 4°, but if they contain cysteines and are to be modified, this should be done immediately after reconstitution to avoid oxidation of the sulfhydryl group.

Modification with Azido Phenacyl Bromide (APB)

In contrast to the method for attaching an EDTA derivative to cysteines, in which the refolded octamer is first modified and then reconstituted into nucleosomes, in the procedure outlined here the target cysteine is modified with the sulfhydryl-specific photoreactive reagent APB only after assembly into a nucleosome. We have attempted to modify the free octamer first, but this led to its disintegration as evidenced by a complete loss of the octamer peak on the Superdex gel-filtration column and failure to assemble into nucleosomes. We have determined that this is due to modification of the only native cysteine in the octamer at position 110 in histone H3 (S. R. Kassabov and B. Bartholomew, unpublished data, 2000). Interestingly, when modification is carried out after octamer assembly into a nucleosome, the native H3 C110 is completely protected and not modified whereas the more accessible target cysteines are modified with high efficiency (Fig. 2B). We have not attempted to modify octamers with an H3 C110A substitution, but this might be a feasible alternative to modification after nucleosome assembly. The procedure outlined below does not include a reduction step before modification because we have determined that even after prolonged storage of the histone octamer under the

[11] Y. Lorch, J. W. LaPointe, and R. D. Kornberg, *Cell* **49**, 203 (1987).

conditions described above, there is no significant oxidation of the target cysteine residues employed in this method.

Because APB is photoreactive, the modification and all subsequent manipulations are carried out under reduced lighting with 20-W incandescent light bulbs. Modification is initiated by adding 3 μl of 80% glycerol and 0.7 μl of freshly made 10 mM APB (Fluka, Buchs, Switzerland) stock in high-quality anhydrous dimethylformamide (Sigma-Aldrich, Milwaukee, WI) to 60 μl of the nucleosome reconstitution and the reaction is incubated for 2 to 3 h at room temperature. Final concentrations in the modification reaction are as follows: 2.0–3.0 μM unique cysteine, 50–100 μM 2-mercaptoethanol (depending on the actual amount of octamer used in the reconstitution), 220 μM p-azidophenacyl bromide, 20 mM Tris-HCl (pH 7.5), 1% dimethylformamide, and 5% glycerol. Because the molar concentration of the cysteine is relatively low, the actual molar ratio of APB to free SH groups is determined by the concentration of 2-mercaptoethanol in the reaction and should be in the range of 2:1 to 10:1. After the modification, excess APB can either be removed by dialysis using microdialyzers (Pierce Biotechnology, Rockford, IL) or quenched by addition of 2-mercaptoethanol to 1 mM (both methods have worked equally well). The efficiency and selectivity of the reaction can be monitored on a Triton–acid–urea (TAU) gel, on which the mobility of the modified histone is altered with respect to the unmodified histone (Fig. 2B).

Alternatively, the derivatization efficiency can be assessed by postmodification labeling with sulfhydryl specific N-[ethyl-1-^{14}C] maleimide reagent and followed by SDS–polyacrylamide gel electrophoresis (PAGE) and fluorography.[8] We have relied primarily on a more functionally relevant test of the modification efficiency afforded by analysis of the DNA cross-linking efficiency following UV irradiation as detailed below. Modified nucleosomes can be stored at 4° for several weeks. Modified nucleosomes can be further purified by sedimentation on a sucrose gradient, although removal of free DNA is not technically required for the histone–DNA contact mapping.

UV Irradiation and Analysis of Cross-Linking Efficiency

Reactions containing the photoreactive nucleosomes (up to 50 μl in volume; larger reactions should be split in multiple tubes or placed in a shallow vessel) are irradiated directly in microcentrifuge tubes with a UV transilluminator equipped with 312-nm UV-B bulbs. Short-wave UV light should be avoided because it significantly damages DNA, leading to increased background. The opened tubes are placed beneath the transilluminator, which is turned face down, and irradiated for 1 min at a distance of

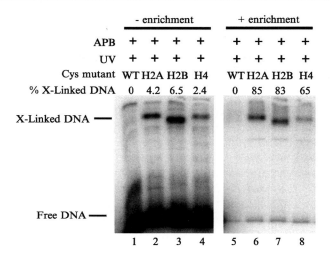

FIG. 3. Analysis of the level of cross-linking efficiency and enrichment for cross-linked DNA. Reactions containing APB-modified nucleosomes assembled with wild-type or Cys substitution mutant histones as indicated were UV irradiated and 2-μl aliquots were taken out for unenriched controls (lanes 1–4). The rest of the reactions were enriched for cross-linked DNA, using a phenol extraction/washing procedure as described and equivalent 2-μl aliquots were mixed with 20 μl of SDS–PAGE loading buffer, heated at 37° for 15 min, and resolved on a 17% SDS–polyacrylamide gel (lanes 5–7). *Note:* Samples should not be heated at 90° to avoid complications due to DNA denaturation/renaturation.

10 cm. A small portion of the irradiated reaction is loaded onto a 17% SDS–polyacrylamide gel to analyze the efficiency of cross-linking (Fig. 3). As can be seen from Fig. 3, the H2A C45 and H2B C53 derivatives cross-link DNA with, respectively, 2- and 3-fold higher efficiency than the H4 C47 derivative. The higher cross-linking efficiency boosts dramatically the signal-to-noise ratio obtained with these new derivatives, making possible detection of minor translational positions that cannot be detected with the H4 C47 derivative alone.

Separation of Cross-linked and Un-Cross-Linked DNA

Important for obtaining a good signal-to-noise ratio is to separate cross-linked DNA from the bulk of non-cross-linked nucleosomal and free DNA (if present), because of the relatively low efficiency of DNA cross-linking. Cross-linked DNA can be enriched for by a simple phenol extraction procedure. This method is based on the tendency of the protein–DNA adduct to partition to the interphase while the non-cross-linked DNA partitions in the aqueous phase.[12] Alternatively, samples can be separated on a preparative SDS–polyacrylamide gel and the cross-linked DNA extracted from the gel.[8] Both methods have produced comparable results in our hands, but we prefer the phenol extraction procedure because it is faster and entails significantly lower loss of signal compared with the SDS–PAGE method.

The volume of the UV-irradiated reactions is brought to 300 μl with buffer containing 20 mM Tris-HCl (pH 8.0) and 50 mM NaCl, and 3 μl of 10% SDS is added to a final concentration of 0.1%. Samples are heated at 70° for 20 min, and after cooling to room temperature 300 μl of phenol–chloroform (4:1) is added and the samples are vortexed for 30 s. Samples are then spun at maximum speed for 2 min and most of the aqueous phase (approximately 280 μl) is removed while taking care not to remove the interphase. A small amount (20–30 μl) of the aqueous phase should be left behind in order not to disturb the interphase. Wash buffer [280 μl of 1% SDS, 1 M Tris-HCl (pH 8.0)] is added to the organic phase and interphase and the samples are vortexed and centrifuged as before. The majority of the aqueous phase is removed (again leaving behind 20–30 μl) and the wash step is repeated three more times. To each sample is added 30 μl of sodium acetate (pH 5.2), 10 μg of sonicated salmon sperm DNA (as carrier), and 750 μl of 100% ethanol and the samples are precipitated by placing them on ice for 2 h to overnight. Precipitation at the standard −20° is not necessary and often leads to coprecipitation of large amounts of salts or SDS. Comparison of cross-linked samples before and after the enrichment procedure in Fig. 3 illustrates its effectiveness for removal of non-cross-linked DNA.

Heat- and Alkali-Induced DNA Cleavage and Determination of Cross-Linked Nucleotide

The procedure for base elimination and DNA cleavage at the cross-linked nucleotide is based on the method developed by Pendergrast et al.[12] The ethanol-precipitated samples containing the cross-linked protein–DNA complex are spun at maximum speed at 4° for 30 min and washed twice with 75% ethanol. The pellets are air dried (approximately 10 min) and resuspended in 100 μl of 2% SDS, 20 mM ammonium acetate, and 0.1 mM Na-EDTA buffer (maximal resuspension can sometimes require mild vortexing for up to 30 min). Drying of the pellets in a centrifugal vacuum device is not recommended because it can lead to desiccation and difficulties in resuspending the sample. The samples are spun at maximum

---

[12] P. S. Pendergrast, Y. Chen, Y. W. Ebright, and R. H. Ebright, *Proc. Natl. Acad. Sci. USA* **89,** 10287 (1992).

speed at room temperature for 10 min to remove any insoluble aggregates and the supernatant is transferred to new microcentrifuge tubes. The samples are then incubated at 90° for 20 min, 5 μl of 2 M NaOH is added, and the samples are heated for another 40 min at 90°. Alkaline cleavage is stopped by addition of an equal volume of 20 mM Tris-HCl (pH 8.0) and 6.0 μl of 2 M HCl. To each sample is added 2 μl of 1 M MgCl$_2$ and 470 μl of ice-cold 100% ethanol and the reactions are precipitated overnight at −20°. Precipitated DNA is recovered by spinning at maximum speed for 30 min at 4° and the pellets are washed once with ice-cold 75% ethanol, air dried, and resuspended in 5 μl of formamide loading buffer [950 μl of deionized formamide (ultrapure grade; USB, Cleveland, OH), 15 μl of 2.5% bromphenol blue, 15 μl of 2.5% xylene cyanol, and 20 μl of 0.5 M EDTA]. Care should be taken that all signal is fully resuspended because the pellet is sometimes smeared along the tube wall. Samples should be spun down after resuspension to check whether any signal is left behind in the tube after removing the supernatant. On occasion, a small gummy pellet containing a significant proportion of the signal is observed, which is difficult to dissolve in the formamide loading buffer. This pellet can be readily resuspended directly in 1 μl of H$_2$O and added back to the rest of the sample in the formamide buffer. Samples are heated for 2 min at 90° and loaded immediately on a 6.5% denaturing 8 M urea–polyacrylamide gel (40 cm × 20 cm × 0.1 mm, 20:1 acrylamide-to-bisacrylamide ratio in 1× TBE buffer) and run at 40–45 W until the bromphenol blue dye reaches approximately 5 cm from the bottom of the gel (about 1.5 h). All electrophoresis reagents are ultrapure grade from USB. Determination of the precise positions of the DNA cuts is done by comparison with a dideoxy sequencing ladder (Sequenase quick denature kit; USB).

Using Site-Directed Histone–DNA Mapping to Characterize Remodeled Nucleosomal State

The site-directed histone–DNA mapping procedure described in the preceding sections was used in our laboratory to map the translational repositioning/sliding and conformational changes in nucleosomes remodeled by two different chromatin-remodeling complexes from yeast: ISW2 and SWI/SNF.[9,13] The results revealed that the two enzymes generate distinct remodeling products. ISW2 was shown to slide nucleosomes from the DNA ends to a single central translational position without altering the core nucleosome structure. The maintenance of the canonical structure of

[13] S. R. Kassabov, B. Zhang, J. Persinger, and B. Bartholomew, *Mol. Cell* **11**, 391 (2003).

the slid/remodeled nucleosome was clearly demonstrated by the preservation of the exact distance between all six modified residue contacts with DNA after sliding to the new central translational position. These results demonstrated that the cysteine replacement and the attachment of the photoreactive aryl azide group to all three sites in the octamer do not affect the activity of the remodeling complex or the outcome of remodeling. In contrast with ISW2, the SWI/SNF complex was shown to translocate the histone octamer up to 50 bp beyond the DNA ends and to alter dramatically the path of the DNA around the histone octamer, forming an intranucleosomal entry/exit site DNA loop. Here we discuss the analysis of the SWI/SNF remodeling reaction because it exemplifies well the capability of the site-directed histone–DNA mapping method to detect changes in both the translational positioning and in the structure of the remodeled nucleosome.

Mapping of Dyad Axis Positions in Remodeled Nucleosomes with H4 C47 Reveals Displacement of Histone Octamer beyond DNA Ends

Determining the H4 C47–DNA contacts revealed the positions of the histone octamer dyad axis on the DNA before and after remodeling with SWI/SNF. In good correlation with the site-directed hydroxy radical cleavage obtained with the same cysteine replacement mutant by Flaus et al.,[7] each of the two modified H4 C47 residues in the octamer cross-linked both DNA strands, resulting in two cuts on each strand per dyad, one two nucleotides 5' of the dyad axis and the second seven or eight nucleotides 3' of the first cut site. As evidenced by the two sets of 7- to 8-bp related cross-links on each DNA strand, there are two translational positions on the 183-bp DNA template used in this experiment with dyads at bp 84 and 104 representing nucleosomes bound near either end of the DNA (Fig. 4A, lanes 2 and 4; and Fig. 4B). In keeping with the results in Fig. 3, generation of specific cross-links was dependent on the selective modification of the engineered cysteine with APB because in control reactions nucleosomes lacking such cysteine and modified analogously failed to produce any distinct cuts (Fig. 4A, lanes 1 and 6). After remodeling, the cross-linking at the original dyad positions was almost completely abolished and new cross-links corresponding to dyad positions near both DNA ends were observed (Fig. 4A, lanes 3 and 5). On the upper strand, multiple, relatively weak new cross-links were generated in the region between bp 130 and 165, indicating the presence of several closely overlapping positions of octamers slid from 22 to 50 bp beyond the right DNA end (Fig. 4B). On the lower DNA strand, there were only two new, distinct 8-bp related cross-links after remodeling at C17 and A25 defining a single dyad position of an octamer slid by 50 bp beyond the left DNA end. Thus the mapping of the H4 C47–DNA contacts provided two pieces of information about the remodeled nucleosomal state. First, the very fact that the H4 C47 residue still cross-links DNA after remodeling and that the characteristic pattern of 8-bp distance and higher intensity of the 5′ over the 3′ cut is preserved indicated that the local histone–DNA interface organization around the dyad axis is maintained in the SWI/SNF-remodeled nucleosome. Second, the positions of the remodeled dyads demonstrated that the histone octamers in the remodeled nucleosomes are translocated up to 50 bp beyond both DNA ends, indicating that SWI/SNF catalyzes a major disruption of histone–DNA contacts in the nucleosome.

Correlation of Remodeled Nucleosome Dyad Axis and H2A–H2B
Dimer–DNA Contacts Reveals an Altered Nucleosomal Conformation

The results of the H4 C47 site-directed mapping raised several important questions regarding the remodeled nucleosomal state: (1) What is the

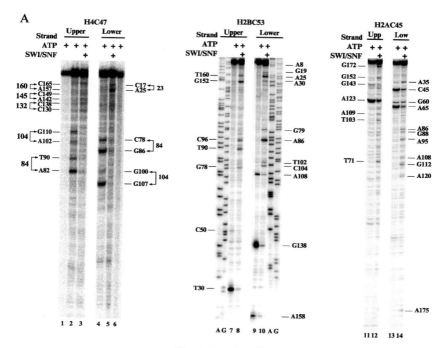

FIG. 4. (continued)

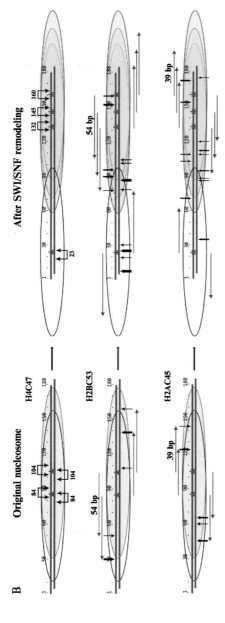

FIG. 4. (A) High-resolution mapping of specific histone–DNA contacts in the nucleosome before and after SWI/SNF remodeling. Remodeling reactions (25 μl) containing nucleosomes with modified mutant (lanes 2–5 and 7–14) or wild-type octamer lacking cysteine (lanes 1 and 6) (5 nM) assembled on DNA radiolabeled at the 5' end of either the upper or lower strand were incubated with or without SWI/SNF (10 nM) as indicated in the presence of ATP (200 μM) for 30 min at 30°. DNA fragments resulting from cleavage of nucleosomal DNA cross-linked to the modified histone residues were resolved on 6.5% sequencing gels. A and G sequencing ladders (lanes A and G) were loaded for reference (only the ladders for the H2B C53 samples are shown). (B) Summary of histone–DNA contacts mapped before and after remodeling with SWI/SNF. Vertical arrows represent the cut sites identified in Fig. 4A. The thickness of the arrow corresponds to the relative efficiency of cutting at the corresponding nucleotide. Asterisks indicate the dyad positions mapped with the H4 C47 mutant and ovals represent the boundaries corresponding to those nucleosome positions. Horizontal arrows represent the expected positions for cuts on the respective DNA strand generated by each of the two symmetrically disposed H2B C53 and H2A C45 residues in the octamer, based on the dyad axis positions mapped with H4 C47 and the crystal structure of the nucleosome. Reprinted from S. R. Kassabov, B. Zhang, J. Persinger, and B. Bartholomew, *Mol. Cell* **11**, 391 (2003), with permission from Elsevier.

fate of the exposed H2A/H2B dimer which lacks interactions with DNA? (2) What is the DNA path around the rest of the DNA-bound octamer surface? (3) Are there major structural alterations in the histone octamer itself? These questions were addressed by correlating the H4 C47–DNA contacts close to the octamer dyad axis with the H2B C53– and H2A C45–DNA contacts in the H2A–H2B dimer–DNA interface in the remodeled nucleosome. Before remodeling, cross-linking with the H2B C53 and H2A C45 derivatives produced single strong cuts on each DNA strand, respectively, 54 bp 5' and 39 bp 3' of the dyad axis positions mapped with the H4 C47 derivative (Fig. 4A, lanes 7, 9, 11, and 13; and Fig. 4B). The precise symmetry and correlation of the distance between DNA cut sites from all three derivatives with the crystal structure demonstrated that the cysteine replacements and covalent attachment of aryl azide to those residues does not affect the structure or the translational position of the nucleosome. Thus, both H2B C53 and H2A C45 derivatives can be used to accurately map nucleosomal translational positions in addition to H4 C47. In addition, the strict strand specificity of cross-linking with the H2B C53 and H2A C45 derivatives provides a sensitive assay for detecting small topological changes of the nucleosomal DNA that could alter the relative orientation of the modified residues with respect to the two DNA strands.

After remodeling the original H2B C53 cross-linking was reduced by nearly 90% and multiple new cross-links were generated in three distinct regions along the length of the DNA template, one at each end of the DNA and one in the middle (Fig. 4A, lanes 8 and 10). The cross-links in the central region of both DNA strands are generally consistent with sliding of the histone octamer off the ends of the DNA, as detected by the H4 C47 dyad mapping (Fig. 4B). However, a careful correlation of the major remodeled dyad position at bp 23 with the H2B C53 cross-linking in the central region on the lower DNA strand revealed that, instead of being at bp 77, the corresponding major H2B C53 cut was shifted by 9 bp to bp 86. Thus, the distance on DNA between the H2B C53 residue and the dyad axis is altered in the remodeled nucleosome from the canonical 54 to 63 bp, resulting in an extra 9 bp of DNA between these two points. A small fraction of the nucleosomes may be maintaining or reverting to near canonical spacing, generating the less efficient cut site at bp 79, with other minor cut sites at bp 102 and 104 being likely due to minor remodeled translational positions that were not detected with the less sensitive H4 C47 mapping.

The presence of H2B C53 cuts near the DNA ends in addition to the central region was unexpected and suggested that the displaced H2A–H2B dimer does not dissociate after remodeling, but instead regains contact with DNA via a major change in the histone octamer structure or in the path of the DNA on the octamer surface or both. Further experiments demonstrated that chemical cross-linking of the histones in the octamer before remodeling does not alter the H2B C53 cutting pattern, ruling out the possibility that the displaced H2A–H2B dimer could be undergoing a major structural rearrangement or could be first dissociating and then rebinding the DNA at a new position as a separate entity. Furthermore, the strict strand specificity of cross-linking with H2B C53 near the DNA ends suggested that, as in the case of H4 C47, the local histone–DNA interface was not altered significantly. An alignment of the free spooled-out DNA ends with the exposed octamer surface in the remodeled nucleosome showed a good correlation of the actual cross-linking positions at the DNA ends with those expected in all remodeled translational frames. This suggested that the spooled-out free DNA end could be binding to the exposed H2A–H2B dimer of the same remodeled nucleosome, resulting in the formation of an entry/exit site DNA loop traversing the dyad axis. Alternatively, the free DNA end could be binding the exposed octamer surface in a different remodeled nucleosome, leading to dinucleosome formation. Gel shift analysis demonstrated that there is little or no formation of slower migrating complexes under our experimental conditions, thus favoring the intramolecular model.

In contrast with H2B C53, there were only weak cross-links generated by H2A C45 after remodeling, indicating that this residue is no longer in close contact with the nucleosomal DNA (Fig. 4A, lanes 12 and 14). Furthermore, the positions of some cuts after remodeling, such as A109 and T103 on the upper strand and A175 on the lower strand, indicate a loss of the original strand specificity of cross-linking from this modified residue. These data indicate a topological change in the histone–DNA interface in this region resulting in a loss of the proximity and relative orientation of the H2A45 residue with respect to the DNA, a finding consistent with the presence of an extra 9 bp of DNA between the H2B C53 contact and the dyad axis.

Taken together, the mapping results with all three derivatives demonstrate that SWI/SNF catalyzes the translocation of the histone octamer off the ends of DNA, while promoting the spooled-out free DNA end in the same nucleosome to associate with the newly exposed H2A–H2B dimer surface via an intranucleosomal entry/exit site DNA loop. These findings were further confirmed and extended by results obtained by other techniques such as gel shift, sedimentation analysis, and restriction endonuclease mapping.

Special Applications of the Site-Directed Histone–DNA
    Contact Mapping Method

Histone–DNA Contact Mapping of Gel-Purified Remodeled Nucleosomal Conformations

Although one of the main strengths of the method presented here lies in its capability to analyze complex mixtures such as nucleosome populations with multiple overlapping positions, it could be advantageous to first separate and then analyze the components of such mixtures individually. For instance, in the analysis of the SWI/SNF-remodeled nucleosome we took advantage of its faster electrophoretic mobility to map the H4 C47– and H2B C53–DNA contacts directly in the stable altered species extracted from the native gel. The nearly identical cross-linking pattern obtained from samples in solution and from the gel-isolated remodeled nucleosomes confirmed the identity of the species with faster electrophoretic mobility, demonstrating the stability of the remodeled nucleosomal conformation and ruled out the formal possibility that the H2B C53 cuts near the ends of the DNA could be generated by the formation of unstable dinucleosomes that dissociate on gel electrophoresis.[13] In the following, we outline the protocol used in these experiments.

Other groups have reported that nucleosomes can be UV irradiated either in solution before electrophoresis or in the gel after resolution with equivalent results, but we have observed that irradiation in the gel produces higher background with our derivatives and therefore preferred the irradiation in solution approach. Fourfold scaled up remodeling reactions (100 μl) with the derivatized photoreactive nucleosomes are UV irradiated as described (after splitting the reactions in two Eppendorf tubes), loaded on a preparative (1.5 mm × 20 cm × 20 cm) 4% native polyacrylamide gel (acrylamide–bisacrylamide, 39:1) with 2-cm-wide wells, and resolved at 200 V in 0.5× TBE at 4° under reduced light. After electrophoresis, one of the gel glass plates is carefully removed and the gel remaining on the other plate is covered with Saran Wrap, placed on ice, and exposed on a PhosphorImager plate for 15–30 min. After phosphoimaging, the faster migrating band containing remodeled nucleosomes is excised from the gel and the cross-linked DNA is electroeluted from gel slices with a Bio-Rad (Hercules, CA) 422 electroelutor and a 0.5× TBE, 0.1% SDS buffer system over 3–5 h at 10 mA per sample/glass tube. The majority of the un-cross-linked DNA elutes significantly earlier than the cross-linked DNA, so that care should be taken that the electroelution proceeds to completion by checking the gel slices after the run. The eluted sample (approximately 400 μl in volume) is then heated at 90° for 20 min and an equal volume of phenol–chloroform (4:1) is added and the samples are processed as described earlier, adjusting for any volume differences.

Rapid Kinetic Analysis of Remodeling Reaction Using Site-Directed Histone–DNA Contact Mapping

In addition to analysis of the stable products of remodeling illustrated in the previous sections, the photochemical histone–DNA contact mapping method can be effectively employed to study the dynamics of the remodeling reaction. Because the assay follows the changes in the contacts of specific histone residues with DNA, it allows for direct and simultaneous monitoring of the individual rates of conversion to and between different states and intermediates during remodeling. Furthermore, the half-life of the reactive nitrene species generated on UV irradiation is in the nanosecond range, permitting ultrafast kinetic analysis provided an appropriate mixing device (such as an adapted stopped-flow system) and a high-intensity laser light source are available. Experiments with resolution on the scale of several seconds, however, can easily be performed with a standard UV transilluminator and laboratory equipment. With this approach, we have been able to demonstrate that the SWI/SNF remodeling reaction occurs much more rapidly and involves the hydrolysis of a significantly lower number of ATP molecules than was previously estimated in other assays.[13] In the following, we outline the protocol used in these kinetic experiments.

Remodeling reactions with a saturating molar ratio of the remodeling complex to nucleosomes (5:1 or higher) are assembled and incubated for 10 min at 30° in the absence of ATP to allow complete binding. ATP is added and after the appropriate remodeling time (5 s to several minutes), the samples are irradiated for 5 s by placing the open tubes in direct contact with the transilluminator glass plate for maximal UV light intensity. Irradiation of a 25-μl sample from this distance for 5 s with the 90-W Spectroline UV transilluminator (model TVD-1000R; Spectronics, Westbury, NY) used in our laboratory results in approximately 60% of the maximum cross-linking efficiency, which is achieved in 10 s or more. The irradiated samples are processed and analyzed as described earlier. The rate of remodeling can then be calculated from the rate of disappearance of the preremodeling histone–DNA contacts or the rate of increase in the new remodeled contacts over time. For estimation of the amount of ATP hydrolyzed in a remodeling "event," an ATPase assay is performed under equivalent conditions and the rate correlated with the results of the histone–DNA contact mapping assay. The remodeling activity is stopped in the ATPase assay at the appropriate time points by addition of 2% SDS and 20 mM EDTA.

[14] Site-Specific Attachment of Reporter Compounds to Recombinant Histones

By MICHAEL BRUNO, ANDREW FLAUS, and TOM OWEN-HUGHES

Introduction

X-ray crystallographic studies of nucleosome core particles provide a detailed description of the structure of the fundamental subunit of eukaryotic chromatin.[1-5] However, under physiological conditions nucleosomes are not static structures and are subject to continuous rearrangement. One of the ways in which this occurs involves the positioning of nucleosomes on DNA being altered. This can occur as a result of thermal nucleosome redistribution.[6-9] Although nucleosomes assembled onto many DNA sequences exhibit some thermal mobility under physiological conditions, it has also been found that many ATP-dependent chromatin-remodeling activities can accelerate this process. Studies of the *Drosophila* ISWI-containing complexes NURF,[10] CHRAC,[11] together with the yeast SWI/SNF complex,[12,13] provided some of the first evidence that these complexes can alter the positions of nucleosomes along DNA. Subsequently, it has been found that many other enzymes including RSC,[14] Mi-2,[15] *Xenopus* ISWI,[16]

[1] K. Luger, A. W. Mader, R. K. Richmond, D. F. Sargent, and T. J. Richmond, *Nature* **389**, 251 (1997).

[2] C. A. Davey, D. F. Sargent, K. Luger, A. W. Maeder, and T. J. Richmond, *J. Mol. Biol.* **319**, 1097 (2002).

[3] J. M. Harp, B. L. Hanson, D. E. Timm, and G. J. Bunick, *Acta Crystallogr. D Biol. Crystallogr.* **56**, 1513 (2000).

[4] C. L. White, R. K. Suto, and K. Luger, *EMBO J.* **20**, 5207 (2001).

[5] R. K. Suto, M. J. Clarkson, D. J. Tremethick, and K. Luger, *Nat. Struct. Biol.* **7**, 1121 (2000).

[6] P. Beard, *Cell* **15**, 955 (1978).

[7] W. O. Weischet, *Nucleic Acids Res.* **7**, 291 (1979).

[8] S. Pennings, G. Meersseman, and E. M. Bradbury, *J. Mol. Biol.* **220**, 101 (1991).

[9] G. Meersseman, S. Pennings, and E. M. Bradbury, *Proc. Natl. Acad. Sci. USA* **89**, 8626 (1992).

[10] A. Hamiche, R. Sandaltzopoulos, D. A. Gdula, and C. Wu, *Cell* **97**, 833 (1999).

[11] G. Langst, E. J. Bonte, D. F. Corona, and P. B. Becker, *Cell* **97**, 843 (1999).

[12] I. Whitehouse, A. Flaus, B. R. Cairns, M. F. White, J. L. Workman, and T. Owen-Hughes, *Nature* **400**, 784 (1999).

[13] M. Jaskelioff, I. M. Gavin, C. L. Peterson, and C. Logie, *Mol. Cell. Biol.* **20**, 3058 (2000).

[14] Y. Lorch, M. Zhang, and R. D. Kornberg, *Mol. Cell* **7**, 89 (2001).

[15] D. Guschin, P. A. Wade, N. Kikyo, and A. P. Wolffe, *Biochemistry* **39**, 5238 (2000).

[16] D. Guschin, T. M. Geiman, N. Kikyo, D. J. Tremethick, A. P. Wolffe, and P. A. Wade, *J. Biol. Chem.* (2000).

and human Brg1[17] also perform this reaction, which may be a common feature of many ATP-dependent remodeling enzymes. In addition to alterations to nucleosome positioning, a subset of ATP-dependent remodeling enzymes are also capable of distorting nucleosomes in ways that appear different from simple changes in the translational positioning of nucleosomes.[12,14,18–22] All these alterations affect access to the underlying DNA and so have the potential to affect any process that requires access to DNA.

A system for the expression of recombinant histone proteins has been developed by the Richmond Laboratory and provides the possibility of manipulating the histone proteins by site-directed mutagenesis. This can be used to study the roles of specific regions of the histone proteins in chromatin function[23,24] and to introduce amino acids at specific sites within nucleosomes for the attachment of reporter compounds. Here we describe the use of this approach to attach compounds that facilitate the study of the dynamic properties of chromatin.

Preparation of Core Histones for Chemical Probe Attachment

Background

The ability to reconstitute nucleosomes of uniform composition from recombinantly expressed materials was a major advance in our ability to study the biochemical and biophysical properties of chromatin.[25] The central techniques for assembling such recombinant nucleosomes were described by Luger et al.[26–28]

[17] S. Aoyagi and J. J. Hayes, *Mol. Cell. Biol.* **22,** 7484 (2002).

[18] Y. Lorch, B. R. Cairns, M. Zhang, and R. D. Kornberg, *Cell* **94,** 29 (1998).

[19] G. Schnitzler, S. Sif, and R. E. Kingston, *Cell* **94,** 17 (1998).

[20] J. R. Guyon, G. J. Narlikar, E. K. Sullivan, and R. E. Kingston, *Mol. Cell. Biol.* **21,** 1132 (2001).

[21] T. Owen-Hughes, R. T. Utley, J. Cote, C. L. Peterson, and J. L. Workman, *Science* **273,** 513 (1996).

[22] Y. Lorch, M. Zhang, and R. D. Kornberg, *Cell* **96,** 389 (1999).

[23] A. Hamiche, J. G. Kang, C. Dennis, H. Xiao, and C. Wu, *Proc. Natl. Acad. Sci. USA* **98,** 14316 (2001).

[24] C. R. Clapier, G. Langst, D. F. Corona, P. B. Becker, and K. P. Nightingale, *Mol. Cell. Biol.* **21,** 875 (2001).

[25] D. Rhodes, *Nature* **389,** 231 (1997).

[26] K. Luger, T. J. Rechsteiner, A. J. Flaus, M. M. Waye, and T. J. Richmond, *J. Mol. Biol.* **272,** 301 (1997).

[27] K. Luger, T. J. Rechsteiner, and T. J. Richmond, *Methods Mol. Biol.* **119,** 1 (1999).

[28] K. Luger, T. J. Rechsteiner, and T. J. Richmond, *Methods Enzymol.* **304,** 3 (1999).

Core Histone Expression Clone Sequences

Luger et al.[26] described pET3-derived plasmids for high-level expression of core histones H2A, H2B, and H3 based on cloned *Xenopus laevis* genes.[29,30] The expression plasmid for histone H4 encodes the native protein sequence, but using optimal *Escherichia coli* codons, because its expression level is typically lower and more variable. For subsequent cloning manipulations and primer design, the indirect information available on these clones is difficult to interpret. We have sequenced samples derived as directly as possible from the original plasmid stocks prepared by Luger et al.[26] and submitted them with appropriate annotations to the EMBL sequence database with the following accession numbers: H2A AJ556870, H2B AJ556871, H3 AJ556872, H4 AJ556873.

Mutagenesis

On removing the single native cysteine in H3 at residue 110 by mutation to alanine (i.e., H3 C110A), it is possible to subsequently introduce a unique cysteine by site-directed mutagenesis of any of the four core histone expression plasmids. Cysteine residues are convenient sites for the attachment of chemical probes containing activated thiol and maleimide-derived reagents, which are highly specific for the cysteine thiol in aqueous solution at neutral pH.

The Quickchange site-directed mutagenesis kit (Stratagene, La Jolla, CA) provides a rapid and highly efficient method of introducing cysteine residues into the histone-coding plasmids. We have also used it successfully to make insertions of 9 bp and deletions of up to 80 bp in these plasmids.

Histone Preparation

Protocols for the preparation of milligram quantities of individual histone proteins have been previously described.[27,28] Here we describe a number of modifications to these methods.

First, we have observed that improvements in yield can be obtained for some histones by expression in BL21* strains of *E. coli* containing the pRIL tRNA supplementation plasmid. The BL21* strain facilitates longer mRNA half-life through a mutation in the mRNA-degrading RNAse E gene (Invitrogen, Carlsbad, CA), and the pRIL codon-supplementing plasmid (Stratagene) increases the normal levels of arginine, isoleucine,

[29] R. W. Old, H. R. Woodland, J. E. Ballantine, T. C. Aldridge, C. A. Newton, W. A. Bains, and P. C. Turner, *Nucleic Acids Res.* **10,** 7561 (1982).

[30] P. C. Turner and H. R. Woodland, *Nucleic Acids Res.* **10,** 3769 (1982).

and leucine tRNAs, which are used less frequently in *E. coli* but that do occur in the histone genes.

Second, we have found that significant DNA contamination can occur if lysed cells are not extensively and efficiently sonicated. The presence of DNA can result in binding of aggregates to the S200 gel-filtration column matrix, broadening of the histone-containing peak during this chromatography step, and precipitation during subsequent dialysis of pooled histone-containing fractions into water.

Third, for carefully prepared inclusion body preparations, it is frequently unnecessary to undertake a subsequent ion-exchange step. However, when ion exchange is needed, we have found that the histones can elute in several peaks during the salt gradient in 7 M urea-containing buffer on TSK SP 5PW, Pharmacia Source 15S, and Poros HS resins. The proteins in each peak are indistinguishable by SDS–PAGE and mass spectrometry, implying that they represent different folding or aggregation states. To simplify purification we employ a stepwise elution from 50 to 600 mM NaCl in 7 M urea-containing buffer.

Preparation of DNA Fragments for Nucleosome Assembly

Preparative PCR

The efficient preparation of sufficient quantities of linear DNA fragments suitable for assembly into nucleosomes enables biochemical analysis of the effect of DNA length and sequence on a broader scale. We have found that scaling up and optimization of individual polymerase chain reactions (PCRs) in standard 96-well plates now enables convenient preparation of $>200 \mu$g $(>1$ nmol) of purified 150- to 500-bp DNA at reasonable cost.

Preparative PCR Method

1. Prepare 5 ml of a stock PCR mixture on ice:

Plasmid template	5 μl
3 mM dNTPs	200 μl
50 mM MgCl$_2$	200 μl
10× PCR buffer	500 μl
100 μM forward oligonucleotide primer ($T_m > 55°$)	50 μl
100 μM reverse oligonucleotide primer ($T_m > 55°$)	50 μl
Distilled H$_2$O	1 ml
Taq polymerase	25 μl

2. Pipette 50 μl per well into a 96-well low-profile PCR plate (ABgene, Epsom, UK), using an eight-channel multipipetter, and seal the PCR plate with PCR plate sealer (ABgene).

3. Subject the plate to PCR in a preheated Eppendorf Mastercycler PCR machine, using 2 min of denaturation at 95° followed by 25 cycles of 30-s steps consecutively at 95°, 50°, and 72°.

4. Pool the products from the wells in a 50-ml conical tube and ethanol precipitate by adding 500 μl of 3 M sodium acetate and 12.5 ml of ethanol and then centrifuging for 15 min at 5000g.

5. Remove the supernatant carefully and resuspend in 500 μl of Tris–EDTA [TE(10/0.1)].

6. Purify by injecting onto a preequilibrated 1- to 2-ml Source 15Q chromatography column (Pharmacia, Uppsala, Sweden) running at 2 ml/min, using buffer A [20 mM Tris (pH 7.5), 0.1 mM EDTA] and buffer B [20 mM Tris (pH 7.5), 2 M NaCl, 0.1 mM EDTA] according to the following steps expressed in column volumes (CV):

    a. Equilibrate with buffer A for 4 CV
    b. Inject sample in 500 μl of TE(10/0.1)
    c. Wash with buffer A for 4 CV
    d. Elute with a gradient from 0 to 40% buffer B over 15 CV
    e. Clean with 100% buffer B for 2 CV
    f. Reequilibrate the column in buffer A

7. Pool the DNA fragment fractions that elute at approximately 700 mM NaCl. Note that preceding large peaks contain, respectively, unused dNTPs and primers. Ethanol precipitate the pooled material and resuspend in 30 μl of TE(10/0.1).

For maximal yields, it is necessary to optimize the amount of template in the PCR for each plasmid template prepared by standard Qiagen maxipreparations (Qiagen, Hilden, Germany). Oligonucleotide primers (Thermo Electron Molecular Biology, Needham Heights, MA) can contain 5′ or internal modifications such as fluorescent dyes, depending on their chemical stability. We have noticed significant variations in product yield depending on the source of both dNTPs and Taq polymerase. In our hands, the highest yields are obtained with dNTPs from Bioline (London, UK), and Taq polymerase prepared in-house[31] or BioTaq obtained commercially from Bioline compared with other commercially available products. PCR buffer and MgCl$_2$ are used as supplied with the BioTaq polymerase or prepared according to the same specifications. This 10× NH$_4$ buffer comprises

[31] D. R. Engelke, A. Krikos, M. E. Bruck, and D. Ginsburg, *Anal. Biochem.* **191,** 396 (1990).

160 mM ammonium sulfate, 670 mM TrisCl (pH 8.8 at 25°), and 0.1% Tween 20.

Improved Methods for Site-Directed Hydroxyl Radical Mapping of Nucleosome Positions

Background

The use of site-directed generation of diffusible hydroxyl radicals to map nucleosome positions at base pair resolution was reported by Flaus et al.[32] This technique is based on methods developed by the laboratories of Tullius[33,34] (using iron chelated by EDTA to footprint DNA nucleosomes), Dervan (using EDTA derivatives tethered to DNA-binding proteins during peptide synthesis),[35] and Ebright and Fox (using thiol-reactive EDTA derivatives on DNA-binding factors).[36,37] A closely related use of the techniques has also been reported by Hayes for linker histones.[38,39]

Briefly, site-directed hydroxyl radical mapping of nucleosomes involves attaching a thiol-reactive EDTA derivative to the free thiol of a cysteine introduced by mutating residue 47 from serine in recombinantly expressed histone H4. Iron is allowed to bind to the chelator, and then ascorbate and hydrogen peroxide are supplied to facilitate cycling of the localized iron between ferrous and ferric forms with the production of hydroxyl radicals. Because of the proximity and symmetry, the hydroxyl radicals cleave the DNA backbone on each strand 2 base pairs upstream, and 4 and 5 base pairs downstream, of the nucleosomal dyad. This gives a characteristic mapping pattern of a predominant strong cut followed by two weaker cuts 7 and 8 base pairs away.

Detailed protocols for this technique have previously been published.[40,41] Here we report several improvements to these protocols.

[32] A. Flaus, K. Luger, S. Tan, and T. J. Richmond, *Proc. Natl. Acad. Sci. USA* **93,** 1370 (1996).

[33] J. J. Hayes, T. D. Tullius, and A. P. Wolffe, *Proc. Natl. Acad. Sci. USA* **87,** 7405 (1990).

[34] T. D. Tullius and B. A. Dombroski, *Science* **230,** 679 (1985).

[35] J. P. Sluka, S. J. Horvath, M. F. Bruist, M. I. Simon, and P. B. Dervan, *Science* **238,** 1129 (1987).

[36] Y. W. Ebright, Y. Chen, P. S. Pendergrast, and R. H. Ebright, *Biochemistry* **31,** 10664 (1992).

[37] M. R. Ermacora, J. M. Delfino, B. Cuenoud, A. Schepartz, and R. O. Fox, *Proc. Natl. Acad. Sci. USA* **89,** 6383 (1992).

[38] J. J. Hayes, *Biochemistry* **35,** 11931 (1996).

[39] D. R. Chafin and J. J. Hayes, *Methods Mol. Biol.* **148,** 275 (2001).

[40] A. Flaus and T. J. Richmond, *Methods Mol. Biol.* **119,** 45 (1999).

[41] A. Flaus and T. J. Richmond, *Methods Enzymol.* **304,** 251 (1999).

Commercial Availability of Reagents

A completely equivalent thiol-reactive EDTA reagent is now available commercially from Toronto Research Chemicals (North York, ON, Canada; http://www.trc-canada.com/) as [*S*-(2-pyridylthio)cysteaminyl]-ethylenediamine-*N,N,N′,N′*-tetraacetic acid. This reagent is identical to the (EDTA-2-aminoethyl)2-pyridyl disulfide (EPD) that has been attached to catabolite activator protein (CAP), staphylococcal nuclease, and other DNA-binding proteins by the groups of Ebright and Fox.[36,37] It differs slightly from the reagent originally used for nucleosome mapping because the activated thiol leaving group is pyridine rather than nitrophenol. However, the resulting coupled reagent–histone is identical and we have successfully used the reagent from Toronto Research Chemicals for some time.

An alternative class of reagents are composed of an EDTA derivative with a reactive haloacetamide, such as (*S*)-1-(*p*-bromoacetamidobenzyl)-ethylenediaminetetraacetate (FeBABE),[42] and are available from a number of commercial sources. However, it is important to note that a significant potential side reaction of haloacetamides is with the primary amines, and that histone octamers contain some 114 lysines, the majority of which are on the highly accessible tails.

Improved Reaction Conditions

Reaction conditions for site-directed hydroxyl radical mapping have been optimized (data not shown). An investigation over a wide range of reaction pH values revealed that buffering in the pH 7.0–8.0 range gives optimal amounts of cutting. The exact identity of the buffer was not significant, with BES, TES, Tris, and HEPES all supporting reaction near their pK_a. Increasing the amount of iron salt added also increases the total amount of mapping cuts up to a maximum level, followed by a rapid increase in background cutting. This presumably represents saturation of the site-specific chelator followed by excess iron then becoming closely associated with the DNA backbone as a counterion. The optimal amount of iron is related to the amount of reagent-derivatized octamer and needs to be determined empirically. Altering the amount of hydrogen peroxide did not significantly affect reactions. However, increasing the amount of ascorbic acid improved the reaction rate significantly, but only if the buffer concentration was also increased to mask the increased acidity of the resultant solution.

[42] D. P. Greiner, R. Miyake, J. K. Moran, A. D. Jones, T. Negishi, A. Ishihama, and C. F. Meares, *Bioconjug. Chem.* **8**, 44 (1997).

It is also possible to perform the mapping reaction in the presence of buffers containing 200 mM NaCl or more, and up to 1–5% glycerol. The presence of up to 2 mM MgCl$_2$ is also not significantly inhibitory, presumably because the affinity of Fe^{2+} for chelation is many orders of magnitude higher than that of Mg^{2+}.

A revised description of the mapping reaction conditions is as follows.

1. Prepare 2 pmol of labeled nucleosome in a volume of 10 μl of 50 mM TrisCl (pH 7.5) on ice.

2. Add 1 μl of 20 μM ammonium ferrous sulfate and allow it to equilibrate for 15 min on ice.

3. Add 5 μl of 22 mM ascorbic acid in 100 mM Tris-HCl (pH 7.5) and then 5 μl of 0.67% hydrogen peroxide. React for 45 min on ice.

4. Stop the reaction by addition of 2 μl of stop solution (50 mM EDTA, 1 M thiophenol).

5. Phenol extract and ethanol precipitate the DNA.

We have also observed that polypropylene tubes from some manufacturers inhibit the mapping reaction although they do not alter the behavior of nucleosomes as assayed by native gels. We have had reliable results using 1.5-ml tubes from Eppendorf (Hamburg, Germany) and Sarstedt (Nümbrecht, Germany), and 0.2-ml tubes from Axygen Scientific (Union City, CA).

Attachment of Fluorescent Dyes to Recombinant Histones

Background

Although the previous section described the attachment of an artificial nuclease to histones, recombinant histone technology can be used for the attachment of a wide variety of compounds to specific sites on nucleosomes. Other compounds that have been attached to nucleosomes include cross-linking reagents[43] and photoactivatable cross-linking compounds.[44]

Recombinant histone technology can also be used to provide a means of attaching fluorescent dyes. Techniques for the detection of fluorescent dyes have increased greatly, enabling detection down to the level of single molecules. To date, the major use of fluorescent technology to study chromatin structure has stemmed from the expression of green or yellow fluorescent protein (GFP or YFP, respectively) histone fusions *in vivo*. These have provided a powerful system for the study of the dynamic properties

[43] K. M. Lee, D. R. Chafin, and J. J. Hayes, *Methods Enzymol.* **304**, 231 (1999).
[44] S. M. Sengupta, J. Persinger, B. Bartholomew, and C. L. Peterson, *Methods* **19**, 434 (1999).

of chromatin *in vivo*.[45] However, the GFP and YFP proteins themselves have a molecular mass of ~30 kDa, which may affect the properties of the histone proteins, which range from 8 to 12 kDa. The attachment of small synthetic dyes to specific sites on histone octamers has the potential to enable the use of powerful fluorescence techniques without so greatly altering the mass of the histone proteins. One such technique is the application of the phenomenon of fluorescence resonance energy transfer (FRET), which since its discovery at the beginning of the century has found widespread application in structural determinations.[46–54] This is a consequence of the discovery by Theodor Förster in 1948 that long-range dipole–dipole interactions between two chromophores were dependent on the inverse sixth power of the interchromophore distance.[55] These interactions occur in the 10- to 100-Å range, representing a sensitive technique for inter- and intramolecular distance determinations. The use of fluorescence to study chromatin dynamics *in vivo* has grown, but applying the phenomenon of FRET to chromatin *in vitro*[56,57] has not yet taken full advantage of the X-ray crystallographic structure of the nucleosome core particle and the ability to produce recombinant histones.[1,26] These developments have allowed the design and implementation of a novel FRET-based assay to analyze nucleosome repositioning. Thus, from the basic technology of site-specific protein labeling using fluorescent dyes, it has been possible to devise a strategy that has the potential to provide greater detail than existing chromatin structural assays in solution. In addition to this, specific fluorescence labeling of histones has enabled the development of a simple gel fluorescence assay for H2A–H2B dimer loss or exchange, a phenomenon that is becoming increasingly important.[58]

[45] H. Kimura and P. R. Cook, *J. Cell Biol.* **153,** 1341 (2001).

[46] R. M. Clegg, *Methods Enzymol.* **211,** 353 (1992).

[47] B. Wieb Van Der Meer, G. Coker, III, and S.-Y. Simon Chen, "Resonance Energy Transfer: Theory and Data." John Wiley & Sons, New York, 1994.

[48] R. M. Clegg, *Curr. Opin. Biotechnol.* **6,** 103 (1995).

[49] M. Yang and D. P. Millar, *Methods Enzymol.* **278,** 417 (1997).

[50] J. R. Lakowicz, Ed., "Principles of Fluorescence Spectroscopy." Kluwer Academic/Plenum, New York, 1999.

[51] P. R. Selvin, *Nat. Struct. Biol.* **7,** 730 (2000).

[52] D. M. Lilley and T. J. Wilson, *Curr. Opin. Chem. Biol.* **4,** 507 (2000).

[53] A. Hillisch, M. Lorenz, and S. Diekmann, *Curr. Opin. Struct. Biol.* **11,** 201 (2001).

[54] T. Heyduk, *Curr. Opin. Biotechnol.* **13,** 292 (2002).

[55] E. V. Mielczarek, E. Greenbaum, and R. S. Knox, "Biological Physics." pp. 148–160, American Institute of Physics, New York, 1993.

[56] L. A. Boyer, X. Shao, R. H. Ebright, and C. L. Peterson, *J. Biol. Chem.* **275,** 11545 (2000).

[57] K. Toth, V. Sauermann, and J. Langowski, *Biochemistry* **37,** 8173 (1998).

[58] M. L. Kireeva, W. Walter, V. Tchernajenko, V. Bondarenko, M. Kashlev, and V. M. Studitsky, *Mol. Cell* **9,** 541 (2002).

Materials and Instrumentation

Fluorescence Labeling of Recombinant Histone Octamer

Oregon Green 488 maleimide (Molecular Probes, Eugene, OR)
Tris(2-carboxyethyl)-phosphine hydrochloride (TCEP·HCl) (Pierce Biotechnology, Rockford, IL)
Dithiothreitol (DTT) (Melford Laboratories, Ipswich, UK)
NuPAGE precast 12% Bis-Tris gels (Invitrogen)
20× NuPAGE MES SDS running buffer (Invitrogen)
Centricon YM-30 concentrators (Millipore, Bedford, MA)
Gel fluorescence: Observed using an FLA-2000 fluorescence image analyzer or LAS-1000 CCD camera (Fujifilm) with suitable filters

Oligonucleotides

54-bp upstream nucleosome A extension:
   5′-TATGTAAATGCTTATGTAAACCA-3′
   5′-Cy3-TATGTAAATGCTTATGTAAACCA-3′
18-bp downstream nucleosome A extension:
   5′-TACATCTAGAAAAAGGAGC-3′
   5′-Cy5-TACATCTAGAAAAAGGAGC-3′
Primers generating nucleosome A 147-bp core positioning sequence:
   5′-ATCTGCAACAGTCCTAAC-3′ (forward)
   5′-Cy5-ATCAAAACTGTGCCGCAG-3′ (reverse)
Modified (i.e., Cy5 or Cy3 labeled) and unmodified oligonucleotides: Purchased from MWG Biotech (Ebersberg, Germany) or Thermo Electron Molecular Biology

Dye-Labeled and Unlabeled DNA Fragments

PCR: Carried out with buffers and *Taq* polymerase available from Bioline
Source 15Q ion-exchange resin (Amersham Biosciences, Piscataway, NJ)

Nucleosome Reconstitution

Salt gradient dialysis buffers:
   0.85 M NaCl, 10 mM Tris (pH 7.5), 0.1 mM EDTA
   0.65 M NaCl, 10 mM Tris (pH 7.5), 0.1 mM EDTA
   0.50 M NaCl, 10 mM Tris (pH 7.5), 0.1 mM EDTA
 10 mM Tris (pH 7.5), 0.1 mM EDTA

Spectra/Por 1 dialysis membrane, MWCO 6000–8000 (Spectrum Laboratories, Rancho Dominguez, CA)

Gel Mobility Shift Assay

50 mM NaCl, 10 mM Tris (pH 7.5), 0.1 mM EDTA
Eppendorf Mastercycler gradient PCR machine (Eppendorf)
0.2-ml thin-wall PCR tubes (ABgene)
0.2× TBE: 18 mM Tris-borate, 0.4 mM EDTA (pH 8.0)
40% acrylamide–bisacrylamide (49:1; Sigma, St. Louis, MO)
Ammonium persulfate (Sigma)
TEMED (Sigma)

FRET Measurements

Steady state fluorescence measurements were made with an F-4500 fluorescence spectrophotometer (Hitachi) with quartz cuvette (Hellma, Müllheim, Germany)

Dimer Exchange

λ DNA/HindIII markers (0.5 μg/μl; available from Promega, Madison, WI)
Novex gel cassettes (10 × 10 cm, 1 mm thick; available from In-vitrogen)

Methods

FRET Strategy

Fluorescence resonance energy transfer occurs when the energy dipoles of two chromophores overlap, resulting in a nonradiative transfer of energy from one dye (donor) to the other (acceptor). This process will take place only under certain conditions.

1. The emission spectrum of the donor must have some overlap with the excitation spectrum of the acceptor. This degree of overlap, expressed as an integral, determines the distance between the two dyes needed for 50% energy transfer efficiency, denoted R_0.
2. Both dyes must have sufficient freedom of rotation because FRET depends on favorable mutual orientation between donor and acceptor.
3. The dyes must be in close proximity (distance range, 10–100 Å).

Using the X-ray crystallographic structure of the nucleosome core particle,[21] one amino acid residue was selected for replacement by a cysteine, which would then be a donor dye labeling site. The molecular modeling program, RASMOL, was used to project the distance between this site and an acceptor labeling site on the 5' end of the DNA when the nucleosome repositions itself upstream as a consequence of thermal shifting. Using this distance and knowing the R_0 between the Oregon Green (donor) and Cy3 (acceptor) dye pair chosen for this study, the energy transfer efficiency could be estimated.

Fluorescence Labeling of Recombinant Histone Octamer

Recombinant histones from *Xenopus laevis* were expressed, purified, and assembled into histone octamer according to Luger *et al.*[26] The resulting histone octamer contained the mutations S113C on H2A and C110A on H3, thus generating only two cysteine residues available for fluorescent dye labeling by a sulfhydril-reactive maleimide derivative.

1. Oregon Green 488 maleimide is dissolved in dimethylformamide (DMF), dispensed into 200-nmol aliquots, lyophilized, and stored at $-20°$ with dessicant.

2. Free dye is electrophoresed on a NuPAGE precast 12% Bis-Tris gel to establish the linear range from which accurate quantitation of dye fluorescence can be taken. Oregon Green fluorescence is found to be linear from 0.8 to 800 pmol, using a fluorescence image analyzer and AIDA software.

3. Recombinant histone octamer [60 μl of 30 μM in 2 M NaCl, 10 mM Tris (pH 7.5), 0.1 mM EDTA] is mixed with a 5-fold molar excess of TCEP (this reducing agent does not interfere with the labeling reaction) relative to the number of cysteines present in the octamer in 120 μl of 2 M NaCl, 20 mM HEPES (pH 7.2), 0.1 mM EDTA at room temperature and left for 20 min.

4. To one 200-nmol aliquot of Oregon Green 488 maleimide is added 20 μl of DMF. From this 10 mM stock, a 10-fold molar excess of dye to reduced cysteines is then thoroughly mixed with the above-described octamer solution; the remainder of the dye is stored at $-20°$. All reactions containing dyes are protected from light.

5. Labeling conditions should be optimized over a time course in the presence of Oregon Green maleimide in order to ascertain the time required for completion of cysteine conjugation while avoiding nonspecific labeling of other histones. Aliquots can be removed from the reaction at various times and the labeling reaction stopped by addition of a 100-fold molar excess of DTT.

6. One microliter (approximately 20–30 pmol of labeled histone octamer) of the labeling reaction from each time point is run on a

NuPAGE precast 12% Bis-Tris gel to assess the labeling by fluorescence detection, using a fluorescence image analyzer and AIDA software. Figure 1 shows an example of such a time course. We have found that typically labeling is complete after 20 min, but that prolonged incubations result in labeling of cysteine-free histones.

7. Once appropriate labeling times have been established, a preparative reaction is performed at the same scale. After the appropriate time, 7 μl of 0.5 M DTT is added (approximately 100× molar excess of DTT to dye) to stop the reaction. The reaction mixture is placed on ice.

8. The labeled octamer–free dye mixture is diluted in 1.5 ml of cold 2 M NaCl, 10 mM Tris (pH 7.5), 0.1 mM EDTA and subsequently dispensed into a precooled YM-30 Centricon concentrator, which is then spun according to the manufacturer's instructions at 4°. After concentration, the sample is diluted and reconcentrated twice, resulting in an approximately 3500-fold dilution of free dye. Seventy to 80% of the octamer can be recovered in a volume of 50 μl. The concentration of labeled histone octamer is determined by absorbance measurement. Specific labeling and removal of unincorporated dye can be assessed by gel electrophoresis as in step 6 above.

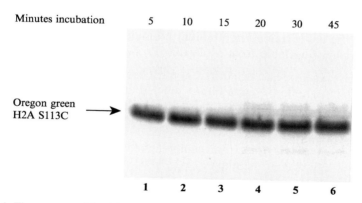

Fig. 1. Time course of the labeling reaction between Oregon Green 488 maleimide and recombinant *Xenopus* histone octamer containing the mutations H3 C110A and H2A S113C. Labeling was stopped by the addition of a 100× molar excess of DTT to dye at 5, 10, 15, 20, 30, and 45 min (lanes 1–6 respectively). Approximately 20 pmol of labeled octamer from each time point was electrophoresed on a NuPAGE pre cast 12% Bis-Tris gel (200 V for 45 min in 1× MES SDS running buffer). The gel was scanned with a fluorescence image analyzer with 480-nm excitation and 520-nm emission filters. Care should be taken to ensure that dyes have not become attached via amine groups to all the histone proteins.

Nucleosome Reconstitution

The nucleosome A core positioning sequence (147 bp) with 54-bp upstream and 18-bp downstream extensions (denoted 54A18), derived from the MMTV LTR promoter (mouse mammary tumor virus long terminal repeat) was generated by the polymerase chain reaction. During this process, oligonucleotides that were either unlabeled or 5'-Cy3 or 5'-Cy5 labeled could be used to incorporate a dye label to the 5', 3', or both ends of this DNA fragment.

1. PCR are precipitated with ethanol and NaCl (sodium acetate is not used as cyanine dyes can be unstable in acidic solutions) and purified as described above. Fragments generated are 54A18-Cy5 and Cy3-54A18-Cy5.

2. Nucleosome reconstitution is carried out by salt gradient dialysis as previously described,[26,59] using the above-labeled DNA fragments and Oregon Green-labeled histone octamer. Nucleosomes are stored in 10 mM Tris (pH 7.5), 0.1 mM EDTA at 4° or in an icebox in a cold room.

3. Nucleosome integrity is assessed by subjecting approximately 4 pmol to native polyacrylamide gel electrophoresis. Briefly, 1 μl of 4 μM nucleosome is diluted with 8 μl of TE(10/0.1) before the addition of 1 μl of 50% (w/v) sucrose. This sample is then loaded onto a 10 × 10 cm (1 mm thick) 5% (w/v) polyacrylamide gel and is run in 1× TBE buffer at 100 V (constant) for 1 h at room temperature. Using a fluorescence image analyzer, the gel is scanned for Cy5 fluorescence, revealing the proportion of free and nucleosomal DNA. When the octamer but not the DNA is labeled, the gel is stained in ethidium bromide or SYBR Green I nucleic acid gel stain (available from Molecular Probes).

At this stage, we had two different nucleosomes: one served as a donor-only-labeled sample (Oregon Green labeled at H2A, 5'-Cy5 at the downstream end) and the other as the donor with acceptor sample (Oregon Green labeled at H2A, 5'-Cy3 at the upstream end and 5'-Cy5 at the downstream end).

Thermally Induced Nucleosome Sliding

Nucleosome repositioning can be induced by thermal incubation and the resulting structural alterations can be observed as a change in electrophoretic mobility on a native polyacrylamide gel.[8] Although Cy5 did not interfere with the energy transfer process between Oregon Green and Cy3 (data not shown), it allowed by selection of the appropriate filters accurate quantitation of thermal shifting on a native polyacrylamide gel

[59] A. Flaus and T. J. Richmond, *J. Mol. Biol.* **275,** 427 (1998).

without interfering fluorescence from the FRET dye pair, because their fluorescence changed as a result of shifting.

Donor-labeled as well as donor- and acceptor-labeled nucleosomes (4 μM) were diluted 10-fold in 50 mM NaCl, 10 mM Tris (pH 7.5), 0.1 mM EDTA in 0.2-ml thin-wall PCR tubes. Using a gradient PCR machine, nucleosomes were incubated at the following temperatures for 1 h and then cooled to 4°: 25, 35.7, 38.5, 46.6, 49.9, 52, 54, and 55.2°.

Gel Mobility Shift Assay

1. Ten microliters of each sample, that is, donor only and donor in the presence of acceptor, representing each temperature listed above, is mixed with 3 μl of 20% (w/v) sucrose and loaded onto a native polyacrylamide gel [5% (w/v) acrylamide–bisacrylamide, 0.2× TBE] that has been prerun and is run at 4° in 0.2× TBE [300 V (constant) for 3.5 h].

2. Gel fluorescence is observed with a fluorescence image analyzer (633-nm excitation and 675-nm emission filter settings) and AIDA software (Fig. 2). There was no difference in mobility or thermal shifting behavior when unlabeled nucleosome made with wild type octamer was compared to labeled nucleosomes (data not shown).

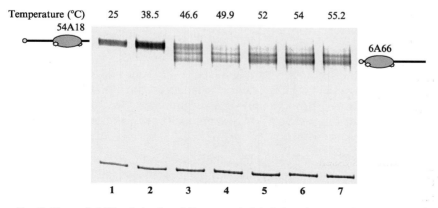

Fig. 2. Thermal shifting behavior of fluorescently labeled nucleosomes in a FRET-based nucleosome sliding assay. Samples containing approximately 4 pmol of Oregon Green 488-labeled (at H2A C113) nucleosome (Cy3 54A18 Cy5) were subjected to thermal shifting, using a gradient PCR machine, allowing a 1-h incubation at 25, 35.7, 38.5, 46.6, 49.9, 52, 54, and 55.2° (lanes 1–7, respectively). Selection of the appropriate excitation (633 nm) and (675 nm) emission filters allowed observation of Cy5 fluorescence without interference from the Oregon Green–Cy3 dye pair. As evident, most of the fluorescent nucleosomes shift 48 bp upstream (to 6A66) of the starting position (54A18). Donor only-labeled or unlabeled nucleosomes behaved similarly (data not shown).

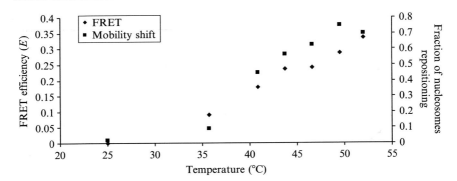

Fig. 3. FRET efficiency (E) correlates with thermally induced nucleosome sliding. Mobility shift (squares) was assessed by calculating the proportion of nucleosomes moving from the starting position (54A18), based on the Cy5 fluorescence intensity of the samples shown in Fig. 2. The donor quenching method was used to calculate the energy transfer efficiency (E) (diamonds), between Oregon Green 488 and Cy3 after incubation for 1 h at each of the temperatures used for samples in Fig. 2.

Fluorescence Measurements and FRET Calculations

FRET was measured by the donor quenching method[46] and energy transfer efficiency E, expressed as a percentage (Fig. 3). The proportion of nucleosomes shifting to within 6 bp from the DNA end or further upstream of the nucleosome was plotted on the same graph.

Dimer Exchange

Nucleosome assembly protein 1 (NAP-1) from *Saccharomyces cerevisiae* facilitates H2A–H2B dimer exchange and it has been possible to monitor this transfer process between recombinant Oregon Green-labeled 219-bp nucleosome and wild-type nucleosome core particle, using a gel-based fluorescence assay.

1. The PCR, octamer-labeling, and reconstitution methods described earlier are employed to generate two nucleosomes of different sizes. The first is composed of Oregon Green-labeled histone octamer (the same octamer used for the previous sliding assay with H2A S113C) and the aforementioned 54A18 DNA fragment, which has a 5'-Cy5 label at the end of the 18-bp downstream extension (denoted 54A18 Cy5). The second nucleosome is essentially wild-type *Xenopus* histone octamer reconstituted onto the 147-bp core nucleosome A positioning sequence with a 5'-Cy5 label at the downstream end of the DNA (denoted 0A0 Cy5).

2. A 0.75-μl volume of yeast NAP-1 [53 μM in 50 mM Tris (pH 7.5), 100 mM NaCl, 5 mM α-mercaptoethanol], 1 μl of the above-described

Oregon Green-labeled nucleosome [4 μM in 10 mM Tris (pH 7.5), 0.1 mM EDTA], 1 μl of wild-type core particle [4 μM in 10 mM Tris (pH 7.5), 0.1 mM EDTA], and 7.25 μl of reaction buffer [50 mM Tris (pH 7.5), 100 mM NaCl] are incubated in a 30° water bath for 2 h. Competitor DNA (250 ng of *Hin*dIII-digested λ DNA) is added and the 30° incubation is continued for a further 10 min before the reaction is placed on ice.

3. The reaction described above contains a 1:1 molar ratio of 54A18 and 0A0 nucleosome. Additional exchange reactions are prepared with molar ratios of 1:2 and 1:4 (54A18:0A0). Control reactions consist of Oregon Green-labeled nucleosome with and without NAP-1, and a 1:1 molar ratio of both nucleosomes in the absence of NAP-1.

4. After placing the reactions on ice, 1 μl of 50% (w/v) sucrose is added to each tube and samples are then subjected to native polyacrylamide gel electrophoresis [10 × 10 cm, 1-mm-thick Novex gel cassette

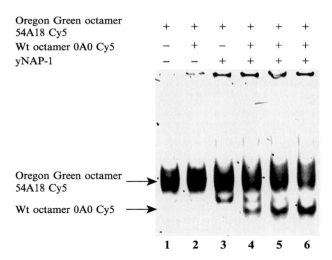

FIG. 4. Histone dimer transfer between nucleosomes can be assessed by gel fluorescence. Oregon Green 488-labeled (at H2A C113) nucleosome (54A18 Cy5, lane 1) was employed to assess the efficiency of dimer transfer to core particle (unlabeled histone octamer on 0A0 Cy5) in the presence of yNAP-1. Each sample contained equivalent amounts of Oregon Green-labeled nucleosome, which were incubated in the presence of yNAP-1 without core particle (lane 3) and with a molar equivalent (lane 4), a 2× molar excess (lane 5), and a 4× molar excess (lane 6) of core particle to Oregon Green-labeled nucleosome. Equimolar quantities of both nucleosomes in the absence of yNAP-1 show no Oregon Green fluorescence transfer (lane 2). The native acrylamide gel was scanned with a CCD camera with filters enabling passage of Oregon Green fluorescence without interference from Cy5 fluorescence, which in a separate scan correlated energy transfer with nucleosome position.

containing 5% (w/v) acrylamide–bisacrylamide (49:1) in 1× TBE, 100 V (constant) for 1 h at room temperature].

5. After electrophoresis, the gel is scanned first with a fluorescence image analyzer with 633-nm excitation and 675-nm emission filter settings, allowing the sole observation of Cy5 fluorescence. Subsequently, the transfer of Oregon Green fluorescence is observed with an LAS-1000 CCD camera (1-min exposure) with a blue light-emitting diode as the excitation source and a 515-nm emission filter (Fig. 4).

[15]   Histone MacroH2A Purification and Nucleosome Reconstitution

By LAKSHMI N. CHANGOLKAR and JOHN R. PEHRSON

Introduction

MacroH2As are unusual core histone variants that contain an N-terminal H2A region and a large C-terminal nonhistone domain.[1] The amino acid sequence of the H2A region is about 65% identical to a full-length conventional H2A and macroH2As can substitute for conventional H2As in the nucleosome.[1,2] Most of the nonhistone region appears to be derived from a gene of unknown function that has an evolutionary history that predates the emergence of eukaryotes.[3] MacroH2As are highly conserved among vertebrate organisms ranging from humans to fish, but are absent from all invertebrate genomes that have been sequenced to date.

There are three known macroH2A variants. MacroH2A1.1 and macro-H2A1.2 are formed by alternate splicing and are identical in sequence except for a segment of about 30 amino acids in the nonhistone domain.[4,5] MacroH2A2 is virtually identical in size and basic architecture to the macro-H2A1 variants, but is only about 70% identical in amino acid sequence and is encoded by a separate gene.[6,7] The pattern of macroH2A variants is different in different cell types and changes during development.[4,7]

[1] J. R. Pehrson and V. A. Fried, *Science* **257**, 1398 (1992).
[2] L. N. Changolkar and J. R. Pehrson, *Biochemistry* **41**, 179 (2002).
[3] J. R. Pehrson and R. N. Fuji, *Nucleic Acids Res.* **26**, 2837 (1998).
[4] J. R. Pehrson, C. Costanzi, and C. Dharia, *J. Cell. Biochem.* **65**, 107 (1997).
[5] T. P. Rasmussen, T. Huang, M. A. Mastrangelo, J. Loring, B. Panning, and R. Jaenisch, *Nucleic Acids Res.* **27**, 3685 (1999).
[6] B. P. Chadwick and H. F. Willard, *Hum. Mol. Genet.* **10**, 1101 (2001).

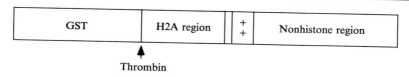

Fig. 1. Diagram of glutathione *S*-transferase (GST)–macroH2A1.2 fusion protein. The H2A region consists of residues 1–122 of rat macroH2A1.2, the basic region (++) includes residues 133–159, and the nonhistone region spans residues 160–370. The arrow marks the site of thrombin cleavage. Reprinted with permission from L. N. Changolkar and J. R. Pehrson, *Biochemistry* **41**, 179 (2002). Copyright (2002) American Chemical Society.

The high evolutionary conservation of macroH2A variants among vertebrates indicates that they have functions that are distinct from conventional H2As. In mammals, macroH2As are preferentially concentrated in the inactive X chromosome, indicating they may have a role in producing transcriptionally repressed chromatin structures.[8] One approach for examining the effects of macroH2As on chromatin structure and function is to reconstitute nucleosomes with purified macroH2A proteins. In this chapter, we describe a method for expressing macroH2A1.2 in bacteria, purifying the recombinant protein, and using this protein to reconstitute nucleosomes.[2]

Expression of MacroH2A1.2 in Bacteria

Rat macroH2A1.2 is expressed as a glutathione *S*-transferase fusion protein (Fig. 1) in *Escherichia coli* BL21, using the expression vector pGEX-4T-1 (Amersham Biosciences, Piscataway, NJ). Cleavage of this fusion protein with thrombin yields a protein that is identical to rat macro-H2A1.2 except for two additional residues on the N terminus, a glycine and a serine. Expression should be done at room temperature because the fusion protein is largely insoluble when expressed at 37°. The amount of soluble protein that is expressed and purified even at room temperature is relatively low. The volumes given are those we use for a typical preparation that should yield one to a few milligrams of purified protein.

Method

1. Grow bacteria containing the expression plasmid in 300 ml of Luria–Bertani medium containing ampicillin (50 μg/ml) for 2–3 days at room temperature to an A_{600} of ~1.5.

[7] C. Costanzi and J. R. Pehrson, *J. Biol. Chem.* **276**, 21776 (2001).
[8] C. Costanzi and J. R. Pehrson, *Nature* **393**, 599 (1998).

2. Expand the culture to 1800 ml of Luria–Bertani medium containing ampicillin (50 μg/ml) and grow for 1 h at room temperature.

3. Induce expression of fusion protein by adding isopropyl-β-D-thiogalactopyranoside to 0.1 mM and culture the cells for 3 h at room temperature.

Purification of Recombinant MacroH2A Protein

All steps are performed at 4° or on ice. Figure 2 shows the protein composition of samples removed at different stages of the purification.

Method

1. Collect the bacteria by centrifugation and resuspend in a 1/10th volume (180 ml) of 20 mM Tris (pH 8.0), 100 mM NaCl, 1 mM ethylenediaminetetraacetic acid (EDTA), 0.5% Igepal CA-630 [Sigma (St. Louis, MO), used as a substitute for Nonidet P-40], 5 mM dithiothreitol (DTT), and 1 mM phenylmethylsulfonyl fluoride (PMSF).

2. Sonicate the bacterial suspension until there is a noticeable decrease in turbidity. Avoid excessive sonication because it may reduce the yield of

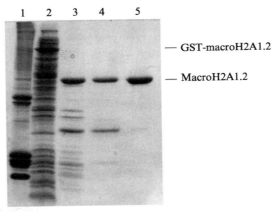

FIG. 2. Purification of macroH2A1.2. The protein composition of samples taken from various stages of purification was analyzed by electrophoresis in a 15% acrylamide–SDS gel. Lane 1, an extract of rat liver nuclei used to mark the location of macroH2A1.2; lane 2, an extract of *E. coli* expressing the GST–macroH2A1.2 fusion protein; lane 3, proteins released from glutathione–Sepharose affinity beads by cleavage with thrombin; lane 4, pooled fractions eluted from a Macro Prep High S column; lane 5, peak tube from a hydroxyapatite column. Adapted with permission from L. N. Changolkar and J. R. Pehrson, *Biochemistry* **41**, 179 (2002). Copyright (2002) American Chemical Society.

soluble macroH2A protein. Remove insoluble material by centrifugation at 10,000g (r_{max}) for 15 min.

3. Mix the supernatant with a 1/10th volume (18 ml) of glutathione–Sepharose 4B beads (Amersham Biosciences) and rock overnight. Collect the beads by centrifugation; wash four times by centrifugation with 80 ml of 0.15 M NaCl and 10 mM Tris (pH 7.5) (TBS), once with 80 ml of 0.35 M NaCl and 10 mM Tris (pH 7.5), followed by four times with 80 ml of TBS. Resuspend the beads in 18 ml of TBS.

4. Add 45 NIH units of bovine thrombin (T 7513; Sigma) to the bead suspension and rock for 4.5 h. If the thrombin cleavage is inadequate, little macroH2A1.2 will be released from the glutathione–Sepharose. Thrombin cleavage can be monitored by sodium dodecyl sulfate (SDS) gel electrophoresis of the proteins released from the beads and of a small sample of the beads.

5. After thrombin treatment, adjust the volume to 60 ml with TBS and add 12.5 ml of 5 M NaCl; this brings the NaCl concentration to 1 M, which improves the extraction of macroH2A1.2 from beads. Rock the bead suspension for 15 min. Collect the supernatant, diluted 2-fold with 10 mM Tris (pH 7.5); add PMSF to 0.2 mM and DTT to 1 mM (the beads can be reused after recycling as described by the manufacturer).

6. Apply the solution to a column containing 5 ml of Macro Prep High S (Bio-Rad, Hercules, CA) that is equilibrated with 10 mM Tris (pH 7.5), 0.5 M NaCl, 1 mM DTT, and 0.2 mM PMSF. Wash the column with 300 ml of 10 mM Tris (pH 7.5), 0.6 M NaCl, 0.2 mM PMSF, and 1 mM DTT. Elute the column with a 300-ml gradient from 0.6 to 2.0 M NaCl, with both solutions containing 10 mM Tris (pH 7.5), 0.2 mM PMSF, and 1 mM DTT. Measure the absorbance of the fractions at 280 nm and analyze the fractions around the major peak (peak absorbance, approximately 0.1) by electrophoresis in a 15% acrylamide–SDS gel.[9] Note that the UV absorbance of macroH2As is low because of a low content of tyrosine and tryptophan residues.

7. Pool the fractions that contain appreciable amounts of macro-H2A1.2 and apply them to a column containing 5 ml of hydroxyapatite that is equilibrated with 1 M NaCl, 1 mM NaPO$_4$ (pH 7.0), 0.2 mM PMSF, and 1 mM DTT. Wash the column with 300 ml of the equilibration buffer and then elute with a 160-ml gradient from 1 mM NaPO$_4$ to 100 mM NaPO$_4$; both solutions contain 1 M NaCl, 0.2 mM PMSF, and 1 mM DTT and the phosphate is at pH 7.0. Measure the absorbance of the fractions at 280 nm and analyze the fractions around the main peak by electrophoresis in a 15% acrylamide–SDS gel.

[9] U. K. Laemmli, *Nature* **227,** 680 (1970).

The purified macroH2A1.2 should be stable when stored for a few weeks at $4°$. Some preparations show a slow accumulation of degradation products. This proteolytic activity is inhibited by the addition of the protease inhibitor cocktail Complete (Roche Applied Science, Indianapolis, IN). To minimize potential contamination with proteases, we autoclave all our solutions, tubes, and beakers. Any relevant equipment that cannot be autoclaved should be soaked in a solution of SDS and then thoroughly rinsed with deionized water.

Preparation of Core Histones

We prepare H2A–H2B dimers and $(H3–H4)_2$ tetramers by salt elution of chicken erythrocyte chromatin bound to hydroxyapatite, following the procedure of Simon and Felsenfeld.[10] Core histones prepared from rat liver chromatin are unsatisfactory because of the presence of protease(s) that degraded macroH2A. Rat liver chromatin also contains a protease that degrades conventional H2A.[11] We have not tried using recombinant core histones. Amounts of material used or purified from one of our typical preparations are included as a guideline. All steps are performed at $4°$ or on ice, unless indicated otherwise.

Method

1. Collect erythrocytes from chicken blood (Pel-Freez Biologicals, Rogers, AR) by low-speed centrifugation, wash once with phosphate-buffered saline (PBS) containing 1 mM EDTA, pellet at low speed, aliquot into tubes, freeze in a dry ice ethanol bath, and store at $-70°$.

2. Thaw and resuspend the erythrocytes (approximately 13 ml) in 30 ml of buffer A [0.4 M mannitol, 60 mM KCl, 15 mM NaCl, 0.15 mM spermine, 0.5 mM spermidine, 2 mM EDTA, 0.5 mM EGTA, 15 mM triethanolamine, 1% thiodiglycol, 0.2 mM PMSF (pH 7.4)] that contains 0.2% Triton X-100.

3. Treat the suspension with 15 strokes in a Dounce homogenizer (Wheaton, Millville, NJ), pellet the nuclei at 1500g (r_{max}), resuspend the suspension in buffer A with 0.2% Triton, repeat the treatment with the Dounce homogenizer, and pellet again.

4. Resuspend the pellet in 18 ml of buffer A, add 66.5 ml of buffer A that contains 2 M sucrose (in place of the mannitol), and mix. Pour into ultracentrifuge tubes and underlay each sample with 5 ml of buffer A containing 2 M sucrose. Centrifuge at 115,000g (r_{max}) for 45 min.

---

[10] R. H. Simon and G. Felsenfeld, *Nucleic Acids Res.* **6**, 689 (1979).
[11] T. H. Eickbush, D. K. Watson, and E. N. Moudrianakis, *Cell* **9**, 785 (1976).

5. Resuspend the pellet in buffer A and estimate the DNA concentration by dissolving 20 μl of the suspension in 1 ml of 2% SDS followed by vigorous vortexing to disperse the DNA; read the absorbance at 260 nm (1 A_{260} equals $\sim$50 $\mu g/ml$). The approximate yield for this size preparation is 50 mg.

6. Pellet the nuclei and resuspend them in 35 ml of 1 mM EDTA, pH 8.0. Leave on ice for 30 min and then sonicate until most of the chromatin has dispersed and dissolved. Spin at 10,000g (r_{max}) for 15 min and collect the supernatant.

7. Add 5 M NaCl to a final concentration of 0.63 M and 0.5 M KPO$_4$ (pH 6.7) to a final concentration of 0.1 M. Add these salts with stirring. Chromatin will first precipitate and then redissolve. After addition, spin the sample at 10,000g (r_{max}) for 10 min to remove insoluble material.

8. Apply the supernatant to a 2.5 × 18 cm hydroxyapatite column that has been previously equilibrated with 0.63 M NaCl and 0.1 M KPO$_4$ (pH 6.7). Wash the column with 300 ml of the equilibration buffer and monitor the protein content of the fractions by measuring the absorbance at 230 nm; the A_{230} should level off before proceeding to the next step.

9. Elute the H2A–H2B dimers with 0.93 M NaCl, 0.1 M KPO$_4$ (pH 6.7). Monitor the elution by measuring the absorbance of fractions at 230 nm. Continue the elution until the absorbance falls and levels off.

10. Wash the column with a 300-ml gradient from 0.93 M NaCl to 1.2 M NaCl containing 0.1 M KPO$_4$ (pH 6.7). This removes residual H2A–H2B dimers before elution of (H3–H4)$_2$ tetramers.

11. Elute (H3–H4)$_2$ tetramers with 2 M NaCl, 0.1 M KPO$_4$ (pH 6.7).

The purity and concentration of the histone fractions should be monitored by SDS gel electrophoresis. For reconstituting with macroH2A, it is necessary to separate H2B from H2A. This is done by gel filtration.[12]

Method

1. Pool the H2A–H2B-containing hydroxyapatite fractions and precipitate the proteins by adding trichloroacetic acid to 20% and rocking the sample for 1 h at 4°. Collect the precipitate by centrifugation, wash the pellet with ethanol containing 1% sulfuric acid and then with ethanol, and briefly air dry.

2. Dissolve the pellet in 4 ml of 7 M urea, 20 mM HCl and apply to a 100 × 2.5 cm column of Bio-Gel P-100 (Bio-Rad) equilibrated with 300 mM NaCl and 20 mM HCl. Run this column at room temperature.

[12] C. von Holt and W. F. Brandt, *Methods Cell Biol.* **16,** 205 (1977).

3. Monitor column fractions by absorbance at 230 nm and analyze fractions around the two major absorbance peaks by SDS gel electrophoresis in a 15% acrylamide gel. The resolution of H2B from H2A may be incomplete and, therefore, fractions used for reconstitution must be carefully selected on the basis of the gel results.

Isolation of Core Particle DNA

We use DNA isolated from nucleosome core particles for reconstituting nucleosome core particles.[13] The advantage of this approach is that it avoids micrococcal nuclease digestion of macroH2A-containing nucleosomes. The preparations of micrococcal nuclease we have used contain protease(s) that degrade macroH2A but have little effect on the other core histones. Most of this protease activity is inhibited by Complete or α_2-macroglobulin (Roche Applied Science), but traces of residual activity make it difficult to purify mononucleosomes without noticeable degradation of macroH2A1.2.

Method

1. Homogenize rat livers (Pel-Freez Biologicals) in buffer A, using a Waring blender. Filter the homogenate through cheesecloth and spin down the crude nuclear pellet by centrifugation at $1400g$ (r_{max}) for 6 min. Isolate the nuclei by ultracentrifugation through a cushion of $2\ M$ sucrose as described above for chicken erythrocyte nuclei. Resuspend the nuclei in buffer A and estimate the DNA concentration as described above.

2. Resuspend the nuclei with 25 mM KCl, 1 mM CaCl$_2$, 1 mM MgCl$_2$, and 50 mM triethanolamine (pH 7.5), pellet at $1400g$ (r_{max}) for 6 min, resuspend them in the same buffer at a concentration of 2 mg/ml, and digest with micrococcal nuclease (20 units/ml) at 37° for 15 min.

3. Pellet the nuclei at $10,000g$ (r_{max}) for 10 min, resuspend the pellet in 1 mM EDTA (pH 8.0) at a concentration of approximately 1 mg/ml, and incubate on ice for 1 h. Centrifuge at $10,000g$ for 10 min and collect the supernatant.

4. Add Tris (pH 7.5) to 10 mM, NaCl to 50 mM, and a 1/10th volume of CM Sepharose beads (Amersham Biosciences) to remove histone H1.[14] Rock this mixture at 4° for 1 h, and then remove the beads by centrifugation and collect the supernatant. In our experience, this treatment removes most, but not all, of the H1.

[13] K. Tatchell and K. E. van Holde, *Biochemistry* **16,** 5295 (1977).
[14] L. J. Libertini and E. W. Small, *Nucleic Acids Res.* **8,** 3517 (1980).

5. Add $CaCl_2$ to a final concentration of 2 mM and digest with micrococcal nuclease. Test digestions with different concentrations of nuclease should be used to optimize the yield of nucleosome core particles while minimizing the amount of compact dinucleosomes and subnucleosomal DNA fragments. In a typical preparation we use 17 units/ml for 30 min at 37°. Stop the digestion by adding EDTA to 5 mM. Some insoluble material will be present after the digestion, possibly the result of incomplete removal of H1. Remove this material by centrifugation and discard.

6. Add SDS to 1% and NaCl to 1 M, incubate at 55° for 20 min, and extract twice with phenol–chloroform–isoamyl alcohol (25:24:1). Precipitate the DNA with 2 volumes of ethanol, wash once with 70% ethanol, briefly air dry, dissolve in TE [10 mM Tris(pH 7.5), 1 mM EDTA], and then dialyze against TE. The DNA concentration can be estimated by A_{260} and by electrophoresis in a 2% agarose gel containing ethidium bromide.

Reconstitution of MacroH2A-Containing Nucleosomes

We reconstitute macroH2A-containing nucleosomes by a salt dialysis procedure similar to that developed by Tatchell and van Holde.[13] DNA and histones are mixed to a final DNA concentration of ~150 μg/ml and a final histone concentration of ~80 μg/ml. The core histones are mixed in equal proportions as estimated by Coomassie Brilliant Blue R-250 staining of bands separated by electrophoresis in 15% acrylamide–SDS gels. The macroH2A1.2 band is adjusted to be approximately 2.5 times darker than H2B in order to compensate for the difference in molecular weight. Histone concentrations are estimated by comparison with Coomassie blue-stained histone bands in an SDS gel of a rat liver chromatin standard of known DNA concentration; it is assumed that the total concentration of histone in this standard is equal to the concentration of DNA. Before mixing with the other histones, the H2B fraction is neutralized with Tris base. After the mixture of histones and DNA is prepared, PMSF is added to 0.2 mM and DTT to 1 mM. All dialysis solutions contain 10 mM morpholinepropanesulfonic acid (MOPS, pH 7.0), 0.5 mM EDTA, and 10 mM 2-mercaptoethanol. Samples are initially dialyzed against 2 M NaCl for 6 h and then overnight against 1.5 M NaCl. This is followed by dialysis for 4 h against 1 M NaCl, 4 h against 0.75 M NaCl, 3 h against 0.5 M NaCl, and overnight against 0.1 M NaCl.

Confirmation of Reconstitution

Nuclease digestion and sedimentation in sucrose gradients can be used to confirm the formation of nucleosomes. When long DNA fragments are used for reconstitution, the reconstituted material can be digested with micrococcal nuclease to show that nucleosome size DNA fragments are protected from digestion. Complete and α_2-macroglobulin should be included in the digestion to inhibit the degradation of macroH2A1.2. Nucleosomes reconstituted with conventional H2A or from native sources can serve as a positive control. The protected fragments for nucleosomes reconstituted with macroH2A1.2 and conventional H2A are indistinguishable when resolved in a 2% agarose gel.[2] Reconstitutions that have no H2A, but include the other core histones and DNA, can serve as a negative control.

To purify reconstituted nucleosome core particles, we use sedimentation in 5–20% sucrose gradients that contain 0.1 M NaCl, 10 mM MOPS (pH 7.0), 1 mM EDTA, 3 mM DTT, and 0.2 mM PMSF. Gradients are prepared in 36-ml tubes for the AH-629 rotor (Du Pont Sorvalli Kendro Laboratory Products, Asheville, NC). Approximately 3–4 ml of reconstituted core particles is concentrated to ~1.5 ml by centrifugal ultrafiltration with Centriplus YM-100 filters (Millipore, Bedford, MA) and then layered onto a sucrose gradient. The gradients are spun for 21 h at 25,000 rpm (115,000g, r_{max}). Fractions (1.5 ml) are collected by puncturing the bottom of the tube. The fractions are analyzed for mononucleosomal DNA by electrophoresis in 2% agarose gels stained with ethidium bromide or by measuring the A_{260}. Before electrophoresis, the DNA is dissociated from the histones by adding SDS to 0.2% and incubating for 20 min at 37°.[15] The protein composition of the fractions is analyzed by electrophoresis in 15% acrylamide–SDS gels. Proteins are concentrated by precipitation with 20% trichloroacetic acid before electrophoresis. H1-free rat liver mononucleosomes from an S1 chromatin fraction[16] are used as a standard in the sedimentation analysis. We also include a positive control of nucleosomes reconstituted with conventional H2A and a negative control reconstituted with no H2A. Figure 3 shows a typical sedimentation profile. The sedimentation of reconstituted macroH2A1.2-containing nucleosome core particles is similar to that of conventional mononucleosomes, suggesting that the basic core structure of macroH2A1.2-containing nucleosomes is similar to that of conventional nucleosomes. The reconstitution mixtures contain some material that sediments more rapidly than mononucleosomes. This is seen with nucleosomes reconstituted with conventional H2A as well as macroH2A1.2, but more is present in the mixture reconstituted with macroH2A1.2. This material contains only mononucleosomal DNA and therefore appears to arise from associations between mononucleosomal particles.[13]

---

[15] J. C. Hansen, J. Ausio, V. H. Stanik, and K. E. van Holde, *Biochemistry* **28**, 9129 (1989).
[16] S. M. Rose and W. T. Garrard, *J. Biol. Chem.* **259**, 8534 (1984).

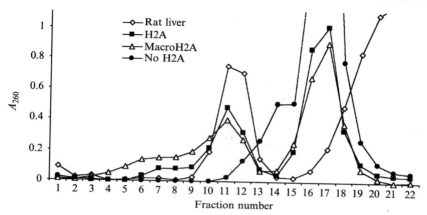

Fig. 3. Sedimentation analysis of nucleosomes reconstituted with nucleosome core DNA. Nucleosomes were reconstituted with nucleosome core DNA and a mixture of histones H2B, H3, and H4 and either conventional H2A or macroH2A1.2; a no-H2A mixture was included as a negative control. After reconstitution the mixtures were sedimented in sucrose gradients, fractions were collected, and the absorbance at 260 nm of each fraction is plotted. A gradient run with rat liver mononucleosomes served as a marker and the mononucleosome peak was at fraction 11. The shoulder at fraction 14 of the no-H2A sample appears to be due to the association of (H3–H4)$_2$ tetramers with the DNA. Adapted with permission from L. N. Changolkar and J. R. Pehrson, *Biochemistry* **41**, 179 (2002). Copyright (2002) American Chemical Society.

DNase I digestion[17,18] can also be used to examine sucrose gradient-purified reconstituted nucleosome core particles. The DNase I digestion pattern of nucleosomes reconstituted with macroH2A1.2 is similar but not identical to that of rat liver nucleosomes or nucleosomes reconstituted with H2A (Fig. 4).

Method

1. Collect sample from the mononucleosome peak. We use a 100-μl sample that contains ~2 μg of DNA and add protease inhibitors E-64 and Pefabloc SC.

2. End label the samples by adding MgCl$_2$ to 2 mM, 5 units of T4 polynucleotide kinase, and 20 μCi of [γ-^{32}P]ATP. Incubate at 37° for 30 min.

3. Digest with DNase I. The amount of nuclease and length of digestion should be determined empirically. We use pancreatic DNase I

[17] M. Noll, *Nucleic Acids Res.* **1**, 1573 (1974).
[18] L. C. Lutter, *J. Mol. Biol.* **124**, 391 (1978).

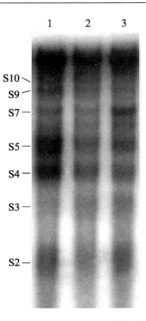

Fig. 4. DNase I digestion patterns of reconstituted nucleosomes. Reconstituted nucleosome core particles were labeled with ^{32}P on their 5' ends and then digested with DNase I. The DNA was denatured and run in an acrylamide gel containing 7 M urea, and the gel was dried and autoradiographed. Reconstituted nucleosome core particles were from fraction 11 of the sucrose gradients shown in Fig. 3. Lane 1, rat liver nucleosome core particles; lane 2, nucleosomes reconstituted with macroH2A1.2; lane 3, nucleosomes reconstituted with H2A. Sites of DNase I cutting are labeled S2, S3, S4, and so on, with the number indicating the number of helical turns of DNA from the labeled 5' end. Adapted with permission from L. N. Changolkar and J. R. Pehrson, *Biochemistry* **41,** 179 (2002). Copyright (2002) American Chemical Society.

(USB, Cleveland, OH) at 25 units/ml, digest at 37°, and remove 20-μl samples at times ranging from 15 s to 3 min.

4. Stop the reaction by adding EDTA to 25 mM and heat in a boiling water bath for 1 min. Digest the samples with proteinase K overnight at 55°. Add glycogen as a carrier and precipitate by adding 2 volumes of ethanol.

5. Dissolve the precipitate in 95% formamide containing 20 mM EDTA, heat in a boiling water bath, and run in a 12% acrylamide gel containing 7 M urea.[19]

[19] J. Sambrook, E. F. Fritsch, and T. Maniatis, "Molecular Cloning: A Laboratory Manual." Cold Spring Harbor Laboratory Press, Plainview, NY, 1989.

[16] Analysis of Histone Variant H2A.Z Localization and Expression during Early Development

By Patricia Ridgway, Danny Rangasamy, Leise Berven, Ulrica Svensson, *and* David John Tremethick

Introduction

In the eukaryotic nucleus, genomic DNA is complexed with histones and histone variants to form a dynamic structure referred to as chromatin. Chromatin also contains a wealth of nonhistone proteins that bind stably or transiently and are required for or regulate DNA-dependent processes such as transcription, DNA replication, and DNA repair.

A chromosome consists of a linear array of nucleosomes folded and compacted to different extents to form a continuum of various "higher order structures." The regulation of higher order chromatin formation is crucial for determining which subset of genes is expressed, in any given higher eukaryotic cell type, by organizing the genome into domains (and subdomains) that are open and transcriptionally active or compacted and repressed. How these architecturally distinct chromatin structures were established during early metazoan development, when the expression patterns of genes were being programmed, is not known.

Covalent modifications of DNA and histones appear to regulate higher order chromatin structures because they generate important landmarks, which distinguish active from inactive chromatin.[1] Potentially, another equally important way to establish specialized chromosomal domains during early development is to alter the biochemical composition of the nucleosome by replacing an individual histone with a histone variant. It has been clearly established that the histone variant H2A.Z performed a crucial role(s) during early metazoan development, but the essential function(s) of H2A.Z remains to be elucidated.[2–4] We demonstrated that H2A.Z can regulate the assembly of higher order chromatin structures *in vitro*.[5]

[1] E. Li, *Nat. Rev. Genet.* **3,** 662 (2002).

[2] A. van Daal, E. M. White, M. A. Gorovsky, and S. C. Elgin, *Nucleic Acids Res.* **16,** 7487 (1988).

[3] X. Liu, J. Bowen, and M. A. Gorovsky, *Mol. Cell. Biol.* **16,** 2878 (1996).

[4] R. Faast, V. Thonglairoam, T. C. Schulz, J. Beall, J. R. Wells, H. Taylor, K. Matthaei, P. D. Rathjen, D. J. Tremethick, and I. Lyons, *Curr. Biol.* **11,** 1183 (2001).

[5] J. Y. Fan, F. Gordon, K. Luger, J. C. Hansen, and D. J. Tremethick, *Nat. Struct. Biol.* **9,** 172 (2002).

To understand the essential role of H2A.Z, it is necessary to investigate the expression and the cellular and nuclear distribution of H2A.Z during early development. In this chapter we describe, using mouse embryos and *Xenopus laevis* as model systems for early development, several procedures to address these important issues.

Mouse Blastocyst Attachment, Outgrowth, and Differentiation *In Vitro*

H2A.Z null mouse embryos die shortly after hatching and implantation, suggesting a critical role for H2A.Z at this stage.[4] To understand the role of H2A.Z at this stage of development, the cellular and nuclear distributions of H2A.Z mRNA and protein were investigated, employing a highly sensitive reverse transcription-polymerase chain reaction (RT-PCR) method and immunofluorescence techniques, respectively.[6] Mouse blastocysts [3.5 days postcoitus (dpc)] were collected and cultured under conditions in which they hatched and attached onto coverslips. The trophoblast spreads out to form an adherent monolayer of cells while the undifferentiated cells of the inner cell mass (ICM) expand and outgrow to form a dense mound of cells on top of the trophoblast layer.[7] Cells located at the outer edge of the ICM mound differentiate into a layer of extraembryonic endoderm, mimicking an early differentiating step during mouse development.[8,9] In the mouse, these extraembryonic cells give rise to tissues including yolk sac and placenta. This *in vitro* system provides an excellent opportunity to study the localization and expression of proteins of interest in totipotent cells and to determine when these cells differentiate into extraembryonic cells. Differentiation *in vitro* of mouse embryonic stem (ES) cells, which are derived from cells of the inner cell mass, also provides a system for study. However, it is not clear whether mouse ES cells grown in culture are identical to cells of the ICM (our unpublished data, 2002).[10]

Using the following approaches, we demonstrated that H2A.Z gene expression is up regulated during the differentiation of ICM cells into extraembryonic cells and that H2A.Z is first targeted to pericentric heterochromatin.[6]

[6] D. Rangasamy, L. Berven, P. Ridgway, and D. J. Tremethick, *EMBO J.* **22**, 1599 (2003).

[7] S. J. Abbondanzo, I. Gadi, and C. L. Stewart, *Methods Enzymol.* **225**, 803 (1993).

[8] A. H. Handyside and S. C. Barton, *J. Embryol. Exp. Morphol.* **37**, 217 (1977).

[9] J. Nichols and R. L. Gardner, *J. Embryol. Exp. Morphol.* **80**, 225 (1984).

[10] A. M. Keohane, L. P. O'Neill, N. D. Belyaev, J. S. Lavender, and B. M. Turner, *Dev. Biol.* **180**, 618 (1996).

Isolation of Inner Cell Mass and Trophoblast Cells

1. Collect mouse blastocysts by flushing them from the uteri of superovulated females, 3.5 days after mating, using bicarbonate-free minimal essential medium (MEM) containing pyruvate (100 μg/ml), gentamicin (10 μg/ml), polyvinylpyrrolidone (PVP, 3 mg/ml), and 25 mM HEPES, pH 7.2 (MEM/PVP). Wash the embryos briefly with 3 or 4 drops of MEM/PVP by gently drawing them through a 25-μl glass mouth-controlled pipette (the tip of which was pulled on a pipette puller and siliconized).

2. Transfer embryos onto coverslips in a 24-well tissue culture plate and allow them to grow in 500 μl of Dulbecco's modified Eagle's medium (DMEM) containing freshly added 1 mM glutamine in an atmosphere of 10% CO_2 at 37° for 4–5 days until embryos hatch, attach, and outgrow onto the coverslips (see note 1 under Notes, below).

3. Wash the embryos once with 500 μl of MEM supplemented with PVP (3 mg/ml).

4. To ensure the complete removal of contaminating trophoblast cells from the ICM (see note 2 under Notes, below), cut the central half of the ICM cone (one-third to one-half) from the embryos, using a pair of 18-gauge needles in a scissor-like action under an inverted microscope at ×100 magnification. Remove the ICM tissue with a mouth-controlled fine pipette (as in step 1). Wash the isolated ICM with 1 or 2 drops of MEM/PVP, by gently pipetting up and down, before placing the ICM in 100 μl of lysis buffer [100 mM Tris-HCl (pH 7.4), 4 M guanidine-HCl, 1 M 2-mercaptoethanol].

5. Remove the remaining ICM cone surgically before isolating adherent trophectoderm cells (see note 3 under Notes, below). Remove trophectoderm cells by gentle scraping, using a siliconized fine glass pipette. Wash the cells with 1 or 2 drops of MEM/PVP before transferring them to 100 μl of lysis buffer.

Option. To perform cell counts, a fraction of each cell population can be removed before lysis, stained with Hoechst dye (0.5 μg/ml) for 2 min, and visualized under a microscope. Although the number of cells recovered per sample varies widely, for RNA isolation, roughly 86–92 ICMs or 124–137 trophectoderm samples give approximately the same number of cells as 30 whole embryos.[11]

Isolation of Total RNA

1. After the addition of lysis buffer to the isolated samples, vortex vigorously for 2 min. After this, add 10 μg of yeast tRNA (Invitrogen Life

[11] D. R. Brison and R. M. Schultz, *Mol. Reprod. Dev.* **44,** 171 (1996).

Technologies, Gaithersburg, MD) as a carrier, vortex again, and freeze once at $-70°$ to facilitate quick lysis.

2. Thaw the tubes on ice; add 0.5 pg of rabbit globin mRNA (Invitrogen Life Technologies) to serve as an internal control for recovery and quantification of RNA needed for RT-PCR (see note 4 under Notes, below). Then, add 8 μl of 1 M acetic acid, 5 μl of 2 M potassium acetate (pH 5.5), and 60 μl of 100% ethanol to precipitate RNA overnight at $-20°$. Spin at 14,000g for 30 min at 4°. Wash once with 70% ethanol, dry (avoid overdrying), and dissolve in 20 μl of TN buffer [40 mM Tris-HCl (pH 8.0), 10 mM NaCl, and 6 mM MgCl$_2$].

3. Add 2 units of DNase I (Promega, Madison, WI) and incubate at 37° for 15 min to remove any residual DNA. Add 50 μl of TN buffer and 70 μl of TE [10 mM Tris-HCl (pH 7.0), 1 mM EDTA]-saturated phenol, vortex briefly, and spin at 14,000g for 3 min.

4. Transfer the aqueous phase into fresh microcentrifuge tubes and precipitate RNA by adding 7 μl of 3 M sodium acetate, pH 7.0, and 155 μl of 100% ethanol. Store overnight at $-20°$ to maximize yields. Spin at 14,000g for 30 min at 4°. Carefully wash the RNA pellets once with 70% ice-cold ethanol, dry the pellets (which will be small, glossy, and often difficult to see), and dissolve in 20 μl of TE buffer with 2 units of RNasin inhibitor (Roche Molecular Biochemicals, Indianapolis, IN). Store the samples at $-70°$.

5. Determine the yield of RNA samples. We use small volumes by employing microcuvettes (5- to 10-μl volume) and a GeneQuant *pro* RNA/DNA Calculator spectrophotometer (Biochrom, Cambridge, UK). TE buffer read at 260 nm serves as a blank. Dilute the RNA sample 1:10 and determine the concentration by using the formula [RNA] = A_{260} (0.04 μg/μl) $\times$ 10.

RT-PCR Assay for H2A.Z and H2A

1. Add the following reagents to thin-walled 200-μl PCR tubes on ice: 3 μl of nuclease-free water, 16 μl of total RNA, and 1 μl of anchored oligo(dT)$_{15}$ (3.5 μM final concentration; Promega). Mix gently and briefly spin to collect all components at the bottom of the tubes. Place the tubes in a thermal cycler at 70° for 10 min to denature the RNA secondary structure for efficient reverse transcription.

2. Remove the tubes, place them on ice, spin them briefly, and then add the following components to each reaction: 29 μl of water, 4 μl of 10$\times$ RT-PCR buffer [500 mM Tris-HCl (pH 8.3), 400 mM KCl, 80 mM MgCl$_2$, and 10 mM dithiothreitol (DTT)], 4 μl of 10 mM dNTP mix, 2 μl of RNase inhibitor (1 unit/μl), and 1 μl of Enhanced Avian Reverse

Transcriptase (2 units/μl; Sigma, St. Louis, MO). Incubate the reactions at 25° for 15 min and then extend the reactions at 42° for 60 min.

3. PCR amplification of the synthesized cDNA is performed by adding the following reagents: 8–10 μl of water, 2 μl of 10× AccuTaq buffer [500 mM Tris-HCl, 150 mM ammonium sulfate (pH 9.3), 25 mM MgCl$_2$, 1% Tween 20], 1 μl of dimethyl sulfoxide (DMSO), 1 μl each of 5′ and 3′ primers (5 μM stock) (see note 5 under Notes, below, for primer sequence), 4–6 μl of cDNA from an RT reaction (varies depending on the yield), and 0.5 μl of JumpStart AccuTaq LA DNA polymerase mix (2.5 units/μl; Sigma). The total volume of each reaction is 20 μl. Mix gently, spin briefly, and place them in the thermal cycler.

4. For amplification of histone genes (see note 5 under Notes, below), the basic PCR program is a 94° soak for 2 min, followed by a 35- to 40-cycle program of 94° for 15 s, 68° for 60 s (annealing/extension), and then a final extension at 68° for 5 min, using a thermal cycler. Amplification parameters will vary for other genes depending on the size of the amplicon and the primers used.

5. After the PCR, chill the tubes on ice and then spin briefly to collect any condensation. Add 3 μl of 6× loading dye [10 mM Tris-HCl (pH 7.5), 50 mM EDTA, 10% Ficoll-400, 0.25% bromphenol blue] containing RNase A (10 μg/ml; Qiagen, Hilden, Germany) to digest any remaining carrier RNA. Load 10 μl of products on a 1.5% agarose gel (Fig. 1) and subsequently quantify the intensity of bands on a PhosphorImager (Amersham Biosciences, Piscataway, NJ).

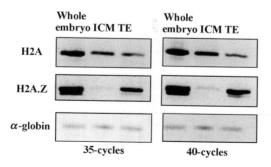

FIG. 1. Analysis of H2A and H2A.Z transcript levels in the early mouse embryo. Transcript levels of histones H2A and H2A.Z were determined by semiquantitative RT-PCR. Shown are 35- and 40-cycle PCRs. Rabbit α-globin mRNA was used as an internal control to correct for differences between samples in RNA recovery and efficiency of RT-PCR. ICM, Inner Cell Mass; TE, trophectoderm cells.

Fluorescence Microscopy

Because the ICM grows as a complex mound of cells, with the surface composed of differentiating endoderm cells, only confocal microscopy can locate proteins in the interior of the ICM cellular mound and in endoderm cells (located at the edge of the ICM) in the same plane of focus. This technique is essential for the resolution of H2A.Z, particularly in pericentric heterochromatin, in differentiating endoderm cells because it can help overcome interference problems due to overlapping layers of cells in the region (Fig. 2). Outgrown embryos, displayed in Fig. 2, are visualized with a Leica confocal laser-scanning microscope (TCS 4D; Leica Microsystems, Bannockburn, IL), equipped with an argon-krypton laser and appropriate filters for the detection of fluorescein isothiocyanate (FITC) and Texas red fluorescence. Whole embryos are imaged with a ×16 oil immersion objective and nuclei are imaged with a ×40 oil immersion objective with a further 5-fold zoom.

Immunostaining of Embryos

1. Fix outgrown embryos with 500 µl of freshly prepared 4% paraformaldehyde for 15 min. Wash the embryos at least five times in phosphate-buffered saline (PBS) at room temperature.

2. Permeabilize the cells with 500 µl of permeabilizing solution [0.1% sodium dodecyl sulfate (SDS), 1% cell culture-grade bovine serum albumin (BSA; CSL, Parkville, Australia) in PBS buffer] for 20 min.

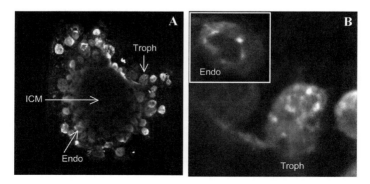

FIG. 2. Fluorescence image of an H2A.Z-stained early mouse embryo. (A) An outgrown blastocyst with trophoblast cells, endoderm cells, and ICM cells in the same plane of focus. H2A.Z staining is detected in trophoblast and endoderm cells, but not in the ICM, as indicated by the arrows. (B) Single trophoblast and endoderm nuclei at higher magnification, showing H2A.Z localized to pericentric heterochromatin.

3. Block the embryos with 500 μl of blocking solution containing 1% BSA in PBS for at least 60 min at room temperature.

4. Incubate the outgrown embryos with 200 μl of blocking solution containing 0.5 μg of anti-H2A.Z[6] overnight at 4° (see note 6 under Notes, below).

5. Wash the coverslips with 500 μl of ice-cold PBS three or four times.

6. Incubate the cells with FITC- or Texas red-conjugated secondary antibodies at a dilution of 1:200 in blocking solution for at least 60 min at room temperature in the dark.

7. Wash the coverslips with 500 μl of PBS three or four times. Care is necessary to avoid removal of the ICM from the attached embryo.

8. Using forceps and an 18-gauge needle with a bent tip, remove the coverslips from the tissue culture plate (blot away excess buffer from the coverslips), invert, and place on a drop of antifade solution (ProLong; Molecular Probes, Eugene, OR) on a slide. Seal the edges of the coverslip with nail varnish.

9. Before examining the slides by fluorescence microscopy, store the slides for 24 h at room temperature, in the dark, to reduce photobleaching.

Notes

1. It is important to monitor the position of the blastocysts on the coverslip until the blastocysts have attached (about 1 day). Unattached blastocysts are easily moved (particularly if the dish is shifted), resulting in blastocysts attaching either too close together or at the edge of the coverslip. Typically, about eight blastocysts are placed in a well, of which at least four to six hatch and attach.

2. To check whether microsurgically isolated ICM samples are free of contaminating trophectoderm cells, the exclusive localization of marker gene(s) can be determined. For example, fibroblast growth factor (Fgf-4) is expressed only in the ICM. Fgf-4 cDNA can be amplified by using 5′-CCGGTGCAGCGAGGCGTGGT and 5′-GGAAGGAAGTGGGT-GACCTTCAT primers.

3. Endo-β-N-acetylglucosaminidase (endo-A) acts as a trophectoderm marker gene that can be amplified by using primers 5′-TGTCGTGTCC-AAGTGAATGGC and 5′-CCATTGGGATATCCCAGATAG.

4. Addition of a known amount of α-globin mRNA before RNA isolation functions as an internal control to normalize the differences between samples in RNA recovery. α-Globin mRNA, which is not present in the embryo at this stage of development, is amplified by using primers 5′-GTGCTGTCTCCCGCTGACAAGACC and 5′-CACCGGGTC-CACCCGCAGCTTGTG.

5. Primers for histone H2A and H2A.Z amplification are, respectively, 5′-GTCTGGACGCGGCAAGCAGGG and 5′-TATTTTCCCTTGGC-CTTGTGG, and 5′-GCTGGCGGTAAGGCTGGAAAG and 5′-AAA-CAGTCTTCTGTTGTC.

6. Antibodies are tested for specificity by incubating the blastocyst either without primary antibody (i.e., secondary antibody alone) or with primary antibody that has been preincubated with a 5-fold molar excess of H2A.Z or H2A protein for 1 h at room temperature.

Xenopus as a Model System for Early Development

Xenopus laevis is an ideal organism to monitor early development because thousands of embryos can be collected, cultured, and manipulated in the laboratory with minimal expense and equipment. Detailed developmental tables[12] combined with the more recent gene expression analysis of embryonic patterning[13] facilitate the analysis of minute disruptions of the normal process of development. In contrast to mouse embryos, where examination of early embryos requires sacrifice of mice and microsurgery, all stages of *Xenopus* development can be monitored with ease from a culture dish, using a stereomicroscope. At the molecular level, *Xenopus* oocytes and embryos have proved to be a powerful tool to define fundamental biochemical aspects of chromatin assembly[14,15] and transcriptional regulation in a chromatin environment.[16]

The specific application of fluorescent proteins as lineage markers in *Xenopus* developmental studies is also well documented.[17] Zernicka-Goetz *et al.*[17] produced a fluorescent protein from injected synthetic mRNA and found it to be a stable, nontoxic marker with which to monitor cell fate and microinjection of overexpressed proteins by coinjection. The use of coinjected fluorescent markers, however, has several disadvantages that can be addressed by a fusion protein strategy. The fusion protein will provide a more accurate localization mechanism for microscopic examination of whole-embryo morphology and for detailed analysis of events occurring at the molecular and cellular levels.

Because core histone–GFP (green fluorescent protein) fusion proteins have provided valuable information about chromatin dynamics in living

[12] P. Nieuwkoop and P. Faber, "Normal Table of *Xenopus laevis*." North-Holland Publishers, Amsterdam, 1967.

[13] V. Gawantka, N. Pollet, H. Delius, M. Vingron, R. Pfister, R. Nitsch, C. Blumenstock, and C. Niehrs, *Mech. Dev.* **77**, 95 (1998).

[14] A. Shimamura, B. Jessee, and A. Worcel, *Methods Enzymol.* **170**, 603 (1989).

[15] A. P. Wolffe and C. Schild, *Methods Cell Biol.* **36**, 541 (1991).

[16] C. C. Robinett and M. Dunaway, *Methods* **17**, 151 (1999).

[17] M. Zernicka-Goetz, J. Pines, K. Ryan, K. R. Siemering, J. Haseloff, M. J. Evans, and J. B. Gurdon, *Development* **122**, 3719 (1996).

mammalian cells,[18] we describe here the application of a fusion protein composed of the H2A.Z histone variant fused at its C terminus to enhanced green fluorescent protein (EGFP) to monitor H2A.Z expression and localization in living embryos and oocytes.

Cloning of H2A.Z–EGFP Fusion Expression Construct

Murine H2A.Z cDNA is fused at its C terminus to EGFP cDNA [derived from BD Biosciences Clontech (Palo Alto, CA) vector pEGFP-N1] by PCR amplification, with the addition of *Bam*HI (5′) and *Xho*I (3′) sites for cloning. The fusion H2A.Z–EGFP cDNA is then cloned into the RN3P vector, which has been specifically modified for protein expression in *Xenopus* after microinjection of synthetic mRNA.[17] It is important to note that factors such as the choice of expression vector, and capping and polyadenylation of the RNA transcript, are critical to the efficient translation and stability of protein expressed from synthetic mRNA in *Xenopus*. For a detailed description of vector design and optimization of translational efficiency, refer to Wormington.[19]

Synthetic mRNA Preparation

Synthetic mRNA is prepared according to standard protocols[19] with some modifications as described below.

1. Cleave H2A.Z–EGFP expression construct DNA with the appropriate restriction enzyme to linearize at the end of the cDNA (*Kpn*I for RN3P derivatives).

2. Spin column purify to remove enzyme and buffers, using a Qiagen QIAquick PCR purification kit.

3. Prepare RNA from 4 μg of linearized DNA, using the appropriate RNA polymerase as in the following example for H2A.Z–EGFP. Combine 4 μg of linearized DNA template in 10μl, 5 μl of an NTP mix (1:10 dilution of 100 mM stock; Amersham Biosciences), 10 μl of a 5× T3 RNA polymerase transcription buffer mix (Fermentas, Hanover, MD), 5 μl of DTT (100 mM stock), 1 μl (40 units) of RNase inhibitor (HPR1; Amersham Biosciences), 3.33 μl of 10 mM CAP analog solution [M7G(5′)ppp(5′)G RNA capping analog; Invitrogen, Carlsbad, CA), and 12.67 μl of sterile MilliQ (Millipore, Bedford, MA) filtered water.

4. Incubate on ice for 10 min and then add 2 μl of T3 RNA polymerase (20 units/μl; Fermentas) to start the reaction.

5. Vortex to mix, centrifuge for 10 s, and then incubate at 37° for 1 h.

[18] T. Kanda, K. F. Sullivan, and G. M. Wahl, *Curr. Biol.* **8**, 377 (1998).
[19] M. Wormington, *Methods Cell Biol.* **36**, 167 (1991).

6. Add an additional 1 μl of T3 RNA polymerase and continue the incubation for an additional 1 h.

7. Add 2 μl of RNase-free DNase I (10 units/μl; Roche Molecular Biochemicals) and incubate at 37° for 30 min to digest the DNA template.

8. Purify and quantify RNA by extracting once with an equal volume of phenol–chloroform–isoamyl alcohol. Remove the upper aqueous phase to a new tube and immediately place on ice.

9. Combine two reactions for each template to give a total volume of approximately 100 μl. Purify, using a Roche Applied Science (Indianapolis, IN) Quick Spin Sephadex G50 spin column to remove unincorporated NTPs.

10. Precipitate with a 2.5× volume of 100% ethanol.

11. Dry the RNA pellet for 5 to 10 min in a SpeedVac (Thermo Savant, Holbrook, NY). An overdried pellet is difficult to resuspend.

12. Resuspend the pellet in a maximum of 10 μl of TE buffer per 4 μg of starting linearized DNA template.

13. To quantify, dilute 1 μl with 9 μl of water. Use 5 μl to test the concentration by determining the optical density at 260 nm (OD_{260}) in a spectrophotometer. Use the other 5 μl to separate on a 1% agarose gel to test the integrity of the RNA product.

14. The remaining RNA is divided into 3-μl aliquots and stored at −70° (after thawing, discard any remaining sample). We routinely obtain a yield of 20 to 50 μg of mRNA from each 4 μg of starting DNA template.

Microinjection of H2A.Z–EGFP into Xenopus Oocytes and Embryos

Preparation of *Xenopus* oocytes and embryos for microinjection of mRNA is done according to standard protocols.[20–22] A Drummond Nanoject injection system (Edwards Instruments, Sydney, Australia) is used for injection of mRNA into the midline of the oocyte animal pole or into one cell of the animal pole of a two-cell stage *in vitro*-fertilized embryo.

For convenience, oocyte mRNA microinjections are used to optimize the H2A.Z–EGFP overexpression system. A range of 1 to 40 ng of mRNA per oocyte is injected into oocytes, depending on experimental conditions. Coinjections with other experimental mRNAs up to a total of 80 ng/oocyte are performed without compromising protein expression levels. Lower amounts are used for embryo overexpression, with 1 ng/embryo giving a bright signal that becomes visible by 5 h postinjection. Although we have expressed up to 10 ng of fusion protein mRNA per embryo, this appears

[20] B. K. Kay, *Methods Cell Biol.* **36,** 663 (1991).
[21] L. D. Smith, W. L. Xu, and R. L. Varnold, *Methods Cell Biol.* **36,** 45 (1991).
[22] J. Heasman, S. Holwill, and C. C. Wylie, *Methods Cell Biol.* **36,** 213 (1991).

to be the upper limit for our H2A.Z–EGFP fusion protein because higher amounts result in aberrant development. It is important to note that optimal amounts of injected synthetic mRNA can vary depending on the protein to be expressed.

In the whole oocyte or embryo, EGFP expression is monitored with a Leica MZ FLIII fluorescence stereomicroscope with a Leica DC 500 digital color camera and a Leica GFP Plus filter set (excitation filter, 480/40 nm; barrier filter, 510 nm). This filter efficiently detects the maximal wavelength of 508 nm for EGFP. Figure 3 shows the localization of the H2A.Z–EGFP fusion protein in the nuclei of *Xenopus* oocytes.

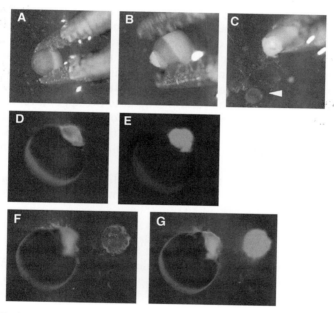

Fig. 3. Nuclear localization of overexpressed H2A.Z–EGFP in *Xenopus laevis* oocytes. (A–C) Manual isolation of nuclei from *Xenopus laevis* oocytes. (A) Oocyte is held in place with forceps. (B) Animal pole of an oocyte is cut, using the point of a needle to make an incision for nuclear removal. (C) Oocyte is squeezed with forceps to remove nucleus. Clear nucleus is indicated by the arrowhead. Nucleus is immediately transferred to a tube on ice. (D–G) Oocytes were injected into the cytoplasm with H2A.Z–EGFP synthetic mRNA (10 ng/ oocyte) and incubated for 18 h for *in vivo* protein expression and accumulation as described. (D) Bright-field color image of incision in animal pole of oocyte. (E) Fluorescence image, using Leica GFP Plus filter set, corresponding to (D). Brilliant green nucleus can be seen through the incision. (F) Bright-field color image of clear extracted nucleus. (G) Fluorescence image, using Leica GFP filter set (with some bright-field light to show location of remainder of oocyte), corresponding to (F). Brilliant green of H2A.Z–EGFP fusion protein accumulated in nucleus is illustrated. (See color insert.)

Analysis of H2A.Z–EGFP Fusion Protein Overexpression by Western Blotting

To assess H2A.Z–EGFP fusion protein expression, crude protein extracts are prepared from *Xenopus* oocytes according to standard protocols as outlined below. Figure 4 shows a Western blot analysis of H2A.Z–EGFP fusion protein expressed in *Xenopus* oocytes.

1. Eighteen hours after injection of H2A.Z–EGFP mRNA, collect 50 healthy oocytes in a 1.5-ml microcentrifuge tube and wash gently several times with oocyte extract buffer [10 mM HEPES (pH 7.5), 70 mM KCl, 5% sucrose] containing protease inhibitors [leupeptin (10 μg/ml), pepstatin (10 μg/ml), 1 mM DTT, phenylmethylsulfonyl fluoride (PMSF, 100 μg/ml)]. Remove any remaining buffer and add, per oocyte, 2.5 μl of fresh oocyte extract buffer containing protease inhibitors.

2. Homogenize the oocytes by pipetting up and down gently 15–20 times. Keep the lysate on ice at all times during homogenization.

3. Centrifuge at full speed in a microcentrifuge for 10 min at 4° to pellet debris to the bottom layer and yellowish lipids to the upper layer of the tube.

4. Carefully, insert a pipette tip from the top through the lipid layer, or a needle into the side of the tube, and remove the clear protein lysate layer.

5. Freeze 25-μl aliquots in liquid nitrogen and then store immediately at −70°.

6. For subsequent Western analysis of H2A.Z–EGFP fusion protein expression, electrophorese 15 μl of crude protein lysate on an SDS–15% polyacrylamide gel (Fig. 4).

Isolation and Fluorescence Analysis of Minichromosomes from Oocytes

The biochemical characterization of minichromosomes, transcriptionally active or repressed, assembled from template DNA injected into *Xenopus* oocytes, would provide new insights into the structure and function of chromatin. However, this has proved to be extremely difficult and complicated by contaminating yolk proteins in the oocyte (over 80% of total protein),[23] the need to separate minichromosomes from endogenous chromatin, and the small amounts of DNA assembled into chromatin in an oocyte.

We have developed a rapid protocol to isolate minichromosomes from oocyte proteins, including yolk proteins, and utilize a sucrose gradient step to separate minichromosomes from bulk endogenous chromatin. To detect the small amounts of chromatin assembled from exogenous DNA, and to facilitate the analysis of the assembly of H2A.Z into chromatin, we have established a simple fluorescence protocol using the H2A.Z–EGFP

---

[23] H. S. Wiley and R. A. Wallace, *J. Biol. Chem.* **256,** 8626 (1981).

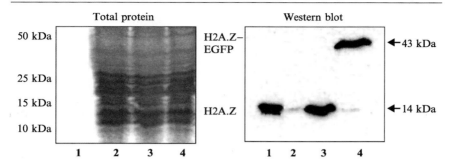

FIG. 4. Western blot analysis of *in vivo*-translated H2A.Z–EGFP fusion protein. Crude protein was extracted from noninjected oocytes and oocytes injected with 40 ng of H2A.Z mRNA or 40 ng of H2A.Z–EGFP mRNA as described. Proteins were separated on an SDS–15% polyacrylamide gel, transferred to a reinforced nitrocellulose 0.45-μm membrane (Schleicher & Schuell BioScience, Keene, NH) and resolved with a polyclonal H2A.Z antibody (1:1250 dilution). *Left:* Total protein detected by Ponceau staining of membrane. *Right:* Western blot. Lane 1, recombinant H2A.Z as a positive control (25 ng); lane 2, endogenous H2A.Z (no injection control); lane 3, H2A.Z mRNA-injected oocyte extract; lane 4, H2A.Z–EGFP mRNA-injected oocyte extract. In lanes 2, 3, and 4, 15 μl of protein extract was loaded in each lane. Approximate sizes of H2A.Z protein (14 kDa) and H2A.Z–EGFP fusion protein (43 kDa) are shown.

fusion protein. This method overcomes the need for immunodetection of limited amounts of chromatin or radioisotope labeling of histones.

1. Oocytes are prepared for injection by standard techniques and injected with up to 40 ng of H2A.Z–EGFP synthetic mRNA in a 23-nl volume.

2. Oocytes are incubated for 18 h to allow translation of mRNA and overexpression of the fusion protein. The maximal fluorescent signal in total protein extracts is detected from 18 to 24 h, but begins to diminish with incubations over 24 h.

3. Single- or double-stranded DNA template (we use 5 ng/oocyte in 23 nl of an M13 derivative DNA for this purpose) is injected into the oocyte nucleus for assembly into chromatin.

4. Oocytes are incubated for at least 5 h for chromatin assembly; however, this time can be varied depending on the experimental questions being addressed.[24]

5. Healthy oocytes are collected and nuclei are removed by manual isolation as previously described (see Fig. 3).[25] Once removed, the nuclei are immediately collected in a 1.5-ml microcentrifuge tube on ice. The bright green fluorescence of the nuclei, when viewed under a fluorescence

[24] M. Ryoji and A. Worcel, *Cell* **37**, 21 (1984).
[25] J. P. Evans and B. K. Kay, *Methods Cell Biol.* **36**, 133 (1991).

stereomicroscope, facilitates their accurate and rapid isolation. Approximately 100 nuclei are collected for each preparation.

6. The volume of the nuclei is raised to 200 μl with OR2 buffer [5 mM HEPES (pH 7.8), 87 mM NaCl, 2.5 mM KCl, 1 mM Na$_2$H-PO$_4$·2H$_2$O, 1 mM MgCl$_2$, and 0.05% polyvinyl pyrrolidone (pH 7.8), filter sterilized] containing protease inhibitors [leupeptin (10 μg/ml), pepstatin (10 μg/ml), 1 mM DTT, and PMSF (100 μg/ml)]. Nuclei are fragile and there is no need to homogenize the nuclei by pipetting.

7. Nuclear lysate is carefully layered onto the top of a 0 to 40% (w/v) sucrose gradient prepared in OR2 buffer with protease inhibitors in Beckman Coulter (Fullerton, CA) centrifuge tubes (11 × 34 mm, polyallomer). Technical details of sucrose gradient centrifugation are given elsewhere.[26] All centrifuge tubes and solutions are kept on ice throughout the procedure.

8. Gradients are centrifuged at 12,000 rpm for 19 h at 4° in a Beckman Coulter TLS-55 swinging bucket rotor in an Optima Max Ultracentrifuge.

9. After centrifugation, 200-μl fractions are carefully removed from the top of the gradient and placed into prechilled microcentrifuge tubes. The bulk of the oocyte chromatin pellets at the bottom of the tubes. The assembled minichromosome is isolated in the 40% sucrose fraction (control free DNA can be isolated in 15 to 20% sucrose fractions). Free H2A.Z–EGFP protein remains at the top of the gradient. A 50-μl volume is processed for DNA extraction and analysis by supercoiling and 150 μl is used for fluorescence analysis for assembly of H2A.Z–EGFP fusion protein into chromatin.

10. In a 96-well plate, 25 μl of each fraction is diluted to a total volume of 200 μl in OR2 buffer containing protease inhibitors. The plate is scanned with a CytoFluor II fluorescence multiwell plate reader (Applied Biosystems, Foster City, CA) with excitation (485/420 nm) and emission (508/520 nm) filters (Applied Biosystems). The detection of a fluorescent signal in the 40% sucrose gradient fraction indicates the assembly of H2A.Z–EGFP fusion protein into minichromosomes.

Acknowledgments

We are grateful to K. Matthaei and H. Taylor for providing us with mouse blastocysts. We also thank G. Almouzni and J.-P. Quivy for invaluable discussions and advice on chromatin assembly in the *Xenopus laevis* system. This work was supported by a Human Frontiers Science program grant to David Tremethick, an NHMRC grant (Grant 179823) to Patricia Ridgway and David Tremethick, and an Australian Academy of Science "Scientific Visits to Europe Program" grant to Patricia Ridgway.

[26] H. Noll and M. Noll, *Methods Enzymol.* **170,** 55 (1989).

[17]   Histone Variant CENP-A Purification,
Nucleosome Reconstitution

By KINYA YODA, SETSUO MORISHITA, and KEIJI HASHIMOTO

Introduction

CENP-A is a centromere-specific histone H3[1,2] and is widely conserved among species, from *Saccharomyces cerevisiae*[3] to *Schizosaccharomyces pombe,*[4] to *Caenorhabditis elegans,*[5] to *Drosophila*[6] to mammals.[7] The C-terminal two-thirds of CENP-A is highly homologous to histone H3, but the remaining N-terminal one-third is unique. The putative histone fold domain located in the C-terminal region is essential for targeting to the centromeric region.[2] In higher eukaryotes, the centromeric DNAs are usually organized with a highly repetitive sequence, but their sequences and unit lengths are totally species specific.[8] It is reported that centromere could be formed even on unique sequences on rare occasions.[9] On the other hand, the mechanism of chromosome segregation is universal to all through the eukaryotes; kinetochores are formed at the centromeric region where mitotic spindles are attached. Thus, CENP-A plays a key role that links the centromeric DNAs, reflecting the variety of eukaryotic species, to the universal mechanism of chromosome segregation.

It is widely accepted that formation of CENP-A nucleosomes defines the active centromeric region.[7] Therefore, to know the structure of CENP-A nucleosomes and the mechanism of CENP-A nucleosome formation is critically important for revealing the structure and function of the active centromere. We have shown that CENP-A nucleosomes are selectively formed on I-type α-satellite arrays in centromeres of all human chromosomes except the Y chromosome.[10]

---

[1] D. K. Palmer, K. O'Day, H. L. Trong, H. Charbonneau, and R. L. Margolis, *Proc. Natl. Acad. Sci. USA* **88,** 3734 (1991).

[2] K. F. Sullivan, M. Hechenberger, and M. Khaled, *J. Cell Biol.* **127,** 581 (1994).

[3] S. Stoler, K. C. Keith, K. E. Curnick, and M. Fitzgerald-Hayes, *Genes Dev.* **9,** 573 (1995).

[4] K. Takahashi, E. S. Chen, and M. Yanagida, *Science* **288,** 2215 (2000).

[5] B. J. Buchwitz, K. Ahnad, L. L. Moore, M. B. Roth, and S. Henikoff, *Nature* **401,** 547 (1999).

[6] S. Henikoff, K. Ahmad, J. S. Platero, and B. V. Steensel, *Proc. Natl. Acad. Sci. USA* **97,** 716 (2000).

[7] K. F. Sullivan, *Curr. Opin. Genet. Dev.* **11,** 182 (2001).

[8] H. F. Willard, *Trends Genet.* **6,** 410 (1990).

[9] D. Sart *et al., Nat. Genet.* **16,** 144 (1997).

[10] S. Ando, H. Yang, N. Nozaki, T. Okazaki, and K. Yoda, *Mol. Cell. Biol.* **22,** 2229 (2002).

In this chapter, we describe methods for purification of CENP-A and for *in vitro* CENP-A nucleosome reconstitution using NAP-1.[11]

Purification of CENP-A from HeLa Cells

Reagents

Note: Asterisks indicate a solution is to be made fresh before use.
* Chromosome isolation buffer (CIB): 3.75 mM Tris-HCl (pH 7.4), 20 mM KCl, 0.5 mM EDTA, 0.5 mM dithiothreitol (DTT), 0.05 mM spermidine, 0.125 mM spermine, 0.1 mM phenylmethylsulfonyl fluoride (PMSF)
* 2× washing buffer (2× WB): 40 mM HEPES-Na (pH 8.0), 40 mM KCl, 1.0 mM EDTA, 1.0 mM DTT, 0.2 mM PMSF, pepstatin (1.0 μg/ml), leupeptin (4 μg/ml)
* Digitonin (biochemical grade; Wako, Kyoto, Japan): 10% in CIB
* 2× extraction buffer (2× EB): 20 mM HEPES-Na (pH 8.0), 30% glycerol, 0.5 mM DTT, 0.5 mM EDTA, 0.5 mM PMSF, pepstatin (0.5 μg/ml), leupeptin (2.0 μg/ml)
* 2× SDS buffer: 0.1 M Tris-HCl (pH 8.0), 50 mM DTT, 2% sodium dodecyl sulfate (SDS), 0.04% bromphenol blue (BPB), 30% glycerol

Preparation of 0.6 M NaCl Nuclear Extract and Nuclear Pellet

1. Thaw frozen HeLa cell pellets (1×10^{10} cells; stocked at $-80°$) in a $37°$ water bath with frequent mixing in order to avoid local temperature increases. Unless otherwise indicated, all subsequent procedures should be performed on ice.

2. Add CIB containing 0.1% digitonin to each cell pellet, to a final concentration of 2×10^7 cells/ml (500 ml; $\sim$50 ml $\times$ 10 disposable centrifuge tubes).

3. Homogenize the cell suspensions with 10 up-and-down strokes in a Dounce homogenizer (Wheaton, Millville, NJ) with a tight-fitting pestle. Be careful not to make air bubbles, especially on the down strokes, because small air bubbles break down the nuclear membrane, which causes loss of products owing to formation of nuclear aggregates.

4. Centrifuge the homogenates at 300g for 5 min at $4°$ with a swing rotor.

[11] K. Yoda, S. Ando, S. Morishita, K. Houmura, K. Hashimoto, K. Takeyasu, and T. Okazaki, *Proc. Natl. Acad. Sci. USA* **97,** 7266 (2000).

5. Discard the supernatants and repeat steps 2–5 twice. Transfer the pellets to two new 50-ml disposable centrifuge tubes.

6. Wash the pellets twice, each time with 50 ml of WB.

7. Suspend each pellet with EB to 12.5 ml ($\times$ 2 tubes) and add an equal volume of EB–1.2 M NaCl with continuous mixing (final concentration, 0.6 M NaCl), and let stand for 1 h at $0°$.

8. Centrifuge the pellet suspensions in 0.6 M NaCl at 100,000g for 30 min at $4°$. Separate the supernatant (0.6 M NaCl nuclear extract; this fraction contains most of the centromere proteins containing CENP-B and CENP-C but not CENP-A, and is also used for purification of Topo-I, as described in a later section) from the pellet (0.6 M NaCl nuclear pellet) and store at $-80°$ after quick freezing in liquid N_2. CENP-A is in the nuclear pellet.

Preparation of Chromatin Pellet

1. Crush the frozen nuclear pellets (1×10^{10} nuclear equivalents), using a mortar and pestle in liquid N_2.

2. Suspend the nuclear powder in 200 ml of EB–1.0 M NaCl and completely homogenize by using, alternatively and repeatedly, a Dounce homogenizer and a sonicator.

3. Add HCl to the homogenates to a final concentration of 0.25 M and let stand for 30 min at $0°$.

4. Centrifuge the homogenates in 0.25 M HCl at 10,000g for 10 min at $4°$.

5. Save the supernatant to new centrifuge tubes ($\sim$50 ml in each of four tubes) and add TCA to final concentration of 5% (w/v). Let stand for 1 h at $0°$.

6. Centrifuge the samples in trichloroacetic acid (TCA) at 10,000g for 10 min at $4°$ and remove the supernatants.

7. Wash the pellets twice with 83.3% acetone and dry the pellets by airflow (chromatin pellet).

First HPLC

1. Dissolve each chromatin pellet (1×10^{10} nuclear equivalents) with 50 ml of 30% acetonitrile–0.1% trifluoroacetate (TFA) and filter through a 0.5-μm pore size membrane for organic solvent (Columnguard-LCR$_{13}$; Millipore, Bedford, MA).

2. Apply 50 ml of the sample solution to a reversed-phase HPLC column [μBondasphere 5 μ C$_{18}$-300 Å, 19 mm $\times$ 15 cm (Nihon Waters, Tokyo, Japan); detection, 280 nm; full scale, 2; flow rate, 5 ml/min; fractionation, 10 ml/fraction]. Wash the column with 50 ml of 30%

acetonitrile–0.1% TFA and elute with 300 ml of a 30–90% acetonitrile linear gradient containing 0.1% TFA.

3. Take 0.1 μl and/or a 1-μl aliquot of each fraction and electrophorese through SDS–12.5% polyacrylamide gels, one for Coomassie Brilliant Blue stain (for detection of core histones) and the other for Western blotting with anti-centromere anti-antibody (ACA) (for detection of CENP-A). Results are shown in Fig. 1.

4. Add 100 μg of Polybrene to the CENP-A fraction (fraction 23 in Fig. 1) in order to make the pellet more soluble to buffer for the second HPLC. Dry the fractions of core histones and CENP-A by vacuum centrifugation. Store these fractions at $-80°$ until use.

Second HPLC

1. Dissolve the CENP-A fraction (total, 3–5 × 10^{10} nuclear equivalents) with 30% acetonitrile–0.1% TFA to a final concentration of 1 × 10^{10} nuclear equivalents/ml. Filter through a Columnguard-LCR$_{13}$ membrane (Millipore).

2. Apply 3–5 ml of the CENP-A fraction to the second HPLC column [μBondasphere 5 μ C$_{18}$-300 Å, 3.9 mm × 15 cm (Nihon Waters); detection, 280 nm; full scale, 0.3; flow rate, 1 ml/min; fractionation, 0.5 ml/tube]. Wash the column for 15 min with 30% acetonitrile–0.1% TFA and elute with a 30–90% acetonitrile–0.1% TFA linear gradient for 30 min (Fig. 2A).

3. Take a 5-μl aliquot of each fraction, dry by vacuum centrifugation, resuspend with 20 μl of 2 × SDS buffer, and boil for 5 min. Electrophorese 18- and 2-μl samples through SDS–13% polyacrylamide gels: one for silver staining (Fig. 2B) and the other for Western blotting (Fig. 2C).

4. Pool CENP-A fractions (Fractions 14–16 in Fig. 2A) and dry by vacuum centrifugation.

SDS–PAGE

Final purification step is SDS–13% PAGE. Purified CENP-A is extremely sticky and apt to be lost. Therefore, it is recommended that an equal amount of histone H4 be added to CENP-A as far as possible and to increase the handling amount of CENP-A as much as possible (2–4 × 10^{10} nuclear equivalents). By these revisions, we obtain a satisfying recovery ($\sim$50%) of CENP-A. The CENP-A fraction after the second HPLC still contains histones H3 and H2A and histone H3 is banded near to the CENP-A band (Fig. 3A). Therefore, it is necessary to set up optimal conditions to completely separate these two bands. Another important point is to use a handmade apparatus for electroelution of protein from SDS gels as shown in Fig. 3B. An advantage of this tool is that CENP-A can

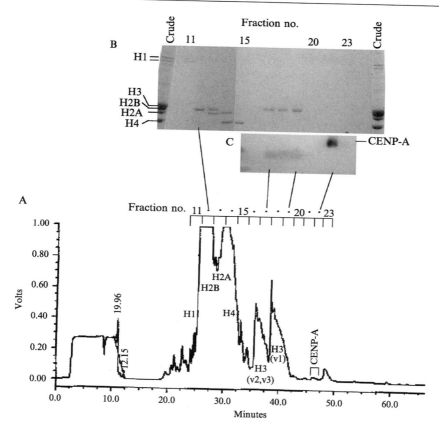

Fig. 1. First HPLC of chromatin proteins. (A) Chromatography profile of chromatin proteins extracted from 1×10^{10} HeLa cells and monitored for absorbance at 280 nm. Chromatin proteins were applied to a C_{18} reversed-phase column (19 mm $\times$ 15 cm) and eluted with 300 ml of a 30–90% acetonitrile linear gradient containing 0.1% TFA. (B) An SDS–12.5% polyacrylamide gel of fractions 11–24. Proteins were detected by Coomassie Brilliant Blue (CBB) staining. (C) Detection of CENP-A by Western blotting of fractions 11–23, using ACA serum. CENP-A can be detected in fraction 22.

be concentrated as well as eluted from the gel. Loss of CENP-A owing to its dilution can be minimized with this tool.

1. Suspend the CENP-A fraction with 1 ml of 1 $\times$ SDS buffer containing ~20 μg of histone H4 and incubate at 65° for 15 min.

2. Prepare an SDS–13% polyacrylamide gel (14 $\times$ 14 $\times$ 0.1 cm) one night before use. Use 99.9% acrylamide for preparation of the gel. A comb

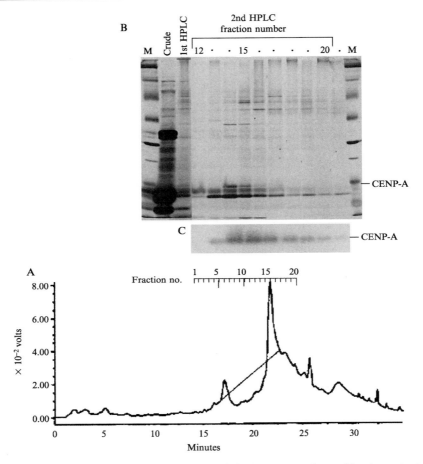

FIG. 2. Second HPLC of chromatin proteins. (A) Chromatography profile of proteins in first HPLC fraction 22 monitored for absorbance at 280 nm. Fraction 22 was applied to a C_{18} reversed-phase column (3.9 mm × 15 cm) and eluted with 30 ml of a 30–90% acetonitrile linear gradient containing 0.1% TFA. (B) An SDS–12.5% polyacrylamide gel of fractions 12–21. Proteins were detected by CBB staining. (C) Detection of CENP-A by Western blotting of fractions 12–21, using ACA serum. CENP-A was recovered in fractions 14–16.

is not used. Let stand overnight after preparation of the gel to complete the chemical reaction. To avoid degradation of protein owing to ions, add 0.1 mM sodium thioglycolate to the cathode buffer as an ion scavenger. Electrophorese the CENP-A fraction (at least 2×10^{10} cell equivalents) at 200 V for 140 min.

A   Copper stain after SDS-PAGE

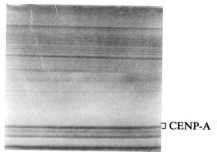

⊐ CENP-A

B   Handmade apparatus for electroelution

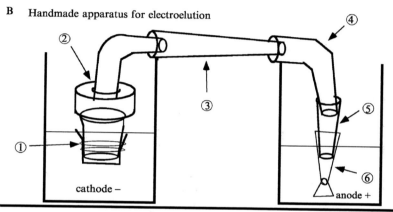

cathode −   anode +

C

4 cm

FIG. 3. Purification of CENP-A by SDS–13% PAGE. (A) CENP-A can be separated from histone H3 by SDS–13% PAGE. The CENP-A fraction from the second HPLC, purified from $3–5 \times 10^{10}$ cells, was applied to an SDS–13% polyacrylamide gel (14 cm × 14 cm × 1 mm). Proteins were detected by copper staining. (B) Illustration of a handmade tool for electroelution. See text for details. The gel tips containing CENP-A are put into 1 and electroeluted at 200 V for 12 h in 1× SDS buffer. (C) Photograph of the tool.

3. Stain the gel after electrophoresis with copper (copper stain and destain kit; Bio-Rad, Hercules, CA) (Fig. 3A) and carefully excise the CENP-A band.

Electroelution

Preparation of Apparatus for Electroelution

The elution tool consists of six parts (**1–6**; Fig. 3B), consisting of a 15-ml centrifuge tube (polypropylene), 1-ml tips (blue tips), a dialysis tube (8/32 in.), a dialysis membrane (5 × 5 cm), and a rubber band. This tool is a closed system and CENP-A in the gel in **1** is electrically eluted from the gel, carried to a dialysis tube (**6** in Fig. 3B), and concentrated in the same dialysis tube. This tool is bent at **2** and **4** (Fig. 3B) and thus changed from a vertical to a horizontal tool. Advantages of the horizontal tool are that we can take each sample on and off separately and freely.

1: A reservoir of gel slices: Cut a 15-ml polypropylene centrifuge tube 2–3 cm from the cap site. Wrap the bottom of the centrifuge tube with dialysis membrane and seal with a rubber band.

2: A cap for reservoir **1** with a link to a bridge **3**: Make a hole in the center of the cap of a 15-ml centrifuge tube (polypropylene tube with a plug seal cap; Corning Life Sciences, New York, NY) with a cork borer (diameter, 8 mm). To make air tight at the junction of the cap and a 1-ml tip, wrap Teflon seal tape around the base of the 1-ml tip and insert the 1-ml tip through the hole of the cap and push the tip to the last part of the base. Heat the 1-ml tip two-thirds from the top and bend ~60°, using a gas burner. Be careful not to squash the inside of the tip at the bend site, which is a pathway of electric current. Cut the tip 15 mm from the top.

3: A bridge, which links cathode to anode: Use a 1-ml tip. Cut the 1-ml tip 23 mm from the top.

4: A link to anode: Use a 1-ml tip. Heat the 1-ml tip two-thirds from the top and bend ~60°, using a gas burner. Be careful not to squash the tube. Cut the tip 15 mm from the top.

5: A support for the dialysis tube: Use a 1-ml tip. Cut the 1-ml tip 21 mm from the base and 20 mm from the top.

6: A dialysis tube (8/32 in.): Make two knots at one end of the ~10-cm dialysis tube. Insert support **6** approximately 2 cm into the dialysis tube. One to 2 cm of the dialysis tube should protrude from the support, where the eluted protein is recovered and concentrated.

Setting Up Electroelution Apparatus

Reagents

1 × running buffer: 25 mM Tris base, 250 mM glycine, 0.1% SDS
Dialysis buffer: 10 mM Tris-HCl (pH 8.0), 1.0 mM EDTA, 0.1% SDS,
1.0 mM DTT, 0.1 mM PMSF

1. Wrap the bottom of **1** (Fig. 3B) with dialysis membrane (5 × 5 cm) and seal with a rubber band. Add 1 × running buffer to **1** and make sure that the buffer does not leak.

2. Cut the gel containing CENP-A every ∼5 mm and place into **1** (Fig. 3B). Add an equal amount of histone H4 (in gel or in buffer). Cover **1** with a cap (**2**) and fill with buffer, using a Pasteur pipette (be careful to remove air bubbles).

3. Fill **6–5** (Fig. 3B) with 1 × running buffer and link **4** and **3** to **6–5** in order. Fill buffer in each step and be careful not to make air bubbles.

4. Link **1–2** to **6–5–4–3** (Fig. 3B), taking care to remove air bubbles. Place the tool on buffer tanks, so that the gel in **1** comes to a cathode and the dialysis tube **6** comes to an anode (Fig. 3B). Be careful the edge of dialysis tube **6** is located above the buffer surface as illustrated in Fig. 3B. (When the edge of the dialysis tube is immersed under the buffer, protein leaks out between the dialysis membrane and the tube.)

5. Add sodium thioglycolate to cathode buffer to a final concentration of 0.1 mM.

6. Electrophorese at 200 V for 24–30 h at 4°.

7. At the end of electrophoresis, reverse the polarity and electrophorese for 30 s at 200 V to detach proteins from the dialysis membrane.

8. Disconnect **5–6** from **4** (Fig. 3B) and carefully remove the buffer in **5**. Usually, protein does not stay in **5**, but in some cases part of the protein leaks in to this area. So it is recommended to save this fraction until recovery of the protein has been measured.

9. Mix the buffer well by rubbing dialysis tube **6** and recover the CENP-A/H4 mixture to a new Eppendorf tube.

10. Dialyze the recovered sample with 500 ml of dialysis buffer overnight at 4° to remove glycine.

11. Transfer the dialysate (∼200 μl) to a new Eppendorf tube and add 5 volumes of acetone (precooled at −20°). Centrifuge at 25,000g for 10 min at 4° after standing overnight at −20°. Remove the supernatant and wash the pellet with 83.3% acetone (precooled at −20°). Dry the pellet with airflow.

12. Dissolve the pellet with 100 μl of 10 mM HCl and repeat the acetone precipitation (step 11).

13. Dissolve the pellet with 100 μl of 10 mM HCl. Quickly freeze the sample with liquid N_2 and store at $-80°$ until use.

Purity and recovery of CENP-A in each step are summarized in Table I. It is possible to obtain more than 95% pure CENP-A by this method. Although CENP-A is purified in a denatured state, we can detect nucleosome formation activity in the mixture of purified CENP-A, histone H4, H2A, and H2B.[11]

Purification of Recombinant CENP-A

As the *Escherichia coli* expression system (pET) scarcely expresses CENP-A, we use the baculovirus system. CENP-A gene tagged with sequence encoding six histidines at the amino terminus is cloned into the pFastBac vector (Invitrogen, Carlsbad, CA) recloned and into the baculovirus genome using the Bac-to-Bac system, and the virus genome is transfected into Sf-9 cells according to the manuals from Invitrogen. Core histones containing CENP-A are recovered by HCl extraction from the baculovirus-infected cells and resuspended with buffer containing 6 M urea, and His_6-CENP-A is purified by passage through a nickel column (data not shown).

Reagents

A-buffer: 20 mM Tris-HCl (pH 8.0), 0.5 M NaCl, 6 M urea
B-buffer: 20 mM Tris-HCl (pH 8.0), 0.5 M NaCl, 6 M urea, 1 M imidazole
10% Triton X-100

TABLE I
SUMMARY OF PURIFICATION OF CENP-A FROM HeLa CELLS

	Total protein	Yield (%)	CENP-A (μg)	Purification (fold)
0.6 M NaCl nuclear pellet	—	—	33	—
Chromatin protein	120 mg	90	30	1
First HPLC	360 μg	45	15	170
Second HPLC	90 μg	30	10	440
SDS–PAGE/electroelution (10^{10} cells)	5 μg	15	5	>3800

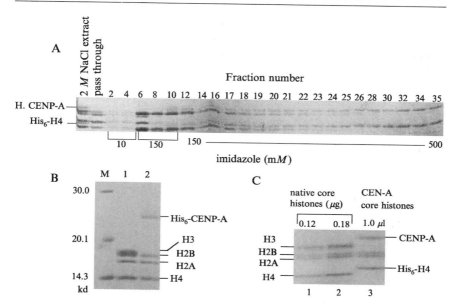

FIG. 4. Purification and reconstitution of CENP-A core histones. (A) Purification of CENP-A/His$_6$-histone H4 in native state. CENP-A and His$_6$-histone H4 were coexpressed in baculovirus-infected Sf-9 cells. The chromatin proteins extracted with 2 M NaCl from the purified nuclei were applied to a nickel column and eluted with 30 ml of a 150–500 mM histidine linear gradient. Proteins in each fraction were separated by SDS–12.5% PAGE. Fraction numbers are indicated at the top. The concentration of imidazole in each fraction is indicated at the bottom. To remove H3/H4 complex from Sf-9 cells fractions 31–35 were pooled as CENP-A/His$_6$-H4 complex. (B) Reconstitution of His$_6$-CENP-A/H4/H2A/H2B: 1.2 μg of native core histones purified from HeLa nuclei (lane 1) and 10 μl of the reconstituted His$_6$-CENP-A core histones 1 (see Table II) (lane 2) were electrophoresed through an SDS–12.5% polyacrylamide gel and histone bands were detected by CBB staining. This result shows an equal ratio of the His$_6$-CENP-A core histones and the protein concentration was calculated as 0.24 μg/μl. Lane M, protein molecular mass marker. (C) Reconstitution of CENP-A/His$_6$-H4/H2A/H2B: 0.12 μg (lane 1) and 0.18 μg (lane 2) of native core histones and 1.0 μl of the reconstituted CENP-A core histones 2 (see Table II) (lane 3) were electrophoresed through an SDS–10–20% polyacrylamide gel and histone bands were detected by CBB staining. An equimolar ratio of CENP-A/His$_6$-H4 and H2A/H2B was confirmed from this result and the protein concentration of the reconstituted CENP-A core histones 2 was calculated as 0.14 μg/μl.

Purification of His$_6$-CENP-A

1. Suspend Sf-9 cells infected with baculovirus containing the His$_6$-CENP-A gene with 1 × EB–1 M NaCl to 2 × 10^7 cells/ml.
2. Add a 1/100th volume of 10% Triton X-100.
3. Homogenize the lysed cell mixture by extensive sonication.

4. Centrifuge the sample at 16,000g for 15 min at 4°.

5. Transfer the supernatant to a new plastic tube and add 6 M HCl to a final concentration of 0.25 M and let stand at 0° for 30 min.

6. Centrifuge at 10,000g for 15 min at 4°.

7. Save the supernatant to a new plastic tube and add 100% (w/v) TCA to a final concentration of 5% and let stand for 60 min at 0°.

8. Centrifuge at 16,000g for 15 min at 4°.

9. Discard the supernatant. Wash the pellet twice with 83.3% acetone precooled to −20°. Dry the pellet with airflow.

10. Suspend the pellet with EB–0.5 M NaCl–6 M urea and dialyze the suspension against A-buffer containing 5 mM imidazole.

11. Apply the sample to a nickel column and elute His$_6$-CENP-A with a 0.01 to 0.5 M imidazole linear gradient (1–50% B-buffer).

12. Pool the CENP-A fractions and dialyze against 0.1% SDS–0.15 M NaCl–20 mM Tris-HCl (pH 8). Measure the protein concentration of the dialysate.

13. Add an approximately equal amount of histone H4. Confirm that the CENP-A:H4 ratio is equal to ~1 by SDS–PAGE each time, because the amount of CENP-A decreases in every procedure.

14. Recover the CENP-A/H4 mixture by acetone precipitation.

Renaturation Process

Although the purified CENP-A is soluble in 10 mM HCl, it becomes insoluble on neutralization even when it contains an equal amount of histone H4. An equimolar mixture of CENP-A and histones H4, H2A, and H2B in 10 mM HCl shows partial nucleosome formation activity.[11] But the activity is unstable and it is difficult to perform quantitative analysis. By denaturation–renaturation treatment using 7 M guanidine-HCl as described below, the CENP-A/H4 mixture partially regains its solubility in neutral buffer and the reconstituted CENP-A core histones show efficient nucleosome formation activity (Fig. 5, lanes 7–9).

Reagents

Note: Asterisks indicate a solution is to be made fresh before use.
   * Denaturation buffer: 7 M guanidine-HCl, 50 mM Tris-HCl (pH 8.0), 10 mM DTT, 2 mM EDTA
   * Renaturation buffer: 10 mM Tris-HCl (pH 8.0), 1 mM DTT, 1 mM EDTA, 2 M NaCl, 0.2 mM PMSF

1. Dissolve the His$_6$-CENP-A/H4 pellet with denaturation buffer at a protein concentration of ~1 μg/μl and let stand for 1 h at room temperature.

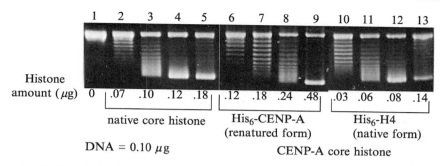

Histone amount (μg)   0   .07   .10   .12   .18   .12   .18   .24   .48   .03   .06   .08   .14

native core histone   His$_6$-CENP-A   His$_6$-H4
                  (renatured form)   (native form)

DNA = 0.10 μg   CENP-A core histone

FIG. 5. Quantitative Topo-I analyses of the efficiency of nucleosome reconstitution, using H3 core histones and CENP-A core histones. pUC119α 11-mer DNA (0.1 μg) and variable amounts of H3 core histones (lanes 2–5; amount in each lane is indicated at the bottom) and CENP-A core histones 1 (CENP-A/H4 complex was renatured) and CENP-A core histones 2 (native form) (see Table II) were subjected to nucleosome reconstitution reaction as described in text. The maximum efficiency is 0.12 μg for H3 core histones, 0.48 μg for CENP-A core 1, and 0.08 μg for CENP-A core 2 (native form). These results show that the increase in linking number in every nucleosome formation is the same for H3 nucleosome and CENP-A nucleosome. Formation of active CENP-A/H4 complex by the renaturation process is estimated to be 20–30% by comparing lanes 6–9 with lanes 10–13.

2. Dialyze against the renaturation buffer for 1 h at room temperature and for several hours at 4°. Although insoluble materials appear after dialysis, do not remove this by centrifugation.

3. Dispense the dialysate to small aliquots and stock at −80° after quick chilling with liquid N$_2$.

Purification of Native CENP-A/His$_6$-H4 Complex

Figure 5, lanes 6–9, shows that the purified CENP-A efficiently forms nucleosomes together with histone H4, H2A, and H2B. But in comparing the stoichiometry of DNA protein ratio with native core histones (Fig. 5, lanes 2–5), where approximately equal amounts of DNA and core histones are needed for maximum nucleosome formation (Fig. 5, lane 4), a five times greater amount of reconstituted CENP-A core histones is needed for maximum nucleosome formation (Fig. 5, lane 9). Considering the possibility that the ratio of proper CENP-A/H4 association may be low by the renaturation process, we plan to form CENP-A/H4 complex *in vivo* and to purify it under native conditions. Using the pFastBacDUAL vector (Invitrogen), the CENP-A gene is cloned under the control of the polyhedrin promoter and the histone H4 gene is tagged with His$_6$ at the amino terminus under the control of the p10 promoter. CENP-A/His$_6$-H4 complex is purified by passage through a nickel column as follows (Fig. 4A).

Reagents

A-buffer: 20 mM Tris-HCl (pH 8.0), 2 M NaCl
B-buffer: 20 mM Tris-HCl (pH 8.0), 2 M NaCl, 1 M imidazole

1. Isolate the nuclei from Sf-9 cells (1.6 × 10^9) infected with baculovirus containing CENP-A/His$_6$-H4 genes by the methods described in the previous section.

2. Suspend the purified nuclei with 1× EB to a final concentration of 4 × 10^8 nuclear equivalents/ml (40 ml) and mix well, until the nuclei become a homogeneous suspension.

3. Add an equal volume of 1 × EB–4 M NaCl with continuous mixing (final concentration, 2 M NaCl) and let stand for 1 h at 0°. (The nuclear suspension becomes extremely viscous because of the spreading out of the chromosomal DNA. Do not use a sonicator. Otherwise it becomes difficult to isolate CENP-A/His$_6$-H4 complex, for reasons not understood.)

4. Centrifuge at 80,000g for 60 min at 4°. Save the supernatant to a new plastic tube.

5. Add imidazole to the supernatant to a final concentration of 5 mM and apply to a nickel column (1 ml).

6. Wash the column with 5 column volumes each of 1% B-buffer and 15% B-buffer (fractionation, 1 ml/tube; flow rate, 0.5 ml/min).

7. Elute with 20 column volumes of a 15–50% linear gradient of B-buffer (fractionation, 0.5 ml/tube; flow rate, 0.5 ml/min).

8. Electrophorese 10 μl of each fraction through an SDS–12.5% polyacrylamide gel and stain with CBB. Results are shown in Fig. 4A. Fractions 31–35 in Fig. 4A are pooled and put into a dialysis tube.

9. To concentrate the proteins by ~5-fold, sprinkle powdered polyethylene glycol 20,000 over the dialysis tube, let stand for a few hours at 4°, and dialyze against 1 × EB–0.5 M NaCl for several hours.

Reconstitution of CENP-A Core Histones

1. Measure the protein concentration of the renatured His$_6$-CENP-A/H4 mixture or of native CENP-A/His$_6$-H4 complex.

2. Add an equal amount of native histone H2A/H2B complex, which is purified from HeLa cells. Confirm that the CENP-A:H4:H2A:H2B ratio is ~1:1:1:1 by SDS–PAGE, by comparing with core histones from HeLa cells as shown in Fig. 4B and C.

3. Dialyze the CENP-A/H4/H2B/H2A mixture against 1 × BB (see later)–0.5 M NaCl.

4. Dispense the dialysate to small aliquots and stock at −80° after quick chilling with liquid N$_2$.

Nucleosome Reconstitution

The salt dialysis method is generally used for nucleosome formation, but the efficiency of nucleosome formation is poor when CENP-A is used instead of histone H3. The low efficiency of nucleosome formation may be due to formation of CENP-A aggregates. When CENP-A core histones are reconstituted, insoluble materials appear at neutral pH. But these insoluble materials can be used as substrate for nucleosome reconstitution when NAP-1 is used. Here we introduce the NAP-1 method, which is efficient for CENP-A nucleosome formation. The advantages of this method compared with salt dialysis are that nucleosome formation can be performed on numerous samples at a time, on a small scale ($\sim$40 μl), in a short period of time ($\sim$1 h).

Purification of NAP-1

To form nucleosomes efficiently, enough NAP-1 is needed. We express NAP-1 using *E. coli* cells transfected with mouse NAP-1 gene cloned into a pET vector. The protein is purified by passage through Q-Sepharose and MonoQ columns.[12]

Purification of Topo-I

Topo-I is purified from HeLa cells.[13] NaCl nuclear extract (0.6 M) is diluted to 0.3 M NaCl and applied to a Q-Sepharose column. The flow-through fraction is then applied to a heparin–Sepharose column and Topo-I is eluted with a 0.4–1.0 M NaCl linear gradient.

Nucleosome Reconstitution

The reaction consists of two steps.

1. Formation of complex with NAP-1 and core histones: Variable amounts of H3 core histones or CENP-A core histones are incubated with 2 μg of NAP-1 in 20-μl reaction cocktails containing 10 mM Tris-HCl (pH 8.0), 0.15 M NaCl, 10% glycerol, 1 mM EDTA, 1 mM DTT, and 0.05% NP-40 for 15 min at 37°.

2. Formation of nucleosomes and Topo-I reaction: Add 0.1 μg of pUC 119-α-satellite 11-mer DNA (relaxed close circular form) and 2 units of Topo I in 20 μl of buffer containing the same reagents as described above and incubate for 45 min at 37°. α-Satellite 11-mer DNA originates from the centromere region of human chromosome 21 and contains five CENP-B boxes.[14]

---

[12] T. F. Nakata, Y. Ishimi, A. Okuda, and A. Kikuchi, *J. Biol. Chem.* **267**, 20980 (1992).

[13] K. Ishii, T. Hasegawa, K. Fujisawa, and T. Ando, *J. Biol. Chem.* **258**, 12728 (1983).

[14] M. Ikeno, H. Masumoto, and T. Okazaki, *Hum. Mol. Genet.* **3**, 1245 (1994).

As an example, we show a Topo-I analysis in which the formation of nucleosomes can be detected by the introduction of superhelical turns detected by DNA ladders in an agarose gel. We perform the quantitative analysis of core histones needed for maximum nucleosome formation with a fixed amount of DNA (Fig. 5).

Reagents

Note: Asterisks indicate a solution is to be made fresh before use.
   * 2× BB: 20 mM Tris-HCl (pH 8.0), 20% glycerol, 2 mM EDTA, 2 mM DTT, 0.1% NP-40
   * 0.5 M NaCl
   Native core histones: 0.12 μg/μl (stocked at $-80°$)
   CENP-A core histones 1 (His$_6$-CENP-A, renatured form): 0.24 μg/μl (stocked at $-80°$)
   CENP-A core histone 2 (His$_6$-H4, native form): 0.14 μg/μl (stocked at $-80°$)

Procedure

1. Mix the reagents in each sample tube as shown in Table IIA.
2. Incubate the tubes for 15 min at 37°.
3. Add 20 μl of the DNA mixture as shown in Table IIB to each tube and mix gently.
4. Incubate the tubes for 45 min at 37°.
5. Add 1 μl of 10% SDS and 1 μl of proteinase K (20 mg/ml) and incubate at 37° for 5 min.
6. Extract the DNA in each tube with phenol and succeding Ethanol precipitation and resuspend with 10 μl of TE buffer.
7. Electrophorese the samples through a 0.7% agarose gel (14 × 14 × 1 cm) without ethidium bromide for 60 min at 150 V. Results are shown in Fig. 5.

Conclusions

We describe methods for purifying CENP-A from HeLa cells. Because of its high hydrophobicity, CENP-A was difficult to solubilize and liable to be lost during purification. We have succeeded in CENP-A purification only under denaturing conditions. The mixture of purified CENP-A and histone H4 was partially reactivated by the renaturation procedure. The CENP-A/H4 complex was purified under native conditions using the baculovirus system. We show that CENP-A can replace histone H3 and efficiently form nucleosomes together with histones H4, H2A, and H2B *in vitro* by using NAP-1. Structures of the reconstituted CENP-A

TABLE II
PREPARATION OF NAP-1 AND DNA MIXTURES

A. Preparation of NAP-1 Mixtures

	Tube no.												
	1	2	3	4	5	6	7	8	9	10	11	12	13
2× BB (μl)	10	10	10	10	10	10	10	10	10	10	10	10	10
Distilled water (μl)	4	4	4	4	4	4	4	4	4	4	4	4	4
H3 core (μl)		.6	.8	1.0	1.5								
CENP-A core 1 (μl)						.5	.7	1.0	2.0				
CENP-A core 2 (μl)										.2	.4	.6	1.0
0.5 M NaCl (μl)	5.0	4.4	4.2	4.0	3.5	4.5	4.3	4.0	3.0	4.8	4.6	4.4	4.0
NAP-1 (2 μg/μl) (μl)	1	1	1	1	1	1	1	1	1	1	1	1	1
Total volume (μl):	20	20	20	20	20	20	20	20	20	20	20	20	20

B. Preparation of DNA Mixtures

	×1	×14
2× BB (μl)	10	140
pUC119α 11-mer (0.1 μg/μl, relaxed form) (μl)	1.0	14
Distilled water (μl)	2.0	28
0.5 M NaCl (μl)	6.0	84
Topo-I (2 U/μl)	1.0	14
Total volume (μl):	20	280

nucleosomes are shown to be basically the same as those of H3 nucleosomes, but some differences are also suggested in nucleosome shape and length of the nucleosomal DNA unit.[11] Although the efficiency of nucleosome formation is high by the NAP-1 method, spaces between nucleosomes are heterogeneous. CENP-A nucleosomes are formed selectively on I-type α-satellite arrays *in vivo*. We performed competitive CENP-A nucleosome formation *in vitro* between pUC119 DNA and pUC119-α-satellite 11-mer DNA and found no preference for I-type α-satellite DNA under the present conditions. The results might be reasonable considering that overexpression of CENP-A results in distribution of CENP-A over the chromosomes.[2] The remodeling complex that forms CENP-A nucleosomes remains to be found. This method for CENP-A nucleosome formation represents a useful tool in searching for remodeling factors; the final goal of the reconstitution system will be to reconstitute CENP-A nucleosomes that reflect *in vivo* events.

[18]   Immunological Analysis and Purification of Centromere Complex

By Kinya Yoda *and* Satoshi Ando

Introduction

Centromeric DNA regions in mammalian cells are usually constituted of highly repetitive sequences ranging from 0.5 to a few megabases.[1] It is important to examine the centromeric structure as DNA–protein complex forms as a result of DNA–protein and/or protein–protein interactions. To date, many centromeric structural proteins have been reported, such as CENP-A,[2,3] CENP-B,[4] CENP-C,[5] CENP-H,[6] CENP-I/h. Mis-6,[7,8] and h.Mis-12.[9] These proteins seem to be components of the centromere, because they are detected at the centromere region all through the cell cycle. Among them, CENP-A is a variant of histone H3[2,3] and it has been widely accepted that formation of CENP-A nucleosomes defines the active centromere region.[10] Therefore, we reason that if we isolate the chromatin containing CENP-A nucleosomes, then we have isolated the centromere chromatin complex. We have raised monoclonal antibodies against CENP-A peptides (from the NH_2 terminus, amino acids 3–19) that recognize CENP-A with high specificity (available at MBL Co. Ltd., Japan).[11] In this chapter, we describe methods for isolating CENP-A chromatin complex from interphase HeLa cells.

The method was originally developed for isolating CENP-A, CENP-B, and CENP-C chromatin complex and for analyzing its molecular structures.[11] The methods for chromatin immunoprecipitation (ChIP) described

[1] H. F. Willard, *Trends Genet.* **6,** 410 (1990).

[2] D. K. Palmer, K. O'Day, H. L. Trong, H. Charbonneau, and R. L. Margolis, *Proc. Natl. Acad. Sci. USA* **88,** 3734 (1991).

[3] S. Stoler, K. C. Keith, K. E. Curnick, and M. Fitzgerald-Hayes, *Genes Dev.* **9,** 573 (1995).

[4] W. C. Earnshaw, K. F. Sullivan, P. S. Machlin, C. A. Cook, D. A. Kaiser, T. D. Pollard, N. F. Rothfield, and D. S. Cleaveland, *J. Cell Biol.* **104,** 817 (1987).

[5] H. Saitoh, J. Tomkiel, C. A. Cooke, H. RatrieIII, M. Maurer, N. F. Rothfield, and W. C. Earnshaw, *Cell* **70,** 115 (1992).

[6] N. Sugata, S. Li, W. C. Earnshaw, T. J. Yen, K. Yoda, H. Masumoto, E. Munekata, P. E. Warburton, and K. Todokoro, *Hum. Mol. Genet.* **9,** 2919 (2000).

[7] K. Takahashi, E. S. Chen, and M. Yanagida, *Science* **288,** 2215 (2000).

[8] A. Nishihashi, T. Haraguchi, Y. Hiraoka, T. Ikemura, V. Regnier, H. Dodson, W. C. Earnshaw, and T. Fukagawa, *Dev. Cell* **2,** 463 (2002).

[9] G. Goshima, Y. Kiyomitsu, K. Yoda, and M. Yanagida, *J. Cell Biol.* **160,** 25 (2003).

[10] K. F. Sullivan, *Curr. Opin. Genet. Dev.* **11,** 182 (2001).

here consist of two steps; solubilization of bulk chromatin and immunoprecipitation of the CENP-A/B/C complex using anti-CENP-A antibody beads. Because we aim to analyze the protein components as well as the DNA, this method differs from that frequently used for analyses in the *Saccharomyces cerevisiae* or *Schizosaccharomyces pombe* system[12] in the following points: (1) the amount of starting materials is increased as much as possible; (2) a cross-linker like HCOOH is not used; (3) bulk chromatin is solubilized not by sonication but only by micrococcal nuclease (MNase) digestion under the 0.3 M NaCl condition; and (4) the antibodies used for immunoprecipitation are chemically linked to a solid support (Sepharose 4B; Pharmacia, Uppsala, Sweden) in order to repress IgG contamination to the least level during recovery of the proteins from the antibody beads.

Coupling Reaction with CNBr-Activated Sepharose Beads and Anti-CENP-A (or Preimmune) IgG

This method is essentially according to the protocol by Pharmacia, with slight modifications. Anti-CENP-A IgG is purified with protein G–Sepharose (Pharmacia) from the culture medium after incubation of hybridoma cells producing the antibody or of mouse ascites containing anti-CENP-A IgG. A human IgG mixture of antibodies against CENP-A, CENP-B, and CENP-C is purified by the same method from anti-centromere antibody (ACA) serum (AI). Mouse preimmune IgG is purchased from Chemicon International, Temecula, CA. To retain the maximum binding capacity of CNBr-activated Sepharose before coupling the ligand, use cold (0–4°) solutions. The time interval between washing and coupling must be minimized; therefore preparation of all required solutions just before coupling is recommended.

Reagents

HCl buffer: 1 mM HCl
Coupling buffer (C-buffer): 0.1 M NaHCO$_3$ (pH 8.3), 0.5 M NaCl
Acidic buffer (A-buffer): 0.1 M CH$_3$COO-Na (pH 4.0), 0.5 M NaCl
Basic buffer (B-buffer): 0.1 M Tris-HCl (pH 8.0), 0.5 M NaCl

Protocol

1. Weigh 1 g of powdered CNBr-activated Sepharose 4B (Pharmacia) into a 50-ml disposable centrifuge tube. Wash for 15 min with 200 ml of

[11] S. Ando, H. Yang, N. Nozaki, T. Okazaki, and K. Yoda, *Mol. Cell. Biol.* **22,** 2229 (2002).
[12] A. Hecht and M. Grundstein, *Methods Enzymol.* **304,** 399 (1999).

ice-cold HCl buffer: Wash four times with 50 ml of HCl buffer. Remove the supernatant, using an aspirator, after centrifugation at $30g$ for 5 min at $4°$. Transfer the Sepharose beads to a new 15-ml disposable centrifuge tube.

2. Wash the HCl-treated beads once with 10 ml of C-buffer and resuspend the pellet with 5 ml of the same buffer.

3. Add 5–15 mg of anti-CENP-A (or preimmune IgG or IgG from ACA serum), dissolved in phosphate-buffered saline (PBS), to the bead suspension. Aliquot 10 μl of the supernatant as the IgG solution before the coupling reaction. CNBr-activated Sepharose reacts with the $-NH_2$ group of the ligand. Therefore, reagents containing the $-NH_2$ group (e.g., Tris-HCl buffer) should be avoided as the ligand buffer. C-buffer is recommended for the ligand buffer in the original protocol, but PBS can also be used.

4. Mix gently by upside-down rotation for a few hours at room temperature or overnight at $4°$.

5. Centrifuge the bead suspension at $30g$ for 5 min and remove the supernatant to a new tube as the IgG solution after the coupling reaction. Measure the protein concentration of the IgG solution before and after the coupling reaction. When the coupling reaction goes well, the protein concentration of the supernatant will change from $\sim$1 mg/ml to <0.1 mg/ml (undetectable). We use the Bradford method: 1- to 10-μl samples are mixed with 300 μl of Coomassie Brilliant Blue G250 (Bio-Rad, Hercules, CA) and left to stand for 15–30 min at room temperature. Measure the absorbency at 595 nm. One to 10 μg of protein is measurable by this method.

6. After confirming the successful coupling reaction, wash away excess ligand with at least 5 volumes of C-buffer (10 ml, twice).

7. Block any remaining active group with 10 ml of 0.1 M Tris-HCl (pH 8.0) for 2 h at room temperature.

8. Wash the product with at least three cycles of altering pH buffers, A-buffer (pH 4.0) and B-buffer (pH 8.0).

9. Suspend the IgG beads with PBS containing 0.1% NaN_3 or thymerosal and stock on ice.

Isolation of CEN Complex from Interphase HeLa Nuclei

Reagents

Note: Asterisks indicate a solution is to be made fresh before use.

* Chromosome isolation buffer (CIB), 2× washing buffer (2 × WB), and digitonin described in [17] in this volume[13]

[13] K. Yoda, S. Morishita, and K. Hashimoto, *Methods Enzymol.* **275**, [17], 2004 (this volume).

  * Micrococcal nuclease (MNase) (nuclease S7; Roche Molecular Biochemicals, Indianapolis, IN): 10 U/ml in 50% glycerol
    $CaCl_2$: 1 M
    EGTA: 0.2 M
    NP-40: 10%

Isolation of Nuclei

1. Thaw frozen HeLa cell pellets (1×10^9 cells; stocked at $-80°$) in a $37°$ water bath with frequent mixing in order to avoid local increases in temperature. Unless otherwise indicated, all subsequent procedures should be performed on ice.

2. Add CIB containing 0.1% digitonin to a final concentration of 2×10^7 cells/ml to the cell pellet (50 ml; ~50 ml × 1 with a 50-ml disposable centrifuge tube).

3. Homogenize the cell suspension with 10 up-and-down strokes in a Dounce homogenizer (Wheaton, Millville, NJ) with a tight-fitting pestle. Be careful not to make air bubbles, especially on the down strokes, because small air bubbles break down the nuclear membrane, which causes loss of products owing to formation of nuclear aggregates.

4. Centrifuge the homogenate at $300g$ for 5 min at $4°$ with a swinging rotor.

5. Discard the supernatant and repeat steps 2–5 twice. Transfer the pellet to a new 15-ml disposable centrifuge tube.

Solubilization of Bulk Chromatin with MNase Digestion of Nuclei in 0.3 M NaCl

We show here the protocol for the kinetic analysis of MNase digestion and recovery of CENP-A/B/C complex by chromatin immunoprecipitation (ChIP) using anti-CENP-A beads.

6. Wash the pellet with 1 × WB to a final concentration of 2×10^7 nuclear equivalents/ml (50 ml).

7. Suspend the pellet with 1 × WB–0.3 M NaCl to a final concentration of 1×10^8 nuclear equivalents/ml (10 ml).

8. Centrifuge the nuclear suspension at $500g$ for 10 min at $4°$ with a swinging rotor and discard the supernatant.

9. Suspend the nuclear pellet with 1 × WB–0.3 M NaCl to a final concentration of 2×10^8 nuclear equivalents/ml (5 ml) and add $CaCl_2$ to a final concentration of 3 mM. Dispense the sample to five Eppendorf tubes (~1 ml each).

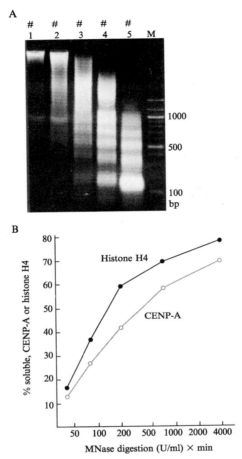

Fig. 1. (A) Size distribution of DNA fragments from bulk chromatin after MNase digestion. HeLa nuclei were digested with MNase as shown in step 10 (see Protocol). Fifty microliters of each MNase supernatant fraction (step 12) was aliquoted and the fragmented DNA in each fraction was extracted with phenol. One-fiftieth of each sample was electrophoresed through a 1% agarose gel. DNA was detected by ethidium bromide staining. Sample numbers, shown at the top of each lane, correspond to those shown in step 10. Positions of a DNA size marker are indicated at the right. (B) Solubility of CENP-A chromatin and bulk chromatin (represented by histone H4). Two sets of 10-μl MNase supernatant fractions and pellet fractions at step 12 were subjected to SDS–12.5% PAGE. Pellet fractions were prepared as follows: the pellets were suspended each with 1 ml of 2× SDS buffer with extensive sonication and boiled for 5 min. Proteins in one gel were stained

10. Add MNase and incubate at 37° as follows:

Sample no.	MNase (U/ml)	Incubation time (min)	(U/ml) × min[a]
1	20	2	40
2	20	4	80
3	40	5	200
4	60	10	600
5	80	45	3600

[a] (U/ml) × min represents the extent of MNase digestion. The extents of DNA digestions are reproducible as long as judged by nucleosome ladders in agarose gels.

11. Stop the reaction by adding 0.2 M EGTA to a final concentration of 5 mM and quickly chilling to 0°.

12. Centrifuge the MNase digests at 13,000g for 10 min at 4°. Transfer the supernatant to new centrifuge tubes and again centrifuge at 13,000g for 10 min at 4° in order to remove the residual pellet. Transfer the supernatant to new centrifuge tubes (MNase supernatant fractions). Figure 1A shows the extent of DNA cleavage in each sample by agarose gel electrophoresis of the extracted DNA.

Immunoprecipitation of centromere chromatin with anti-CENP-A IgG beads

13. Add NP-40 to each tube to a final concentration of 0.1% (10%, 10 μl).

14. Add 50 μl of IgG beads, which are prewashed with WB–0.3 M NaCl–0.1% NP-40.

---

with Coomassie Brilliant Blue. Proteins in the other gel were electrically transferred to a PVDF membrane and CENP-A was detected by immunostaining, using ACA serum (AI). Intensities of histone H4 in the gel and of the CENP-A bands in the membrane were measured using the NIH Image digitizer. Relative amounts of CENP-A or histone H4 in the soluble fractions were calculated as $s/(s + p) \times 100$, where s and p represent the intensities of CENP-A (or histone H4) in the soluble and insoluble fractions, respectively, and the solubility of each sample is plotted against MNase digestion represented as (U/ml) × min. The results indicate that CENP-A chromatin is solubilized at approximately the same rate as bulk chromatin by MNase digestion of HeLa nuclei.

15. Mix gently by upside-down rotation for 1 h at room temperature or for several hours to overnight at 4°.

16. Centrifuge the samples at 30g for 5 min at 4°. Transfer the supernatant to new 1.5-ml Eppendorf tubes. Save the IgG beads as a negative control for step 19.

17. Pour each of the supernatants into new 1.5-ml centrifuge tubes containing 50 μl of anti-CENP-A IgG beads, which are prewashed with WB–0.3 M NaCl–0.1% NP-40, and gently mix by upside-down rotation overnight at 4°.

18. Centrifuge the samples at 30g for 5 min at 4°. Remove the supernatant and save the pellets.

19. Wash the nonimmune IgG beads and anti-CENP-A IgG beads after immunoreaction with 1.0 ml of ice-cold WB–0.3 M NaCl–0.1% NP-40 three times.

20. Add an equal volume of 2× SDS buffer (50 μl) to each of the bead pellets and boil for 5 min.

21. Centrifuge the bead suspensions at 13,000g for 1 min and save the supernatant to new Eppendorf tubes. Freeze the samples with liquid N_2 and stock at $-80°$ until use.

Results and Discussion

CENP-A chromatin can be solubilized by MNase digestion of HeLa interphase nuclei to approximately the same extent as bulk chromatin (Fig. 1B). Recovery of CENP-A chromatin in the soluble fraction is increased up to ~70% (Fig. 1B), but recovery of CENP-A/B/C complex is decreased by extensive MNase digestion (Fig. 2), for DNA cleavage cuts out CENP-B and CENP-C as well as CENP-A/B/C complex.[10] Recovery of CENP-A/B/C complex reaches a maximum at a solubility of CENP-A chromatin of 30–50% (Fig. 2B). Solubility of chromatin is highly dependent on salt concentration. The higher the salt concentration, the more soluble chromatin becomes,[10] but CENP-B is detached from chromatin at 0.5–0.6 M NaCl without MNase digestion. Therefore, we chose 0.3 M NaCl as the salt condition for MNase digestion. Chromatin isolated with anti-CENP-A antibody contains CENP-H, CENP-I (human homolog of *Sch. pombe* Mis-6), and human Mis-12 as well as CENP-A, CENP-B, and CENP-C (Genes to cells impress 2003), but not CENP-G.[11] Thus, the isolated centromere chromatin contains all centromeric proteins known as constitutive proteins. These results, together with the information that CENP-A/B/C chromatin forms the

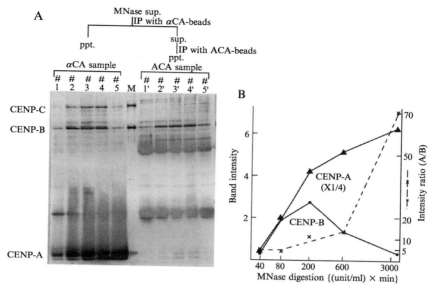

FIG. 2. Kinetic study of the amounts of CENP-A, CENP-B, and CENP-C in A/B/C chromatin complexes precipitated by anti-CENP-A antibodies. (A) Western blotting with ACA serum. Five MNase digests at step 13 were immunoprecipitated with anti-CENP-A antibody beads (steps 13–21) [lanes 1–5, 12-μl samples of anti-CENP-A antibody (αCA)]. The supernatants were subsequently precipitated with ACA serum (lanes 6–12, 12-μl samples of ACA) to isolate and detect the residual CENPs. The precipitated proteins were resolved by SDS–12.5% PAGE and immunoblotted with ACA serum. The diffuse bands between CENP-A and CENP-B in the ACA samples (lanes 6–12) were human IgG released from the ACA–Sepharose beads. Lane M shows an A/B/C marker mixture. Lanes 1'–5' show that immunoprecipitation (IP) with anti-CENP-A beads can deplete most of the CENP-A chromatin from the MNase supernatant fractions. (B) Kinetic analysis of relative amount of CENP-A and CENP-B measured from the intensity of each band shown in (A). The intensity of each band of CENP-A (solid triangles) and CENP-B (solid circles) is measured by the NIH Image Digitizer and the A/B intensity ratio is calculated (dashed line). Amounts of CENP-A continue to increase as the extent of MNase digestion goes up, while those of CENP-B increase first (data points 1 and 2), come to maximum (data point 3), and decrease (data points 4 and 5). The A/B intensity ratio converges to a constant value as MNase digestion weakens; its value might represent the molar ratio of CENP-A/CENP-B in the CENP-A/B/C complex.

unit structure reflecting the repetitive structure of I-type α-satellite DNA arrays,[11] suggest that the centromere complex fragmented by weak MNase digestion may well preserve all the components of the centromere complex.

[19] Purification and Analyses of Histone H1 Variants and H1 Posttranslational Modifications

By Craig A. Mizzen

Introduction

Before the advent of recombinant DNA technology, surveys of histone expression using specialized electrophoretic and chromatographic techniques revealed that amino acid sequence variants (subtypes) of all histones except H4 are coexpressed in many eukaryotes. Depending on the species and tissue analyzed, as many as eight forms of H1 and five, three, and three forms of H2A, H2B, and H3, respectively, have been resolved that are not accounted for by posttranslational modification.[1–6] Typically, these forms are expressed in tissue-specific fashion to similar degrees in different individuals of a species, suggesting that they represent nonallelic variants rather than allelic variants. This distinction is important to note because allelic variants of H1 and core histones have also been resolved by some of the techniques described in this chapter. However, unless noted otherwise, discussion here is limited to analyses of nonallelic variants of H1.

Subsequent analyses have revealed details of the organization, structure, and regulation of histone genes in a variety of species.[7–10] In addition to confirming the presence of genes encoding nonallelic variants of H1, H2A, H2B, and H3, this work, together with protein sequence data, enables the variants separated by a given technique to be correlated with

[1] A. Zweidler, *in* "Histone Genes" (G. S. Stein, J. L. Stein, and W. F. Marzluff, eds.). John Wiley & Sons, New York, 1984.

[2] R. D. Cole, *Anal. Biochem.* **136,** 24 (1984).

[3] R. D. Cole, *Int. J. Pept. Protein Res.* **30,** 433 (1987).

[4] R. W. Lennox and L. H. Cohen, *in* "Histone Genes" (G. S. Stein, J. L. Stein, and W. F. Marzluff, eds.). John Wiley & Sons, New York, 1984.

[5] K. E. van Holde, "Chromatin." Springer-Verlag, New York, 1989.

[6] A. P. Wolffe, "Chromatin." Academic Press, San Diego, CA, 1998.

[7] G. S. Stein, J. L. Stein, and W. F. Marzluff, "Histone Genes." John Wiley & Sons, New York, 1984.

[8] T. Nakayama, S. Takechi, and Y. Takami, *Comp. Biochem. Physiol.* **104,** 635 (1993).

[9] Z. F. Wang, A. M. Sirotkin, G. M. Buchold, A. I. Skoultchi, and W. F. Marzluff, *J. Mol. Biol.* **271,** 124 (1997).

[10] D. Doenecke, W. Albig, C. Bode, B. Drabent, K. Franke, K. Gavenis, and O. Witt, *Histochem. Cell Biol.* **107,** 1 (1997).

specific genes.[11] The amino acid sequences of mouse H1.1–H1.5, variants that are expressed in many non-germ-line tissues (hence, "somatic" variants), and the testis-specific H1t are shown in Fig. 1A. As has been found in other species, sequence corresponding to the globular domain is more highly conserved than that of the flanking N- and C-terminal domains, where significant numbers of conservative and nonconservative substitutions involving charged residues and potential sites of phosphorylation (S/T) are found. In Fig. 1B, the amino acid sequence of the most divergent variant in mice, H1^0, is shown aligned with that of H1.2. In this case, extensive variation is found throughout the molecule. Such molecular differences, together with evidence of tissue-specific patterns of expression and differences in variant metabolism from chromatographic and electrophoretic analyses, support the notion that H1 variants acquired specialized functions in chromatin during evolution even though direct evidence of this is often lacking. To aid investigators unfamiliar with this area of research, this chapter reviews the capabilities of conventional approaches to this problem and describes techniques for resolving H1 variants. Current knowledge and techniques for analyzing posttranslational modifications of H1 are also described.

Resolution of H1 Variants by Chromatographic Methods

Cation-Exchange Chromatography

Species-specific and tissue-specific heterogeneity of H1 was first discovered using Bio-Rex 70 (Amberlite IRC-50) chromatography.[12–14] Four

---

[11] A standardized nomenclature for histone variants has not been adopted. Core histone variants are usually identified by appended numerals (e.g., H3.1–H3.3) based on acid urea Triton gel electrophoretic mobility, as proposed by S. G. Franklin and A. Zweidler, *Nature* **266,** 273 (1977). Generally, these fractions appear to be homogeneous and this nomenclature has been applied to the corresponding genes on their identification. This nomenclature has also been applied to gene sequences reported for H1 variants in molecular databases, but it should be noted that in some instances, these have not been correlated to the H1 proteins resolved by various techniques. Chromatographic and electrophoretic fractions of H1 are often denoted as H1a–H1e or H1-1–H1-5, based on mobility or order of elution. In many instances, the purity of these fractions is unknown or they are known to be heterogeneous. Moreover, the degree of resolution achieved by chromatographic and electrophoretic methods varies significantly among species due to interspecies sequence variation of H1. At present, the genes corresponding to electrophoretic or chromatographic fractions of H1 are known for mice, chickens, and humans.

[12] J. M. Kinkade, Jr. and R. D. Cole, *J. Biol. Chem.* **241,** 5790 (1966).

[13] M. Bustin and R. D. Cole, *J. Biol. Chem.* **243,** 4500 (1968).

[14] J. M. Kinkade, Jr., *J. Biol. Chem.* **244,** 3375 (1969).

A

```
H1.1 (a)  SETAPVAQAA STATEKPAAA KKTKKPAKAA APRKKPAGPS VSELIVQAVS SSKERSGVSL AALKKSLAAA GYDVEKNNSR
H1.2 (c)  **A**A*P** APPA**AP*K **AA*K--P* GV*R*AS**P *****TK**A A********** ******A**** **********
H1.3 (d)  *****A*P** PAPV**TPVK *A**TG-** **AR*A--*G GAKR*AS**P *****TK**A A********** ******A**** **********
H1.4 (e)  *****A*P** PAPA**TPVK **AR*A--*G GAKR*TS**P *****TK**A A********** ******A**** **********
H1.5 (b)  *****AET** PAPV**SP*K **T*K--*G *AKR*AT**P *****TK*** A****G**** P****A***G **********
H1.t      *****A*SST LVPAPVEKPS S*RRGKKPGL **AR*R*F* **K**PE*L* T*Q**A*M** ******A****


          IKLGLKSLVN KGTLVQTKGT GAAGSFKLNK KAES-----K AITTKVSVKA KASGAAKKPK KTAGAAAKKT --VKTPKKPK
          *********S **I******* **S******* **A*GEAKPQ *KKAGAAKAK *PA******* *AT***TP*K AAK*****A*
          *********S ********** **S******* **A*GEAKP* *KKAGAAKAK *PA******* *AT***TP*K TAK*****A*
          *********S ********** **S******* **A*GEAKP* *KRAGAAKAK *PA******* *AA*T*TA*K STK*****A*
          *****A*R** **V******* **S******S* **A*GNDKG* *KK*GAAKAK *PA*T--** *PKKT*GA*K TVK*****A*
                                                     GKKSASAKAK *MGLPR---- ASRSPKSS** KA**K**ATP


          KPAVSKKTSK SPKKPK---V VKAKKVAKSP AKAKAVKPKA SKAKVTKPKT PAKPKKAAPK KK-----   212
          **AAAV*K*  VA*------ -SP**AKVTK P*KVKSAS** V*P*AA***V AKAK*V**K* *-----     211
          ***AAAGAK* VS*S*KVKA A*P**A**** *****T***** A*P*AS***A TKAK*A*PR* *-----     220
          ***AAAGAK* AKSPK*AKAT K-****AP*** *****T***** A*P*TS***A AKPK*T**K* *-----     218
          ***AAGVKKV AKSPK*AKAA A*P*A***** **P****S** **P******* AKPKAAK*K* AVSKKK      222
          TK*SGSGRKT KGA*G----- ---VQQR**** ***R*AN*NS G***MVMQ** DLRKAAGRK-  ------     207
```

B

```
H1.2 (c)  SEAAPAPAA APPAEKAPAK KKAAKKPAGV RRKASGPPVS ELITKAVAAS KERSGVSLAA LKKALAAAGY DVEKNNSRIK
H1°       T*NSTS***** K*------- ----*RAKAS KKSTDH*KY* DM*VA*IQ*E *N*A*S*RQS IQ*YIKSHY* VG*NA**Q**


          LGLKSIVSKG ILVQTKGTGA SGSFKLNKKA ASGEAKPQAK KAGAAKKPKKA TGAATPKKAA KKTPKKAKKP
          *SI*R*TT* V*K****V** ****R*A*GD EPKRSVAFK* TKKEV*KVAT PKK*A***** ASK*PS**PK ATPV*****-


          AAAAVTKKVA KSPKKAKVTK PKKVKSASKA VKPKAAKPKV AKAKKVAAKK K   211
          -KP*A*F*K* *K**VV**-* *V*----***P K*A*TV***A KSSA*RGS** *      193
```

to six components are usually apparent when metazoan H1 is chromatographed on this weak cation exchanger, using shallow gradients of guanidine hydrochloride (GuHCl) at neutral pH.[15] Although this method surpasses conventional reversed-phase high-performance liquid chromatography (RP-HPLC) for the resolution of H1 variants from certain species (e.g., chicken; see below), several disadvantages have led to the increased use of more convenient HPLC techniques. Because such shallow gradients are required, Bio-Rex 70 H1 fractionations lasting 1 week or more have been reported. Typically, the method is performed at room temperature, providing ample opportunity for proteolysis to occur. Moreover, because histones are not readily detected at 280 nm because of a paucity of aromatic residues, analyses of small samples employing detection at lower wavelengths is hampered by the strong absorbance of solutions of guanidine hydrochloride below 230 nm.

Other cation-exchange media and buffer systems have been used to purify H1 proteins, but these procedures do not resolve H1 variants as well as Bio-Rex 70 chromatography.[16–19] The efficacy of Bio-Rex 70 chromatography for resolving H1 variants may be due in part to the effects of GuHCl on the conformation of H1 proteins because four peaks were partially resolved within 30 min when a gradient of GuHCl was used to elute chicken erythrocyte H1 from a MonoS column.[20]

[15] R. D. Cole, *Methods Enzymol.* **170**, 524 (1989).
[16] N. Harborne and J. Allan, *FEBS Lett.* **194**, 267 (1986).
[17] M. Garcia-Ramirez, F. Dong, and J. Ausio, *J. Biol. Chem.* **267**, 19587 (1992).
[18] G. A. Rice and R. D. Cole, *Protein Expr. Purif.* **1**, 87 (1990).
[19] L. C. Garg and G. R. Reeck, *Protein Expr. Purif.* **14**, 155 (1998).
[20] M. F. Shannon and J. R. Wells, *J. Biol. Chem.* **262**, 9664 (1987).

FIG. 1. Amino acid sequences of mouse linker histones. The amino acid sequences of nonallelic variants of mouse H1 predicted from nucleotide sequences contained in the nonredundant set of the histone sequence database as of May 2000 are shown in one-letter code after alignment using CLUSTAL W 1.8 [D. G. Higgins, J. D. Thompson, and T. J. Gibson, *Methods Enzymol.* **266**, 383 (1996)]. (A) The sequences of the "somatic" and the testes-specific H1 variants are compared. (B) The sequence of the "extreme" variant H1° is compared with that of the shortest somatic variant, H1.2. Genes are identified according to the nomenclature (H1.1–H1.5) proposed previously [B. Drabent, K. Franke, C. Bode, U. Kosciessa, H. Bouterfa, H. Hameister, and D. Doenecke, *Mamm. Genome* **6**, 505 (1995)]. The corresponding proteins are identified by the nomenclature (a–e) based on 2-D electrophoresis [R. W. Lennox and L. H. Cohen, *J. Biol. Chem.* **258**, 262 (1983)] as described previously [Z. F. Wang, A. M. Sirotkin, G. M. Buchold, A. I. Skoultchi, and W. F. Marzluff, *J. Mol. Biol.* **271**, 124 (1997)]. The total number of predicted residues is listed at the end of each sequence. Asterisks indicate residues identical to the corresponding position in the first sequence. Hyphens indicate gaps introduced to maximize homology. The approximate location of the conserved globular domain is indicated by the line above the sequences.

Reversed-Phase Chromatography

The majority of reports using HPLC to fractionate histones have employed RP-HPLC. However, the resolution of H1 variants by RP-HPLC, in some cases, is lower than that achieved by Bio-Rex 70 chromatography or electrophoretic methods. H1 samples exhibiting electrophoretic heterogeneity eluted from RP-HPLC columns as a single peak,[21,22] although two to five peaks have been resolved depending on the sample and the chromatographic method.[23–27] Comparison of published reports suggests that the efficacy of RP-HPLC for resolving H1 variants may be species dependent. Mouse,[27–29] rat,[23,26,30] and human H1[25] have been resolved into four or more components (excluding H1⁰) by RP-HPLC. However, calf thymus H1 manifesting six components in Bio-Rex 70 chromatography,[3,31] was resolved into only two components by RP-HPLC.[24] Similarly, chicken H1, which has been resolved into four[20] or five[14,32] components by cation-exchange chromatography, eluted as a single peak in an optimized RP-HPLC method.[22] Comparison of the gene-predicted protein sequences of mouse and chicken H1 variants (Figs. 1 and 2) reveals that relatively few amino acid sequence differences distinguish the chicken variants from each other compared with the mouse variants. Taken together, these observations suggest that RP-HPLC is likely to be inferior to other methods for resolving H1 variants from species displaying minimal degrees of amino acid sequence divergence. A more promising approach in such instances is the variant of cation-exhange chromatography described below.

Cation-Exchange Hydrophilic Interaction Chromatography

Hydrophilic interaction chromatography (HILIC) is a variant of normal phase chromatography in which biomolecules are separated on the basis of polar and hydrophilic interactions with polar neutral and ionizable

[21] L. R. Gurley, D. A. Prentice, J. G. Valdez, and W. D. Spall, *J. Chromatogr.* **266,** 609 (1983).

[22] W. Helliger, H. Lindner, S. Hauptlorenz, and B. Puschendorf, *Biochem. J.* **255,** 23 (1988).

[23] M. Kurokawa and M. C. MacLeod, *Anal. Biochem.* **144,** 47 (1985).

[24] H. Lindner, W. Helliger, and B. Puschendorf, *Anal. Biochem.* **158,** 424 (1986).

[25] Y. Ohe, H. Hayashi, and K. Iwai, *J. Biochem. (Tokyo)* **100,** 359 (1986).

[26] J. C. Tchouatcha-Tchouassom, J. H. Julliard, and B. Roux, *Biochim. Biophys. Acta* **1009,** 121 (1989).

[27] V. Giancotti, A. Bandiera, L. Ciani, D. Santoro, C. Crane-Robinson, G. H. Goodwin, M. Boiocchi, R. Dolcetti, and B. Casetta *Eur. J. Biochem.* **213,** 825 (1993).

[28] T. Wurtz, *Eur. J. Biochem.* **152,** 173 (1985).

[29] W. Helliger, H. Lindner, O. Grubl-Knosp, and B. Puschendorf, *Biochem. J.* **288,** 747 (1992).

[30] H. Lindner, W. Helliger, and B. Puschendorf, *Biochem. J.* **269,** 359 (1990).

[31] M. J. Smerdon and I. Isenberg, *Biochemistry* **15,** 4233 (1976).

[32] T. Dupressoir and P. Sautiere, *Biochem. Biophys. Res. Commun.* **122,** 1136 (1984).

```
      SETAPAPAAE AAPAAAPA-P --AKAA-AKK PKKAAGGAKA RKPAGPSVTE LITKAVSASK ERKGLSLAAL KKALAAGGYD
H1a
H1a'  A*****-**-* A-****-*** ********** ********** ********** ********** ********** **********
H1b   A****V---- ***-DVA*A* TP****P*** ********** ********** ********** ********** **********
H1c   A****-**** ***-*VAX-* A-****-*** ********** ********** ********** ********** **********
H1c'  A*****-**- ***-D***-* G-****-*** *********P ********** ********** ********** **********
H1d   A*****V--- ***-*VS*-* G-****-*** ********** ********** ********** ********** **********


      VEKNNSRIKL GLKSLIVSKGT LVQTKGTGAS GSFRLSKKPG EVKEKAPKKK ASAAKPKKPA AKKPAAAAKK PKKAVAVKKS
H1a   ********** ********** ********** ********** *GL******* ********-A* ********** **********
      **S******* ********** ********** *********S* D********** TP******** ********** **********
      ********** ********** ********** ********** ******-R*R TP******** *******S*** ****A*A***
      ********** ********** ********** *****N**** ******-R*R *T******** ********** ****A****
      ********** ********** ********** ***K*N**** *T****T*** -P******** ********** ****A****


      PKKAKKPAAS ATKKSAKSPK KVTKAVKPKK AVAAKSPAKA KAVKPKAAKP KAAKPKAAKA KKAAAKKK  224
      *********** *****V**** *AA*--*** ***V****** ********** **T******* ******** 218
      **********A ****A***** *A**A***** ***V****** ********** **T******* *****P*** 223
      **********A ****A***** *A**A***** *AT******* ********** ********** ******** 219
      **********A ****A***** *AA**GR*** *--******* ********** **T******* **T***** 218
      **********A ****A***** *A**GR*** T--**GR*** *********S ********** **T***** 217
```

FIG. 2. Amino acid sequences of chicken H1 variants. The protein sequences predicted from the genes described previously [L. S. Coles, A. J. Robins, L. K. Madley, and J. R. Wells, *J. Biol. Chem.* **262,** 9656 (1987)] are shown aligned for maximum homology. Variants are named according to AU gel mobility [M. F. Shannon and J. R. Wells, *J. Biol. Chem.* **262,** 9664 (1987)]. Asterisks indicate residues identical to the H1a sequence; gaps introduced for alignment are indicated by hyphens. The predicted protein lengths are indicated at the end of each sequence. The approximate location of the conserved globular domain is indicated by the line above the sequences.

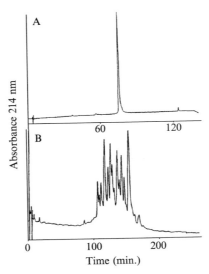

FIG. 3. Resolution of pooled chicken erythrocyte H1 variants by RP-HPLC and CX-HILIC. Identical samples (100 μg) of pooled chicken erythrocyte H1 were chromatographed by RP-HPLC (A) and CX-HILIC (D). (A) A Chromegabond MC-18 column (4.6 mm i.d. × 250 mm; ES Industries, West Berlin, NJ) was eluted with a 2-h linear gradient from 5 to 60% (v/v) acetonitrile in 0.1% (v/v) TFA. (B) A PolyCAT A column (4.6 mm i.d. × 200 mm; 100-nm-diameter pores) was eluted with a 4-h linear gradient from 380 to 590 mM $NaClO_4$ in 10 mM propionic acid (pH 6.5), 70% (v/v) acetonitrile. The flow rate was 0.8 ml/min and detection was at 214 nm in both cases.

constituents of stationary phases in hydrophobic mobile phases.[33] We and others have adapted this methodology for the separation of histone variants and modified forms, using a weak cation-exchange support. Both electrostatic and hydrophilic interactions contribute to the retention of H1 variants under the conditions described here and we refer to the method as cation-exchange hydrophilic interaction chromatography (CX-HILIC).[34] Whereas H1 from pooled chicken blood elutes as a single peak in conventional RP-HPLC (Fig. 3A), a total of 21 peaks representing all 6 nonallelic variants and allelic forms of 5 of these variants are resolved for the same sample in less than 4 h by CX-HILIC (Fig. 3B). This method is considerably more rapid and more sensitive than Bio-Rex 70 chromatography and can also be used to resolve posttranslationally modified forms of both linker and core histones after appropriate modification.[35,36]

[33] A. J. Alpert, J. Chromatogr. 499, 177 (1990).
[34] C. A. Mizzen, A. J. Alpert, L. Lévesque, T. P. A. Kruck, and D. R. McLachlan, J. Chromatogr. 744, 33 (2000).

Procedure for CX-HILIC of Chicken H1 Variants

Chicken erythrocyte nuclei are isolated as described previously.[37] Crude linker histones are prepared by either direct extraction of nuclei with 5% perchloric acid or perchloric acid fractionation of total histone extracted with 0.4 N H_2SO_4.[34] Because the elution profile of forms of H5 resolved by CX-HILIC overlaps partially with that of forms of H1 under the conditions employed here, bulk H1 and H5 are first separated by RP-HPLC[22] or size-exclusion chromatography on Bio-Gel P-100[34] before CX-HILIC.

Chromatographic Method

Chromatography is performed at ambient temperature (22°) on a binary gradient chromatography system at 0.8 ml/min with absorbance detection at 214 nm. The column should be well equilibrated (i.e., 5 column volumes) with 100% buffer A before sample injection.

Columns

4.6 mm i.d. × 200 mm packed with PolyCAT A based on 5-μm-diameter silica particles with average pore diameters of 30 and 100 nm (PolyLC, Columbia, MD)

Buffers

Buffer A: 0.2 M NaClO$_4$, 10 mM propionic acid, 70% (v/v) CH_3CN adjusted to pH 6.5 with NaOH

Buffer B: 0.8 M NaClO$_4$, 10 mM propionic acid, 70% (v/v) CH_3CN adjusted to pH 6.5 with NaOH

A (typical) gradient is as follows:

Time (min)	% B
0–5	0
5–10	0–30 (linear)
10–250	30–70 (linear)
250–260	70–100 (linear)
260–270	100
270–275	100–0 (linear)

[35] H. Lindner, B. Sarg, C. Meraner, and W. Helliger, *J. Chromatogr.* **743**, 137 (1996).

[36] H. Lindner, B. Sarg, and W. Helliger, *J. Chromatogr.* **782**, 55 (1997).

[37] C. A. Mizzen, J. E. Brownell, R. G. Cook, and C. D. Allis, *Methods Enzymol.* **304**, 675 (1999).

Comments

CX-HILIC is performed with columns packed with PolyCAT A, a silica-based weak cation exchanger available with different average particle and pore diameters. The pore diameter of the silica matrix can affect the resolution of H1 variants when PolyCAT A is used for conventional cation-exchange chromatography or CX-HILIC.[34] In general, the best resolution is achieved with media based on 30- or 100-nm pore diameter silica. Although we found that the elution profiles of H1 from pooled chicken blood on 30- and 100-nm pore PolyCAT A were overtly similar, acid urea gel analyses of the peaks resolved by each column revealed differences in the resolution of certain variants.[34] Thus, the resolutions achieved on 30- and 100-nm pore diameter supports should be compared when optimizing separations of H1 from other sources.

Analyte retention in CX-HILIC is sensitive to minor changes in mobile phase composition and it is necessary to adopt a standard method for buffer preparation to obtain good chromatographic reproducibility. $NaClO_4$ and propionic acid are first dissolved in the desired volume of acetonitrile, using a magnetic stirrer, and water is added to bring the volume to 95% of the intended final volume. The solutions are allowed to come to room temperature and the pH is adjusted with a glass combination electrode. Water is then added to give the final intended volume and buffers are vacuum-filtered, using a 0.45-μm-pore diameter Durapore membrane (Millipore, Bedford, MA) before use. Buffer reservoirs are sealed with only a small vent to the atmosphere to minimize evaporation during use.

The salt concentration required to elute H1 proteins from PolyCAT A in CX-HILIC depends on the acetonitrile concentration employed.[34] We have found that a mobile phase containing 70% acetonitrile provides the highest resolution of chicken H1 at pH 6.5 and that H1 variants elute between 400 and 600 mM $NaClO_4$ without detectable losses during chromatography under these conditions. Although chicken H1 remains highly soluble in 90% CH_3CN in the presence of $NaClO_4$, these proteins begin to precipitate within column frits when CH_3CN concentrations greater than 72% are employed. Resolution is decreased when less than 70% CH_3CN is employed. The optimum CH_3CN concentration depends on the species of origin of the H1 to be separated and must be established empirically.

Recovery of H1 proteins from CX-HILIC eluents by trichloroacetic acid (TCA) precipitation is not effective unless the concentration of CH_3CN is first significantly reduced by dialysis or evaporation in a Speed-Vac (Thermo Savant, Holbrook, NY). Cellulose-based dialysis tubing is compatible with CX-HILIC eluents and fractions can also be recovered by lyophilization after dialysis against a suitable buffer. Alternatively, fractions sufficiently depleted of CH_3CN can be desalted by RP-HPLC with volatile buffers and the H1 proteins recovered by vacuum drying.

Resolution of H1 Variants by Electrophoretic Methods

SDS–PAGE

Electrophoretic techniques differ in their ability to resolve H1 variants from different species. Excluding species such as *Tetrahymena* and *Drosophila melanogaster,* which appear to express only a single H1 polypeptide,[38,39] and in the absence of posttranslational modifications, H1 from most eukaryotes migrates as a set of closely spaced bands in sodium dodecyl sulfate (SDS) gels. In general, the resolution of H1 variants by SDS–polyacrylamide gel electrophoresis (PAGE) is not complete: only two or three bands are resolved for H1 prepared from tissues such as calf thymus and chicken erythrocytes known to express six H1 subtypes.[3,40] In avian species, the erythrocyte-specific linker histone H5 is sufficiently smaller in size that it is well resolved from the common H1 subtypes on SDS gels, as are the related $H1^0$ proteins found in certain mammalian tissues.[41]

AU/AUT–PAGE

H1 variants are usually better resolved on acid–urea (AU) and acid–urea–Triton (AUT) gels compared with SDS gels. In contrast to the core histones, Triton X-100 does not appear to affect the resolution of H1 variants.[42] However, modifications of H1 such as phosphorylation and poly(ADP-ribosylation) that affect net charge lead to altered mobility in both AU and AUT gels[43] and appropriate controls are required to distinguish mobility differences attributable to modifications from those due to amino acid sequence heterogeneity. The extent to which H1 variants are resolved on AU gels varies with the species and tissue analyzed and methodological details such as the concentrations of acrylamide and urea

[38] M. Wu, C. D. Allis, R. Richman, R. G. Cook, and M. A. Gorovsky, *Proc. Natl. Acad. Sci. USA* **83,** 8674 (1986).

[39] G. E. Croston, L. A. Kerrigan, L. M. Lira, D. R. Marshak, and J. T. Kadonaga, *Science* **251,** 643 (1991).

[40] L. S. Coles, A. J. Robins, L. K. Madley, and J. R. Wells, *J. Biol. Chem.* **262,** 9656 (1987).

[41] B. J. Smith and E. W. Johns, *FEBS Lett.* **110,** 25 (1980).

[42] R. W. Lennox and L. H. Cohen, *Methods Enzymol.* **170,** 532 (1989).

[43] R. Balhorn and R. Chalkley, *Methods Enzymol.* **40,** 138 (1975).

employed. In some cases, the number of bands resolved for H1 on AU gels is less than the number of components apparent in Bio-Rex 70 chromatography. Although five variants of calf thymus H1 are partially resolved by Bio-Rex 70 chromatography, the two or three bands resolved in AU gels are so closely spaced that they can appear to be a single broad band.[31] In contrast, AU–PAGE, under optimal conditions, separates six well-resolved bands for chicken erythrocyte H1[20] whereas lengthy chromatography on Bio-Rex 70 incompletely resolves only five peaks.[14,32]

Two-Dimensional Electrophoresis

Two-dimensional electrophoresis has been employed to analyze both linker and core histones. The extremely basic nature of the histones (pI 10–12) precludes the use of conventional isoelectric focusing (IEF) because the histones migrate into the cathodal electrolyte during focusing. Although conditions can be employed in nonequilibrium pH gradient electrophoresis (NEPHGE) that keep histones on these gels, the resolution of nonallelic histone variants and their modified forms achieved by IEF–SDS and NEPHGE–SDS procedures is less than that achieved in AU and AUT gels alone.[44] Thus, alternative systems combining AU or AUT with SDS gels (AU–SDS or AUT–SDS) have proved more useful for histone work with variations including AU–AUT, SDS–AU, and SDS–AUT gels applied in specific instances.[45] A technique in which the second dimension AU gel includes the cationic detergent cetyltrimethylammonium bromide (CTAB) (AUC gel) has also been described.[46] Lennox and Cohen used AU–SDS gels to systematically analyze the expression and metabolism of H1 variants in mice.[4,47–49] Depending on the sample, as many as 8 nonphosphorylated and 15 phosphorylated components of mouse H1 were resolved in these studies. Changes in the expression and turnover of the nonphosphorylated components that were apparent in electropherograms of various tissues taken from mice at different stages of development have provided compelling, albeit indirect, evidence that nonallelic variants of H1 are functionally distinct.

[44] P. Z. O'Farrell, H. M. Goodman, and P. H. O'Farrell, *Cell* **12**, 1133 (1977).
[45] J. R. Davie, *Anal. Biochem.* **120**, 276 (1982).
[46] W. M. Bonner, M. H. West, and J. D. Stedman, *Eur. J. Biochem.* **109**, 17 (1980).
[47] R. W. Lennox, R. G. Oshima, and L. H. Cohen, *J. Biol. Chem.* **257**, 5183 (1982).
[48] R. W. Lennox and L. H. Cohen, *J. Biol. Chem.* **258**, 262 (1983).
[49] R. W. Lennox and L. H. Cohen, *Dev. Biol.* **103**, 80 (1984).

Capillary Electrophoresis

Capillary electrophoresis (CE), an instrumental form of electrophoresis in which proteins are separated after application of voltage along capillary tubing filled with buffer or sieving matrices, has been applied to histone analysis in a few reports. This methodology is potentially more rapid and more sensitive than conventional electrophoretic techniques because proteins are detected in real time as they migrate past absorbance or fluorescence detectors. However, because fused silica tubing is most commonly employed, strategies that minimize or prevent adsorption of such highly basic proteins to capillary walls due to interactions with acidic silanol groups are necessary to succesfully analyze histones by CE. Lindner and colleagues, using acidic buffers in combination with the dynamic coating agent hydroxypropylmethyl cellulose and specific cations to suppress silanol ionization and decrease the rate of endoosmotic flow of electrolyte within capillaries during electrophoresis, have developed conditions that permit rapid, high-resolution separations of H1 variants from a number of species.[50–52]

We have developed a neutral pH method that relies on the interaction of the quaternary ammonium-containing polymer Polybrene with silanols to prevent the adsorption of H1 proteins to capillary walls.[53] Treatment with Polybrene before electrophoresis renders the normally negatively charged capillary walls cationic and thus reverses the direction of endoosmotic flow so that it opposes the direction of electromigration of the histone molecules. Although this results in lengthier analysis times compared with methods employing acidic buffers, this approach has the advantage that when reagents with multiple anionic groups such as EDTA are included in the electrophoresis buffer, they assemble on the cationic Polybrene layer to form a dynamic cation-exchange layer that provides additional selectivity for the resolution of H1 variants and modified forms. As shown in Fig. 4, all six nonallelic variants and one or more allelic forms present in a sample of pooled chicken erythrocyte H1 are resolved by this method.

[50] H. Lindner, W. Helliger, A. Dirschlmayer, H. Talasz, M. Wurm, B. Sarg, M. Jaquemar, and B. Puschendorf, *J. Chromatogr.* **608,** 211 (1992).
[51] H. Lindner, M. Wurm, A. Dirschlmayer, B. Sarg, and W. Helliger, *Electrophoresis* **14,** 480 (1993).
[52] H. Lindner, W. Helliger, B. Sarg, and C. Meraner, *Electrophoresis* **16,** 604 (1995).
[53] C. A. Mizzen and D. R. McLachlan, *Electrophoresis* **21,** 2359 (2000).

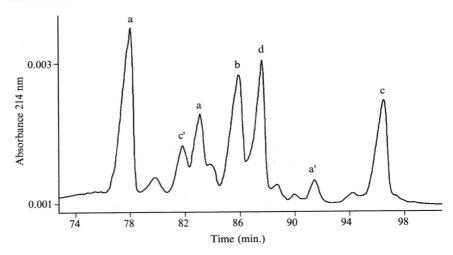

FIG. 4. CE of pooled chicken erythrocyte H1 variants in a Polybrene-coated capillary. Chicken H1 (0.2 mg/ml in water, 2.0-s injection) was electrophoresed at 10 kV in a 75-μm i.d. × 57 cm capillary (previously coated with Polybrene) in 10 mM Na-EDTA (pH 6.8), 0.001% (w/v) Polybrene at 30°. The separation of all six known nonallelic variants and two allelic variants of H1a (letters) was established by coelectrophoresis of H1 variants purified by CX-HILIC.

Procedure for Capillary Electrophoresis of Chicken H1 Variants

Polybrene (1,5-dimethyl-1,5-diazaundecamethylene polymethobromide) is obtained from Sigma (St. Louis, MO). All other chemicals are ACS grade or better. Deionized water prepared with a Milli-Q system (Millipore) is used throughout. H1 from pooled chicken blood and free of H5 is prepared as described above.

Separations are performed on a P/ACE 2000 system under computer control, using System Gold software (Beckman Courter, Fullerton, CA). Fused silica tubing (75 μm i.d. × 375 μm o.d.; total length, 57 cm; Polymicro Technologies, Phoenix, AZ) is contained within a Beckman Coulter capillary casette such that the inlet-to-detector length is 50.0 cm. Capillary treatment solutions and sample injections are introduced by positive pressure at the capillary inlet. Capillaries are washed with 1.0 M NaOH and cathodal endoosmotic flow is assessed at 10 kV in 0.1 M sodium phosphate, pH 6.8, before initial use and between trials with different electrolytes. Injections (typically 2 s) are made from a 0.2-mg/ml solution of H1 in water. Electrolyte solutions containing EDTA are prepared with the disodium salt and adjusted to the desired pH with sodium hydroxide. Solutions containing Polybrene are prepared by diluting a 10% (w/v) stock solution in water. Capillaries are flushed with 1% Polybrene in water and then with running buffer [10 mM EDTA (pH 6.8) containing 0.001% (w/v) Polybrene] for 10 min each before sample injection. Analyses of H1 are performed at 10 kV and 30° with detection at 214 nm.

Comments

Polybrene binds silica tubing avidly and CE can be performed in capillaries previously coated with Polybrene without the addition of Polybrene to the running buffer. However, we prefer to add a low level (0.001%, w/v) of Polybrene to the running buffer to replenish any molecules lost from capillary walls during electrophoresis and ensure that the rate of endoosmotic flow, and hence protein migration, remains consistent among runs. Both the rate of endoosmotic flow and the resolution of H1 variants are sensitive to the level of Polybrene employed and the nature and concentration of buffer agents employed. The effects of different buffer compositions on the resolution of chicken H1 variants have been described previously.[53]

Other Analyses of H1 Variants

Mass Spectrometry

Mass spectrometry (MS) has been used to establish the purity of chromatographic fractions of H1 prepared from calf thymus[54] and to determine the molecular masses of mouse liver H1 variants separated by RP-HPLC and confirm identification of the corresponding gene sequences.[9] MS techniques have also been used to characterize posttranslational modifications of H1 variants as discussed below.

Analyses of H1 Variant Function

Typical of early attempts to obtain direct evidence of functional differences among H1 variants are reports demonstrating that H1 fractions separated by Bio-Rex 70 chromatography differed in their ability to condense naked DNA or dinucleosomes previously depleted of H1 in vitro.[55,56] More recently, the affinity of individual mouse H1 variants separated by RP-HPLC for mononucleosomes reconstituted on a 210-bp fragment of the mouse mammary tumor virus (MMTV) promoter has been

[54] R. G. Berger, R. Hoffmann, M. Zeppezauer, W. Wagner-Redeker, L. Maljers, A. Ingendoh, and F. Hillenkamp, J. Chromatogr. 711, 159 (1995).

[55] L. W. Liao and R. D. Cole, J. Biol. Chem. 256, 6751 (1981).

[56] L. W. Liao and R. D. Cole, J. Biol. Chem. 256, 10124 (1981).

compared.[57] Strikingly, these authors noted that the affinity of H1b for this model nucleosome was less than one-half that of the other variants and that the binding of both H1b and the testis-specific variant H1t was less cooperative than that of other variants, suggesting the possibility that the degree of compaction achieved in chromatin containing these variants on higher order folding is likely to differ from that of bulk chromatin. This model is consistent with earlier reports describing differences in the distributions of H1 variants in chromatin fractions differentially solubilized by nuclease digestion of isolated nuclei[58,59] and differences in the solubility and hydrodynamic properties of chromatin oligomers reconstituted with different linker histone fractions.[60,61] However, because H1 can exchange between chromatin fragments *in vitro* during fractionation procedures[62] and the binding of H1 to chromatin is dynamic *in vivo*,[63,64] it remains to be resolved whether H1 variants are distributed nonrandomly *in vivo* such that enrichment of particular variants (or possibly H1 depletion) localizes to chromatin domains differing from bulk chromatin in higher order folding or transcriptional activity. Differences in the staining of cultured fibroblasts analyzed by immunofluorescence, using four antisera reported to be specific for different human H1 variants, suggested unique distributions for some of these proteins although the degree to which these correlate with differences in chromatin condensation or activity has not been determined.[65,66]

Molecular genetic approaches have provided direct evidence of functional differences between H1 variants *in vivo*. Regulated overexpression of mouse $H1^0$ in 3T3 cells caused transient inhibition of G_1- and S-phase progression whereas overexpression of H1c to comparable levels had no effect on cell cycle progression.[67] Quantitation of mRNAs corresponding to several cell cycle-regulated and housekeeping genes revealed that expression of all these genes was significantly reduced in cells overexpressing

[57] H. Talasz, N. Sapojnikova, W. Helliger, H. Lindner, and B. Puschendorf, *J. Biol. Chem.* **273,** 32236 (1998).

[58] C. Gorka and J. J. Lawrence, *Nucleic Acids Res.* **7,** 347 (1979).

[59] H. C. Huang and R. D. Cole, *J. Biol. Chem.* **259,** 14237 (1984).

[60] R. Hannon, E. Bateman, J. Allan, N. Harborne, and H. Gould, *J. Mol. Biol.* **180,** 131 (1984).

[61] S. Nagaraja, G. P. Delcuve, and J. R. Davie, *Biochim. Biophys. Acta* **1260,** 207 (1995).

[62] J. O. Thomas and C. Rees, *Eur. J. Biochem.* **134,** 109 (1983).

[63] M. A. Lever, J. P. Th'ng, X. Sun, and M. J. Hendzel, *Nature* **408,** 873 (2000).

[64] T. Misteli, A. Gunjan, R. Hock, M. Bustin, and D. T. Brown, *Nature* **408,** 877 (2000).

[65] M. H. Parseghian, R. F. Clark, L. J. Hauser, N. Dvorkin, D. A. Harris, and B. A. Hamkalo, *Chromosome Res.* **1,** 127 (1993).

[66] M. H. Parseghian, D. A. Harris, D. R. Rishwain, and B. A. Hamkalo, *Chromosoma* **103,** 198 (1994).

[67] D. T. Brown, B. T. Alexander, and D. B. Sittman, *Nucleic Acids Res.* **24,** 486 (1996).

H1⁰. On the other hand, overexpression of H1c resulted in either no significant change or dramatic increases in the levels of these transcripts. These
data provide strong evidence that functional differences exist between
these two H1 variants. Subsequent analyses of the effects of overexpression
of chimeric molecules containing N-terminal, C-terminal, and globular
domains from either H1c or H1⁰ suggested that the differential effect of
overexpression of H1c and H1⁰ on transcription of the genes analyzed
was due primarily to differences in their globular domains.[68] Given the
degree of sequence divergence in the globular domain of H1⁰ compared
with that of H1c, evidence of functional differences between H1c and
H1⁰ is perhaps not surprising and raises a question concerning whether
functional differences exist between the "somatic" H1 variants, which display higher degrees of sequence conservation (Fig. 1). However, experiments involving targeted disruption of H1 variants in mice and chickens
have provided evidence of both differences and redundancy in the function
of somatic H1 variants in transcriptional regulation.[69–71]

Analyses of H1 Modifications

Phosphorylation

Clear evidence that phosphorylation of H1 is involved in transcriptional
regulation comes from work in which the effects of mutagenesis of known
phosphorylation sites in macronuclear H1 of *Tetrahymena* were assayed
in vivo[72–74] and evidence of a similar role in higher eukaryotes has also
been obtained.[75–78] Much earlier work demonstrated that H1 variants are
progressively phosphorylated at multiple sites during cell cycle progression
in mammalian cells.[79,80] Thus H1 phosphorylation appears to be involved

[68] D. T. Brown, A. Gunjan, B. T. Alexander, and D. B. Sittman, *Nucleic Acids Res.* **25,** 5003 (1997).
[69] Y. Takami, R. Nishi, and T. Nakayama, *Biochem. Biophys. Res. Commun.* **268,** 501 (2000).
[70] R. Alami, Y. Fan, S. Pack, T. M. Sonbuchner, A. Besse, Q. Lin, J. M. Greally, A. I. Skoultchi, and E. E. Bouhassira, *Proc. Natl. Acad. Sci. USA* **100,** 5920 (2003).
[71] Y. Fan and A. I. Skoultchi, *Methods Enzymol.* **377,** [5] (2004).
[72] Y. Dou, C. A. Mizzen, M. Abrams, C. D. Allis, and M. A. Gorovsky, *Mol. Cell* **4,** 641 (1999).
[73] Y. Dou and M. A. Gorovsky, *Mol. Cell* **6,** 225 (2000).
[74] Y. Dou and M. A. Gorovsky, *Proc. Natl. Acad. Sci. USA* **99,** 6142 (2002).
[75] G. C. Banks, L. J. Deterding, K. B. Tomer, and T. K. Archer, *J. Biol. Chem.* **276,** 36467 (2001).
[76] R. N. Bhattacharjee, G. C. Banks, K. W. Trotter, H. L. Lee, and T. K. Archer, *Mol. Cell. Biol.* **21,** 5417 (2001).
[77] P. J. Horn, L. M. Carruthers, C. Logie, D. A. Hill, M. J. Solomon, P. A. Wade, A. N. Imbalzano, J. C. Hansen, and C. L. Peterson, *Nat. Struct. Biol.* **9,** 263 (2002).
[78] R. Koop, L. Di Croce, and M. Beato, *EMBO J.* **22,** 588 (2003).

in remodeling chromatin in relation to transcription and changes in chromatin condensation that occur during the cell cycle. Although direct evidence is lacking at present, data suggest that H1 phosphorylation also plays roles in apoptosis and DNA replication and repair.[81,82]

Sites of H1 phosphorylation *in vivo* have been determined systematically only in *Tetrahymena,* an organism that expresses a single H1 protein.[83] Given that the sequence of this H1 is somewhat divergent from that of the H1 proteins expressed in other eukaryotes and that H1 phosphorylation is involved in multiple important processes, it is unfortunate that sites of phosphorylation have not been better characterized in other eukaryotes. Presumably, this reflects the difficulties encountered in resolving the H1 variants and the array of differently phosphorylated forms of these variants that can occur in other species. However, application of some of the chromatographic and electrophoretic techniques described above may help to rectify this situation. Previously, estimates of the stoichiometry of H1 phosphorylation were inferred from the degree to which the migration of proteins was retarded on AU gels.[43,84] Lindner's group have demonstrated that HILIC and CE can both be used to resolve the human H1.1 fraction (arbitrary designation) eluting as a single peak in RP-HPLC into non-phosphorylated and mono-, di-, and triphosphorylated components.[36] Although it has not been reported whether the components resolved in either case are homogeneous with respect to phosphorylation sites, it seems likely that manipulation of the separation conditions could enable the resolution of at least some phosphorylation site isomers as described previously for multiply phosphorylated forms of *Tetrahymena* H1.[85]

MS has also been employed to assess the phosphorylation stoichiometry of H1 variants.[27,75] Although not yet reported in the literature, the application of MS/MS techniques potentially represents an approach for mapping phosphorylation sites in individual H1 variants that would not rely on the prior separation of variants to the same degree as conventional biochemical approaches such as microsequencing ^{32}P-labeled

[79] L. R. Gurley, J. A. D'Anna, S. S. Barham, L. L. Deaven, and R. A. Tobey, *Eur. J. Biochem.* **84,** 1 (1978).

[80] K. Ajiro, T. W. Borun, and L. H. Cohen, *Biochemistry* **20,** 1445 (1981).

[81] H. Talasz, W. Helliger, B. Sarg, P. L. Debbage, B. Puschendorf, and H. Lindner, *Cell Death Differ.* **9,** 27 (2002).

[82] C. Y. Guo, C. A. Mizzen, Y. Wang, and J. M. Larner, *Cancer Res.* **60,** 5667 (2000).

[83] C. A. Mizzen, Y. Dou, Y. Liu, R. G. Cook, M. A. Gorovsky, and C. D. Allis, *J. Biol. Chem.* **274,** 14533 (1999).

[84] H. Talasz, W. Helliger, B. Puschendorf, and H. Lindner, *Biochemistry* **35,** 1761 (1996).

[85] M. Lu, C. A. Dadd, C. A. Mizzen, C. A. Perry, D. R. McLachlan, A. T. Annunziato, and C. D. Allis, *Chromosoma* **103,** 111 (1994).

phosphopeptides. MS data demonstrating that hormonal stimulation affects the phosphorylation of H1 variants differentially[75] highlight the need for alternate site-mapping techniques.

N-Terminal Acetylation

Acetylation of the amino terminus of H1 proteins has been documented in a number of protein-sequencing studies.[86] Although little is known of the functional significance of this modification, for either H1 or the wide variety of other eukaryotic proteins known to be similarly modified, it has been reported that H1 variants differ with regard to their N-terminal acetylation status and that this acetylation can affect only a portion of specific variants.[25,87–89] Lindner and colleagues have used HILIC and CE to show that the accumulation of N-terminal acetylated H1[0] in rat liver and brain is age related.[87] Similar accumulations are not detected for other H1 variants in these tissues, nor are they detected when H5 from young and adult chickens is compared, suggesting the possibility that accumulation of N-terminal acetylated H1[0] may be related to age-related decreases in transcription described previously for some tissues.[89]

Poly(ADP-Ribosylation)

Poly(ADP-ribosylation) of a wide variety of chromatin proteins, but particularly H1, occurs to a small extent in various types of cells[5] and is strongly induced in response to DNA strand breaks.[90] Typically, the modification is detected by monitoring the incorporation of [^{32}P] NAD$^+$ into poly(ADP-ribose) added to proteins in cell extracts or permeabilized cells or through the use of antisera to poly(ADP-ribose). Despite intensive research, the functional significance of this modification remains obscure.[91] It is not clear how many sites in H1 are poly(ADP-ribosylated) and whether any of these are unique to certain variants or functionally distinct, as the only data available come from early work implicating several glutamate residues and the C-terminal carboxyl residue[92,93] and possibly

[86] C. Von Holt, W. F. Brandt, H. J. Greyling, G. G. Lindsey, J. D. Retief, J. D. Rodrigues, S. Schwager, and B. T. Sewell, *Methods Enzymol.* **170**, 431 (1989).
[87] H. Lindner, B. Sarg, B. Hoertnagl, and W. Helliger, *J. Biol. Chem.* **273**, 13324 (1998).
[88] H. Lindner, B. Sarg, H. Grunicke, and W. Helliger, *J. Cancer Res. Clin. Oncol.* **125**, 182 (1999).
[89] B. Sarg, W. Helliger, B. Hoertnagl, B. Puschendorf, and H. Lindner, *Arch. Biochem. Biophys.* **372**, 333 (1999).
[90] F. R. Althaus and C. Richter, *Mol. Biochem. Biophys.* **37**, 1 (1987).
[91] G. DeMurcia and S. Shall, "From DNA Damage and Stress Signaling to Cell Death: Poly(ADP-ribosylation) Reactions." Oxford University Press, New York, 2000.

phosphoserine[94] and arginine residues.[95] Poly(ADP-ribosylation) of H1 correlates with internucleosomal DNA fragmentation during apoptosis, suggesting that this modification is likely to affect chromatin accessibility.[96] Poly(ADP-ribose) polymerase I (PARP I), the enzyme responsible for poly(ADP-ribosylation), has been shown to bind H1 *in vitro*. Comparison of phosphopeptide maps monitoring the phosphorylation of H1 by cdc2 *in vitro* suggests that H1–PARP I interaction exposes new phosphorylation sites in H1.[97] Although it is not known whether similar interaction can occur *in vivo,* this finding suggests the possibility that pathways regulated by DNA damage may converge on H1 to change chromatin accessibility through coordinated H1 phosphorylation and poly(ADP-ribosylation).

Ubiquitination

Although data implicating monoubiquitination of histones H2A and H2B with gene expression have long existed,[5] evidence of H1 ubiquitination was reported only more recently. Pham and Sauer have reported that the $TAF_{II}250$ subunit of *Drosophila* TFIID mediates H1 ubiquitination *in vitro* and *in vivo* and that the ubiquitin-conjugating activity of $TAF_{II}250$ is required for activated transcription of two genes controlled by the Dorsal activator.[98] Human and *Drosophila* $TAF_{II}250$ have also been shown to possess kinase and acetyltransferase activity,[99,100] suggesting that multiple activities may play roles in coactivation of transcription by TFIID. As there are currently no reports demonstrating that H1 is ubiquitinated in any species apart from *Drosophila,* it remains to be determined how widespread the occurrence of this signaling pathway is.

Concluding Remarks

Any attempt to summarize the current state of knowledge regarding a field as broad and complex as that of the nature and function of H1 variants and their modified forms is likely to fall short of this goal in one or more

[92] P. T. Riquelme, L. O. Burzio, and S. S. Koide, *J. Biol. Chem.* **254,** 3018 (1979).

[93] N. Ogata, K. Ueda, H. Kagamiyama, and O. Hayaishi, *J. Biol. Chem.* **255,** 7616 (1980).

[94] J. A. Smith and L. A. Stocken, *Biochem. Biophys. Res. Commun.* **54,** 297 (1973).

[95] Y. Tanigawa, M. Tsuchiya, Y. Imai, and M. Shimoyama, *Biochem. Biophys. Res. Commun.* **113,** 135 (1983).

[96] Y. S. Yoon, J. W. Kim, K. W. Kang, Y. S. Kim, K. H. Choi, and C. O. Joe, *J. Biol. Chem.* **271,** 9129 (1996).

[97] P. I. Bauer, K. G. Buki, and E. Kun, *Int. J. Mol. Med.* **8,** 691 (2001).

[98] A. D. Pham and F. Sauer, *Science* **289,** 2357 (2000).

[99] R. Dikstein, S. Ruppert, and R. Tjian, *Cell* **84,** 781 (1996).

[100] C. A. Mizzen, X. J. Yang, T. Kokubo, J. E. Brownell, A. J. Bannister, T. Owen-Hughes, J. Workman, L. Wang, S. L. Berger, T. Kouzarides, Y. Nakatani, and C. D. Allis, *Cell* **87,** 1261 (1996).

aspects. Moreover, given the impressive advances made in this area, some of the findings discussed here may eventually be found to be erroneous. Regardless of these caveats, creative application and revision of the techniques described here should permit greater understanding of this important problem.

[20] HMGA Proteins: Isolation, Biochemical Modifications, and Nucleosome Interactions

By RAYMOND REEVES

Introduction

The mammalian HMGA family of high mobility group (HMG) nonhistone proteins [formerly called HMG-I(Y) proteins[1]] has been referred to as a "hub" of nuclear function because of its intimate involvement in a wide variety of biology processes.[2] An extensive literature suggests that HMGA proteins act as architectural transcription factors that function by recognizing and modulating the structure of both DNA and chromatin substrates and, as a consequence, have the ability to participate in a diverse array of nuclear events ranging from DNA replication and repair to the regulation of gene transcription and mRNA splicing and the control of integration of retroviruses into the genome (reviewed in Ref. 3). HMGA proteins have also attracted considerable medical interest because their aberrant expression or overexpression has been demonstrated to lead to neoplastic transformation of cells and to promote metastatic progression of cancers (reviewed in Refs. 3–6).

HMGA Proteins Exhibit Unusual Properties

In both mice and humans, the principal members of the HMGA protein family (i.e., HMGA1a, HMGA1b, and HMG2) are produced by translation of alternatively spliced mRNA transcripts encoded by two different

[1] M. Bustin, *Trends Biochem Sci.* **26,** 152 (2001).
[2] R. Reeves, *Gene* **277,** 63 (2001).
[3] R. Reeves and L. Beckerbauer, *Biochim. Biophys. Acta* **1519,** 13 (2001).
[4] J. L. Hess, *Am. J. Clin. Pathol.* **109,** 251 (1998).
[5] G. Tallini and P. Dal Cin, *Adv. Anat. Pathol.* **6,** 237 (1999).
[6] R. Reeves, *Environ. Health Perspect.* **108,** 803 (2000).

genes, *HMGA1* and *HMGA2* (reviewed in Refs. 6 and 7). HMGA proteins share a number of common biochemical, biophysical, and biological properties that distinguish them from other nuclear proteins. The major HMGA protein species are small (~10.6–12 kDa), have unusually high concentrations of basic, acidic, and proline residues, are soluble in dilute acids, and are resistant to inactivation by heating. Furthermore, they have little (if any) detectable secondary structure while free in solution but assume specific conformations when bound to DNA substrates,[8] and possess three copies of a highly conserved DNA-binding motif called the "A/T hook"[9] that preferentially bind to the minor groove of short stretches (4–6 bp) of AT-rich regions of B-form DNA.[10–12] Nevertheless, the HMGA proteins do not bind to all AT-rich stretches equally well or with equal affinity, suggesting that they recognize structural features of DNA rather than nucleotide sequence.[8,9,13] Selective substrate structural recognition is also demonstrated by the fact that HMGA proteins preferentially bind to non-B-form DNA structures found in supercoiled plasmids,[14] to locally distorted regions of DNA found on the surface of nucleosome core particles,[13,15] and to synthetic four-way junction (FWJ) DNAs *in vitro*.[16,17] In addition to substrate structural recognition, HMGA proteins also possess the capacity to alter the structure of DNA after binding as evidenced by their ability to bend, straighten, unwind, and introduce supercoils in various kinds of DNA molecules *in vitro* without a requirement for ATP hydrolysis, an aptitude that is thought to underlie many of their biological functions *in vivo* (reviewed in Ref. 2).

Production and Purification of Recombinant HMGA Proteins

Wild-type, mutant, and truncated forms of recombinant HMGA proteins produced in either bacteria or insect cells have been widely used in studies to elucidate their *in vitro* biochemical, biophysical, and biological

[7] M. Bustin and R. Reeves, *Prog. Nucleic Acid Res. Mol. Biol.* **54**, 35 (1996).
[8] J. R. Huth *et al.*, *Nat. Struct. Biol.* **4**, 657 (1997).
[9] R. Reeves and M. S. Nissen, *J. Biol. Chem.* **265**, 8573 (1990).
[10] M. Solomon, F. Strauss, and A. Varshavsky, *Proc. Natl. Acad. Sci. USA* **83**, 1276 (1986).
[11] T. S. Elton, M. S. Nissen, and R. Reeves, *Biochem. Biophys. Res. Commun.* **143**, 260 (1987).
[12] J. E. Disney *et al.*, *J. Cell Biol.* **109**, 1975 (1989).
[13] R. Reeves and A. P. Wolffe, *Biochemistry* **35**, 5063 (1996).
[14] M. S. Nissen and R. Reeves, *J. Biol. Chem.* **270**, 4355 (1995).
[15] R. Reeves and M. S. Nissen, *J. Biol. Chem.* **268**, 21137 (1993).
[16] D. A. Hill and R. Reeves, *Nucleic. Acids. Res.* **25**, 3523 (1997).
[17] D. A. Hill, M. L. Pedulla, and R. Reeves, *Nucleic Acids. Res.* **27**, 2135 (1999).

characteristics. Unfortunately, however, it is often difficult to isolate appreciable quantities of recombinant HMGA proteins because their levels of production in many host cell expression systems are frequently low. Among the factors that contribute to this low level of expression are the fact that overexpression of HMGA proteins is generally toxic to the host cells, the proteins themselves are highly prone to proteolytic degradation, and (at least in the case of bacteria) the codon bias of host expression systems is against the tRNA species that recognize the codons for several of the amino acids found in abundance in HMGA proteins. A number of strategies have been employed to try and overcome these difficulties. Cellular toxicity problems have been dealt with by inducing the expression of recombinant HMGA proteins for only a limited amount of time in "non-leaky" strains of host cells in which expression of transgenes is under tight transcriptional control. Minimization of intracellular degradation problems has usually been addressed by employing host cell strains that are deficient in various proteolytic enzymes and/or by expressing the proteins as "tagged" fusion derivatives (e.g., hexahistidine tagged, hemagglutinin tagged, or glutathione *S*-transferase tagged) and the tags then used as an aid in the rapid purification of intact, undegraded recombinant proteins. Problems with codon bias have been largely ameliorated by the use of genetically engineered strains of bacteria containing transgenes encoding tRNA species that are rarely used in bacteria but are frequently employed by mammalian cells.

Although the tag-linked recombinant HMGA fusion proteins have proved useful in their purification, for many structural and biophysical studies it has been necessary to produce HMGA proteins without any additional amino acid sequences. For this purpose, we have found the pET bacterial expression vector systems, originally developed by Studier and colleagues[18,19] and now commercially available from Novagen (Madison, WI), to be especially useful.[8,20,21] By employing standard *in vitro* mutagenesis techniques and recombinant DNA procedures,[22] full-length, mutant, or truncated forms of the cDNAs encoding either the HMGA1a, HMGA1b,[23] or HMG2[24] proteins are subcloned into the appropriate pET expression vector, which is then transformed into the double *lon/ompT*

[18] A. H. Rosenberg *et al., Gene* **56,** 125 (1987).
[19] F. W. Studier *et al., Methods Enzymol.* **185,** 60 (1990).
[20] M. S. Nissen, T. A. Langan, and R. Reeves, *J. Biol. Chem.* **266,** 19945 (1991).
[21] R. Reeves and M. S. Nissen, *Methods Enzymol.* **304,** 155 (1999).
[22] F. M. Ausubel *et al.,* "Current Protocols in Molecular Biology." John Wiley & Sons, New York, 1998.
[23] K. R. Johnson, D. A. Lehn, and R. Reeves, *Mol. Cell. Biol.* **9,** 2114 (1989).
[24] U. A. Patel *et al., Biochem. Biophys. Res. Commun.* **201,** 63 (1994).

protease mutant B strain of *Escherichia coli* BL21(DE3)pLysS.[19] Alternatively, to overcome problems of codon bias and greatly increase the yield of recovered proteins, the cDNA-containing pET vectors can be transformed into the BL21-CodonPlus-RP strain of *E. coli* (Stratagene, La Jolla, CA), which has been genetically engineered to express tRNAs that recognize arginine (AGG/AGA) and proline (CCC) codons that are rarely used in bacteria but are common in transcripts encoding HMGA proteins. In either case, transcription of the HMGA transgenes is induced by addition of 0.4–0.6 mM isopropyl-β-D-thiogalactopyranoside (IPTG) to exponentially growing bacterial cultures for a period of 4–6 h (37°) in order to obtain the greatest amount of recombinant HMGA protein while minimizing toxicity and cell death in the culture. After induction, the bacteria are pelleted (10 min at 5000 rpm in a GSA rotor), washed once in 2–3 pellet volumes of RSB [10 mM NaCl, 3 mM MgCl$_2$, 10 mM Tris-HCl, 0.5 mM phenylmethylsulfonyl fluoride (PMSF), pH 7.4], and either immediately extracted as described below or, if desired, frozen at $-70°$ for later extraction.

Isolation of Recombinant HMGA Proteins

Extraction of recombinant HMGA proteins from expressing host cells is greatly facilitated by the fact that, because of their unusual amino acid composition, they are soluble in dilute acids such as 5% perchloric acid (HClO$_4$) or trichloroacetic acid (TCA). Furthermore, because the proteins have little, if any, secondary structure while free in solution,[8] such denaturing extraction conditions do not appear to adversely affect the substrate-binding properties of the proteins. In our laboratory, we generally use the following extraction protocol to isolate recombinant proteins.

1. Add 25 ml of cold (4°) 5% TCA–0.5% Triton X-100 (TX-100) to bacterial pellets obtained from 250-ml cultures and resuspend the pellets by vigorously vortexing/stirring/pipetting until as fine a colloidal suspension as possible is produced. It is important to include TX-100 [or some other detergent, such as 0.1% sodium dodecyl sulfate (SDS) or 1% Nonidet P-40] with the dilute acid in order to efficiently lyse the bacteria and release the recombinant HMGA protein. Incubate the suspension at 4° for at least 30 min with continuous stirring to maximize protein extraction. The solution can also be slowly mixed on a shaker table overnight at 4° with no adverse effects. Pellet acid-insoluble material from the extract by centrifugation (10 min, 10,000 rpm in a GSA rotor) and collect the supernatant.

2. Precipitate recombinant HMGA from the supernatant by adding ice-cold TCA to a final concentration of 25% (i.e., 0.34 volume of 100% TCA solution is added to the supernatant) and place the mixture on ice for

at least 1 h with occasional agitation. It is convenient to perform the precipitation directly in a centrifuge bottle.

3. The precipitated crude recombinant HMGA protein is recovered by centrifugation at 10,000 rpm in a GSA rotor for 15 min at 4° and the supernatant is decanted. Dissolve the protein pellet in 15–20 ml of distilled water and transfer to a 30-ml Corex glass centrifuge tube. The recombinant protein is precipitated as before with TCA (a 30-min incubation on ice is sufficient) and pelleted by centrifugation (10,000 rpm for 15 min at 4°). The supernatant is discarded and the pellet is then carefully washed (without dislodgement) once with ice-cold, acidified (containing 10 mM HCl) acetone and then twice with ice-cold acetone alone. The washed protein pellet is dried under vacuum and stored at −20°.

Chromatographic Purification of Recombinant HMGA Proteins

Recombinant HMGA proteins are purified from the dissolved crude bacterial extracts by cation ion-exchange chromatography on a Macro-Prep 50-S column (1 × 20 cm; Bio-Rad, Hercules, CA) employing a gradient of potassium salts for elution. Protein elution peaks are monitored by absorbance at 220 nm.[20]

The solutions used for column elution are as follows:
Buffer A: 25 mM KH$_2$PO$_4$, pH 7.0 (HPLC-grade phosphate salt)
Buffer B: Buffer A plus 1.0 M KCl

Elution Steps

1. The Macro-Prep column is first equilibrated in 5% buffer B (i.e., 50 mM KCl) and then loaded with approximately 2 mg of crude recombinant protein dissolved in the same solution.

2. For elution, the KCl concentration is initially raised to 300 mM in 5 min (increasing 5% of buffer B per minute).

3. The elution gradient is then increased from 300 to 550 mM over a period of approximately 70–75 min at a flow rate of 1.0 ml/min. Absorbance at 220 nm is used to monitor proteins eluted in the fractions.

Identification of the recombinant HMGA protein and degradation products in various chromatographic elution fractions is monitored by Western blot analysis using specific anti-HMGA antibodies[21] and the purity of the fractions is monitored by either SDS–polyacrylamide gel electrophoresis (SDS–PAGE)[25] or acid/urea (AU)–PAGE[26] using known HMGA proteins as reference standards. Such independent assessments

[25] U. K. Laemmli, *Nature* **227**, 680 (1970).
[26] S. Panyim and R. Chalkley, *Arch. Biochem. Biophys.* **130**, 337 (1969).

are necessary because there is often considerable degradation of the HMGA proteins inside bacterial cells, even in protease-deficient strains such as *E. coli* BL21. For example, chromatographic profile analyses of recombinant proteins produced in this bacterial strain usually reveal two major peaks of HMGA1 proteins, one corresponding to the full-length protein and the other to a C-terminal truncated form.[21]

Because the HMGA1 proteins lack aromatic amino acids, estimations of their presence and concentrations in chromatographic fractions are conveniently determined by measuring the absorbance of the proteins in water or elution buffers at 220 nm[9] rather than at the more usual 280 nm. The extinction coefficient for the HMGA1a isoform protein is $\varepsilon_{220} = 73,500$ liters/mol·cm or 6.23 ml/mg·cm in water whereas for the shorter HMGA1b (Fig. 2) isoform $\varepsilon_{220} = 67,500$ liters/mol·cm or 6.37 ml/mg·cm.[21] In contrast to the HMGA1 proteins, both the human and mouse HMGA2 proteins contain a single tryptophan residue that allows accurate determination of their concentration in purified solutions by monitoring absorbance at 280 nm.

Isolation and Purification of HMGA Proteins from Mammalian Cells and Tissues

Two different methods have been widely used to extract HMGA proteins from the nuclei and chromatin of mammalian cells. Either they can be eluted by extraction with dilute acids, as described above for recombinant proteins, or they can be extracted with a buffered solution containing 0.3–0.4 M NaCl.[27] There are advantages and disadvantages to both methods. Although salt extraction is mild and allows for relatively efficient recovery of native, nondenatured HMGA proteins, it results in the solubilization of a complex mixture of nuclear proteins from which the desired HMGA proteins must subsequently be purified by chromatographic methods. Use of the salt extraction method is usually restricted to isolation of HMGA proteins from embryonic, cancerous, or immortalized cells, which contain exceptionally high concentrations of these proteins; in normal cells and tissue, the levels of these proteins are low and often undetectable by this extraction method.[7] Salt extraction methods also have the disadvantage that they involve the handling of large volumes of dilute protein solutions and, because considerable protein degradation occurs due to the presence of contaminating proteases in the extracts, high concentrations of expensive protease inhibitors are required in all steps of the isolation protocol.

[27] T. Lund *et al.*, *FEBS Lett.* **152**, 163 (1983).

Extraction of HMGA proteins from mammalian cells with dilute acids has numerous advantages. For example, it greatly simplifies the composition of proteins obtained from isolated nuclei and chromatin, selectively solubilizing total HMG proteins (e.g., members of the HMGA, HMGB, and HMGN families[1]), various histone H1 variants, and only a few other proteins,[28] thus facilitating the subsequent isolation of individual protein species. Dilute acids can also be used to extract HMGA proteins from whole, unfractionated cells, and even from intact tissues and organs, with the complexity of the resulting extracts still being considerably less than is observed with salt-extracted nuclei. Because acid extraction methods always include a protein precipitation step, they are also advantageous because they efficiently concentrate HMGA proteins from the large volume of a dilute solution. Acid extraction methods also minimize possible proteolytic degradation of HMGA proteins during the isolation procedure and are usually used to isolate these proteins directly from frozen samples during the thawing process. Disadvantages of acid extraction procedures include a somewhat lower yield of recovered HMGA protein compared with salt isolation methods and the fact that TCA exhibits strong absorbance at 220 nm, thus prohibiting determination of the concentration of HMGA proteins (which lack aromatic amino acids that absorb at 280 nm) by spectrophotometric methods at this wavelength. Dilute perchloric acid, which does not have strong absorbance at 220 nm, is therefore an alternative reagent of choice for such extractions. A combination of the salt extraction and acid fractionation methods is sometimes used to maximize HMGA recovery and facilitate protein purification from mammalian cells. Detailed experimental protocols for extracting HMGA proteins from cells and tissues by both salt and acid extraction methods have previously been published in this series.[21]

Purification of HMGA Proteins Isolated from Mammalian Cells

Although a number of chromatographic procedures have been successfully used to purify HMGA proteins eluted by either the salt or acid extraction method, we routinely employ ion-pair reversed-phase high-performance liquid chromatography (RP-HPLC) for such separations because it provides a fast and convenient method for obtaining highly purified forms of each of the different members of the HMGA protein family.[21,29] As illustrated by the chromatographic scan shown in Fig. 1B of acid-extracted proteins isolated from mammalian cells growing in culture,

[28] E. E. Johns, ed., "The HMG Chromosomal Proteins." Academic Press, New York, 1982.
[29] T. S. Elton and R. Reeves, *Anal. Biochem.* **157**, 53 (1986).

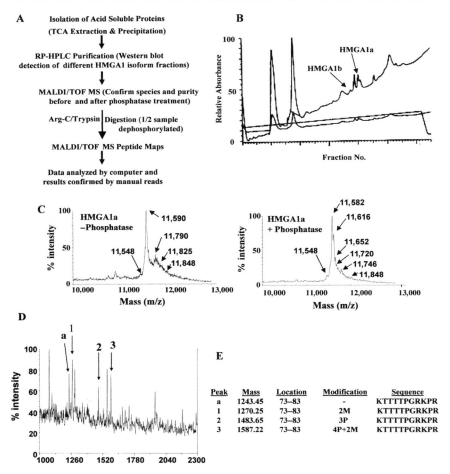

FIG. 1. Purification scheme and strategy for identification of posttranslational modifications (PMTs) on native HMGA1 proteins, using MALDI/TOF mass spectrometry. (A) Outline of the overall strategy for isolation, purification, and analysis by mass spectrometry of *in vivo* biochemically modified HMGA proteins. (B) Profile of total acid-soluble proteins isolated from mammalian cells and chromatically separated on an RP-HPLC column. Peaks corresponding to the HMGA1a and HMGA1b proteins are identified by arrows. (C) MALDI/TOF spectra of the full-length HMGA1a protein before (*left*) and after (*right*) it has been dephosphoryated by exposure to shrimp alkaline phosphatase. (D) A restricted region of the MALDI/TOF MS spectrum of Arg-C-derived peptides of native HMGA1a proteins. (E) Table of the masses, sequences, and PTMs of peptide fragments detected by MALDI/TOF MS in (D). The numbers in (C) correspond to molecular mass (*m/z*) values. In (D) the letter "a" indicates a peptide fragment without any PTMs and the numbers indicate peptide fragments with one or more posttranslational modifications.

RP-HPLC chromatography can be used to separate the HMGA1a and HMGA1b isoform proteins not only from each other but also from other contaminating protein species in the extracts.

RP-HPLC fractionation of HMGA proteins can be performed on either Vydac C_4 or C_{18} silica gel (Microsorb; Rainin Instruments, Oakland, CA) columns (4.6 × 250 mm i.d.; 5-μm beads, 300-Å pore size). Dried HMGA protein samples are dissolved in solvent A [0.1% (w/v) trifluoroacetic acid (TFA) in H_2O, pH 2.14] containing 0.1 M dithiothreitol (DTT), centrifuged to remove insoluble particulates, and injected onto the column. A linear gradient of solvent A to 0.1% TFA in 95% acetonitrile (solvent B), with a flow rate of 1 ml/min over a period of about 75 min, is used to separate HMGA proteins from other contaminants (Fig. 1B). For the reasons previously described, the presence and concentration of proteins in the chromatographic fractions are monitored by absorbance at 220 nm, the purity of the proteins is determined by either SDS– or AU–PAGE, and the identification of HMGA proteins in the fractions is assessed by Western blot analysis using specific anti-HMGA antibodies (Fig. 1A).

In Vivo Biochemical Modifications of HMGA Proteins

HMGA proteins isolated from living cells exhibit complex patterns of postsynthetic phosphorylations, acetylations, and methylations, making them among the most highly modified proteins in the mammalian nucleus (reviewed in Refs. 2 and 3). These modifications are dynamic and rapidly change as a function of the cell cycle,[30,31] the state of cellular differentiation,[32] in response to intra- and extracellular signaling events,[32] and during the activation of apoptotic cell death.[33,34]

The diagram in Fig. 2 summarizes the currently known sites of in vivo modification of the mammalian HMGA1 proteins and some of the enzymes that catalyze these modifications either in vitro or in vivo (reviewed in Ref. 2). The enzymes involved include the cell cycle-regulated kinase cdc2,[20,30] the developmentally regulated kinase HIPK,[35,36] the constitutively expressed kinases cdk-1,[37] the environmentally responsive

[30] R. Reeves, T. A. Langan, and M. S. Nissen, *Proc. Natl. Acad. Sci. USA* **88,** 1671 (1991).
[31] R. Reeves, *Curr. Opin. Cell Biol.* **4,** 413 (1992).
[32] G. C. Banks, Y. Li, and R. Reeves, *Biochemistry* **39,** 8333 (2000).
[33] F. Diana et al., *J. Biol. Chem.* **276,** 11354 (2001).
[34] R. Sgarra et al., *Biochemistry* **42,** 3575 (2003).
[35] G. M. Pierantoni et al., *Oncogene* **20,** 6132 (2001).
[36] G. M. Pierantoni et al., *Biochem. Biophys. Res. Commun.* **290,** 942 (2002).
[37] L. Detvaud, G. R. Pettit, and L. Meijer, *Eur. J. Biochem.* **264,** 55 (1999).

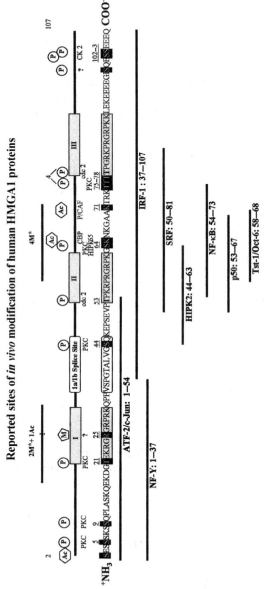

Fig. 2. Diagram of the human HMGA1 protein, showing the known sites of *in vivo* biochemical modifications and the regions of the protein that have been identified as minimal areas required for physical interaction with other proteins (see text for details). *Indicates the exact site of modification has not been determined. Redrawn, with modification, from Reeves and Beckerbauer (2001).[3]

kinases CK2[38,39] and PKC,[32] and the gene transcriptional regulatory acety-transferases CBP and P/CAF.[40] It has also been demonstrated that many of these sites of *in vivo* modification occur in regions of the HMGA1 proteins that have been demonstrated to physically interact with other gene tran-scription regulatory proteins. These include, among others, ATF-2/c-Jun, IRF-1, NF-κB, NF-Y, SRF, and Tst-1/Oct-6 (reviewed in Ref. 3). Not sur-prisingly, therefore, evidence also suggests that secondary biochemical modifications likely modulate both the substrate-binding properties and *in vivo* biological functions of the HMGA proteins.[30,31,40,41]

Determination of *In Vivo* Biochemical Modifications by Mass Spectrometry

Although a variety of procedures have been employed to determine the sites of individual types of *in vivo* biochemical modifications on HMGA proteins, mass spectrometry (MS) techniques have proved to be the most efficient and facile methods for analyzing the complex patterns of multiple modifications that are simultaneously present on these proteins in living cells. Matrix-assisted laser desorption/ionization time-of-flight (MALDI/TOF) MS has been successfully used in our laboratory to analyze the types, and in some cases even the actual locations, of posttranslational modifications (PTMs) on HMGA1 proteins[32,42] isolated from mammalian cells.

Figure 1A outlines one strategy for determining by MALDI/TOF MS the types of *in vivo* PTMs (in this example, phosphorylations) found on native protein.[32] The steps in the procedure are as follows: (1) proteins that are soluble in dilute acid, including HMGA proteins, are extracted from mammalian cells, precipitated, and washed as described above; (2) the acid-soluble proteins are then separated by RP-HPLC (Fig. 1B) and peaks containing HMGA1 proteins are identified by Western blot analysis; (3) the identity and masses of the HMGA1 proteins in the fractions identified in step 2 are then confirmed by MALDI/TOF MS analysis before and after, aliquots of the protein have been digested with shrimp alkaline phos-phatase (SAP) enzyme to assess the degree of phosphorylation on the proteins (Fig. 1C). It is obvious from the MS profiles shown in Fig. 1C that the RP-HPLC-purified endogenous HMGA1a protein, whose molecular mass without any secondary PTMs is 11,548 kDa, is highly modified by

[38] D. Z. Wang, P. Ray, and M. Boothby, *J. Biol. Chem.* **270,** 22924 (1995).

[39] D. Wang *et al., J. Biol. Chem.* **272,** 25083 (1997).

[40] N. Munshi *et al., Mol. Cell* **2,** 457 (1998).

[41] J. S. Siino, M. S. Nissen, and R. Reeves, *Biochem. Biophys. Res. Commun.* **207,** 497 (1995).

[42] D. D. Edberg and R. Reeves, unpublished data (2003).

phosphorylation (in addition to other modifications) *in vivo* and that this modification can be substantially removed by SAP treatment; (4) the untreated (Fig. 1C, left) and SAP-treated (Fig. 1C, right) samples are then digested with either trypsin or Arg-C protease and the resulting peptide mixtures are analyzed by MALDI-TOF MS to generate diagnostic "maps" of peptides with differing molecular masses (m/z). Figure 1D shows a small region of a MALDI-TOF MS ion spectrum of Arg-C-derived peptides from HMGA1a proteins that have not been treated with SAP. The m/z peaks in Fig. 1D, identified by lower case letters (i.e., "a"), identify peptide ions without any secondary PTMs whereas those peaks identified by numbers (i.e., 1 . . . 3) indicate peptides that contain additional PTMs; and (5) both computer and manual analyses of multiple MS spectra data [to a mass accuracy of ∼200 parts/million (or greater) per m/z species] are then employed to determine the type(s), and in many cases the actual location(s), of specific PTMs present on both typsin- and Arg-C protease-derived peptide fragments originating from various regions of the HMGA proteins (Fig. 1E).

Enzyme Treatments for MS Analysis of In Vivo-Phosphorylated HMGA Proteins

Dephosphorylation reactions are performed by digesting RP-HPLC-purified HMGA1 proteins with either calf intestinal alkaline phosphatase (CIP) (Roche, Mannheim, Germany) at 37° for 7 h, or with 20 units of shrimp alkaline phosphatase (Sigma, St. Louis, MO) at 37° for 10 h, in 1× reaction buffers supplied by the manufacturer. Purified HMGA1 proteins that act as controls are incubated under the same conditions but without the addition of enzymes. The dephosphorylated proteins are separated from the CIP and SAP enzymes by using the dilute acid HMGA extraction procedure described above. Aliquots of the dephosphorylated and control proteins are then dissolved in 100 mM NH$_4$HCO$_3$, pH 8.0, and digested at 37° with sequencing-grade trypsin (Promega) at a protein-to-enzyme ratio of 33:1 (w/w) for between 6 and 12 h to produce various partial HMGA1 digestion products for subsequent analysis.[32] The resulting tryptic peptide mixtures are then lyophilized and reconstituted in sterile distilled water. For digestions with Arg-C protease (Sigma), the same buffer and reaction conditions are employed except that the protein-to-enzyme ratio is 50:1 (w/w) and the time of incubation is 6 h at 37°.

MALDI/TOF Mass Spectrometry

The HMGA protein samples are dissolved in a 1:1 acetonitrile-to-water, 0.25% TFA solution saturated with either 3,5-dimethoxy-4-hydroxycinnamic acid (for full-length proteins) or α-cyano-4-hydroxycinnamic acid (for tryptic peptides) and the samples are spotted and dried on the stage of a MALDI plate. MALDI/TOF MS is performed in linear mode using an Applied Biosystems (Foster City, CA) Voyager DE-RP instrument equipped with a 337-nm nitrogen laser and run according to published protocols.[43-45] Analysis of full-length protein samples is performed in the positive mode with 256 laser pulses per sample averaged into a single spectrum and with each spectrum being calibrated either internally and/or externally, using commercially available mass standards (Applied Biosystems) to determine time of flight. Mass/charge ratios of full-length proteins are analyzed with the program Data Explorer, version 5.1 (Applied Biosystems). Both positive and negative modes are used for analysis of both tryptic and Arg-C peptide solutions in order to better visualize phosphopeptides.[46] Analyses of peptide digestion pattern spectra, done to determine the sites and species of potential PTMs, are best performed by employing a combination of both manual calculations[47] and the use of publicly available software programs such as Protein Prospector MS-digest program (http://prospector.ucsf.edu/) and the ExPASy FindMod tool (http://us.expasy.org/tools/findmod/). As an additional aid in determining potential PTMs, mass/charge peaks observed in the MALDI/TOF MS spectra are compared with theoretical peptide digestion patterns of posttranslationally modified HMGA1 proteins.

HMGA Proteins Bind to Nucleosome Core Particles Both *In Vitro* and *In Vivo*

Although HMGA proteins have received a considerable amount of attention because of their active role in regulating the transcription of more than 40 different eukaryotic and viral genes by participating in the formation of stereospecific, multiprotein complexes called "enhanceosomes" on their promoter/enhancer regions (reviewed in Refs. 3 and 48), these proteins have long been recognized as being important participants in many other nuclear processes. HMGA proteins are integral components of both metaphase chromosomes[12] and interphase heterochromatin[49] and have

[43] R. S. Annan and S. A. Carr, *Anal. Chem.* **68,** 3413 (1996).

[44] S. A. Carr and R. S. Annan, "Current Protocols in Molecular Biology," pp. 10.22.1–10.21.27. John Wiley & Sons, New York, 1998.

[45] S. D. Patterson, "Current Protocols in Molecular Biology," pp. 10.22.1–10.22.24. John Wiley & Sons, New York, 1998.

[46] S. A. Carr, M. J. Huddleston, and R. S. Annan, *Anal. Biochem.* **239,** 180 (1996).

[47] M. Kinter and N. E. Sherman, "Protein Sequencing and Identification Using Tandem Mass Spectrometry." Wiley-Interscience, New York, 2000.

[48] M. Merika and D. Thanos, *Curr. Opin. Genet. Dev.* **11,** 205 (2001).

HMGA proteins bind to nucleosome core particles

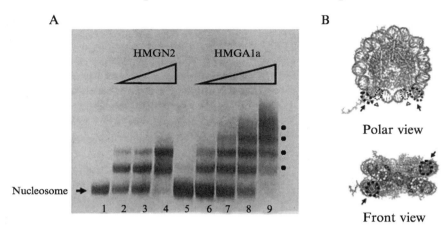

FIG. 3. Nonmodified, recombinant HMGA1 proteins bind to four regions of random sequence nucleosome core particles. (A) The results of EMSA gel assays in which increasing concentrations of either purified nonhistone HMGN2 (which binds to two sites on nucleosome core particles) or recombinant human HMGA1a protein were bound to nucleosome core particles isolated from chicken erythrocytes. (B) Two different views (polar and front) of the X-ray structure of the nucleosome cores particle,[50a] showing the sites of binding of HMGA proteins (dashed circles) determined by DNA footprinting analyses and other techniques (see text for details). (See color insert.)

been postulated to be involved in the marked changes in chromatin and chromosome structure that occur during the cell cycle (reviewed in Refs. 31 and 50). HMGA proteins are also members of a small group of transcription factors that are able to bind to nucleosome core particles both *in vitro* and *in vivo*. For example, HMGA proteins have been found associated with monomer nucleosomes isolated from mammalian cells[51] and they can be cross-linked by chemical reagents to nucleosomal histones *in vivo*.[15] Furthermore, HMGA1 proteins have been demonstrated to bind to specific DNA regions of random-sequence nucleosomes *in vitro*.[15]

As illustrated by the electrophoretic mobility shift assay (EMSA) experiments shown in Fig. 3A, when increasing concentrations of pure recombinant HMGA1 protein are added to nucleosome core particles

[49] M. Z. Radic *et al.*, *Chromosoma* **101,** 602 (1992).

[50] R. Reeves and M. S. Nissen, *Prog. Cell Cycle Res.* **1,** 339 (1995).

[50a] K. Luger *et al.*, *Nature* **389,** 251 (1997).

[51] A. Varshavsky *et al.*, *Cold Spring Harb. Symp. Quant. Biol.* **47,** 511 (1983).

isolated from chicken erythrocytes (lanes 6–9), up to four nucleosome–HMGA1 bands with retarded mobility are observed on the gel. This EMSA pattern stands in contrast to the two retarded bands that are observed when another protein, HMGN, is added to the nucleosomes at the same molar ratios (Fig. 3A, lanes 2–4). Analyses of the HMGA1-binding sites on nucleosomes, employing a variety of techniques (thermal denaturation studies, DNase I digestions, and hydroxyl footprinting methods), revealed that these proteins bound to the DNA as it enters and exits the core particle as well as to distorted regions of DNA on the surface of the nucleosome, about 1.5 helical turns on either side of the dyad axis (Fig. 3B).[13,15,50a] Additional experiments demonstrate that HMGA proteins can also bind to localized regions of A/T-rich DNA sequence located on the surface of strongly positioned nucleosome core particles that have been reconstituted *in vitro* from purified histones and to defined sequences of cloned DNA.[13] In this case, both hydroxyl radical and DNase I footprinting analyses indicated that binding of HMGA1 proteins caused localized changes in the rotational setting of the DNA on the surface of the core particles, that is, they induced a limited form of nucleosome remodeling (see the following).

Importantly, as illustrated by the results of the EMSA gels shown in Fig. 4, the affinity of binding of HMGA1 proteins to both free DNA and

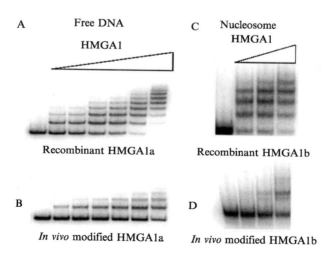

FIG. 4. The binding affinity of HMGA proteins for DNA and nucleosome substrates is reduced by secondary *in vivo* biochemical modifications. Shown are the results of electrophoretic mobility shift assays (EMSAs) in which increasing concentrations of either unmodified recombinant human HMGA1 proteins (A and C) or *in vivo*-modified native HMGA1 proteins (B and D) bound to either radiolabeled free DNA (A and B) or random-sequence nucleosome core particles (C and D). See text for details.

nucleosome core particles *in vitro* is markedly affected by the types of post-synthetic modifications found on the proteins in living cells.[32] Figure 4A shows the numerous retarded bands that are observed when increasing concentrations of unmodified, recombinant HMGA1a protein are added to a naked B-form DNA substrate that contains multiple, short stretches of A/T-rich binding sites for the protein. Figure 4B shows the results of an EMSA experiment using the same DNA substrate and the same concentrations of HMGA1a protein, but in this case the protein has been isolated from mammalian cells and contains numerous postsynthetic modifications. A comparison of Fig. 4A and B clearly demonstrates that the PTMs found on native HMGA1 proteins markedly reduce their affinity for binding to A/T-rich regions of free DNA substrates. A comparison of Fig. 4C and D also indicates a marked reduction of binding affinity of *in vivo*-modified HMGA1 proteins for isolated nucleosome core particles. These results strongly suggest that the PTMs present on HMGA proteins *in vivo* will have similar effects on their interactions with substrates in living cells and, hence, are likely to regulate their biological function(s). For detailed protocols describing methods for analyzing the modes of interaction of HMGA1 proteins with both native and *in vitro*-reconstituted nucleosome core particles the reader is referred to an earlier article in this series.[21]

HMGA1 Binding to Nucleosomes Is Orientation Specific

Interaction of HMGA1 proteins with nucleosomes not only induces localized changes in the rotational setting of DNA on core particles,[13] it also binds to A/T-rich stretches of DNA on the nucleosome surface in a orientation-specific manner.[53] HMGA1 proteins participate in the transcriptional induction of the gene encoding the α subunit of the human interleukin 2 receptor (*IL-2Rα*) in stimulated T lymphocytes by controlling the formation of an enhanceosome on the gene's 5' proximal promoter region.[52–56] Before transcriptional activation and enhanceosome formation, however, in unstimulated lymphocytes it has been demonstrated that the DNA of an important control sequence in the proximal promoter, the positive regulatory region II (PRRII), is incorporated into a stably positioned nucleosome that prevents transcription initiation by occluding the binding site for an essential transcription factor, Elf-1.[53] Immediately after

[52] H. P. Kim and W. J. Leonard, *EMBO J.* **21,** 3051 (2002).
[53] R. Reeves, W. J. Leonard, and M. S. Nissen, *Mol. Cell. Biol.* **20,** 4666 (2000).
[54] S. John *et al., Mol. Cell. Biol.* **15,** 1786 (1995).
[55] S. John, C. M. Robbins, and W. J. Leonard, *EMBO J.* **15,** 5627 (1996).
[56] H. Kim, J. Kelly, and W. J. Leonard, *Immunity* **15,** 159 (2001).

lymphocyte activation, however, it has also been shown that the PRRII-positioned nucleosome is disrupted or "remodeled," thus allowing for Elf-1 binding to the PRRII element, enhanceosome formation, and transcriptional activation of the *IL-2Rα* gene. Detailed studies of this transcriptional activation process have been aided by the fact that, as illustrated in Fig. 5B, cloned fragments of *IL-2Rα* promoter DNA containing the PRRII region can be reconstituted into a stably positioned nucleosome *in vitro* that has the same translational and rotational setting as the nucleosome found in unstimulated lymphocytes. Furthermore, as shown in Fig. 5A, *in vitro* footprinting experiments employing the chemical cleavage reagent 1,10-phenanthroline copper(II) (i.e., OP) covalently attached to the C-terminal end of the HMGA1a protein demonstrate that this protein binds to the surface of the PRRII-positioned nucleosome in an orientation-specific manner.[53]

In vitro DNase I and hydroxyl footprinting experiments have demonstrated that, as illustrated by the diagram to the side of the sequencing ladder in Fig. 5A (lanes 1–4), the three A/T hooks of a single HMGA1 protein molecule bind to a homopolymer stretch of 18 adenine residues (A_{18}) in the PRRII region (Fig. 5A, lanes 1–4), but do not bind to a PRRII element in which this stretch has been mutated to a random sequence of nucleotides (R_{18}; lanes 9–12), regardless of whether the promoter DNA is free in solution or is incorporated into a stably positioned nucleosome.[53] When a single cysteine residue is introduced by *in vitro* mutagenesis techniques at the C-terminal end of a truncated form of HMGA1a and the OP moiety is covalently attached to this residue, the resulting modified protein (OP–HMGA1a) can be used as an "orientation-specific" chemical cleavage agent when it is bound to either free DNA or reconstituted nucleosome substrates. For example, when OP–HMGA1a is bound to reconstituted PRRII nucleosomes containing wild-type A_{18}, or mutant R_{18}, PRRII DNAs that have been radiolabeled at one end and the complex is then exposed to chemical conditions promoting DNA strand cleavage by the OP moiety, it is observed that only nucleosomes reconstituted with A_{18} DNA (Fig. 5A, lanes 7 and 8), and not those reconstituted with R_{18} DNA (Fig. 5A, lanes 5 and 6), exhibit strand scission inside the core particle. In contrast, OP–HMGA1a molecules that bind to A/T stretches in promoter DNA located outside of the PRRII-positioned nucleosome (Fig. 5B) are able to cleave both the A_{18} (Fig. 5A, lanes 7 and 8) and R_{18} (Fig. 5A, lanes 5 and 6) DNAs near the edge of the core particle. These results unequivocally demonstrate that, at least *in vitro*, HMGA proteins can bind to both DNA and nucleosome substrates in a distinctly polar manner. It has been suggested that if similar direction-specific binding of HMGA proteins to the PRRII-positioned nucleosome occurs in unstimulated T cells *in vivo*, this

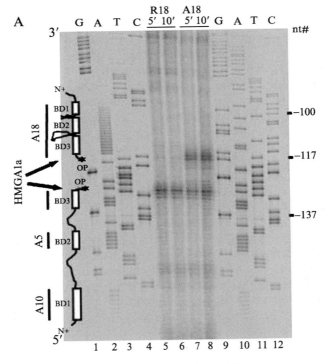

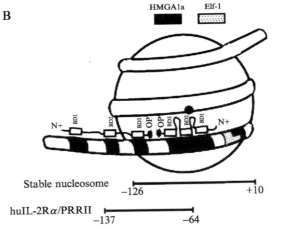

FIG. 5. (continued)

could impart a distinct architectural specificity to the core particle that would uniquely "mark" it for subsequent remodeling after lymphocyte activation.[53]

Directional Binding of HMGA1 to *In Vitro*-Positioned Nucleosomes

Isolation, Mutagenesis, and Radiolabeling of DNA Fragments

A cloned 581-bp *Bam*HI–*Pst*I restriction fragment encompassing nucleotides −472 to +109 of the human *IL-2R*α gene and its 5′ proximal promoter region[57] serves as the starting material for isolating subfragments of the promoter containing the PRRII element for use in *in vitro* chromatin reconstitution experiments. Standard polymerase chain reaction (PCR) techniques[22] are used to amplify a 277-bp promoter fragment that encompasses PRRII and its flanking regions, using the following PCR primer pair: PCR#1 (sense; nucleotides −192 to −173: 5′-CCAGCCCACACCTCCAG CAA-3′ and PCR#2 (antisense; +85 to +65: 5′-CCTCTTTTTGGCATCG CGCCG-3′). The mutant *IL-2R*α promoter construct designated R$_{18}$ (see Fig. 5A) is produced as follows: an oligonucleotide containing 18 randomized residues (along with nonmutagenized *IL-2R*α oligonucleotides flanking either end of the sequence) is chemically synthesized and this fragment is inserted by PCR techniques into the *IL-2R*α promoter DNA, thereby replacing the wild-type homopolymer adenine tract (A$_{18}$) in the PRRII enhancer element. Polynucleotide kinase and [γ-^{32}P]ATP are used to label the 5′ end of the PRRII-containing cloned fragment whereas 3′-end labeling is achieved by filling-in reactions using Klenow polymerase and [α-^{32}P]ATP as described in Ausubel *et al.*[22]

[57] N. H. Ishida *et al.*, *Nucleic Acids Res.* **13**, 7579 (1985).

Fig. 5. Directional binding of HMGA1a proteins to the nucleosome positioned on the PRRII region of the human *IL-2R*α gene promoter. (A) Cleavage of 5′ end-labeled 277-bp wild-type (A$_{18}$) and random mutant (R$_{18}$) promoter fragments assembled into monomer nucleosomes in the presence of a 2-fold molar excess of OP-complexed HMGA1a protein. Sequencing ladders of the wild-type and mutant DNAs are shown in lanes 1–4 and lanes 9–12, respectively. Lanes 5–8 show the products of chemical cleavage reactions (5 and 10 min in duration) resulting from the binding of OP–HMGA1a proteins to positioned nucleosomes reconstituted from wild-type (A$_{18}$) and mutant (R$_{18}$) DNAs (see text for discussion). (B) Diagrammatic representation of the centrally positioned nucleosome on a 277-bp fragment of the *IL-2R*α promoter bound, in an orientation-specific manner, by two HMGA1a proteins that have a copper phenanthroline chemical cleavage reagent attached to their C-terminal ends. The positions of the HMGA1 and Elf-1 protein-binding sites, along with the position of the PRRII enhancer element, are indicated. (A) and (B) reproduced from Reeves *et al.* (2000),[53] with permission.

Creation and Use of OP-Labeled Recombinant Mutant HMGA1a Proteins

Production and purification of the mutant protein HMG-IΔE91, from which the 17 amino acids normally found at the C-terminal end of the wild-type protein are deleted, has been described in detail.[15] Briefly, the 17 highly acidic, glutamic acid-rich residues at the carboxyl end of the HMGA1a protein are removed by *in vitro* mutagenesis of the full-length human HMGA1a cDNA,[23] using standard recombinant DNA procedures.[22,58] The synthetic antisense mutagenesis primer 5'-TCCTCTTCCT CGGATCCAGTTTTTTGGG-3' is used to introduce a *Bam*HI restriction enzyme cut site covering amino acid residues Glu$_{91}$-Lys$_{92}$ (altering these to Gly$_{91}$-Ser$_{92}$) in a previously modified version of the full-length HMG-I(7C)-*Nde*I/*Bam*HI cDNA clone.[20] A 271-bp *Nde*I–*Bam*HI fragment, designated HMG-IΔE91, is isolated from the resulting DNA construct and directionally subcloned, in frame, into a pET expression vector for production of recombinant HMG-IΔE91 proteins as described above. The mutant HMG-IΔE91 protein has a single cysteine residue added to its C terminus that is used as the conjugation site for the chemical nuclease 1,10-phenanthroline copper(II) complex (i.e., OP) used in DNA cleavage footprinting reactions as described by Pan *et al.*[59] The optimum conditions for OP cleavage of naked DNA and each reconstituted chromatin preparation are determined empirically. After the cleavage reactions, labeled DNA fragments are recovered from the reconstituted chromatin substrates by protease digestion and phenol–chloroform–isoamyl alcohol extraction and precipitated with ethanol. Radioactive cleavage products are separated by electrophoresis on either 6 or 8% polyacrylamide sequencing gels and band intensities on the gels are analyzed and quantified with a PhophorImager and ImageQuant software (Molecular Dynamics, Sunnyvale, CA).

Isolation of Nucleosomes, Histone Octamers, and Chromatin Reconstitutions

Packed, frozen chicken erythrocytes in sodium citrate are purchased from various commercial suppliers such as Lampire Biologicals (Pipersville, PA). Trimmed chicken core particles are prepared by micrococcal nuclease digestion of isolated nuclei as previously described.[15,60] Histone

[58] T. A. Kunkel, J. D. Roberts, and R. A. Zakour, *Methods Enzymol.* **154**, 367 (1987).
[59] C. Q. Pan *et al.*, *Proc. Natl. Acad. Sci. USA* **91**, 1721 (1994).
[60] L. J. Libertini and E. W. Small, *Nucleic Acids Res.* **8**, 3517 (1980).

concentrations are determined (ε_{230} = 4.2 liters/g·cm) and purity is monitored by denaturing gel electrophoresis as described above.

Nucleosomes are reconstituted onto radiolabeled DNA fragments in one of two ways[61]: (1) by exchange with isolated chicken erythrocyte nucleosomal core in high salt followed by stepwise dilutions to low salt concentrations, or (2) by dialysis from high-salt to low-salt conditions, using purified histone octamers according to established protocols. Quality control of chromatin reconstitutes is monitored by native nucleoprotein gel electrophoresis [0.8% agarose, 45 mM Tris-borate (pH 8.3), 1 mM EDTA] and the integrity of the core histones is checked before and after reconstitutions by denaturing SDS–PAGE. For typical chromatin assembly reactions, less than 5% of the input labeled DNA fragments should remain as free DNA at the end of the reconstitutions and, when necessary to avoid possible background problems, can be removed by standard purification techniques.[22] Electrophoretic mobility shift assays (EMSAs) employ purified proteins and either radiolabeled DNA substrates or *in vitro*-reconstituted nucleosome core particles and are performed as previously described in this series.[21]

A/T-Hook Peptide Motif Binds Nucleosome Core Particles

Studies of cocomplexes of HMGA1 proteins with synthetic DNA substrates, using solution nuclear magnetic resonance (NMR) techniques, revealed that A/T-hook peptides that interact with the minor groove of A/T-rich sequences of DNA do so in a direction-specific manner,[8] raising a question concerning whether the A/T-hook regions of these proteins are also responsible for mediating the directional interactions with nucleosome core particles. Of even greater biological importance is determining whether or not A/T-hook peptides, which in the context of free HMGA1 proteins in solution are unstructured,[3] are also able to specifically bind to DNA and chromatin substrates in a manner similar to the native proteins when these peptide motifs are incorporated as part of proteins that have a defined secondary structure. These questions are significant for a number of reasons. First, the A/T hook is an evolutionarily highly conserved functional peptide motif dating back to bacteria and is found in a large number of structured proteins in many different organisms,[7] many of which appear to be transcription factors.[62] Second, chromosomal translocations of genes encoding HMGA proteins are among the most common chromosomal abnormalities observed in human neoplasms (reviewed in Refs. 4–6). In

[61] A. P. Wolffe and J. J. Hayes, *Methods Mol. Genet.* **2,** 314 (1993).
[62] L. Aravind and D. Landsman, *Nucleic Acids Res.* **26,** 4413 (1998).

the majority of cases, these chromosomal translocations result in the creation of chimeric genes encoding hybrid proteins in which the three A/T-hook motifs derived from one of the HMGA gene loci are fused "in frame" to the N-terminal end of ectopic peptide sequences derived from a number of other genes. Such translocations are the most common chromosomal abnormalities observed in benign mesenchymal tumors such as lipomas, leiomyomas, fibroadenomas, pleiomorphic adenomas, aggressive angiomyxomas, and pulmonary hamartomas. But the function of these A/T hook-containing chimeric fusion proteins, several of which are predicted to have well-defined secondary structures, in these tumors is unknown. One suggestion, however, is that the attachment of A/T-hook motifs to other proteins or peptides changes the substrate-binding characteristics of the hybrid proteins so that they are now "mistargeted" to sites within the nucleus.[6,63]

To experimentally determine whether the A/T-hook peptide is able to retain its normal DNA and nucleosome substrate specificities when it is placed in the restricted confines of a structured protein fusion partner, we performed the *in vitro* "domain swap" experiments diagrammed in Fig. 6A.[64] In these experiments, the second and strongest DNA-binding A/T-hook motif of the HMGA1 protein was inserted in place of a corresponding length peptide at the N terminus of the second DNA-binding "B-box" domain of the human HMGB1 protein (e.g., HMG-1[1]). The HMGB1 B-box peptide was chosen as a "model" fusion partner for the domain swap experiments because (1) NMR studies demonstrated that the B-box peptide is highly structured in solution, having an "L" or boomerang-like shape consisting of three α helices and an extended N-terminal peptide terminus[65,66]; (2) the extended N-terminal region of the B-box peptide has extensive sequence homology to the HMGA1 A/T-hook motif (Fig. 6A) and yet, unlike the A/T hook, the wild-type B-box peptide does not selectively bind to either the minor groove of A/T-rich sequences or to nucleosome core particles *in vitro;* and (3) an extensive literature exists on the probable biological function(s) of proteins that contain one or more B-box DNA-binding motifs (reviewed in Refs. 7 and 67).

The results of these domain swap experiments demonstrated that the hybrid B-box:A/T-hook peptide correctly refolds into an L-shaped

[63] T. H. Rabbits, *Nature* **372**, 143 (1994).
[64] G. C. Banks, B. Mohr, and R. Reeves, *J. Biol. Chem.* **274**, 16536 (1999).
[65] C. M. Read *et al., Nucleic Acids Res.* **21**, 3427 (1993).
[66] H. Weir *et al., EMBO J.* **12**, 1311 (1993).
[67] J. J. Turchi, K. M. Henkels, I. L. Hermanson, and S. M. Patrick, *J. Inorg. Biochem.* **77**, 83 (1999).

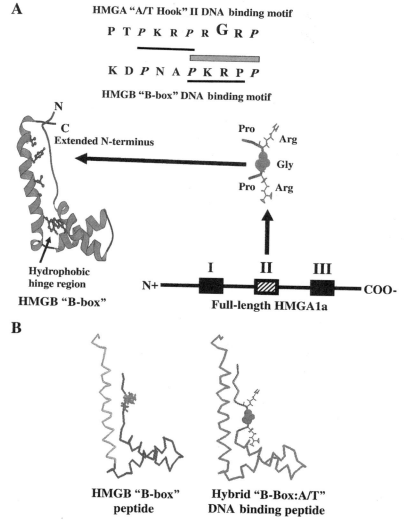

Fig. 6. (A) Schematic diagram of a domain-swap experiment in which a single A/T-hook motif (DNA-binding domain 2 from the human HMGA1 protein) is exchanged for the extended N-terminal peptide segment of the second B box of the HMGB1 protein. Also shown is a sequence comparison between the two exchanged peptide segments. (B) Computer-generated models of the wild-type B-box peptide and the hybrid B-box:A/T-hook peptides, showing the relative orientation of the side chains of the P-K-R-P-P residues in the wild-type B box and the P-R-G-R-P residues in the hybrid protein. See text for discussion. Redrawn from Reeves (2000),[6] with modifications. (See color insert.)

structure resembling a typical B box (Fig. 6B). Furthermore, unlike the wild-type B box, the chimeric molecule now exhibits the substrate-binding specificity of the A/T-hook peptide in that it has the ability to selectively bind to the minor groove of A/T-rich DNA sequences and also binds to random-sequence nucleosome core particles in a manner similar to the wild-type HMGA1 proteins.[64] Thus, the A/T-hook peptide motif has the ability to confer its DNA- and chromatin-binding substrate specificities onto structured proteins of which it is a part.

Construction of Hybrid B-Box:A/T-Hook Plasmid Expression Vectors and Recombinant Protein Production

A plasmid expressing vector encoding the wild-type (WT) second B box of the HMGB1 protein, corresponding to amino acid residues 89–164 of the 214-amino acid full-length HMGB protein, is constructed by PCR amplification of this region of the human HMGB1 cDNA,[68] using the following wild-type primers: 5'-GST-WT1 (sense), 5'-ATGGATCCAAGGACCC TAATGCA-3'; 3'-GST-WT2 (antisense), 5'-TGAATTCTTTTGCACGG TATGC-3'. Starting with this wild-type B-box construct, the hybrid B-box:A/T-hook chimeric expression vector is created in several steps, the first of which is to generate a plasmid encoding a B box that lacks the 10-amino acid (residues 89–98) N-terminal extended segment (B-box w/o N-term). The wild-type B-box vector is used as the template for constructing the B-box w/o N-term vector with the following PCR primers: 5'-GST-WT3 (sense), 5'-CGTGGTCGACCTTCGGCATTCTTC-3'; 3'-GST-WT2 (anti-sense). In the second step, the A-T hook II DNA-binding motif, corresponding to amino acid residues 52–61, PTPKRPRGRP, is amplified by PCR, using the human HMGA1a (clone 7C) cDNA[23] as template and PCR primers with the following sequence: 5'-GST-HOOK1 (sense), 5'-CTGGATCCCCAACACCTAAG-3'; 3'-GST-HOOK2 (antisense), 5'-GAAGAATGCCGAAGGTCGACCACGAGG-3'. In the last step, construction of the hybrid B-box:A/T-hook expression vector is completed in the following manner. The purified PCR constructs corresponding to the A/T hook II DNA-binding motif and the B-box w/o N-term DNA are annealed and extended by PCR, using the 5'-GST-HOOK1 (sense) and 3'-GST-WT2 (antisense) primers. This PCR amplification results in the creation of the completed hybrid expression vector because the 5' region of the B-box w/o N-term [which corresponds to the 5'-GST-WT3 (sense) primer] contains sequence complementary to the 3' portion of A-T hook II, and the 3' region of the 3'-GST-HOOK2 (antisense) primer. The final

[68] L. Wen, J. K. Huang, B. H. Johnson, and G. R. Reeck, *Nucleic Acids Res.* **17,** 1197 (1989).

amplified PCR product is purified by preparative gel electrophoresis, ligated into pGEX-2T plasmids (Amersham Biosciences) at *Bam*HI and *Eco*RI restriction sites and transformed into *E. coli* XL-1 Blue. After confirming the DNA sequence of the B-box:A/T-hook construct inserted into the vector, the plasmid is transformed into either BL21(DE3)pLysS or BL21-CodonPlus-RP bacteria for subsequent production and purification of recombinant hybrid protein as described above.

Refolding of Hybrid B-Box:A/T-Hook Proteins

Recombinant hybrid proteins are purified from bacterial inclusion bodies and, therefore, must be renatured before use. Renaturation is achieved by dissolving the RP-HPLC-purified recombinant proteins in 10 mM Tris-HCl buffer (pH 7.8) containing 1 mM DTT and 3 M guanidine-HCl (Sigma), followed by slow (overnight at room temperature) dialysis of the solution against a large excess of the same buffer lacking guanidine-HCl until equilibrium is reached. Confirmation of correct refolding of the hybrid protein is confirmed by both circular dichroism (CD) and intrinsic fluorescence (IF) determinations of protein structure. In CD analyses, the absorption spectrum of recombinant protein preparations should be measured at least three times, with the average being used to calculate the molar ellipticity. The CD spectrum is measured under reducing conditions in 20 mM KH$_2$PO$_4$ (pH 6.0) and 1 mM Tris-(2-carboxyethyl)-phosphin hydrochloride (Molecular Probes, Eugene, OR). CD data should be collected with nitrogen-purging instruments, such as the AVIV 62DS (AVIV, Lakewood, NJ), that register wavelengths below 190 nm. IF measurements of the hybrid proteins can be performed on instruments such as the Shimadzu 5300 spectrofluorometer (Shimadzu, Kyoto, Japan). For comparative purposes, equal concentrations of both a correctly folded hybrid B-box:A/T-hook protein and a mutant hybrid B-box:A/T-hook protein that is incapable of refolding[64] are dissolved separately in a buffer containing 10 mM Tris-HCl (pH 7.8) plus 1 mM DTT and their IF spectra are read, using an excitation wavelength of 280 nm and an emission wavelength of 290 to 400 nm.

DNA and Nucleosome Binding by Hybrid B-Box:A/T-Hook Proteins

A number of methods for analyzing the A/T-DNA-binding specificity of hybrid recombinant proteins by both DNase I and hydroxyl radical footprinting techniques have been previously published.[13,64,69] Likewise, detailed protocols are available for monitoring the interaction of hybrid

---

[69] T. D. Tullius *et al., Methods Enzymol.* **155,** 537 (1987).

recombinant proteins with isolated nucleosomes, using both EMSA[15,64] and various footprinting techniques.[13,21,70]

A/T Hooks Are Essential Components of Many Chromatin-Remodeling Complexes

The observation that HMGA proteins bind to nucleosome core particles via their A/T-hook motifs and, thereby, induce localized changes in the structure of DNA on the surface of core particles has taken on unanticipated significance. It is now becoming apparent that A/T hook-containing proteins are essential components of many of the multi-protein, ATP-dependent chromatin-remodeling complexes (or "machines"; CRMs) found in yeast, *Drosophila,* and mammalian cells (reviewed in Refs. 2 and 3). For example, mammalian CRM complexes contain one or the other of two essential and closely related ATPases, known as brm/SNF2α and BRG-1/SNF2β,[71,72] each of which contains a single AT-hook motif in its C-terminal region. Interestingly, Bourachot and colleagues[73] have demonstrated that when the AT hook is deleted from brm/SNFα, the CRM complex loses its functional chromatin-remodeling activity in cells and also can no longer bind to chromatin substrates *in vitro*. Likewise, Xiao *et al.*[74] discovered that the N-terminal end of the largest subunit of the *Drosophila* CRM complex, NURF301, contains two AT hooks that are responsible for binding to nucleosomes *in vitro* and also demonstrated that when the two AT hooks of this region are deleted, the ability of the truncated protein to both bind ("tether") to core particles and induce chromatin remodeling (i.e., sliding) is greatly inhibited. Available *in vivo* and *in vitro* evidence, therefore, strongly supports an active role for the AT-hook motifs found in various CRMs in both nucleosome binding and in ATP-dependent chromatin sliding/remodeling. This information, when considered in the context of activation of the human *IL-2Rα* gene discussed above, has led to a proposal that the directional binding of HMGA1 proteins (via their AT-hook motifs) to the nucleosome positioned on the PRRII promoter region in unstimulated T cells likely acts as a "marker" or "placeholder" for binding by the AT hooks of CRM proteins during the subsequent ATP-dependent disruption of the core particle that occurs during transcriptional activation of the gene *in vivo*.[3,53]

[70] D. Pruss *et al., Science* **274,** 614 (1996).
[71] W. Wang *et al., EMBO J.* **15,** 5370 (1996).
[72] M. Vignali, A. H. Hassan, K. E. Neely, and J. L. Workman, *Mol. Cell. Biol.* **20,** 1899 (2000).
[73] B. Bourachot, M. Yaniv, and C. Muchardt, *Mol. Cell. Biol.* **19,** 3931 (1999).
[74] H. Xiao *et al., Mol. Cell* **8,** 531 (2001).

[21]   Preparation and Functional Analysis of HMGN Proteins

By Jae-Hwan Lim, Frédéric Catez, Yehudit Birger,
Yuri V. Postnikov, and Michael Bustin

High-mobility group N (HMGN, formerly named HMG-14/-17[1]) is a family of nuclear proteins that binds transiently to nucleosomes, changes the local architecture of the chromatin fiber, and impacts various DNA-related activities such as transcription, replication, and DNA repair. HMGN proteins are found in the nuclei of all mammalian and most vertebrate cells and contain three functional domains: a bipartite nuclear localization signal, a nucleosome binding domain, and a chromatin unfolding domain. The HMGN family consists of five closely related proteins with a molecular weight of ~10 kDa: HMGN1 (formerly HMG-14), HMGN2 (formerly HMG-17), HMGN3a, HMGN3b, and HMGN4. HMGN3b is a splice variant of HMGN3a that lacks the C-terminal, chromatin unfolding domain.[2]

In the nucleus, HMGN proteins are highly mobile and their interaction with chromatin is transient. HMGN proteins interact with the amino terminal tail of histone H3, compete with histone H1 for nucleosome binding sites, and induce architectural changes that reduce the compaction of the chromatin fiber. By decreasing the compaction of the chromatin fiber, HMGN proteins enhance the accessibility of nucleosomes to chromatin-modifying enzymes and to DNA binding regulatory factors. Thus, together with many other nuclear factors, HMGN proteins impart flexibility to the chromatin fiber and modulate DNA-related activities that ultimately affect the cellular phenotype. Studies with HMGN proteins emphasize the important role of nonhistone structural proteins in chromatin structure and function. Several recent reviews describe the properties and mechanism of action of these proteins in detail.[2,3]

In this chapter, we summarize the methods used to purify the proteins from various sources, present the main approaches used for studying the interaction of the proteins with chromatin, and describe experimental approaches suitable for elucidating their cellular function.

---

[1] M. Bustin, *Trends Biochem. Sci.* **26,** 152 (2001).
[2] M. Bustin, *Trends Biochem. Sci.* **26,** 431 (2001).
[3] M. Bustin, *Mol. Cell. Biol.* **19,** 5237 (1999).

METHODS IN ENZYMOLOGY, VOL. 375

Isolation and Purification of HMGN and HMGN-Containing Multiprotein Complexes

General Considerations

The method of choice for isolating HMGN from various sources, including cultured cells, whole tissue homogenates and bacteria, is direct extraction with cold 5% perchloric acid (PCA) (Fig. 1). With purified nuclei, this procedure yields a simple mixture of proteins consisting mainly of histone H1 and all the members of the HMG superfamily (HMGB, HMGA, and HMGN). This protein extract can be used directly for western analysis. Direct extraction of HMGN proteins from whole tissues is extremely convenient, rapid, and prevents protein degradation. Whole-tissue extracts contain all of the 5% PCA soluble nuclear proteins, and additional cytoplasmic proteins with similar properties. Although the PCA procedure is harsh and involves treatment with 25% trichloroacetic acid (TCA) and 0.3 M HCl, there is no indication that the proteins extracted by this procedure differ from those obtained by milder procedures. However, until this question is clarified it is prudent to minimize the time of exposure to the strong acids and work at low temperatures.

A milder procedure for isolating HMGN proteins is extraction of purified nuclei with 0.35 M NaCl. This procedure allows the isolation of

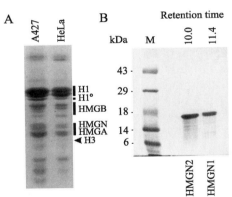

FIG. 1. Purification of HMGN proteins from 5% PCA extracts. (A) Proteins extracted from the nuclei of HeLa or A427 cells with 5% PCA and fractionated on 18% polyacrylamide gels containing sodium dodecyl sulfate. Note the prominent H1 doublet in the gels containing the various HMG proteins is indicated. The bands seen are not necessarily these proteins. The extract does not contain histone H3; the arrow indicates the mobility of H3 in these gels. (B) Pure mouse HMGN1 and HMGN2 proteins. Fractions eluted from an HPLC column at the time indicated at the top of the columns were resolved on SDS-containing 15% polyacrylamide gels. Note: The difference in the electrophoretic mobilities of mouse HMGN1 and HMGN2 is smaller than between the proteins isolated from human tissues.

nondenatured proteins and multiprotein complexes containing HMGN proteins (see later). However, this procedure requires isolation of nuclei, and the extracts contain numerous additional nuclear components including proteolytic enzymes that readily degrade HMGN proteins, necessitating the inclusion of protease inhibitors. Precipitation of bulk proteins from these extracts with 2% TCA leaves HMGN, together with all the other HMGs and a small amount of basic proteins in the supernatant. This extract contains very little histone H1 and is suitable for western analysis. Additional procedures for isolating HMG proteins are described elsewhere.[4,5]

Regardless of the extraction procedure used, to obtain pure HMGN protein the extracts must be fractionated by chromatographic techniques. High-resolution HPLC yields pure proteins; however, due to posttranslational modifications and other components in the extracts, the elution profile is complex. The purified proteins can be identified by immunological techniques.

Recent studies suggest that a large fraction of the HMGN proteins are associated with other proteins into metastable macromolecular complexes.[6] These complexes may enhance the interaction of the proteins with nucleosomes and target them to specific sites. Later we outline two methods suitable for initial purification and detection of HMGN-containing multiprotein complexes. One of the methods involves extraction with low-salt solutions; the second method involves affinity purification of recombinant tagged HMGN protein expressed in cells.

Preparation of HMGN Proteins from Cells Grown in Tissue Culture by Extraction with 5% PCA

1. Harvest cells (no trypsin if possible; if small volumes, the entire procedure can be done in an Eppendorf tube).
2. Wash cells twice with ice cold phosphate-buffered saline (PBS) to remove serum components.
3. Add 10 volumes of cold H_2O (with optional protease inhibitors) to swell the cells and nuclei.
4. Homogenize or sonicate briefly to break the cells (avoid overheating).
5. Make the solution 5% PCA by adding the appropriate volume of 60% PCA (HMGs and H1 remain in the supernatant).
6. Leave on ice for 20 min with occasional vortexing (maximizes extraction).

[4] E. W. Johns, "The HMG Chromosomal Proteins." Academic Press, London, 1982.
[5] R. Reeves and M. S. Nissen, *Methods Enzymol.* **304**, 155 (1999).
[6] J. H. Lim, M. Bustin, V. V. Ogryzko, and Y. V. Postnikov, *J. Biol. Chem.* **277**, 20774 (2002).

7. Spin 10,000g for 10 min at 4°, etc. (removes insoluble materials).
8. Save supernatant and add (slowly, with mixing) 3 volumes of 100% TCA to a final concentration of 25% (HMGN and H1 are insoluble in 25% TCA).
9. Leave at least 1 h at 4° (to fully precipitate the HMG proteins).
10. Spin 10,000g for 10 min at 4° and discard supernatant.
11. Resuspend the pellet in 5 volumes (volume of pellet) of 0.3 N HCl (pellet is not soluble and is difficult to resuspend).
12. Spin 10,000g for 10 min at 4° and save the supernatant.
13. To the supernatant, add 6 volumes of cold acetone. Leave overnight at −20°.
14. Spin 10,000g for 10 min at 4° and discard supernatant.
15. Wash pellet twice with 90% cold acetone and twice with 100% cold acetone (to remove all acids).
16. Air dry.

Modification. To avoid 25% TCA treatment, it is possible to modify the procedure as described below. This procedure involves manipulations of larger volumes.

Proceed to step 7 of the procedure described earlier. Then:

8. Make the supernatant 0.3 N in HCl by adding appropriate volume of 6 N HCl.
9. Proceed to step 12 in original procedure.

Extraction from Nuclei. For a cleaner preparation, purify nuclei from the tissue culture cells, and extract the proteins from these nuclei.

1. Spin down the cultured cells at 1200g for 10 min at 4°. Measure the volume of pellet (the following procedure is exemplified for 0.5 ml of pelleted cells).
2. Wash cells twice with ice cold PBS.
3. Add 5 ml of ice cold solution A (50 mM Tris-HCl, pH 7.5, 25 mM KCl, 0.25 M sucrose, 1 mM PMSF, or other protease inhibitors) plus 0.5% NP-40 and scrape.
4. Spin at 3000g for 10 min at 4°.
5. Suspend the nuclei in 5 ml of ice cold solution A (without NP-40).
6. Homogenize 10 times with a loose pestle.
7. Centrifuge at 1500g for 10 min at 4° in 15 ml polystyrene tube.
8. Repeat steps 5–7 once.
9. Suspend nuclei in 5 ml of solution B (1 mM EDTA, 10 mM Tris-HCl, pH 7.5, 1 mM PMSF, and additional protease inhibitors).
10. Homogenize 15 times with tight pestle to break up nuclei.

11. Make the solution 5% PCA by adding the appropriate volume of 60% PCA (HMGs and H1 remain in the supernatant).
12. Proceed to step 6 in original procedure.

All of the 5% PCA preparations contain large amounts of histone H1, all the HMGs, and additional components. To obtain pure HMGN it is necessary to fractionate them further by HPLC (Fig. 1).

Preparation of HMGN from Tissues

Crude preparations of HMGN proteins can be obtained by extracting tissues directly with 5% PCA. In this procedure, washed tissues are homogenized with a tissue grinder or blender, in 10 volumes of water, the extract is made 5% PCA by the addition of appropriate amount of 60% PCA, and the procedure described earlier for preparation of the proteins from tissue cultures scaled up and followed closely.

However, with tissues it is best to isolate HMGN from nuclei prepared as follows:

1. Rinse tissues (1–2 ml) in 25 ml buffer A (15 mM HEPES, pH 7.5, 60 mM KCl, 15 mM NaCl, 2 mM EDTA, 0.5 mM EGTA, 0.34 M sucrose, 10 mM DTT, 0.15 mM spermine (freshly added), 0.5 mM spermidine (freshly added), 0.2 mM PMSF) in a 100-ml glass beaker on ice. Transfer tissues to 100-mm plastic Petri dishes on ice containing 10 ml of buffer A. Mince tissues into chunks several millimeters in size using razor blades or scalpel. Decant each minced preparation into a separate homogenizer. Let the fragments settle and decant. Homogenize tissues with 5–10 strokes of a loose-fitting motor-driven tissue grinder, followed by an additional 5 sharp strokes by hand.
2. Nuclei preparations could be done according to various protocols depending on the tissue studied. We use either detergent lysis[7] or mechanical disruption with a Dounce homogenizer, followed by sucrose cushion ultracentrifugation.[8] The nuclear pellet is suspended in 20 mM HEPES-Na, pH 7.5, 1.5 mM MgCl$_2$, 15 mM NaCl, 60 mM KCl, 0.2 mM EDTA, 0.5 mM DTT, 1× cocktail of proteinase inhibitors (Roche). For storage at −20°, add glycerol up to 50%.
3. HMGN together with other HMGs and large amounts of H1 can be extracted from the nuclei directly with 5% PCA as described earlier. To obtain pure HMGN the extract needs to be purified by HPLC or ion exchange chromatography.

[7] J. Ausio, F. Dong, and K. E. van Holde, *J. Mol. Biol.* **206**, 451 (1989).
[8] D. R. Hewish and L. A. Burgoyne, *Biochem. Biophys. Res. Commun.* **52**, 504 (1973).

Extraction of HMGN from Nuclei with 0.35 M NaCl

This procedure is applicable only to purified nuclei. The procedure is much milder than the 5% PCA approach but yields a very complex mixture of proteins and special precautions are needed to avoid HMGN proteolysis:

1. Mix freshly prepared nuclei with equal volume of 625 mM NaCl solution, pH 7.0 (final NaCl concentration is 350 mM).
2. Extract by over-the-top mixing in a cold room for 20 min.
3. Spin down the pellet at 4000g for 20 min at 4°, and use the supernatant for further purification.

The salt extract could be considerably enriched in HMG proteins by adding TCA to a final concentration of 2%. The HMGN proteins are among the proteins that remain in the supernatant after centrifugation at 5000g for 15 min at 4°. Further details are described elsewhere.[4]

Preparation of Recombinant Proteins

Recombinant HMGN proteins can be produced in bacteria by standard protocols.[9] The HMGNs do not seem to form inclusion bodies. We break the cells, suspended in H$_2$O with either a French press, or by sonication. The proteins are extracted by making the bacterial homogenate 5% in PCA and recovering the proteins as described earlier.

A frequent complication that we encountered in the production of recombinant HMGN proteins is the presence of numerous peptides derived from the HMGN sequence. These can originate from either premature termination or degradation. The use of bacterial codons in the expression vector[10] could minimize some of the problems due to premature termination. We purify the recombinant proteins first by reverse-phase HPLC and then by MonoS ion exchange chromatography (Fig. 2).

Purification of HMGN Proteins by Reverse-Phase HPLC or
   MonoS Ion Exchange Chromatography

Some of these techniques have been described previously.[11] The following is a description of the major steps involved.

HPLC on Reverse-Phase C4 Column

We use HP1100 HPLC chromatography modules, Butyl Aquapore 250 × 4.6 mm columns (Applied Biosystems) with pre-column guard cartridges. The buffers are:

[9] M. Bustin *et al.*, *Nucleic Acids Res.* **19**, 3115 (1991).
[10] S. M. Paranjape, A. Krumm, and J. T. Kadonaga. *Genes Dev.* **9**, 1978 (1995).
[11] T. S. Elton and R. Reeves, *Anal. Biochem.* **144**, 403 (1985).

Buffer A: 0.1% trifluoroacetic acid (TFA) in water.
Buffer B: 0.1% TFA in acetonitrile.
Sample: 0.1–5 mg of the PCA extract dissolved in 0.5 ml buffer A and filtered through 0.22-μm Millex GX C_4 syringe filter.

After purging the system with buffers A and B, we equilibrate the column according to the supplier's recommendations. The proteins are fractionated with a linear gradient from 90% buffer A: 10% buffer B to 10% buffer A: 90% buffer B, for 40 min at 1 ml/min flow rate. The eluate is monitored with a diode array detector at 220 and 280 nm. HMGN proteins do not adsorb at 280 nm, a property that can be used for preliminary identification of peaks of interest.

The fractions are analyzed on 15% SDS-PAGE and if needed, by western analysis.

We have also successfully tried a number of C_3–C_{18} columns from Vydac, Applied Biosystems, Phenomenex, and Waters. All of them purify HMGN proteins up to 75–95% purity in a single run (Fig. 1).

MonoS Cation Exchange Chromatography

We use either FPLC Pharmacia system or HP1100 HPLC chromatography modules. Column: Mono S HR 5/5 (Pharmacia) eluted with buffers degassed and filtered through 0.22-μm filter:

Buffer A: 100 mM sodium phosphate, pH 7.
Buffer B: 100 mM sodium phosphate, pH 7, 1 M NaCl.
Sample: 1–5 mg of the lyophilized preparation, dissolved in 0.5 ml buffer A and filtered through 0.22-μm Millex GX C_4.

After equilibrating the column according to the supplier's recommendations, we apply the sample and elute it with a linear gradient from 100% buffer A to 100% buffer B, for 24 min, at 1 ml/min flow rate. The diode array detector is set at 220 and 280 nm. The fractions are analyzed on 15% SDS-PAGE and if needed, by western blots.

Detection and Initial Isolation of HMGN-Containing Multiprotein Complexes

In HeLa cells, a large fraction of nuclear HMGN proteins, most likely nonchromatin bound, is associated with other proteins in metastable multiprotein complexes.[6] The complexes can be isolated from 0.35 M NaCl nuclear extracts prepared essentially as described.[12]

---

[12] K. A. Lee and M. R. Green, *Methods Enzymol.* **181,** 20 (1990).

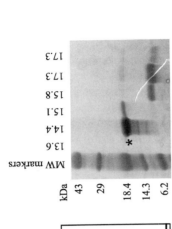

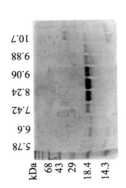

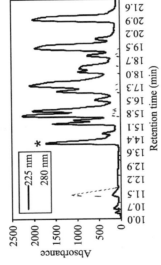

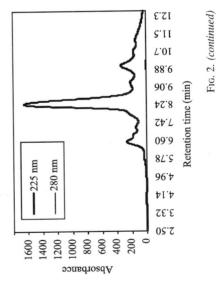

FIG. 2. (continued)

Cells

One liter HeLa cells, grown to 5×10^5 cells/ml in 5% FCS MEM-Joklik. Spin 5 min at 1000g at 4°. One liter of cells yield about 2 g packed cells in a total volume of 2 ml (1 volume).

For large quantities, it is possible to obtain cell pellets from commercial sources.

Nuclear Extract (All Steps at 4°)

1. Add 15 volumes (30 ml) buffer A (10 mM HEPES-Na, pH 7.9, 1.5 mM MgCl$_2$, 10 mM KCl, 0.5 mM DTT, 1 mM AEBSF), suspend on vortex, transfer to 50-ml Falcon centrifuge tubes.
2. Spin 1000g, 10 min, 4°.
3. Resuspend in 4 volumes (8 ml) buffer A.
4. Suspend with 15 strokes with Dounce (loose fit, L type), transfer to 15-ml Falcon centrifuge tubes.
5. Spin down on 1000g, 10 min, 4°. The nuclear pellet is about 1.5 ml.
6. Suspend in 4 volumes (6 ml) Buffer C-350 (20 mM HEPES-Na, pH 7.5, 25% glycerol, 1.5 mM MgCl$_2$, 430 mM NaCl, 0.2 mM EDTA, 0.5 mM DTT, 1 mM AEBSF).
7. Break nuclei with 15 strokes with Dounce (tight fit, T-type) homogenizer.
8. Keep on ice for 20 min.
9. Repeat with 5 strokes with Dounce (tight fit, T-type) homogenizer.
10. Spin in a Sorvall centrifuge, rotor SS-34, 30 min 12,000 rpm, 4°.
11. Supernatant (~7 ml) is "nuclear extract" (S350).
12. Dialyze S350 on Slyde-A-Lyzer 3-10K (Pierce, 3–15-ml cassette) against 50 mM NaP, pH 7.2, 0.2 mM AEBSF, overnight.

Usually some of the proteins precipitate after dialysis. Spin down the pellet in 1.5-ml Eppendorf tubes, 13,000 rpm at 4° for 15 min. Save supernatant. It can be stored frozen but best if it is used fresh.

FIG. 2. Purification of recombinant HMGN proteins. (A) Separation of the 5% PCA extract from BL-21 bacterial culture expressing wild-type HMGN1 protein. The gel on the left depicts the protein profile of a cell lysate and a 5% extract of this lysate fractionated on 15% polyacrylamide gels containing SDS. The gel on the right depicts purified HPLC fractions separated on the same type of gel. * indicates the position of the full-length HMGN1. Fractions containing the protein were pooled and dried twice with SpeedVac. (B) Purification on Mono S HR 5/5 columns. HMGN-containing fractions recovered from the HPLC were dissolved in 0.5 ml of buffer B (see text) and the solution applied to the column and eluted with a linear gradient from 0 to 1 M NaCl in 100 mM sodium phosphate, pH 7, at 1 ml/min. The peaks of interest were loaded on SDS-containing 15% polyacrylamide gels. Note the purity of the final preparation of HMGN1 protein, eluting at 8.5 min.

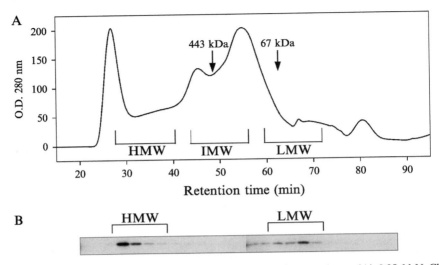

FIG. 3. Detection of HMGN1-containing macromolecular complexes. (A) 0.35 *M* NaCl nuclear extracts from HeLa cells applied to Superose 6 size exclusion columns. HMW, IMW, and LMW denote, respectively, the presence of HMGN in high, intermediate, and low molecular weight fractions as indicated by western analysis (B).

For large quantities, it is possible to obtain nuclear extracts from commercial sources.

Chromatography

1. Run complexes on Zorbax GF450 (or Superose 6) exclusion columns.
2. Test for HMGN proteins in the various fractions by western analysis.

Figure 3 depicts typical results of the initial isolation and detection. Further purification of the complexes can be achieved by fractionation on various columns, such as MonoS, Heparin-Agarose, or MonoQ columns.[6]

It is also possible to isolate HMGN-containing complexes by affinity purification from S3 HeLa cells that stably express HMGN1 or HMGN2 tagged at the C-terminal with both FLAG and HA peptide tags according to established procedures.[13] We have obtained such complexes but did not study them in detail.

---

[13] V. V. Ogryzko *et al.*, *Cell* **94,** 35 (1998).

Analysis of the Interaction of HMGN with Nucleosomes and Chromatin

Mobility Shift Assay to Study In Vitro Interactions of HMGN with Chromatin Subunits

The method of choice for *in vitro* analysis of the interaction of HMGN proteins with nucleosome cores is the mobility shift assay. In this assay, increasing amounts of HMGN protein are added to a constant amount of purified nucleosome core particles and the mixture examined by electrophoresis on 5% native polyacrylamide gels. The interaction of proteins with core particles is ionic strength dependent. At salt concentrations lower than 1.2× TBE (1× TBE; 89 mM Tris, 89 mM borate, 1 mM EDTA, pH 8.4) the dissociation constant is in the order of 10^{-9} M and the binding is noncooperative. Under these conditions, core particles form complexes containing either 1 or 2 molecules of HMGN. In higher ionic strength conditions that are close to physiological, the dissociation constant is about 10^{-7} M and the HMGN proteins bind to the cores cooperatively producing complexes containing two molecules of HMGN per particle. At these physiological ionic strength concentrations, the interactions are highly specific. Point mutations and posttranslational modifications alter measurably the interaction of the proteins with cores. Significantly, both *in vitro* and in nuclei, HMGN proteins form homodimeric complexes containing 2 molecules of one type of HMGN protein (i.e., 2 HMGN1 or 2 HMGN2). The absence of heterodimeric complexes (i.e., two different HMGNs on 1 nucleosome) argues for specificity in the interaction of the proteins with nucleosomes. To date, the nature of this specificity is not fully understood.

More information on the mobility shift assay and additional methods for detail analysis of the interaction of HMGN with isolated core particles, including hydroxyl radical footprinting, DNase I digestion, and thermal melting can be found elsewhere.[14] In the following, we present the protocol for mobility shift assays suitable for analysis of the cooperative interaction of wild-type and mutant HMGN proteins with isolated core particles under physiological salt concentrations:

1. Incubate 5 pmol of core particles with 5–15 pmol of HMGN protein in a volume of 10 μl, at 4° for 5 min in 2 × TBE. Add to each tube 2 μl of 20% Ficoll-400.
2. Pre-electrophoresis a minigel (85 × 70 × 0.7 mm), 5% native polyacrylamide gel (acrylamide: bis = 19:1), containing 2 × TBE for 2 h at 4° at constant 100 V.

---

[14] Y. V. Postnikov and M. Bustin, *Methods Enzymol.* **304**, 133 (1999).

3. Add loading buffer and load the reaction mixtures onto the gel. Two microliters of $1 \times$ DNA loading buffer containing bromophenol blue and xylene cyanol applied to an empty well serves as a control to monitor the electrophoresis progression. Run at a constant 100 V for about 2.5 h, until the xylene cyanol is about two-thirds into the gel length.
4. Stain the gel with ethidium bromide (Fig. 4).

Assays for the Interaction of HMGN Proteins with Chromatin

The most convenient assays for the analyses of the effects of HMGN on chromatin are electron microscopy, sucrose gradient centrifugation, and digestion with various nucleases. The various techniques for analysis of the interaction of HMGN with chromatin have already been described.[14,15] These and additional assays indicate that the binding of HMGN to nucleosomes reduces the compaction of the chromatin fiber. Thus, supercoiling analysis in chloroquine-containing agarose gels and electron microscopy of minichromosomes assembled *in vitro* revealed that HMGN does not affect the rate of nucleosome assembly or the number of nucleosomes in the minichromosomes, but changed the apparent diameter of the minichromosomes. In sucrose gradients, HMGN-containing chromatin sediments slower than HMG-free chromatin, presumably because "decompaction" increases the radius of gyration of the chromatin fiber. Likewise, the correct binding of HMGN to nucleosomes increases the rate at which restriction enzymes, or micrococcal nuclease, digest the DNA in chromatin.

Micrococcal Nuclease Assay for HMGN-chromatin Interactions. Micrococcal nuclease digestion assay is an accepted method to reveal changes in chromatin structure. Micrococcal nuclease cleaves initially in the linker region as an endonuclease, and in a second stage reduces the size of the chromatin fragment by an exonucleolytic digestion proceeding bidirectionally from the initial cleavage point.[16] In the exonucleolytic digestion stage, the histone octamers slide toward each other thereby exposing additional nucleotides to digestion, resulting in a gradual decrease in the size of the oligonucleosome fragment. The presence of HMGN proteins on nucleosomes affects the kinetics of micrococcal nuclease in several ways. By decompacting chromatin, HMGN proteins facilitate access to the linker DNA and increase the rate of the initial endonucleolytic attack. By stabilizing the structure of the nucleosome core, they prevent octamer sliding, protect several bases at the ends of the chromatin particle, and decrease the rate of the exonucleolytic digestion. As a result, digestions of HMGN-containing chromatin yield a "cleaner" nucleosomal ladder with

[15] K. L. West, Y. V. Postnikov, Y. Birger, and M. Bustin, *Methods Enzymol.* **371,** 524 (2003).
[16] K. E. van Holde, "Chromatin." Springer, New York, 1988.

less "smear" between the oliogonucleosomal fragments. In addition, these digestions produce longer mono-, di-, and tri-nucleosomal DNA fragments. These effects are most obvious when comparing *in vitro* assembled templates in either the presence or absence of HMGN.

Nuclease digestion is a major tool in the analysis of the role of HMGN in chromatin structure. The application of the technique to study the role of HMGN using *in vitro* assembled templates has already been described in detail.[14] The use of this approach for studies with HMGN knock-out mice, as a means to study its function in a physiological context, is described in the last section of this chapter.

In Vivo *Interactions Analyzed by Photobleaching Techniques*

Until recently, most of the information on the intranuclear organization of HMGN proteins was obtained by immunofluorescence techniques. Immunofluorescence approaches to chromatin are described elsewhere.[17] Immunofluorescence analysis demonstrated that the proteins are found only in nuclei (1) they colocalize with active sites of transcription,[18] (2) their intranuclear organization changes during the cell cycle,[19] (3) they do not interact with metaphase chromosomes, and (4) the proteins cluster into separate domains, that is, nuclear domains containing either only HMGN1 or only HMGN2.[20]

Immunofluorescence also suggested that the intranuclear organization of the proteins is dynamic, a finding confirmed by more recent photobleaching techniques. HMGN was the prototype used to demonstrate that in living cells, the intranuclear organization of chromatin binding proteins is dynamic, and that these proteins move very fast throughout the nucleus.[21]

Photobleaching techniques with wild-type and mutant HMGN proteins verified that in living cells HMGN binds to chromatin through its nucleosome binding domain and that mutations in this domain decrease its residence time on chromatin. Furthermore, HMGN mutants that do not bind to nucleosomes mislocalize to the nucleolus. These studies with living cells demonstrated that the large body of experimental data obtained *in vitro* reflects the *in vivo* function and organization of the protein.

Photobleaching techniques, in conjunction with microinjection techniques, can be used to study competition between proteins for common chromatin binding sites, in living cells. In these experiments, a nuclear protein is microinjected into the cytoplasm of a cell expressing a GFP-labeled

[17] M. Bustin, *Methods Enzymol.* **170**, 214 (1989).
[18] R. Hock, F. Wilde, U. Scheer, and M. Bustin, *EMBO J.* **17**, 6992 (1998).
[19] R. Hock, U. Scheer, and M. Bustin, *J. Cell Biol.* **143**, 1427 (1998).
[20] Y. V. Postnikov, J. E. Herrera, R. Hock, U. Scheer, and M. Bustin, *J. Mol. Biol.* **274**, 454 (1997).
[21] R. D. Phair and T. Misteli, *Nature* **404**, 604 (2000).

nuclear protein. Photobleaching analysis of the chromatin residence time of the GFP-labeled nuclear protein in control and microinjected cells allows quantitative determination of competition between the labeled protein and the microinjected protein.[22]

Experimental details on the application of photobleaching techniques to studies of chromatin binding proteins are described in this volume (Misteli, this volume). Here we describe fluorescent recovery after photobleaching (FRAP) analysis of HMGN protein.

Fluorescence Recovery After Photobleaching Analysis of the Interaction of HMGN Proteins with Chromatin in Living Cells

The HMGN-GFP proteins are produced from a peGFP-HMGN vector in which the *Hmgn* coding sequence is placed 5' of eGFP coding sequence. Mouse embryonic fibroblasts are cultured in regular conditions (Dulbecco's modified Eagle's medium (DMEM) + 10% serum, 5% CO_2, 37°) and are transfected 24 h before FRAP experiment with peGFP-HMGN vector using classical transfection protocols. Cells should be 25–50% confluent at the time of FRAP experiment. Confocal microscopy should be performed at 37° and under 5% CO_2 atmosphere. If no CO_2 chamber is used, the medium can be buffered with 25 mM HEPES, pH 7.5. FRAP analyses are performed on an inverted confocal microscope with high-speed imaging capabilities (e.g., Zeiss LSM 510) as HMGN proteins are highly mobile. Typically, five images are collected to determine prebleach fluorescence intensity, followed by a short high-power laser pulse (typically 150–300 ms) of 4 μm in diameter, using the 488 nm line of an argon laser. After bleaching, images are acquired at 150-ms interval for at least 20–25 s. Imaging over 20 s can be done at slower rate (e.g., 500 ms or 1 s interval).

Recovery curves are generated from background-subtracted images as described elsewhere[21] and by Misteli in this volume. Total fluorescence of each image is compared to the initial fluorescence to determine the loss of fluorescence during imaging. Fluorescence of the region of interest at each time point is normalized to the initial fluorescence of the same area as: $I_{rel} = (T_0 I_t)/(T_t I_0)$, where T_0 is the prebleach total fluorescence, T_t, the total fluorescence at time point t, I_0, the initial fluorescence in the region of interest, and I_t is the fluorescence of the region of interest at time point t.

Alternatively, transfection of HMGN-GFP can be replaced by microinjection of a fluorescently labeled purified HMGN protein. Briefly, HMGN proteins containing a S88C (HMGN1) or S82C (HMGN2) amino acid modification[19] are labeled with a fluorescent dye (AlexaFluor

[22] F. Catez, D. T. Brown, T. Misteli, and M. Bustin, *EMBO Rep.* **3,** 760 (2002).

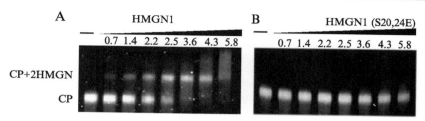

FIG. 4. Electrophoretic mobility shift assays for studies on the interaction of HMGN with core particles (CP). CP were incubated with either recombinant HMGN1 (A) or with an HMGN1 S20,24E double-point mutant (B) that does not bind to chromatin. The reaction mixture were electrophoresed on native 4% polyacrylamide gel in 2× TBE buffer and the gel stained with ethidium bromide. Numbers on top of the columns indicate HMG:core particle in the reaction mixture. Note that increasing concentrations of wild-type HMGN1, but not of the mutant HMGN1, shifted the mobility of the core particles.

488 or Texas Red [Molecular Probes]). Labeled proteins are diluted in 10 mM Tris-HCl, pH 7.5, and microinjected into mouse fibroblast at 0.03–0.05 μM concentration. Labeling and microinjection procedures are detailed elsewhere.[19,22] FRAP experiments should be performed at least 30 min after injection to ensure complete nuclear localization of the injected protein.

Typical recovery curves for wild-type HMGN2 and HMGN2-S24,28E mutant, which does not bind chromatin, are shown in Fig. 5. Note that loss of nucleosome binding (compare to Fig. 4) decreases the length of the time that HMGN resides on chromatin and increases its apparent intranuclear mobility. This protein also mislocalizes to the nucleolus.

Analysis of the *In Vivo* Function of HMGN Proteins

Phenotypic analysis of $Hmgn^{-/-}$ mice is the most adequate method for elucidating the true biological function for HMGN proteins

Comparison of transcription profiles of cells and tissues derived from wild-type and HMGN knock-out mice could identify transcriptional targets of HMGN proteins. The chromatin structure of potential genes targets in knock-out and wild-type cells, by the methods described here and elsewhere, can provide information on the true mechanism of action of these proteins. Methods for production and analysis of these mice are beyond the scope of this chapter. A recent manuscript describes one of the phenotypes observed in $Hmgn1^{-/-}$ mice.[23]

A major question of general interest is whether changes observed in the transcription levels or chromatin structure of a gene are a direct effect of

[23] Y. Birger *et al.*, *EMBO J.* **22**, 1665 (2003).

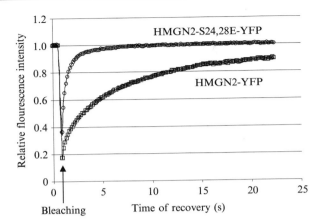

Fig. 5. FRAP analysis of HMGN-chromatin interaction in living cells. Shown is a quantitative FRAP analysis of wild-type HMGN2-YFP (yellow fluorescent protein) and mutant HMGN2-S24,28E-YFP, which *in vitro* does not bind to nucleosomes (Fig. 4). A 4-μm wide spot was bleached in the nucleus and the relative intensity of the bleached area was measured before and after bleaching. Fluorescence intensity was normalized to the prebleach intensity of the same region and plotted as indicated in the text. The curves are averages generated from 10–12 cells for each experiment.

the absence of HMGN protein. Insights into these questions can be obtained by chromatin immunoprecipitation (ChIP) approaches.

Immunoprecipitation of HMGN1-Containing Chromatin (ChIP)

1. Grow mouse embryo fibroblast (MEF) cells, $Hmgn1^{+/+}$ and $Hmgn1^{-/-}$, to 90% confluence in DMEM with 10% (v/v) fetal bovine serum.

2. Add formaldehyde (Sigma) directly to the cell culture medium at a final concentration of 1% and incubate the plate for 10 min ($37°$) to cross-link the HMGN1 to chromatin.

3. Rinse the cells in the plate once with PBS, and PBS-containing protease inhibitors (1 μg/ml pepstatin A, 1 μg/ml leupeptin, 1 μg/ml aprotinin and 1 mM phenylmethylsulfonyl fluoride), HDAC inhibitor (10 mM sodium butyrate), and phosphatase inhibitor (100 μM sodium orthovanadate, 50 nM okadaic acid).

4. Scrape the cells of the plate and recover by centrifugation.

5. Resuspend the cells in lysis buffer (1% SDS, 10 mM EDTA, 50 mM Tris-HCl, pH 8.1, protease inhibitor cocktail [Roche], HDAC inhibitors [10 mM sodium butyrate and 50 nM Trichostatin A], and phosphatase inhibitor cocktails I and II [Sigma-Aldrich]) and incubate 10 min on ice.

6. Sonicate the lysates nine times with 10-s bursts (with 10 s chilling on ice between bursts). We use an Ultrasonics sonicator model W-225R, at setting 3, with a 1/16″ stepped microtip. (*Caution:* The quality of the sonicator tip affects the results.)

7. Spin the sonicated lysate at 10,000g for 10 min, at 4°. The DNA fragments in the supernatant should be about 200–800 base pair long.

8. Dilute the supernatant tenfold with ChIP dilution buffer (0.01% SDS, 1.1% Triton X-100, 1.2 mM EDTA, 16.7 mM Tris-HCl, 167 mM NaCl, 10 mM sodium butyrate, protease inhibitors). Save small amounts (1/20 volume ratio) of the supernatants as input DNA control.

9. Preclear the chromatin solutions with 80 μl of salmon sperm DNA/protein A(G)-agarose 50% gel slurry (Upstate Biotechnology, Lake Placid, NY), preequilibrated in 1 × ChIP dilution buffer, for 4 h at 4° on a rotatory shaker. Filter and save the cleared chromatin solution using 0.45-μm spin filter (Millipore). *Note:* Make sure to select the proper affinity media (protein A versus proteins G) for the antibodies used.

10. Add HMGN1 antibodies or 8 μg of normal nonimmune IgG to the precleared chromatin solutions and rotate overnight at cold room.

11. Add 60 μl of salmon sperm DNA/protein A-agarose slurry and incubate for 2 h.

12. Wash the chromatin-antibody-protein A-agarose complexes (5 min each wash) sequentially with 1 ml each of:

   a. Low-salt buffer (150 mM NaCl, 20 mM Tris-HCl, pH. 8.1, 2 mM EDTA, 0.1% SDS, 1% Triton X-100)
   b. High-salt buffer (low salt buffer plus 500 mM NaCl)
   c. Lithium chloride solution (0.25% LiCl, 10 mM Tris-HCl, pH 8.1, 1 mM EDTA, 1% NP-40, 1% deoxycholate)

13. Wash the precipitates two times with 1× TE buffer.

14. Elute the precipitated DNA twice with freshly made elution buffer (1% SDS and 50 mM NaHCO$_3$) by rotating 30 min at room temperature.

15. Add NaCl to 200 mM final concentration and heat at 65° for 6 h to reverse the formaldehyde cross-linking. Cool and digest with proteinase K for 1 h at 45°.

16. Recover the DNA fragments by phenol/chloroform extraction and ethanol precipitation after treatment.

17. Resuspend the DNA in 50 μl of sterile water. Store overnight before quantification.

18. Quantify the DNA amount with PicoGreen dsDNA Quantitation Kit (Molecular Probes).

As controls, perform parallel reaction with nonimmune IgG and with chromatin isolated from HMGN knock-out mice. The specific sequence DNAs in input DNA and immunoprecipitated DNA samples are quantified by either conventional or the real-time PCR using primers specific for a region of interest (Fig. 6).

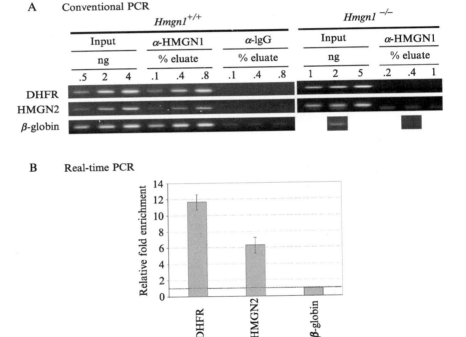

FIG. 6. ChIP analysis of HMGN-containing chromatin. (A) Conventional PCR analysis with the primer sets for the transcribed *Dhfr* and *Hmgn2* and the nontranscribed β-globin genes. As nonspecific binding controls, mock immunoprecipitation was performed with rabbit nonimmune IgG and with chromatin fragments obtained from *Hmgn1*[−/−] cells. Note the lack of signal in the immunoprecipitates from these controls. Since these controls did not contain DNA, comparisons with experimental values were done on the basis of volume rather than amounts of DNA. (B) Quantitation by real-time PCR reaction. Note that the regions amplified from transcribed genes were enriched in HMGN1. (We routinely use several sets of primers to scan a large portion of the gene.)

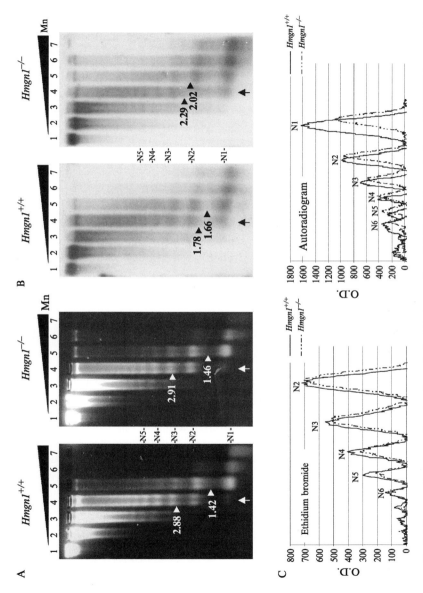

FIG. 7. Micrococcal nuclease digestion for analysis of the interaction of HMG with chromatin. Loss of HMGN1 decreases the accessibility of the *Hmgn2* gene. Micrococcal nuclease digestions of nuclei isolated from the livers *Hmgn1*[−/−] and *Hmgn1*[+/+] mice.

Micrococcal Nuclease Digestion

In living cells HMGN are highly mobile and the amount present in an average nucleus is sufficient to bind to about 1–2% of the nucleosomes. It is highly likely that most of the HMGN-induced changes in chromatin structure are local and temporary. Therefore, it is not reasonable to expect to detect HMGN-induced global changes in the chromatin structure. On the other hand, it is possible that these proteins would affect the structure of selected genes. Indeed, using HMGN1 knock-out mice we have detected differences in the chromatin structure of the HMGN2 gene (Fig. 7). The following is the outline of the experimental approach to detect HMGN1-induced changes in the chromatin structure of a specific gene:

1. Isolate nuclei from tissue of interest from wild-type and knock-out mice.
2. Digest nuclei with various concentration of micrococcal nuclease. The exact conditions of digestion need to be calibrated for each micrococcal nuclease lot and for each nuclear preparation. The goal is to generate a complete kinetic picture, from undigested chromatin to fully digested chromatin. In a good preparation, the nucleosomal ladder extends to decamers.
3. Extract the DNA and run it on 1.5% agarose gels in 1× TBE alongside appropriate molecular weight markers. Stain the gel with ethidium bromide, photograph, and save the picture.
4. Blot the gel onto a membrane using a standard Southern procedure and probe the blot for a gene of interest. Develop the blot.
5. Quantify the area under each of the peaks observed in the nucleosomal ladders in the ethidium bromide-stained gels and in the probed membranes. Usually it is difficult to get good resolution beyond hexanucleosomes.
6. Calculate the average nucleosomal length in a digest using the formula: $L_a = \sigma N1{-}Nx \, (P_N)$, where L_a, is the average nucleosome length, $N1$ to Nx is the oligonucleosome size (i.e., mono, di), and P_N is the fraction of a particular oligonucleosome size out of the total scan.

(A) Ethidium bromide stain prior to Southern transfer. (B) Southern analysis with *Hmgn2*-specific probe. (C) Scans of lane 4 (arrow). The numbers adjacent to the arrowheads indicate the length of the average oligonucleosomes (L_a) in the digest calculated as indicated in the text. The increased average oligonucleosome length in the autoradiogram of the *Hmgn1*$^{-/-}$ cells indicates slower rate of digestion of the *Hmgn2* chromatin. Lanes 1–7 correspond, respectively, to 0, 0.05, 0.3, 2, 11, 65, and 400 units micrococcal nuclease in the digest. N1–N5 denotes the oligonucleosomal size. Figure reproduced from West *et al.*[15] and Birger *et al.*[23]

Section III

Molecular Cytology of Chromatin Functions

[22] Methods for Visualizing Chromatin Dynamics in Living Yeast

By Florence Hediger, Angela Taddei,
Frank R. Neumann, and Susan M. Gasser

Introduction

It is now well established that chromatin is not randomly organized within the nuclei of eukaryotic cells. Its arrangement varies broadly among species and cell types, yet conserved aspects of chromosomal organization are thought to facilitate common nuclear functions, such as transcription, replication, and repair. One obvious example of a subnuclear domain with a specialized activity is the nucleolus, a prominent nuclear substructure that serves as the unique site of ribosomal RNA transcription, processing, and assembly into ribosomal particles. Subnuclear positioning may also contribute to the establishment of differentiated patterns of gene expression or to the maintenance of heritable epigenetic controls over expression states.[1,2]

In higher eukaryotic cells and particularly in primate species, chromosomes occupy discrete territories, and their placement tends to correlate with the density of genes and timing of replication: gene-rich, early-replicating chromosomes are generally more central in the nucleus than gene-poor, late-replicating chromosomes.[3] Moreover, within chromosome territories, specialized subdomains like centromeres and telomeres, or Giemsa-dark bands and R-bands, often occupy distinct subdomains. In some cells, a radial organization can be detected within chromosome territories, with heterochromatin being frequently associated with the nuclear periphery.[4] Despite evidence suggesting certain reproducible patterns of nuclear organization, the mechanisms that move chromosomes in interphase and govern their positioning remain largely a mystery.

Budding and fission yeast genomes are nearly 200-fold less complex than those of vertebrate and mammalian species, and both yeasts have been extremely useful for correlating function with the position of chromatin domains. For example, in both organisms centromeres and telomeres are clustered in distinct subnuclear sites, each reflecting a different

[1] M. Cockell and S. M. Gasser, *Curr. Opin. Genet. Dev.* **9**, 199 (1999).
[2] J. Baxter, M. Merkenschlager, and A. G. Fisher, *Curr. Opin. Cell Biol.* **14**, 372 (2002).
[3] L. Parada and T. Misteli, *Trends Cell Biol.* **12**, 425 (2002).
[4] D. L. Spector, *Annu. Rev. Biochem.* **72**, 573 (2003).

mechanism of subnuclear localization. Budding yeast telomeres cluster at the nuclear periphery in groups of four to six, forming discrete subcompartments in which histone-binding silencing factors concentrate.[5] Centromeres, on the other hand, cluster around the spindle pole body, being held in place by microtubules.[6–8] Genetic manipulations have revealed functional links between the formation and maintenance of silent chromatin and the positioning of telomeres, and artificial means to anchor DNA domains at the yeast nuclear envelope (NE) have helped implicate subnuclear position in the regulation of nuclear functions.[9] It was shown, for instance, that late-replicating DNA is found preferentially at the nuclear periphery in G_1 phase, although this localization need not persist into S phase to maintain late replication status.[8] Consistently, the anchoring of an origin at the periphery through a transmembrane ligand will not necessarily render it late-firing, although the nucleation of silent chromatin will.[10] Nevertheless, tethering a potential silencer element at the nuclear periphery does enhance its ability to repress transcription.[9]

The notion that chromatin is generally immobile, and thus can provide anchorage sites for various nuclear processes, appears repeatedly in the literature on nuclear organization despite an absence of supporting evidence. In fact, a large body of data from live imaging of tagged chromosomal loci in interphase nuclei of fly, mammalian, and yeast cells shows the contrary: chromatin is highly dynamic with specific tagged loci moving in a rapid, random walk, which in some cases is coupled with a slow directional movement of the chromosomal domain.[11–13] Determining the exact nature of chromatin movement and its impact on nuclear functions is an area of intense pursuit. Similar efforts focus on identifying the proteins that anchor certain specialized domains of chromosomes, such as telomeres and centromeres in yeast,[14] and other periodic anchorage sites along chromosomes in flies.[15] Only with the identification of *bona fide* structural

[5] M. Gotta and S. M. Gasser, *Experientia* **52**, 1136 (1996).

[6] V. Guacci, E. Hogan, and D. Koshland, *Mol. Biol. Cell* **8**, 957 (1997).

[7] Q. Jin, E. Trelles-Sticken, H. Scherthan, and J. Loidl, *J. Cell Biol.* **141**, 21 (1998).

[8] P. Heun, T. Laroche, M. K. Raghuraman, and S. M. Gasser, *J. Cell. Biol.* **152**, 385 (2001).

[9] A. D. Andrulis, A. M. Neiman, D. C. Zappulla, and R. Sternglanz, *Nature* **394**, 592 (1998).

[10] J. B. Stevenson and D. E. Gottschling, *Genes Dev.* **13**, 146 (1999); D. C. Zappulla, R. Sternglanz, and J. Leatherwood, *Curr. Biol.* **12**, 869 (2002).

[11] W. F. Marshall, *Curr. Biol.* **12**, R185 (2002).

[12] S. M. Gasser, *Science* **296**, 1412 (2002).

[13] J. R. Chubb and W. A. Bickmore, *Cell* **112**, 403 (2003).

[14] F. Hediger, F. R. Neumann, G. Van Houwe, K. Dubrana, and S. M. Gasser, *Curr. Biol.* **12**, 2076 (2002).

proteins can the mechanisms of chromosome positioning and the functional implications of anchoring and dynamics be rigorously tested.

The protocols included here describe common techniques for the visualization of chromatin in living cells (primarily in budding yeast), while pointing out pitfalls and artifacts that can arise during live cell imaging. We also present analytical tools that have been developed for the quantitation of data generated by digital imaging. These tools allow us to define a new field of quantitative analysis: the dynamic behavior of DNA in real time. As these approaches are new, we expect there to be ongoing developments both in analytical tools and in high-resolution fluorescence microscopy. Half a century after the model was proposed for the structure of the DNA helix, and 10 years after the first eukaryotic chromosome was sequenced (i.e., Chr 3 of yeast), it is hoped that spatial and dynamic analyses may uncover novel mechanisms working to control the expression, replication, and repair of the genome.

Visualization of Specific Chromosomal Loci in Yeast

Tagging Genomic Loci

The precise identification of specific chromosomal loci in living cells was initially rendered possible by the development of a green fluorescent protein (GFP)-tagged lac repressor–operator system for site recognition.[16] This system exploits the high affinity and highly specific interaction of the bacterial lac repressor (lac^i) for its recognition sequence called the lac operator (lac^{op}).[17] Directed insertion of an extended array of lac^{op} (usually 256 copies or roughly 10 kb) and expression of a fusion construct between lac^i and a fluorescent protein like GFP, enables *in vivo* visualization of a defined DNA locus (Fig. 1A and B). Naturally, the specificity of targeting depends on the ability of the organism to carry out site-specific homologous recombination, an event that is extremely efficient in budding yeast.

To visualize the lac^i repressor in yeast, its gene is fused in frame to sequences encoding a nuclear localization signal, and the S65T derivative of natural GFP, which has a red-shifted excitation spectrum and higher emission intensity. Fusions to optimized forms of cyan fluorescent protein (CFP) or yellow fluorescent protein (YFP) (or ECFP and EYFP; Fig. 2)

[15] W. F. Marshall, A. F. Dernburg, B. Harmon, D. A. Agard, and J. W. Sedat, *Mol. Biol. Cell* **7,** 825 (1996).

[16] C. C. Robinett, A. Straight, G. Li, C. Willhelm, G. Sudlow, A. Murray, and A. S. Belmont, *J. Cell. Biol.* **135,** 1685 (1996).

[17] A. F. Straight, A. S. Belmont, C. C. Robinett, and A. W. Murray, *Curr. Biol.* **6,** 1599 (1996).

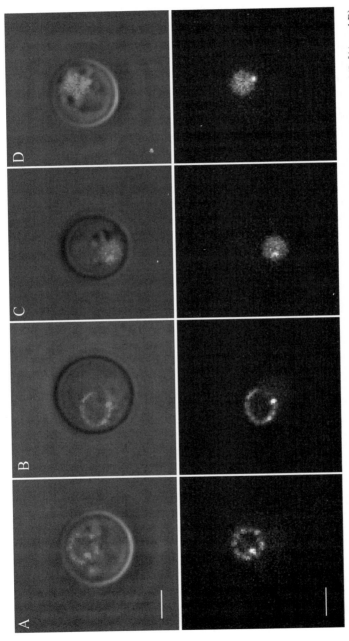

FIG. 1. Representative images of G_1-phase budding yeast cells in which a lac^{op} array has been inserted near telomere 14L [(A and B); Hediger *et al.*[14]] or near *MAT*a, the mating type locus at Chr 3 [(C and D); K. Dubrana, unpublished data, 2003]. The cells express the lac$^i_-$-GFP fusion and either a Nup49–GFP fusion (A and B) or a Tet^R–GFP fusion (C and D). Nup49–GFP gives a ringlike signal that allows precise determination of the nuclear volume, but interferes with detection of the tagged DNA locus when the latter is near the nuclear envelope (B). Tet^R–GFP fusion gives uniform diffuse nuclear staining, which is easily discernible from the spot signal [compare (B) and (D)]. The nuclear center can be accurately determined from the Tet^R–GFP signal, although a precise determination of the nuclear periphery is not possible. Scale bars: 2 μm.

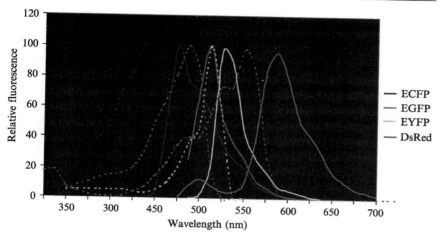

FIG. 2. Excitation and emission spectra of enhanced cyan, green, yellow, and red fluorescent proteins. Excitation spectra are represented as dashed lines, excitation spectra as unbroken lines. ECFP, EGFP, and EYFP are all variants of the green fluorescent protein from *Aequorea victoria*. DsRed is derived from a coral of the *Discosoma* genus (for more information, see www.clontech.com). (See color insert.)

are also successfully used. The lac repressor itself is modified to prevent tetramerization, thus minimizing artifactual higher order interactions between lac[op] sites. The minimal dectectable cluster of lac[op] sites is less than 24 operators,[18] although longer time-lapse series are facilitated by the insertion of larger arrays. Long arrays of lac[op] sites can be assembled into nucleosomes, yet these are largely disordered. Mild micrococcal nuclease analysis of these sequences reveals primarily monomeric and dimeric nucleosomal patterns (F. R. Neumann, unpublished data, 2002). This underscores the fact that lac[op] inserts may not form normal nucleosomal arrays, yet in our hands 10-kb arrays did not grossly affect the timing of replication of either late- or early-replicating domains.[8]

The repetitive 10-kb lac[op] arrays are difficult to propagate both in bacteria and in yeast. Amplification of these constructs in *Escherichia coli* often results in plasmids that have lost many repeats. To avoid this, it is recommended that growth temperatures be reduced to 24 or 30°, storage of colonies be minimized, and recombination-deficient bacteria strains be used. Even after integration in the yeast genome, lac[op] repeats are subject to recombination events that shorten or even eliminate the array. To have sufficiently strong fluorescent signals, we usually screen individual transformants for the brightest signal and freeze these isolates immediately. When

[18] A. S. Belmont, *Trends Cell Biol.* **11,** 250 (2001).

strains are recovered from frozen stocks, they are grown on selective media if possible, to avoid recombination events that would eliminate both the lac[op] repeats and the integration marker.

When comparing the position or mobility of two different genomic loci, one should avoid tagging both with the same repeats. It has been shown that two integrated identical arrays undergo an ill-defined pairing event that, at least in the case of the Tet system, depends on the expression of the repressor.[19,20] These may interfere with the positioning or dynamics of normal chromosomal loci. By using the Tet repressor (Tet[R])–operator combination for one site, and lac[i]–lac[op] sequences for the second, the risk of spurious pairing is eliminated.[19] We have successfully used fusions between either the Tet[R] or lac[i] with EYFP or ECFP and found no interference for binding or fluorescent signals (K. Bystricky, personal communication, 2002). A monomeric red fluorescent protein has become available (mRFP1),[21] which is functional when fused to lac[i] and which can be easily distinguished from the other fluorescent proteins.

We note that in contrast to the lac[i]–GFP fusion, Tet[R]–GFP gives a high and generally diffuse background of nuclear staining in yeast. This has two consequences: it can render the detection of a small array more difficult to detect, but at the same time, it can be useful for identifying the global nuclear volume and for calculating the center of the nuclear sphere (see below and Fig. 1C and 1D).

Nuclear Detection

To correct for any movement of the cell or the nucleus, or for mechanical vibration due to the instrumentation, it is necessary to determine the precise coordinates of the nuclear position in each acquired frame. Two alternative approaches have been used with success. In the first, a Nup49–GFP fusion is introduced into the yeast genome, generating a characteristic perinuclear ring due to the regular distribution of nuclear pores within the NE.[22] The second makes use of a Tet[R]–GFP fusion in the presence or absence of an integrated Tet[op] array. Tet[R]–GFP produces a uniform low-intensity fluorescence throughout the yeast nucleoplasm, including the nucleolus. This background allows for a precise calculation of the nuclear center within each frame, and permits ready detection of a tagged locus near the nuclear periphery. Its disadvantage is that it does

[19] L. Aragon-Alcaide and A. V. Strunnikov, *Nat. Cell Biol.* **2,** 812 (2000).
[20] J. Fuchs, A. Lorenz, and J. Loidl, *J. Cell Sci.* **115,** 1213 (2002).
[21] R. E. Campbell, O. Tour, A. E. Palmer, P. A. Steinbach, G. S. Baird, D. A. Zacharias, and R. Y. Tsien, *Proc. Natl. Acad. Sci. USA* **99,** 7877 (2002).
[22] N. Belgareh and V. Doye, *J. Cell Biol.* **136,** 747 (1997).

not provide a precise determination of the nuclear limit. The GFP–Nup49 fusion, on the other hand, defines the nuclear periphery clearly, but renders detection of perinuclear GFP-tagged loci difficult (Fig. 1). This can be overcome by using a CFP–Nup49 fusion, although this requires double imaging of the nucleus at each time point.

Growth Conditions

Strains are grown to an early exponential phase of growth (not more than 0.5×10^7 cells/ml) in synthetic or YPD medium. Cells are usually washed once before observation if grown in YPD, because this medium gives an autofluorescent signal.

We observe highly significant differences in the dynamics of internal DNA loci under different growth conditions. As little as one doubling of the culture (i.e., from a concentration of 1×10^7 to 2×10^7 cells/ml) was shown to produce a sharp drop in the dynamics of an internal tagged locus on Chr 14.[23] Less quantitative analyses suggest that this is true for other sites, as long as they are not tethered due to their proximity to a telomere or centromere. This drop in dynamics occurs just before the so-called diauxic shift, which entails a major reprogramming of the transcriptional pattern of genes involved in glycolytic metabolism, resulting in the induction of genes required for oxidative respiration.[24] It may reflect a general drop in global ATP levels or changes in ratios of other small molecules such as NAD/NADH. Treatment of yeast cells for as little as 15 min with the uncoupler carbonyl cyanide chlorophenyl hydrazone (this depletes both mitochondrial and plasma membrane potentials, inducing a hydrolysis of ATP) also provokes a significant drop in internal chromatin movement.[23]

Growth in medium containing galactose, rather than glucose, as carbon source has also been observed to shift specific loci on Chr 3R to more perinuclear positions (K. Dubrana, M. Gartenberg, and K. Bystricky, personal communication, 2002). It is not known how general this phenomenon may be. These observations nonetheless stress the importance of using identical growth conditions and of including appropriate controls for growth density when pursuing comparative studies of nuclear dynamics.

Cell Preparation

Because we use cultures with a low cell concentration, a gentle centrifugation step (1 min at $\leq 5000g$) is recommended to ensure a convenient cell

[23] P. Heun, T. Laroche, K. Shimada, P. Furrer, and S. M. Gasser, *Science* **294,** 2181 (2001).
[24] J. L. DeRisi, V. R. Iyer, and P. O. Brown, *Science* **278,** 680 (1997).

density on the slide. For short periods of observation (i.e., minutes), cells can be deposited on depression slides filled with 1.4% high-quality agarose dissolved in synthetic medium containing 4% glucose. The agarose is necessary to prevent deformation of the cells due to pressure between coverslip and slide (Fig. 3A). However, these anaerobic conditions and the production of CO_2 from cellular respiration can perturb extended observations. As an alternative, we recommend the use of a cell chamber system, such as the Ludin chamber (Life Imaging Services, Reinach, Switzerland). The chamber consists of a round 18-mm glass coverslip on which cells are deposited, mounted in a stainless steel chamber of 0.75 ml volume. For yeast, this slide must be coated with concanavalin A (1-mg/ml solution in H_2O; Sigma, St. Louis, MO) to allow cells to stick efficiently. The chamber can be closed by an upper glass coverslip and two perfusion lines permit medium changes or drug addition (Fig. 3B). With this system, the distance between the objective and the sample is separated only by the thickness of the coverslip, which improves resolution and prevents loss of focus.

The optimal temperature control can be achieved with an incubator box that encloses not only the specimen, but also the stage and much of the microscope. This allows a tight regulation of the specimen conditions and also avoids physical changes in microscope stand, stage, and objective due to thermal variations. We have not systematically analyzed the effects of temperature on chromatin dynamics, but imaging conditions should be standardized by using a well-thermostatted microscopy room (temperature, $22 \pm 2°$). Work with thermosensitive yeast strains can be performed by locally heating the stage, the specimen, and the slide, if an incubator box is not available.

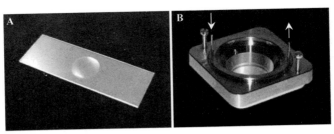

FIG. 3. (A) Glass slides containing a spherical depression are used for short time-lapse imaging. The depression is filled with synthetic medium containing 4% glucose and 1.4% agarose, on which cells are deposited. This avoids cell movement and deformation due to coverslip pressure. (B) The stainless steel Ludin chamber (Life Imaging Services, Reinach, Switzerland) permits deposition of the cells on an 18-mm coverslip precoated with concanavalin A. The chamber is then filled with 0.75 ml of medium. Two perfusion lines (arrows) allow medium or gas exchange.

Time-Lapse Acquisition

General Considerations. Because chromatin movement in yeast can be fast (we frequently detect movements >0.5 μm in less than 10 s), it is important to capture cell images very quickly. A compromise between resolution, the number of z frames, intervals between frames, exposure time, and bleaching must be found. It is also critical to control for laser- or light-induced damage to the organism. This is usually done by comparing the time required for an imaged cell to complete a division cycle with that of a nonimaged cell. We find that for rapid time-lapse imaging, the Zeiss (Thornwood, NY) LSM510 scanning confocal microscope is optimal for achieving fast acquisition, low bleaching, and minimal cell damage. To reduce the risk of damage by illumination, we keep the argon laser transmission as low as possible (0.1–1% for GFP), and image the cell as fast as possible by limiting the acquisition area to a minimal region of interest (ROI). High scanning speeds are used as described below.

The simplest approach is to capture "2D movies" (xy images over time) by manually maintaining the moving spot in focus during the acquisition. Image capture (average of four scans) requires approximately 500 ms for an ROI of 7 × 7 μm or roughly 250 ms for 3 × 3 μm, which is sufficient for a yeast nucleus. Image capture is repeated every 1.5 s. Using this method with minimal laser intensity, we were able to collect time-lapse series over 12 min without detectable effects on cell growth. For 3D time-lapse stacks, we typically take six to eight focal planes in z, each at a spacing of 300 to 450 nm, using an ROI of 3 × 3 μm. The stack is repeated every 1.5 s. Again, we could acquire up to 12-min movies (i.e., 480 stacks) without impairing cell division. The 3D time-lapse information has two advantages: (1) the spot is always present in one of the focal planes, meaning that after a maximal projection of the z axis, a complete 2D time-lapse sequence is obtained without ever losing the focal plane of the GFP spot; and (2) by image reconstruction, one can calculate distances and volumes in three dimensions. Three-dimensional measurements are nonetheless compromised by the reduced optical resolution in z (≥ 0.5 μm for 488-nm wavelength light).

When imaging a DNA locus as a 2D series over time (moving the focal plane when necessary), it should be noted that we generally exclude time-lapse series in which the GFP spot is within 0.5 μm of the top or bottom of the nucleus. In this zone, the spot is not easily distinguishable from GFP–Nup49, nor is the center of the nucleus readily determined. Particularly for measuring radial movements (see below), 2D time-lapse imaging is generally restricting to a 1-μm central zone of these nuclei (diameter, 2 μm).

Zeiss LSM510 Settings. The specific settings for a Zeiss LSM510 on an Axiovert 200M, equipped with Plan-Apochromat × 100/NA 1.4 oil immersion and Plan-Fluar × 100/NA 1.45 oil immersion lenses, an argon laser, and hyperfine motor HRZ 200, are as follows:

Laser: Argon/2: 458-, 488-, or 514-nm tube; current, 4.7 A; output, 25%

Filters: Channel 1, Lp 505 for GFP alone; channel 1, Lp 530; channel 3, Bp 470–500 for YFP/CFP single-track acquisition

Channel setting: Pinhole, 1–1.2 Airy units (corresponding to optical slice of 700 to 900 nm); detector gain, 950 to 999; amplifier gain, 1–1.5; amplifier offset, 0.2–0.1 V; laser transmission AOTF (acousto-optic tuned filter) = 0.1–1% for GFP alone, 5–15% for YFP, and 10–40% for CFP in single-track acquisition. The pinhole must be regularly aligned to use minimal laser transmission

Scan setting: Speed, 10 (0.88 μs/pixel); 8 bits one scan direction; 4 average/line; zoom, 1.8 (pixel size, 100 × 100 nm), using an ROI of 3 × 3 to 4 × 4 μm

z settings: Hyperfine HRZ 200; six to eight optical slices of 300 to 450 nm each

Imaging intervals: 1.5 s

If CFP and YFP signals are weak, one can acquire both signals sequentially on the LSM510 channel 1 (which is more sensitive than the others), using the multitrack mode to allow the use of broader filters. In this case, CFP signal is recovered through long-pass filter Lp 475 and YFP through Lp 530. Alternatively, and to avoid any cross-talk, YFP signal is recovered as before but CFP is recovered on channel 3 through bandpass filter Bp 470–500. Obviously, these latter parameters slow the imaging process.

Wide-Field Microscopy Imaging with Deconvolution. An efficient alternative to confocal imaging of chromatin dynamics is the use of a sensitive monochrome charge-coupled device (CCD) on a wide-field microscope equipped with a piezoelectric translator [PIFOC; Physik Instrumente (PI), Karlesruhe, Germany], xenon light source, monochromator, and rapid imaging software, such as Metamorph (Universal Imaging, Downingtown, PA). The advantage is that the most recent cooled CCD cameras have high sensitivity and speed, allowing for less than 50-ms exposures to acquire images up to 1392 × 1040 pixels. This makes this system convenient for imaging several cells at once. Here, the limiting step is the speed of signal transfer from the CCD chip to the computer, which depends on the image size. The monochromator allows the continuous regulation of incident light wavelength over a larger range of values than does the AOTF and the confocal laser system. Because out-of-focus haze makes wide-field images noisier than confocal images, deconvolution of the 3D stack is recommended to reassign blurred intensities back to their original source. This can be readily performed by Metamorph software, although other deconvolution packages are also available.

The instrumentation that we generally use is as follows: an inverted Olympus IX70 microscope equipped with a piezoelectric translator (PIFOC; PI) placed at the base of a Planapo × 60/NA 1.4 objective, a polychromator (Polychrome II; TILL Photonics, Gräfelfing, Germany) and a CoolSNAP-HQ digital camera (Roper Scientific, Tuczon, AZ). Suppression of stray light may require an additional shutter. The conditions of 3D time-lapse series capture are as follows: 5–11 optical z slices taken every 1 to 4 min; optical sections are 200 to 400 nm in depth and have 50-ms exposure. Using these settings, we are able to capture up to 300 stacks of 5 sections each (1500 frames) at 1-min intervals, which corresponds to 2.5 cell cycles, without affecting cell cycle progression. More rapid sampling with this system, on the other hand, leads to bleaching and potential cellular damage. Until this can be remedied by more rapid CCD cameras, we recommend confocal microscopy for rapid time-lapse imaging (intervals ≤ 2 s) on small regions of interest (typically, one yeast nucleus) and wide-field microscopy for less rapid time-lapse imaging (intervals ≥ 60 s) on larger fields.

Positional Information

Because time-lapse movies are necessarily limited to tens of cells for each condition tested, other more statistically rigorous methods are recommended to obtain an accurate determination of the subnuclear position occupied by a tagged locus with relation to a nuclear landmark or in relation to another site. This is performed by taking one stack of 17 to 19 images (exposure, 200 ms; step size, 200 nm) through a field of yeast cells that are either growing on agar or placed in a Ludin chamber, using the wide-field microscope equipped with a cooled, high-sensitivity CCD camera as described above. Analysis of position is generally performed on 200–300 cells, and cells are classified by their position in the cell cycle.

Position is routinely determined in relation to the nuclear envelope, which is usually tagged with the same fluorescent protein as the locus. In this way, both the DNA and the NE are imaged in each frame of the stack. If a tagged locus is to be localized with respect to a second spot or another specific nuclear landmark [e.g., spindle pole body (SPB) or nucleolus], it is preferable to tag the two with different fluorescent proteins to eliminate confusion during analysis. A bright-field or phase image of each cell is also essential to determine its cell cycle stage.

The position of the GFP-tagged locus is monitored relative to the middle of the Nup49–GFP ring. To do this, we measure from the center

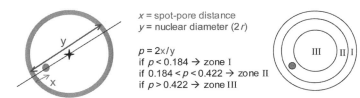

FIG. 4. Schematic representation of the Nup49–GFP and DNA tagged locus fluorescent signals. For hundreds of different cells, the distance from the middle of the spot to the middle of the envelope signal (x), and the nuclear diameter (y), are measured. By dividing x by $y/2$ ($p = 2x/y$), we can classify the spot position into three concentric zones of equal surface. The outermost zone (I) contains peripheral spots ($p < 0.184$). Zone II regroups intermediate positioned spots ($0.184 < p < 0.422$). Zone III contains internal spots ($p > 0.422$). (See color insert.)

of intensity of the GFP spot to the nearest pore signal along a nuclear diameter, as well as measuring the nuclear diameter itself. By dividing the first value by half of the second (i.e., the radius), we can classify each spot falling into one of three concentric zones of equal surface, as depicted in Fig. 4. The most peripheral zone (zone I) is a ring of width $0.184 \times$ the nuclear radius (r). Zone II lies between $0.184r$ and $0.422r$ from the periphery and zone III is a central core of radius $0.578r$. The three zones are of equal surface no matter where the nuclear cross-section is taken. We usually eliminate nuclei in which the tagged locus is at the very top or bottom of the nucleus, because the pore signal no longer forms a ring but a surface.

Once the distribution of a given locus is determined, it can be compared with either the predicted random array or another distribution by χ^2 analysis or by proportional analysis, which compares percentages in one zone for different conditions. Statistical significance is determined using a 95% confidence interval.

Quantitative Approaches to Motion Analysis

Reproducible variations in chromatin mobility have been detected between G_1- and S-phase nuclei, making it important to monitor the cell cycle stage of each imaged cell. In budding yeast, the cell cycle stage is easily determined by monitoring bud presence and size, together with nuclear position, in bright-field images taken before and after the fluorescence imaging series. Such analysis will also confirm that the imaged cell is progressing normally through the cell cycle. Figure 5 summarizes the characteristics used to classify budding yeast cells as G_1, early S, mid-to-late S, G_2, mitosis, and telophase. Unbudded cells are considered G_1 phase; this category includes posttelophase stages in which two equal-sized cells remain attached,

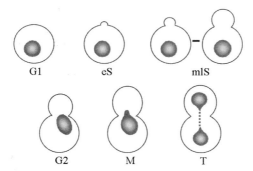

FIG. 5. Nuclear shape, position, and bud size provide elaborate criteria for identifying the cell cycle stage of individual *Saccharomyces cerevisiae* cells. Round nonbudding cells are in G_1 phase; early S phase cells (eS) have a very small bud. Other budded cells are classified into mid-to-late S phase (mlS). We consider cells to be in G_2 phase when they have a large bud (about two-thirds of the mother cell) and when the nucleus has moved to the bud neck. Most of the nuclei in G_2 cells start to elongate. Mitotic cells (M) are characterized by intrusion of the nucleus into the bud. Once the two nuclei are clearly separated but not yet round, and still linked through residual nuclear envelope structures, cells are considered to be in telophase (T). As much of the dynamics and positional studies are based on the assumption that nuclei are round, we usually do not analyze G_2, M, and T cells.

although their nuclei are clearly separated and round. Cells with an emerging bud and small-budded cells are classified as early S phase, and all other budded cells, in which the bud is big enough to form a ring at the neck and the nucleus is still round, are grouped as mid-to-late S phase. Once the nucleus moves to the bud neck and the bud is two-thirds the size of the mother cell, the cells are considered G_2-phase cells. As soon as the nuclear envelope begins to extend into the daughter cell as a result of spindle extension, cells are considered to be mitotic (Fig. 5). Telophase can be defined as a state in which two distinct nuclei are visible, but remain connected via residual NE structures. In addition to cell cycle effects, we observed a reduced frequency of large movements (defined below) in G_1 daughter nuclei compared with G_1 mother nuclei, which can be attributed to the difference in nuclear size (nuclear diameters in G_1 average 1.68 $\pm$ 0.2 μm, whereas it is 1.95 $\pm$ 0.3 μm in mother cells[23]). For this reason, we include only mother cells in our G_1-phase analyses.

Tracking

The detection of chromatin movement is of interest if it can be correlated with physiological changes; therefore, it is necessary to quantify its behavior under different controlled states. To this end, it is important to accurately identify the position of the tagged locus relative to the nuclear center for each frame of a typical time-lapse movie. This laborious task was originally done by hand with simple measurement tools (AIM tool, LSM510 software; Zeiss). It is now greatly simplified by the development of a new tracking algorithm developed in collaboration with D. Sage and M. Unser (Swiss Federal Institute of Technology, Lausanne, Switzerland).[25]

The automated image analysis software consists of three components.

Alignment phase: The first step of the program is an alignment module that compensates for the movement of the nucleus or the cell. This is achieved by a threshold on the image and the extracted points are fit within an ellipse using the least-squares method of Fitzgibbon *et al.*[26] Each image is then realigned with respect to the center of the ellipse.

Preprocessing phase: To facilitate detection of the tagged locus, the images are convolved with a Mexican-hat filter. This processing compensates for background variation and enhances small spotlike structures.

Tracking phase: The final component is the tracking algorithm, which uses dynamic programming to extract the optimal spatiotemporal trajectory of the particle. Detection of the spot is a difficult task mainly because of the low signal-to-noise ratio of the images and the similarity between the DNA spot and the nuclear pore signal. We take advantage of the strong dependency of the spot position in one image on its position in the next: all endogenous chromosomal loci observed so far show a constrained motion within the nucleus and a spot localized at a certain time point is generally not found more than 2 μm away at the next time point (1.5 s). The algorithm evaluates all possible trajectories over the sequence and the one that maximizes an appropriate set of criteria is kept. The criteria used are as follows:

1. Maximum intensity is favored (i.e., the tagged DNA is usually brighter than the pore signal).
2. Smooth trajectories are favored.
3. Positions that are closer to the center of the nucleus are favored. This is because the Nup49 staining can give signal similar to that of

[25] D. Sage, F. Hediger, S. Gasser, and M. Unser, *in* "IEEE International Symposium on Image and Signal Processing and Analysis (ISPA 2003)." Rome, Italy.

[26] A. Fitzgibbon, M. Pilu, and R. B. Fisher, *IEEE Trans. Pattern Anal. Mach. Intell.* **21,** 476 (1999).

the tagged locus when it is located at the nuclear periphery; a bright signal found inside the nucleus, on the other hand, is unlikely to be anything other than the tagged locus.

Importantly, these three parameters can be modulated individually in order to optimize the tracking for different situations (loci that are more mobile, more peripheral, of variable intensity, etc.). The program also has the option of further constraining the optimization by forcing the trajectory to pass through a manually defined pixel. In other words, the user can move or add nodes interactively when the spot is not well defined automatically and the new optimal trajectory is recomputed quasi-instantaneously. This tracking method proves to be extremely robust because of its global approach: the decision for spot definition is based not only on the present and the recent past but on the future as well, taking the data set as a whole into consideration.

This complete system has been implemented as a Java plug-in for the public domain ImageJ software.[27] Application to time-lapse series of *S. cerevisiae* nuclei produces results that are equal to—if not better than—manual tracings, with the enormous advantage of being reproducible and analyzer independent. This new algorithm reduces the tracking time from 10 min for a typical experiment, when it is done manually, to a few seconds (Pentium IV, 1 GHz) if no user intervention is required. The spatiotemporal trajectory is exportable to a spreadsheet and will soon be implemented for 3D image stacks over time. Some software applications commercially available are able to track objects [Imaris (Bitplane, Zurich, Switzerland); Volocity (Improvision, Lexington, MA)], but in addition to being expensive, tracking efficiency is variable and usually requires uniformly high-quality images. Available algorithms are mostly based on threshold priniciples, and are rarely modifiable or interactive.

Controls for Nuclear Rotation

Before the analysis of the tagged locus dynamics, certain artifacts of spatial analysis must be eliminated by control studies. For instance, it is necessary to demonstrate that the yeast nucleus is not rotating on itself, as nuclear movement might be misinterpreted as tagged locus movement. To do this, we track the movement of a fixed point within the nuclear envelope, using an integral component of the spindle pole body (SPB) called Spc42,[28] fused to CFP. Time-lapse analysis of SPB shows little movement over time, and individual steps are generally restricted to changes smaller

[27] W. Rasband, National Institutes of Health, Bethesda, MD, 2003.
[28] A. D. Donaldson and J. V. Kilmartin, *J. Cell Biol.* **132**, 887 (1996).

than the optical resolution size (~0.2 μm).[23] A second control for nuclear rotation uses fluorescence recovery after photobleaching (FRAP). A restricted zone of the NE visualized through the Nup49–GFP signal is bleached by repeated irradiation (50 iterations of 100% power pulses with the 488-nm laser). Subsequent imaging shows that the bleached zone persists in the same position for 2–3 min and is then "invaded" by one or two discrete pores, which diffuse from the edges of the bleached zone. This shows that there is no general rotation of the nucleus, for this movement would have shifted unbleached pores into the photobleached zone. Moreover, pore diffusion is much slower than chromatin movement. Such controls should be routinely performed before attributing movement to a tagged chromosomal locus, rather than to the nucleus in general.

Characteristic Parameters of Movement

Once the *xy* or *xyz* coordinates for each time point of a time-lapse series are exported, which will identify the center of the nucleus and the center of the tagged locus for each frame, the moving particle can be characterized by many different parameters. Those presented here were selected because they accurately discriminate between visually different dynamics. These analyses can be carried out with a variety of programs designed for advanced calculations, including Excel, Mathematica, or MATLAB. Macros for implementation can be obtained by e-mail from our laboratory.

Trace. For each time-lapse series, we sum the total distance traveled by the particle over 5 min of 1.5-s intervals, assuming a straight trajectory between sequential positions (Fig. 6A). This gives the total track length for a given time period, which should be averaged over 8–10 similar movies. Alternatively, the movies and tracks can be "added," such that the sum represents the track length over 40 to 50 min of time-lapse imaging (1600 to 2000 frames). From this, one can calculate an "average velocity" for spot movement with some reliability. Analysis of individual movies (i.e., one cell within a 5-min time period) is too anecdotal for any reliable conclusions to be drawn. Examples of track length variation for internal chromosomal sites versus telomeric or centromeric sites, and an internal site in G_1- versus S-phase cells, are available in Ref. 23.

Step Size Distribution. For small changes in dynamic behavior, the trace measurement is sometimes not sufficiently discriminating and a statistical approach is preferable. Position-to-position distances can be averaged over one or multiple movies and the standard deviation calculated. These parameters can be compared for different groups of time-lapse series with statistical tests (e.g., ANOVA) and even small but reproducible differences

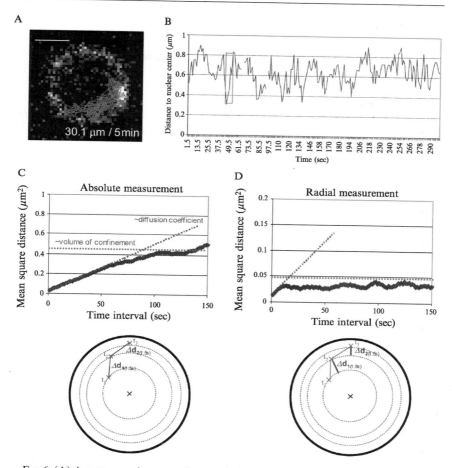

FIG. 6. (A) A representative trace of a tagged telomere on Chr 14L in a G_1-phase yeast cell is shown. Images were captured over 5 min with frames taken every 1.5 s. The position of the spot was tracked after alignment of the nuclei on the basis of their nuclear envelope fluorescence, using the tracking algorithm. The trajectory of the locus is projected in red on a single focal section of the nucleus. The mean length of the path in micrometers for a 5-min movie (200 frames) averaged over 8 movies is indicated. Scale bar: 1 μm. (B) Radial movement of the GFP-tagged telomere 14L relative to the nuclear envelope (NE) was monitored by measuring the distance from the middle of the spot to the middle of the pore signal in each frame. The red box indicates radial movement > 0.5 μm within 10.5 s or 7-frame intervals. (C) Mean square difference (MSD) of actual point-to-point distances is shown for the time-lapse series pictured in (A) and (B). For each time interval (1.5 s to 150 s), the mean of the absolute distance (Δd) covered by the spot is calculated and plotted against time intervals. (D) The MSD of radial distances is shown for the same time-lapse series. For each time interval (1.5 s to 150 s), the mean of the radial distance (Δd) covered by the spot is calculated and plotted against time intervals. Telomere 14L moves significantly along the nuclear envelope and thus appears less

can be documented. Because of the large number of measurements taken into consideration, small differences can nonetheless be highly significant.

Large (Radial) Movements. Some loci do not appear to move with a significantly higher average speed, but the frequency of large rapid jumps (movements $\geq$ 0.5 μm within a few frames; see Fig. 6B) varies greatly among different loci and points in the cell cycle. This criterion has been found to be one of the most useful for distinguishing patterns of mobility. Any threshold over 0.2 μm can be chosen for the definition of a "large" movement, but in our experience a useful criterion for comparison has been the frequency of radial movements $\geq$ 0.5 μm within a 7-frame interval, that is, within 10.5 s. Once radial distances are plotted over time, the scoring of these can readily be done by hand and confirmed by computation, although this analysis is also readily automated. A similar quantitation of the frequency of point-to-point movements over a certain size is also possible, although, in this case, we have no standard to recommend.

Relative Surface Coefficient. The fraction of the nuclear surface occupied by a particle tracked by 2D imaging can be calculated for a given time period. In the case of chromosomal loci, these show a restricted zone of movement, which is significantly different for "tethered" sites, such as telomeres, and internal chromosomal loci. The relative surface coefficient (RSC) defines the minimal ellipse that encloses 95% of the individual particle positions during a given movie. The surface of the ellipse is divided by the average surface of the nucleus. This number must be determined movie by movie, but the RSC of a group of time-lapse series can be averaged. The surface coefficient quantifies graphically the freedom of movement of a given tagged locus, and these values generally agree well with results from the mean square displacement analysis (see below).

Mean Square Displacement Analyses. The observation of a moving DNA locus over time gives information not only about the rate of movement but also about the area or volume that it occupies during this period of time. It has been amply shown that certain chromosomal domains or loci are able to move locally, but remain in a given subvolume of the nucleus.[11–13] The fact that they move within a defined area can be considered to be a constraint on their apparently random local motion. This particular behavior can be analyzed by a calculation based on the random walk model that describes a linear relationship between time intervals and the square of the distance traveled by a particle during this period of time (mean square displacement = MSD or $<\Delta d^2>$, where

constrained when the MSD is calculated from absolute distance as compared with radial distances. The fact that telomere 14L is restricted to a small volume close to the nuclear periphery is shown by the low plateau in the radial MSD graph. (See color insert.)

$\Delta d^2 = \{d(t) - d(t + \Delta t)\}^2$).[29,30] The slope reflects the diffusion coefficient of the particle. The linearity of the curve is lost, however, if the particle is restricted in its freedom of movement (obstructed random walk), or when a global directional movement is observed (random walk with flow). A particle moving in a closed volume (as a cell nucleus) will show a plateau in the MSD analysis, and the height of this plateau is related to the volume in which the particle is restricted (see red dashed line in Fig. 6).

The slope of the MSD relation is directly correlated with the diffusion coefficient (D) :$< \Delta d^2 > = 2nDt^{2/d_w - 1}$ (where n is the number of spatial dimensions and d_w is the anomalous diffusion exponent). d_w can be calculated by plotting $\log(\Delta d^2 / \Delta t)$ versus $\log(\Delta t)$ and is larger than 2 if the diffusion is obstructed. In enclosed systems, the diffusion coefficient decreases with increasing Δt, because of space constraints exerted on the particle dynamics. Nevertheless, the maximal diffusion coefficient can be calculated for short time intervals and reflects intrinsic mobility of particles (see green dashed line in Fig. 6C and D). Diffusion coefficients of nuclear components have been calculated by MSD or FRAP and values can range from ~60 $\mu m^2/s$ for EGFP in hamster cells,[31] to ~$1 \times 10^{-4} \mu m^2/s$ for freely diffusing Cajal bodies in HeLa cells.[32] We observe diffusion coefficients in the range of 1×10^{-2} to 1×10^{-3} $\mu m^2/s$ for chromosomal loci in yeast. Of course, in an obstructed random walk diffusion situation, the calculated D will depend on the time interval used to acquire images, as D decreases with increasing time intervals.

In practical terms, the distances traveled by the spot for each time interval (1.5 s, 3 s, 4.5 s, ...) are calculated and the square of their mean is plotted against increasing time intervals (Fig. 6C and D). The original slope of MSD is equal to $2nD\Delta t$ (D is the diffusion coefficient), such that $n = 1$ for radial measurements and $n = 2$ for absolute measurements. With respect to position-related questions, it is of interest to plot MSD for radial distances. In this case, the MSD plateau will reflect the tendency of the particle to be associated with the nuclear periphery, discriminating loci that are moving freely in the nucleoplasm from loci that move in a restricted zone near the nuclear envelope. The graphs in Fig. 6C and D are based on the same time-lapse data from telomere 14L, but show MSD calculations using

[29] J. Vazquez, A. S. Belmont, and J. W. Sedat, *Curr. Biol.* **11**, 1227 (2001).
[30] W. F. Marshall, A. Straight, J. F. Marko, J. Swedlow, A. Dernburg, A. Belmont, A. W. Murray, D. A. Agard, and J. W. Sedat, *Curr. Biol.* **7**, 930 (1997).
[31] A. B. Houtsmuller, S. Rademakers, A. L. Nigg, D. Hoogstraten, J. H. Hoeijmakers, and W. Vermeulen, *Science* **284**, 958 (1999).
[32] M. Platani, I. Goldberg, A. I. Lamond, and J. R. Swedlow, *Nat. Cell Biol.* **4**, 502 (2002).

absolute distances and radial distances, respectively. The low plateau in Fig. 6D reflects the fact that telomere 14L is constrained close to the NE.

The MSD analysis has been used to characterize the dynamic behavior of chromosomal loci in yeast, *Drosophila,* and mammalian cells, and is a useful means to compare the levels of constraint experienced by chromatin in these different organisms. In other organisms, distances between two separate moving loci have been used, in which case $<\Delta d^2>$ reflects two times the MSD of an individual spot or locus moving relative to a fixed point.[29] For a more theoretical treatment of this parameter, the reader is referred to Ref. 33.

Discussion

We have described techniques that allow a quantitative evaluation of the position and mobility of specific DNA sites in the yeast nucleus and give the details of how we and others analyze such results. These analytical methods are applicable to data from any organism. The most striking conclusion that arises from this type of analysis is that chromatin mobility is nearly identical in the three organisms studied in detail to date. Notably, tagged sites along yeast chromosomes (but not telomeres or centromeres[14,23]), sites on the X chromosome in *Drosophila* spermatocytes[29,30] and various insertions at random positions on human chromosomes[34] show similar dynamics. In general, one detects both continuous small oscillations of 0.2 to 0.4 μm/s in random directions and, less frequently, larger movements of >0.5 μm, which are not strictly random: these are often followed by movements in the opposite direction. In contrast to these chromosomal loci, yeast telomeres and centromeres and inserts near human nucleoli or satellite repeats have significantly fewer large movements and are more spatially constrained.[12,13,23,34]

One might question the significance of motion within this range, as the smaller movements approach the focal resolution of a confocal microscope (0.2 to 0.3 μm). However, these distances are large compared with chromatin structure. A movement of 0.5 μm spans half the radius of a yeast nucleus and is equivalent to about 100 kb of interphase chromatin (based on a linear compaction ratio of ~70-fold[35]). We stress that these hyperfine movements are, to a large degree, sensitive to ATP levels in yeast, and that the conditions that cells are exposed to during imaging may influence

[33] H. C. Berg, "Random Walks in Biology." Princeton University Press, Princeton, NJ, 1993.
[34] J. R. Chubb, S. Boyle, P. Perry, and W. A. Bickmore, *Curr. Biol.* **12**, 439 (2002).
[35] K. Bystricky *et al.*, submitted (2003).

the mobility of a given locus. DNA damage from short-wavelength light may provoke changes in chromatin dynamics as a result of the cellular response to DNA damage. For the analysis of chromatin position within a nucleus, it is strongly advised that time-lapse analyses (which give only a temporally restricted sampling of a single cell, usually averaged over a limited number of movies) be complemented with more accurate analyses of specific cell populations (>200 cells), monitored at a single time point. Both sets of data are necessary to understand chromatin position and behavior in living cells.

Present efforts now focus on linking chromatin mobility with either active transcriptional states or the potential for transcription. The source of movement does not appear to be RNA polymerase II elongation, but possibly the events of chromatin remodeling that render promoters accessible to the machinery of transcription. The ATP dependence of chromatin-remodeling machines (SWI/SNF, etc.) correlates well with the sensitivity of large chromatin movements to ATP depletion and changes correlated with the diauxic shift. Thus, the near future promises at least some enlightenment on the physiological implications of chromatin dynamics in living cells.

Acknowledgments

We are indebted to Drs. Marek Blaszczyk, Kerstin Bystricky, Karine Dubrana, and Thierry Laroche (Gasser Laboratory) and to Drs. Daniel Sage and Michael Unser (EPFL, Lausanne) for help in designing, testing, and revising both the tools of microscopy and the tools of analysis described here. We also thank John Sedat, Julio Vasquez, Andrew Belmont, Marc Gartenberg, and others in the field for helpful advice, and Patrick Heun and Thierry Laroche for establishing this approach in our laboratory. This work is supported by the Swiss National Science Foundation in a grant to S.M.G. and to the National Center of Competence in Research, Frontier in Genetics. A.T. is supported by an EMBO Long-Term Fellowship.

[23]　Direct Visualization of Transcription Factor-Induced Chromatin Remodeling and Cofactor Recruitment *In Vivo*

By Anne E. Carpenter and Andrew S. Belmont

Introduction

The application of chromatin immunoprecipitation procedures, as described elsewhere in this volume, has greatly facilitated *in vivo* measurement of recruitment of various transcriptional activators and cofactors to specific *cis* sequences in a wide range of biological systems. The impact on the field of chromatin structure and gene regulation has been tremendous. However, this method does have two specific limitations. Not all proteins will cross-link efficiently to nearby DNA sequences. More importantly, perhaps, chromatin immunoprecipitation is a biochemical procedure involving averaging among cells in a population, precluding single-cell observations. This is a particular limitation when cell heterogeneity is present and/or when fine temporal ordering of events is important. Similarly, biochemical and molecular probes for chromatin remodeling have been limited to methods that measure averages over cell populations.

Our laboratory and others have begun to develop assays for chromatin remodeling and transcription factor and coactivator recruitment based on direct, microscopic observations within individual live cells. The experimental basis for this approach involves the visualization of chromosome regions containing multiple transgene copies. To date, two approaches have been used. The first uses direct repeats of bacterial operators to bind repressor fusion proteins as a means of tethering specific transcription factors to specific chromosomal sites. The second uses repeats of transgenes containing viral promoters with binding sites for known transcription factors. We anticipate a natural experimental progression will be to combine both of these approaches, using the bacterial operator repeats for tagging the chromosome regions, while using transgene repeats containing specific promoters with known transcription factor-binding sites and reporter constructs for monitoring gene expression.

These methods promise to allow real-time visualization of transcription factor dynamics and their relationship to changes in chromatin structure and gene expression. Here, we review previous work using these methods. We then discuss key methodologies used in these experiments together with ongoing technological developments in our laboratory. We conclude with several example protocols for specific experimental procedures used in these studies.

Review of Previous Work

The ability to tag chromosomes with green fluorescent protein (GFP), using direct repeats of bacterial operators and operator/repressor recognition, has provided a highly sensitive capability for localizing and measuring the dynamics of specific DNA sequences within living cells.[1] First described for the lac operator/repressor system,[2,3] this approach has now been extended to the tet operator as well,[4] and applied to a wide range of organisms, including bacteria, yeast, *Caenorhabditis elegans, Drosophila, Arabidopsis,* and mammalian cells.[1] Although this approach readily allows detection of single-copy insertions of a 256-copy lac operator direct repeat in mammalian cells (and as low as 25 repeats under favorable conditions in *Drosophila;* A. Belmont, unpublished results), it is also possible to create cell lines containing labeled chromosome regions of varying size up to an entire chromosome arm in length, either by selection of cells that have integrated high copy numbers of the transfected plasmid[5] or by a secondary process of gene amplification.[2]

The first demonstration of large-scale chromatin unfolding induced by the tethering of a transcriptional activator used a lac repressor–VP16 AAD (acidic activation domain) fusion protein.[6] The CHO cell line A03_1 contains a gene-amplified chromatin array approximately 90 Mbp in size. This array is formed by repeating blocks of transgene repeats, several hundred kilobase pair blocks in size, spaced by large coamplified regions of genomic DNA, roughly 1 Mbp in size. Each transgene copy contains a 256-copy lac operator direct repeat and a dihydrofolate reductase (DHFR) cDNA construct. This array appears in metaphase as a homogeneously staining region (HSR) and has properties of heterochromatin, forming a condensed mass roughly 1 μm in diameter through most of interphase.[7]

Targeting the lac repressor–VP16 AAD fusion protein resulted in a dramatic uncoiling of the condensed chromatin array into extended, ~80-nm-diameter chromonema fibers up to 25–40 μm in length that accompanied gene activation.[6] In a related experiment, tethering of the

[1] A. S. Belmont, *Trends Cell Biol.* **11,** 250 (2001).

[2] C. C. Robinett, A. Straight, G. Li, C. Willhelm, G. Sudlow, A. Murray, and A. S. Belmont, *J. Cell Biol.* **135,** 1685 (1996).

[3] A. F. Straight, A. S. Belmont, C. C. Robinett, and A. W. Murray, *Curr. Biol.* **6,** 1599 (1996).

[4] C. Michaelis, R. Ciosk, and K. Nasmyth, *Cell* **91,** 35 (1997).

[5] Y. G. Strukov, Y. Wang, and A. S. Belmont, *J. Cell Biol.* **162,** 23 (2003).

lac repressor–VP16 fusion protein to a peripherally located chromosome site containing the lac operator repeats resulted in this spot moving into the nuclear interior.[8] This initial work also demonstrated the ability to use this experimental system to detect *in vivo* recruitment of coactivators by the VP16 acidic activation domain. Histone H3 and H4 hyperacetylation was observed over the chromatin array, as well as recruitment of p300/CBP, PCAF, and GCN5, and splicing factors U2B and Sm100.[6] More recently, this work has been extended by using bead loading, rather than transient transfection, to load GFP–lac repressor–VP16 AAD directly into cell nuclei within several minutes.[9] This method was used to demonstrate asynchronous recruitment of different components of the SAGA histone acetyltransferase complex, and the mammalian SWI/SNF complexes. These results suggested that targeting of chromatin-modifying complexes to condensed chromatin may proceed through initial recruitment of "pioneer factors" consisting of individual components or partial subcomplexes that might act by themselves to produce initial chromatin unfolding.[9,10]

Similar approaches have now been taken using lac repressor fusions to other transcription factors to assay their effect on large-scale chromatin conformation. BRCA1, E2F1, and p53,[11] the estrogen receptor,[12] and several acidic activators including Gal4 and p65[13] have been demonstrated to have large-scale chromatin-unfolding activity. Interestingly, cancer-predisposing mutations in *BRCA1,* mapping to its BRCT 3′ repeats, were found to markedly enhance its chromatin-opening activity. When both the lac operator-tethered transcription factor and a recruited coactivator protein are labeled with color variants of GFP, recruitment can be visualized directly in living cells, as was done for the estrogen receptor and its coactivators SRC-1 and CBP.[12,14] This also allowed the kinetics and dynamics of recruitment to be observed and revealed weaker ligand independent interactions that had been overlooked in biochemical assays.

[6] T. Tumbar, G. Sudlow, and A. S. Belmont, *J. Cell Biol.* **145,** 1341 (1999).

[7] G. Li, G. Sudlow, and A. S. Belmont, *J. Cell Biol.* **140,** 975 (1998).

[8] T. Tumbar and A. S. Belmont, *Nat. Cell Biol.* **3,** 134 (2001).

[9] S. Memedula and A. S. Belmont, *Curr. Biol.* **13,** 241 (2003).

[10] C. L. Peterson, *Curr. Biol.* **13,** R195 (2003).

[11] Q. Ye, Y. F. Hu, H. Zhong, A. C. Nye, A. S. Belmont, and R. Li, *J. Cell Biol.* **155,** 911 (2001).

[12] A. C. Nye, R. R. Rajendran, D. L. Stenoien, M. A. Mancini, B. S. Katzenellenbogen, and A. S. Belmont, *Mol. Cell. Biol.* **22,** 3437 (2002).

[13] A. C. Nye, Ph.D thesis, Effects of transcription activators on a large-scale chromatin structure, Dept. of Cell and Structural Biology, University of Illinois, Urbana-Champaign, Urbana, 2003. *Work from ref. 13 is now in press:* A. E. Carpenter, A. Ashouri, and A. S. Belmont, *Cytometry,* in press.

[14] D. L. Stenoien, A. C. Nye, M. G. Mancini, K. Patel, M. Dutertre, B. W. O'Malley, C. L. Smith, A. S. Belmont, and M. A. Mancini, *Mol. Cell. Biol.* **21,** 4404 (2001).

An improved VP16 AAD targeting system was developed in the Spector laboratory by adding a protein reporter system to monitor gene activation simultaneous with chromatin conformation. This system also separated the lac operator/repressor chromosome visualization tag from the inducible transcription factor-targeting mechanism based on tet operator/repressor binding.[15] In this system, a significant lag between transcriptional initiation and detection of reporter protein exists, because of the time it takes for transcript buildup and protein accumulation. More recently, however, the Spector laboratory has added an RNA *in vivo* detection component to this system, incorporating repeats of a stem–loop structure in the RNA transcript to which bacteriophage MS2 protein binds.[16,17] Using this inducible system, it is now possible to simultaneously read out chromatin conformation and nascent RNA transcript accumulation within the 5–30 min following VP16 AAD targeting.[16] This represents a powerful system to investigate *in vivo* dynamics of the transcriptional machinery.

However, a major limitation of all the systems described above is that they rely on the tethering of a transcription factor to a 100- to 250-multicopy operator repeat. In a second experimental approach, a chromatin array containing multiple copies of a viral promoter was used to monitor changes in large-scale chromatin structure and coactivator recruitment after glucocorticoid receptor (GR) targeting. A GFP–GR fusion protein was expressed in a cell line[18] with ~200 copies of a plasmid containing the mouse mammary tumor virus (MMTV) promoter, with 4–6 GR-binding sites, upstream of the *ras* and bovine papilloma virus (BPV) genes.[19,20] Addition of ligand led to GR translocation to the nucleus and binding to the chromatin array. Recruitment to the array of GR coactivators SRC1 and CBP together with BRG1 and transcription factors NF1 and AP-2 accompanied targeting of GR.[18] Ligand-dependent GR targeting led to a cycle of large-scale chromatin decondensation and then recondensation, which roughly correlated with transcriptional activation and then downregulation.[18] A maximum fold induction of gene expression by GR of between 20 and 40 min correlated with maximum recruitment of GFP–pol2 over the array.[21]

[15] T. Tsukamoto, N. Hashiguchi, S. M. Janicki, T. Tumbar, A. S. Belmont, and D. L. Spector, *Nat. Cell Biol.* **2**, 871 (2000).

[16] D. L. Spector, T. Tsukamoto, E. Bertrand, D. Fusco, R. Singer, and S. M. Janicki, *Mol. Cell. Biol.* **13** (Suppl.), 2a (2002).

[17] E. Bertrand, P. Chartrand, M. Schaefer, S. M. Shenoy, R. H. Singer, and R. M. Long, *Mol. Cell* **2**, 437 (1998).

[18] W. G. Muller, D. Walker, G. L. Hager, and J. G. McNally, *J. Cell Biol.* **154**, 33 (2001).

[19] P. R. Kramer, G. Fragoso, W. Pennie, H. Htun, G. L. Hager, and R. R. Sinden, *J. Biol. Chem.* **274**, 28590 (1999).

[20] D. Walker, H. Htun, and G. L. Hager, *Methods* **19**, 386 (1999).

While providing a more physiological context by allowing analysis of chromatin remodeling and factor recruitment by real promoters, one technical difficulty of this second approach is in visualizing the chromosomal array in the absence of the fluorescently tagged *trans* factor. In the MMTV/GR experiments described above, the chromosomal array could not be examined before GR targeting, and the signal was sufficiently bright to allow detection of the chromosomal array in only a fraction of the cells after GR targeting. Similarly, many coactivators are present in various nuclear compartments and bodies at significant levels; therefore, differentiation between the chromosomal array and other nuclear concentrations of a given cofactor may be difficult. A natural progression for these *in vivo* methods will be to combine the two approaches, using natural promoters together with a direct operator repeat as a tag for the chromosomal array.

Methodology

Basic Approach

Our previous work used the lac operator/repressor system for transcription factor tethering. This required several experimental components. First, we needed to generate and subclone the lac operator repeat. Inverted repeats are highly unstable in bacteria. We therefore used a directional cloning method to force direct repeats of the lac operator, doubling the number of repeats with each cloning cycle.[2] Recombination between repeats was a significant problem in typical host cells. Reaching high copy numbers of lac operator repeats therefore required use of a special recA minus *Escherichia coli* strain designed for increased stability of direct repeats, switching to a moderate or low copy number plasmid at later stages of repeat amplification, and using nonstandard protocols for plasmid preparation. A detailed description of these methods for working with lac operator repeats has been published elsewhere.[22]

Second, we needed GFP–lac repressor fused with transcriptional activation domains. We have typically fused proteins at the 3' end of the GFP–lac rep protein-coding region, mainly for cloning convenience. A unique *Asc*I restriction site was engineered downstream of the lac repressor and before the stop codon to facilitate convenient cloning of PCR products.[12] Caution should be exercised when designing a cloning scheme, as we have discovered that short (~10 amino acid) stretches rich in acidic and

---

[21] M. Becker, C. Baumann, S. John, D. A. Walker, M. Vigneron, J. G. McNally, and G. L. Hager, *EMBO Rep.* **3,** 1188 (2002).
[22] A. S. Belmont, G. Li, G. Sudlow, and C. Robinett, *Methods Cell Biol.* **58,** 203 (1999).

hydrophobic residues that resemble acidic activators are sufficient to direct dramatic large-scale chromatin unfolding.[13] This is consistent with previously published reports that random peptides with such properties can serve as potent transcriptional activators.[23,24] Therefore, linker regions should be avoided, or if used, a GFP–lac repressor fused to the linker alone should be used as control construct.

A significant number of lac repressor fusion proteins are inactive. Therefore, it is critical that functional assays be performed on lac repressor fusion proteins. In the case of transcriptional activators or repressors, the obvious functional assay is a direct measure of activation or repression, using a reporter system. We now typically use a transient transfection assay using a lac operator 8-mer adjacent to a minimal TATA box core promoter driving a luciferase reporter construct.[13] Additional functional criteria that can be used would include correct intranuclear localization, cofactor recruitment, and, where appropriate, normal ligand dose–response curves.[12] The ultimate functional test would be genetic complementation of the wild-type transcription factor. This has been used for a lac repressor–HP1 fusion protein.[25]

The third requirement is the generation of stable cell lines containing integrated plasmid arrays containing the lac operator repeats. In general, we created these cell lines in cells not expressing lac repressor constructs, giving us the flexibility of then expressing a variety of different lac repressor fusion proteins in a particular cell line. Two approaches can be used to generate large chromosomal arrays containing the transfected plasmids. The first is to select stable transformants with large copy numbers of the transfected plasmid. Alternatively, one can start with a given plasmid insertion and then use gene amplification to further increase the size of the chromosomal array. Detailed descriptions and protocols for both methods of cell transformation and cloning have been published elsewhere.[26]

The final step is to express the GFP–lac repressor fusion protein in the cell, detect the chromosomal array, and monitor cofactor recruitment. We typically have used transient transfection to examine steady state levels of large-scale chromatin decondensation and cofactor recruitment.[6,9] For time courses of opening and factor recruitment, we have used ligand-dependent inducible systems[12] or bead loading as a method for rapidly

[23] D. M. Ruden, J. Ma, Y. Li, K. Wood, and M. Ptashne, *Nature* **350,** 250 (1991).

[24] K. Melcher, *J. Mol. Biol.* **301,** 1097 (2000).

[25] Y. Li, J. R. Danzer, P. Alvarez, A. S. Belmont, and L. L. Wallrath, *Development* **130,** 1817 (2003).

[26] Y. G. Strukov and A. S. Belmont, *in* "Live Cell Imaging: A Laboratory Manual" (D. L. Spector and R. D. Goldman, eds.). Cold Spring Harbor Laboratory Press, Cold Spring Harbor, NY, 2003.

introducing a recombinant protein into the cell.[9] Factor recruitment can be done either with antibodies to detect endogenous proteins or with GFP- or epitope-tagged transfected proteins. The use of GFP-tagged coactivator proteins provides the unique capability of monitoring coactivator recruitment in live cells. However, there are always concerns when expressing exogenous proteins, particularly by transient transfection methods, with regard to mislocalization and alterations in binding kinetics due to protein overexpression. We have preferred to conduct initial work with antibodies to monitor the distribution of endogenous proteins, reserving the use of GFP-tagged coactivators for more specific follow-up experiments.

Ongoing Methodology Development

pSP Plasmids. Our laboratory is moving to combine the two approaches used to date for *in vivo* analysis of transcription factor-mediated changes in large-scale chromatin structure and cofactor recruitment. The idea is to separate the chromatin tag, using lac operator or other repeats, from the targeting of transcription factors, allowing the use of natural promoters. At the same time, where appropriate, protein and/or RNA reporter constructs can be included for direct visualization of gene expression,[15,16] as described earlier.

Toward this end, the pSP plasmids, which contain a synthetic 258-bp polylinker, were designed with several intended goals.[13] First, they contain a large number of unique restriction sites, many of them rare, 8-bp recognition sequences, to facilitate cloning of multiple components including upstream regulatory elements, promoter, reporter gene, repetitive DNA-binding sequences (such as lac operators), and a selectable marker. Because of the large number of unique rare restriction sites, the design of the polylinker allows cloning of four components, up to ~20 kb, by simple cloning methods. Although we have designed the polylinker for generating plasmids to make stable cell lines, it is also useful for any cloning project where several large inserts must be combined.

Second, the plasmid contains a combination of restriction sites allowing directional cloning of repeats. In the first step, a DNA sequence of interest is cloned into the blunt *Pme*I site. The *Spe*I and *Nhe*I sites surrounding *Pme*I have complementary cohesive ends that do not recreate either restriction site when ligated together. Multiple copies can be cloned by cutting out the DNA sequence of interest, using *Nhe*I and *Asc*I, and ligating it into the *Spe*I and *Asc*I sites of a separate aliquot of the same plasmid. In this way, the number of DNA sequences of interest can be doubled in each round of cloning, as was done previously to generate 256 tandem repeats of the lac operator sequences.[2] In addition, *Sal*I, *Xho*I, and any nearby unique site can be used for directionally cloning a second set of repeats of a DNA sequence. Alternatively, existing lac operator repeats flanked by *Xho*I and *Sal*I sites can be cloned into these sites.

Third, because the plasmid must stably carry repetitive DNA sequences, it must be present at low copy within bacteria in order to prevent recombination and therefore loss of some repeats. We chose the pMB1 origin of replication, which is used in the plasmid pBR322. This origin of replication includes the *rop* gene, which is responsible for maintaining the plasmid at about 20 copies per cell.[27] Last, because plasmids become more difficult to clone as they approach 20 kb in size, we made the plasmid as small as possible. At under 3 kb, the pSP2 plasmid is able to contain the average 13–16 kbp of inserts required for the typical project.

We have used this pSP plasmid for a number of different constructs containing different combinations of regulatory elements, promoters, and reporter genes, together with lac operator repeats.[13] The mammalian selectable marker chosen for these plasmids was puromycin.[28] Because this is an uncommon marker, it allows other plasmids with different markers to be stably cotransfected into mammalian cells. Puromycin also allows cells to survive even with low-level expression of the puromycin resistance gene, which might prevent bias of clones toward integration sites with high-level expression.

For the repetitive DNA-binding sites, we inserted previously constructed 64- and 256-copy lac repressor-binding sites.[2] Several promoters were inserted: the simian virus 40 (SV40) promoter and enhancer, which has high transcriptional activity; the cytomegalovirus (CMV) core promoter, which has high transcriptional activity, although not as high as the entire CMV promoter; the F9 polyoma promoter, which has moderate activity; and the vitellogenin B1 TATA[29] and the E1b TATA, both of which have low basal activity but are capable of induction when strong activators are tethered nearby. Although we did not directly compare these two TATA-based reporters, preliminary tests indicated that the E1b TATA is capable of about 2-fold higher induction than the vitellogenin B1 TATA.

We chose several reporter genes: CFP-PTS and YFP-PTS are derivatives of the green fluorescent protein that are cyan and yellow, respectively. Fluorescent proteins are also used to label the lac operator array in the nucleus, so as to prevent overlap of the lac operator spots in the nucleus with the reporter genes, CFP-PTS and YFP-PTS are tagged with a

[27] M. Lusky and M. Botchan, *Nature* **293,** 79 (1981).
[28] S. de la Luna, I. Soria, D. Pulido, J. Ortin, and A. Jimenez, *Gene* **62,** 121 (1988).
[29] T. C. Chang, A. M. Nardulli, D. Lew, and D. J. Shapiro, *Mol. Endocrinol.* **6,** 346 (1992).

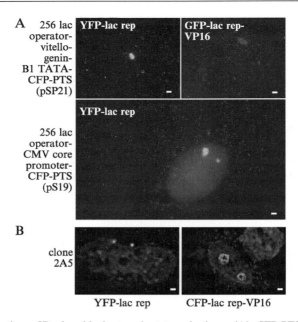

Fig. 1. Testing pSP plasmids in transient transfections. (A) CFP-PTS observed by microscopy. pSP plasmids were transfected into wild-type CHO-K1 cells along with YFP–lac rep or GFP–lac rep–VP16. GFP bleeds through in the CFP channel, so the GFP–lac rep–VP16 protein is not distinctly visible. Single optical sections are shown. Scale bar: 1 μm. (B) Stable cell lines containing pSP plasmids. Chromatin unfolding and reporter activation in response to VP16 activator in stable clone 2A5 with the 256 lac op–vitellogenin B1 TATA–CFP-PTS plasmid (pSP21) integrated into the genome. DNA stained with DAPI is shown in red, the YFP channel is shown in green, and the CFP channel is shown in blue. A single optical section from a deconvolved image series is shown. Scale bar: 1 μm. (See color insert.)

peroxisome-targeting signal.[15,30] This tag directs the fluorescent proteins to the peroxisomes, which appear as a pattern of spots in the cytoplasm. These reporters are compatible with live, individual cell observations by microscopy or flow cytometry. For more routine analysis of cell populations, the luciferase reporter is more convenient.

During plasmid construction, we tested pSP plasmids by transient transfection; this represents a significant time saving over generating stable cell lines. For plasmids with a fluorescent CFP-PTS or YFP-PTS reporter gene, transiently transfected CHO-K1 cells were observed by microscopy (Fig. 1A). For example, the 256 lac op–vitellogenin B1 TATA–CFP-PTS plasmid (pSP21) was found to show low CFP-PTS reporter expression

[30] S. J. Gould, G. A. Keller, and S. Subramani, *J. Cell Biol.* **107,** 897 (1988).

when cotransfected with YFP–lac rep. In contrast, the plasmid showed strong CFP-PTS expression in most transfected cells when cotransfected with GFP–lac rep–VP16 acidic activation domain, a strong transcriptional activator. When the constitutively active 256 lac op–CMV core–CFP-PTS plasmid (pSP19) was transiently cotransfected with YFP–lac rep, nearly 100% of transfected cells expressed CFP-PTS (Fig. 1A).

Interestingly, despite the fact that the plasmid was not integrated into the genome, it was possible to see some structure of the plasmids even in the transiently transfected state. Aggregates of the 256 lac op–vitellogenin B1 TATA–CFP-PTS plasmid (pSP21) detected by YFP–lac rep were typically condensed whereas plasmid aggregates detected by GFP–lac rep–VP16 were usually less condensed. Although not as dramatic, these results mimic the results previously seen with stably integrated plasmids.[6] It is known that transiently transfected DNA takes on some, but not all, properties of plasmids integrated into chromatin.[31]

We also examined the behavior of transiently transfected plasmids by flow cytometry and by luciferase assays, where appropriate.

Generation of Stable Cell Lines. Multiple copy repeats are subject to transgene silencing, a phenomenon still not well understood. Work is in progress to find transformation conditions that minimize the effects of transgene silencing and increase the fraction of chromosome arrays that show inducible expression. In a previous study that used 96 copies of the tet operator to target the VP16 AAD to a chromosome array, only 3 of 71 stable transformant clones showed both good growth and inducible expression of the reporter gene.[15] Using the pSP21 plasmid, containing the 256 lac op–vitellogenin B1 TATA–CFP-PTS plasmid, only 1 of 200 clones showed under basal conditions a large, condensed chromosome array that also visibly unfolded and showed significant inducible expression when transfected with a GFP–lac repressor–VP16 AAD expression vector. This was the 2AS cell line containing two large chromosome arrays per nucleus (Fig. 1B).

The pSP plasmid contains a low-copy origin of replication. The *rop* gene sequences that make the plasmid low-copy are called "poison" sequences because they have been found to reduce expression of trans-fected proteins when transfected into mammalian cells.[27,32] We tested the effects of removal of this sequence, using the 256 lac op–vitellogenin B1 TATA–CFP-PTS plasmid (pSP21). Transfection of linear DNA increased the proportion of stable transformants showing both a large, visible chromosome array and inducible expression after expression of GFP–lac

[31] C. L. Smith and G. L. Hager, *J. Biol. Chem.* **272**, 27493 (1997).
[32] D. O. Peterson, K. K. Beifuss, and K. L. Morley, *Mol. Cell. Biol.* **7**, 1563 (1987).

repressor–VP16 AAD, with 4 of 18 positive colonies from 0.6 μg of linear DNA with the poison sequences removed versus 3 of 50 from 5 μg of intact, supercoiled plasmid.[13]

Therefore, removal of these prokaryotic poison sequences appears to improve efficiency of inducible chromosome array isolation. A second improvement we have found is to cotransfect plasmid DNA with genomic DNA. Using the 256 lac op–CMV core–CFP-PTS plasmid (pSP19), we found that cotransfection of genomic DNA increased CFP-PTS expression. Greater than 80% of stably transfected cells cotransfected with genomic DNA expressed the CFP reporter versus ∼50% without genomic DNA in cells expressing GFP–lac repressor. Of 18 stable colonies with visible chromosomal arrays, 15 showed CFP expression in cells expressing YFP–lac repressor.

Automated Microscopy of Chromatin Structure and Coactivator Recruitment. Our initial work used qualitative or semiquantitative measures of array opening and/or coactivator recruitment. For instance, percentages of cells with small or large array opening or weak or strong coactivator recruitment were typical measures.[6,9,11] Even for these rough classification schemes, it became important to collect statistics on large numbers of cells, which was time consuming and tedious when working with transient transfection efficiencies of 0.5–5%. Later work used quantitative measures of array size to better analyze experimental results.[9,12] To analyze a larger number of proteins and their mutant versions for effects on large-scale chromatin structure, it became clear that manual microscopy was not sufficiently high-throughput.

We therefore have adapted a commercially available motorized microscope and software package to automate data collection and analysis. The program we developed runs unattended, scanning a slide in search of the 0.5–5% of cells expressing the transiently transfected, fluorescently labeled lac repressor.[13] It finds the optimal focus and exposure for the chromosome array, collects images, and measures the array size. We further adapted this automated routine to analyze not only the size of the lac repressor-labeled chromatin array but also the recruitment of a second protein to that region of the nucleus, using two different colors of GFP.[13] An algorithm determines the ratio of fluorescence intensity at the spot versus fluorescence intensity in the remainder of the nucleus to determine quantitatively the extent of colocalization. This automation made possible the analysis of chromatin unfolding and recruitment of 25 different fusion proteins within a matter of weeks, with each data point, duplicated in two independent experiments, consisting of measurements of 100–200 transiently transfected nuclei.[13]

Using such an automated data collection approach should greatly facilitate the adaptation of these cell systems to provide more general assays of chromatin remodeling and coactivator recruitment.

Selected Protocols

Directional Cloning in pSP Plasmids

Note: Tris = 10 mM Tris-HCl, pH 8.0 or 8.5.

1. Check that the promoter of interest lacks *Nhe*I, *Spe*I, and *Asc*I sites. Clone the promoter into the blunt *Pme*I site of pSP12 and check the orientation, noting that the reporter gene runs counterclockwise (3′ to 5′). Clone enhancers of interest into the *Pme*I site of pSP21 or pSP32, which already contain TATA core promoters. Check that the promoter works in this context with a transient transfection.
2. Prepare the vector for cloning, using the plasmid resulting from step 1, which contains one copy of the promoter.

   a. Digest ∼2 μg of the plasmid with *Spe*I (in buffer H from Roche Molecular Biochemicals, Indianapolis, IN) for 2 h at 37°. Check an aliquot on a gel to make sure the digest is complete. *Spe*I is allowed to digest alone before the *Asc*I digest (step 2c) because *Spe*I may not cut well near the end of DNA. That is, if *Asc*I cuts first, the *Spe*I site will be close to the end of the fragment and may not be cut.
   b. Purify DNA on a cleanup column (e.g., QIAquick PCR purification column; Qiagen, Hilden, Germany). Elute in ∼40 μl of Tris.
   c. Digest with *Asc*I (in NEB4 buffer; New England BioLabs, Beverly, MA) overnight at 37°.
   d. Heat kill the *Asc*I at 65° for 20 min.
   e. Reserve 2 μl of phosphorylated vector to ligate as a control. If it is truly cut by both *Spe*I and *Asc*I, it should not yield any colonies. Dephosphorylate the remainder with shrimp alkaline phosphatase (SAP; Roche Molecular Biochemicals) for ∼4 h. Dephosphorylation is theoretically not necessary because the vector should have incompatible ends, but there appears to be always some vector that is cut by only one enzyme.
   f. Purify the DNA on a cleanup column or run on an agarose gel, excise without exposing to UV light (see Generating Stable Cell

Lines, below), and extract with a QIAquick gel extraction column. Elute in 30 μl of Tris.

3. Prepare the insert for cloning, using the plasmid resulting from step 1, which contains one copy of the promoter.

   a. Digest ∼4 μg of the plasmid with NheI and AscI (in NEB4 buffer) for 2 h at 37°.
   b. Run the DNA on an agarose gel, excise the fragment without exposing to UV light, and extract with a QIAquick gel extraction column. Elute in 30 μl of Tris.

4. Run 2 μl of vector and 2 μl of insert on a miniature gel to compare their concentrations.

5. Ligate. As controls, ligate vector plus ligase, and phosphorylated vector plus ligase.

6. Check the colonies for the appropriate insert. Because pSP is a low-copy plasmid, prepare 1.5 ml of bacterial culture and use about half of the 50-μl product for each diagnostic digest.

7. After obtaining a plasmid with two copies of the promoter, repeat the cloning to obtain 4 copies, 8, 16, 32, and so on. To obtain enough DNA for cloning, prepare 5 ml of culture, using one QIAquick spin miniprep column, or use the chloramphenicol amplification procedure.[33] Recombination and loss of repeats are problematic for large numbers of repeats (roughly eight or more), so the chloramphenicol amplification procedure should not be used and STBL2 cells from Invitrogen (Carlsbad, CA) should be used and grown at 30°.

Generating Stable Cell Lines

1. *Test pSP plasmid in a transient transfection.* Transfect the pSP plasmid with transcription factors of interest to make sure that the reporter responds appropriately and strongly, given that most plasmids will have much lower expression when stably integrated than when transiently transfected. pSP19 (CFP-PTS reporter) or pSP32 (luciferase reporter) can be used as a positive control: these plasmids should express the reporter in nearly all transfected cells. Cotransfect a plasmid that allows detection of which cells are transfected, for example, YFP–lac rep or CMV–β-galactosidase.

2. *Determine transfection conditions.* Optimize transfection conditions for the cell line of choice, using a convenient constitutively active

---

[33] J. Sambrook, E. F. Fritsch, and T. Maniatis, "Molecular Cloning: A Laboratory Manual." Cold Spring Harbor Laboratory Press, Cold Spring Harbor, NY, 1989.

reporter (e.g., CMV–β-galactosidase or GFP). Choose the transfection reagent, the amount of reagent and DNA to use, and the size of the dish. We use 60-mm dishes, 2 μg of DNA, 40 μl of FuGENE6 (Roche Molecular Biochemicals), and 300 μl of serum-free medium for the easily transfectable CHO-K1 cell line. Other cell lines would probably require a 75- or 150-cm^2 flask.

3. *Test sensitivity of the cell line of interest to puromycin.* Transfect cells with pPUR (BD Biosciences Clontech, Palo Alto, CA; contains puromycin selectable marker) as a positive control and some other DNA (e.g., pUC19) as a negative control. The day after selection begins, cells can be passaged to a larger flask if necessary. Passaging 1 day after selection seems to get rid of nonexpressing cells quickly, because dying cells do not reattach. The minimal concentration of puromycin should be used because the selection reagent slows the growth even of resistant cells. The concentration selected for CHO-K1 cells was 7.5 μg/ml. Puromycin (Sigma, St. Louis, MO) is dissolved in H$_2$O. Store aliquots at $-20°$.

4. *Prepare pSP DNA for transfection.*
   Make DNA. Preparing DNA of pSP plasmids is not an easy task, because the 256 lac operators are unstable. For this reason, the plasmid is low-copy (to minimize recombination) and is maintained in STBL2 cells (GIBCO/Life Technologies, Grand Island, NY), which exhibit repeat-stabilizing properties when grown at 30°. To grow a culture for DNA preparation, streak from a frozen glycerol stock or a stab culture onto an LB-Amp plate and grow overnight at 30°. The next day around noon, pick 10–20 colonies into 2–3 ml of liquid LB-Amp cultures and grow at 30° overnight. The next day, prepare 1.5 ml of culture, digest 40 of the 50 μl of plasmid DNA in a 50-μl restriction digest that should cut on either side of the lac operator repeats, and run the digests on a gel (with fairly large wells to hold the large volume). Choose the cultures that have full-length repeats and use the leftover "miniprep" culture to inoculate a large culture. Grow at 30° overnight. It is best to choose two different positive clones and do one 60-ml culture for one Bio-Rad (Hercules, CA) "midiprep" for each in case one of the cultures loses its repeats in culture overnight, although this rarely occurs. Elute the midiprep DNA, using 400 μl of hot 10 mM Tris-HCl, pH 8.5. Each midiprep usually yields 100–300 μg of DNA, using the instructions for a low-copy plasmid. Do not use larger culture volumes than suggested.
   Remove poison sequences. If it is desirable to eliminate the poison sequences, digest the pSP plasmid with enzymes cutting at sites on either side of the DNA of interest (e.g., *Bgl*II and *Fse*I, or an *Xmn*I

digest, as long as the enzymes do not cut within the promoter of interest). Run the DNA on a low-percent (0.75%) SeaPlaque GTG gel (Cambrex Bioscience, Rockland, Rockland, ME), which should be poured and allowed to set at 4° because it will be flimsy. The gel can be run at room temperature. Fresh agarose and Tris–acetate–EDTA (TAE) should be used in a clean gel box, to avoid contamination of the sample. It is helpful to cut out a minimal gel slice to aid purification later. Avoid exposing the DNA to UV light or ethidium bromide when cutting out the large band containing the pSP fragment of interest. Once the band has been excised, there are two options.

a. Heat the agarose at 65° for 10 min or until melted and transfect the unpurified DNA. In this case, there is no need to check the concentration if the amount of DNA loaded onto the gel is known and if the majority of the band has been cut.

b. Use a QIAEX II gel extraction kit. Use 30 μl of reagent. Do not vortex the pellet—gently flick to avoid shearing. Use hot elution buffer to elute the large DNA fragment. Elute twice. Because yields of 5–20% are expected for a fragment of this size, this method requires that a great deal of DNA be run on the gel. In addition, the concentration of the DNA fragment should be checked before transfection.

Option (a) is recommended because it is much easier and provides much higher yields of DNA. We obtained similar size arrays with both protocols. There is, of course, the possibility of contamination when transfecting the unpurified melted agarose gel, whereas the QIAEX II protocol includes an ethanol wash. In practice, the nonsterile melted gel does not produce contamination as long as antibiotics are present in the cells' medium.

5. *Transfect cells with the linear fragment or supercoiled plasmid to make a stable cell line.* Use the conditions determined in steps 2 and 3, including positive and negative control transfections.

6. *Screen clones for properties of interest.* Cells can be analyzed as a mixed population (which may rapidly change in properties over time, because some clones grow faster than others), or they can be subcloned to obtain a pure population. We note, however, that subcloned populations do tend to display variety in the number and appearance of chromatin arrays, presumably because of genomic instability, which is particularly evident in CHO-K1 cells. An aliquot of the stably transfected mixed population should be frozen as a backup in case subcloning fails. To subclone, the cells are passaged

and serially diluted, flow sorted, or plated for filter paper cloning in order to obtain individual clones in a 96-well plate. For filter paper cloning, dilute cells are plated in large 150-mm dishes and grown until colonies contain ~20–50 cells. Trypsin-treated bits of sterile filter paper are then placed on well-isolated colonies for several minutes until they detach, and the paper with some attached cells is transferred to a 96-well plate. Further details on cell cloning are provided elsewhere.[26]

Acknowledgments

This work was supported by grants from the National Institutes of Health to A. S. Belmont (R01-GM58460 and R01-GM42516). Anne E. Carpenter is a Howard Hughes Medical Institute Predoctoral Fellow.

[24] Measuring Histone and Polymerase Dynamics in Living Cells

By Hiroshi Kimura, Miki Hieda, and Peter R. Cook

Introduction

In eukaryotic cells, DNA is packaged into nucleosomes by wrapping it around histone octamers; each octamer contains two copies of H2A, H2B, H3, and H4.[1] In dividing mammalian cells, where DNA is made during the S phase, DNA is first wrapped around the $(H3–H4)_2$ tetramer before the addition of two H2A–H2B dimers.[2] Once assembled, these core histones are so tightly bound to DNA that they resist extraction with salt concentrations below $0.63\ M$.[3] Therefore, it is assumed that histone–DNA interactions must be loosened or remodeled to allow access of proteins such as polymerases to DNA.[4,5] Various factors mediating chromatin assembly, disassembly, and remodeling have been identified; some slide nucleosomes along the DNA without dissociating the octamer, others displace some or all of the histones. However, important questions remain as to when,

[1] K. Luger, A. W. Mader, R. K. Richmond, D. F. Sargant, and T. J. Richmond, *Nature* **389**, 251 (1997).
[2] A. Verreault, *Genes Dev.* **14**, 1430 (2000).
[3] R. H. Simon and G. Felsenfeld, *Nucleic Acids Res.* **6**, 689 (1979).
[4] J. L. Workman and R. E. Kingston, *Annu. Rev. Biochem.* **67**, 545 (1998).
[5] A. P. Wolffe and J. J. Hayes, *Nucleic Acids Res.* **27**, 711 (1999).

where, and how such histone exchange occurs in living cells and what role this exchange might play during transcription and replication.

Early studies on histone exchange and deposition in living cells utilized radiochemical labeling. In a seminal series of studies, cells were incubated in radioactive amino acids, and the assembly of the radiolabeled (newly made) histones into nucleosomes was monitored; H2A and H2B exchanged more rapidly than H3 and H4.[6,7] The stable association of H3 with H4 was also demonstrated using radiolabeled arginine.[8] HeLa cells were incubated in [³H]arginine, which was then incorporated preferentially into the arginine-rich histones (i.e., H3 and H4); after fusion with mouse 3T3 cells, some of the nuclei in the resulting heterokaryons entered mitosis to give chimeric daughter nuclei. In these nuclei, autoradiography revealed that ³H remained associated with the HeLa chromosomal territories, showing that the arginine-rich histones remained associated with HeLa DNA over several days.

Studies *in vitro* have also revealed a great deal about how RNA polymerase II transcribes naked DNA templates *in vitro;*[9,10] however, we still know little about how it transcribes natural templates *in vivo.* For example, the TATA-binding protein plays a critical role *in vitro,* but "knockouts" reveal it has little effect on activity *in vivo.*[11] Given this precedent, it seems studies *in vivo* will uncover other surprises.

Fortunately, the dynamics of proteins can be monitored in living cells after tagging them with the green fluorescent protein (GFP). A construct encoding GFP fused to the protein of interest is expressed in a cell, so the resulting fluorescent hybrid can be seen directly.[12] Mutant GFPs with altered fluorescence are available (e.g., enhanced cyan and yellow fluorescent proteins—ECFPs and EYFPs), and one—PAGFP—is photoactivated by irradiation with 413-nm light so its fluorescence (488-nm excitation) increases 100-fold.[13] Therefore, it is now possible to monitor histone and polymerase dynamics in real time, using these GFP tags and photobleaching techniques such as fluorescence recovery after photobleaching (FRAP) and fluorescence loss in photobleaching (FLIP).[14–17] Here, we use

[6] V. Jackson and R. Chalkley, *Biochemistry* **24,** 6921 (1985).

[7] V. Jackson, *Biochemistry* **29,** 719 (1990).

[8] T. Manser, T. Thacher, and M. Rechsteiner, *Cell* **19,** 993 (1980).

[9] R. G. Roeder, *Methods Enzymol.* **272,** 165 (1996).

[10] G. Orphanides and D. Reinberg, *Nature* **407,** 471 (2000).

[11] I. Martianov, S. Viville, and I. Davidson, *Science* **298,** 1036 (2002).

[12] R. Y. Tsien, *Annu. Rev. Biochem.* **67,** 509 (1998).

[13] G. H. Patterson and J. Lippincott-Schwartz, *Science* **297,** 1873 (2002).

[14] J. Ellenberg and J. Lippincott-Schwartz, *Methods* **19,** 362 (1999).

[15] A. B. Houtsmuller and W. Vermeulen, *Histochem. Cell Biol.* **115,** 13 (2001).

our studies on the dynamics of the core histones[18] and RNA polymerase II[19,20] as examples to illustrate these techniques.

Some Problems Associated with Use of GFP-Tagged Proteins

When using a GFP-tagged protein, it is essential to ensure that the tagged protein functions like the untagged protein. For example, when determining subcellular location, the requirement is simply that the protein localizes like the untagged protein. But even though a GFP-tagged protein might localize correctly, its dynamics will differ; it is inevitably 28 kDa larger—the size of the GFP tag—than the normal counterpart and it will diffuse more slowly. And when studying function, it becomes essential to demonstrate that the tag does not interfere with that function. Such a demonstration is best achieved by replacing all copies of the gene of interest with its counterpart encoding the GFP fusion under the control of the natural promoter; then, if the cell grows normally, the hybrid gene must be able to function normally.[21] However, this kind of demonstration proves to be practically difficult in higher eukaryotes. A second-best option is to use mutant cells that are defective in the gene of interest so that the GFP-tagged protein can be demonstrated to function normally, using genetic complementation. For example, the kinetics of the nucleotide excision repair factors have been analyzed with GFP-tagged proteins that complement the ultraviolet (UV)-sensitive phenotype in mutant cells defective in such repair.[22] In our case, the kinetics of transcription were analyzed with a GFP-tagged version of RNA polymerase II, which complemented a temperature-sensitive mutation in the polymerase;[19,20] as the mutant cells died at the restrictive temperature, it must have been the tagged version that kept the cells alive. If mutant eukaryotic cells are not available, it may be possible to demonstrate that the hybrid eukaryotic gene will complement the genetic defect in a homologous yeast mutant, if available. When mutants are unavailable, function may be demonstrable by enzymatic assay.[23] In the case of histones, no mutant eukaryotic cell lines are available, and it is difficult to prove decisively that the ectopically

[16] R. D. Phair and T. Misteli, *Nat. Rev. Mol. Cell. Biol.* **2**, 898 (2001).

[17] G. Carrero, D. McDonald, E. Crawford, G. de Vries, and M. J. Hendzel, *Methods* **29**, 14 (2003).

[18] H. Kimura and P. R. Cook, *J. Cell. Biol.* **153**, 1341 (2001).

[19] K. Sugaya, M. Vigneron, and P. R. Cook, *J. Cell Sci.* **113**, 2679 (2000).

[20] H. Kimura, K. Sugaya, and P. R. Cook, *J. Cell Biol.* **159**, 777 (2002).

[21] Y. Dou, J. Bowen, Y. Liu, and M. A. Gorovsky, *J. Cell Biol.* **158**, 1161 (2002).

[22] A. B. Houtsmuller, S. Rademakers, A. L. Nigg, D. Hoogstraten, J. H. Hoeimakers, and W. Vermeulen, *Science* **284**, 958 (1999).

expressed GFP-tagged protein is fully functional and behaves exactly like the endogenous protein. In this case, a wide range of different biochemical and cytological analyses were used to demonstrate that the endogenous and GFP-tagged proteins behaved similarly, including copurification/ sedimentation with nucleosomes, coextraction with the endogenous counterparts by different salt concentrations or nuclease treatments, and immunoprecipitation of nucleosome-sized DNA, using anti-GFP antibody from micrococcal nuclease-treated nucleosomes.[18,24,25] Even so, there is always the suspicion that the tagged proteins might not be incorporated into nucleosomes exactly like their untagged counterparts.

It is also important to demonstrate that the GFP-tagged protein is expressed at the appropriate level, as both dynamics and function are likely to be affected by the expression level. For example, if the tagged protein is expressed in addition to the untagged protein, the combined concentration is likely to be higher than that normally found, and this will affect dynamics through mass action. Expression levels are particularly difficult to control when the tagged protein is introduced by transiently transfecting cells with the hybrid gene; then, different cells receive different numbers of plasmids and expression levels can vary widely from cell to cell in the population. Therefore, it is generally better to use stable and clonal cell lines so that each cell studied will contain the same copy number. But even then, it is still unlikely that the hybrid gene will be expressed at the same level as the endogenous gene. Just as in proving functional equivalence, proof that expression levels are equivalent is best achieved by replacing all copies of the gene of interest with its counterpart encoding the GFP fusion under the control of the natural promoter,[21] but—as before—this is practically difficult in higher eukaryotes. In the case of the histones, which are present at $>10^7$ molecules in a human cell, an expression of H2B–GFP at 10% of total H2B gives bright fluorescence.[18,24] The amount of the GFP-tagged protein relative to its endogenous counterpart is often determined by "western blotting," as the GFP-tagged protein is 28 kDa bigger than the untagged protein. Although this method can be applied to relatively large proteins, the blotting efficiency of which is less affected by the extra 28 kDa, it cannot be used for GFP–histones because the blotting efficiency of GFP-tagged histone is quite different from that of nontagged histone. Therefore, it is better to purify nucleosomes, run sodium dodecyl-sulfate (SDS)-polyacrylamide gels, and estimate the relative amount of GFP-tagged histone by Coomassie staining.[18,24]

[23] E. A. Reits, A. M. Benham, B. Plougastel, J. Neefjes, and J. Trowsdale, *EMBO J.* **16**, 6087 (1997).

[24] T. Kanda, K. F. Sullivan, and G. M. Wahl, *Curr. Biol.* **8**, 377 (1998).

[25] T. Misteli, A. Gunjan, R. Hock, M. Bustin, and D. T. Brown, *Nature* **408**, 877 (2000).

Constructing Expression Plasmids and Establishing Stable Cell Lines

Choice of End for Attaching GFP-Fusion

Experience shows that the GFP tag can be attached to either the amino- or C-terminal end of many proteins without affecting their function,[12] and this seems to be true of the histones. Fusion at the N terminus might appear attractive as the N termini of all core histones extend from the nucleosome and do not form a defined structure detectable by X-ray crystallography.[1] One might worry that such a tag would affect the posttranslational modifications occurring at this end and that are known to play important roles,[26] but such N-terminal tags have been used with H2B, H2A, and H3 without any obvious problems.[24,27,28] The C termini of H2A and H2B are also attractive candidates because they are also located at the surface of the nucleosome. Indeed, H2B fused with GFP at its C terminus has become the most commonly used GFP-tagged histone, and an appropriate plasmid vector is commercially available from BD Biosciences Pharmingen (San Diego, CA), in which the C terminus of H2B is connected with GFP through six amino acids. In contrast to H2A and H2B, the C termini of H3 and H4 are located at the center of the nucleosome, and we failed to obtain stable HeLa cell clones showing bright fluorescence when using GFP attached through the same six amino acids.[18] However, many stable clones expressing high levels of H3– or H4–GFP were obtained by using a longer linker with 23 amino acids; this suggests the longer linker enabled the GFP moiety to be placed outside the particle. Note, however, that a different six-amino acid linker has been used successfully in *Drosophila* cells.[29,30] In the case of the linker histone H1, the C-terminal GFP fusion is the only one reported[21,25,31] and the linker connecting H1 and GFP can be as short as one alanine residue.[25] In conclusion, both ends of histones can be used for attaching the tag, but the length (and perhaps flexibility) of the linker must be considered for C-terminal GFP fusions with H3 and H4.

[26] B. D. Strahl and C. D. Allis, *Nature* **403**, 41 (2000).

[27] P.-Y. Perche, C. Vourc, L. Konecny, C. Souchier, M. Robert-Nicoud, D. Dimitrov, and S. Khochbin, *Curr. Biol.* **10**, 1531 (2000).

[28] K. Sugimoto, T. Urano, H. Zushi, K. Inoue, H. Tasaka, M. Tachibana, and M. Dotsu, *Cell Struct. Funct.* **27**, 457 (2002).

[29] S. Henikoff, K. Ahmad, J. S. Platero, and B. van Steensel, *Proc. Natl. Acad. Sci. USA* **97**, 716 (2000).

[30] K. Ahmad and S. Henikoff, *Mol. Cell* **9**, 1191 (2002).

[31] M. A. Lever, J. P. Th'ng, X. Sun, and M. J. Hendzel, *Nature* **408**, 873 (2000).

Only one of the subunits of RNA polymerase II has been tagged with GFP—the largest catalytic subunit.[19,32] This subunit was chosen because a cell line with a temperature-sensitive mutation in this subunit was available, and this enabled genetic complementation to be used to demonstrate that the hybrid protein was functional. Here, the N terminus was chosen for tagging, as the C terminus plays such an important role in regulation and message production.[33] Various subunits (i.e., RPA194, RPA43, RPA40, and RPA16) of RNA polymerase I have also been tagged at either the N or C terminus.[34]

Choice of Expression System

GFP expression vectors for mammalian cells are available commercially (e.g., from BD Biosciences Clontech, Palo Alto, CA). We found the use of one of these, in which expression of the fusion protein was under the control of the constitutive elongation factor 1α promoter, led to uneven distribution of H3– and H4–GFP; this was traced to a preferential assembly of nucleosomes during DNA replication.[18] Therefore, the use of a promoter that enables more natural[21] or inducible expression[29] has advantages. For analysis of the largest subunit of RNA polymerase II tagged with GFP, we used the cytomegalovirus (CMV) promoter.[19]

Cell Line

The use of established clonal cell lines rather than transient transfections has many advantages (see above). Their use is essential in the particular case of the core histones, which exchange so slowly that it takes several days before the GFP-tagged histone has fully equilibrated.

FRAP and FLIP

Overview

With the introduction of confocal microscopes and laser illumination, FRAP and FLIP have become popular techniques for analyzing the kinetics of molecules in living cells, and many good reviews are now available.[14–17,35] For FRAP, a small part of a cell expressing the GFP-tagged

[32] M. Becker, C. Baumann, S. John, D. A. Walker, M. Vigneron, J. G. McNally, and G. L. Hager, *EMBO Rep.* **3**, 1188 (2002).
[33] T. Maniatis and R. Reed, *Nature* **416**, 499 (2002).
[34] M. Dundr, U. Hoffmann-Rohrer, Q. Hu, I. Grummt, L. I. Rothblum, R. D. Phair, and T. Misteli, *Science* **298**, 1623 (2002).
[35] J. Lippincott-Schwartz, N. Altan-Bonnet, and G. H. Patterson, *Nat. Cell Biol.* Suppl., S7–14 (2003).

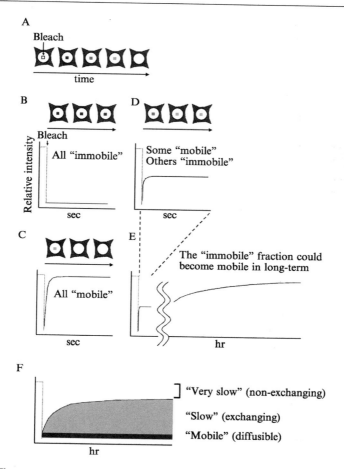

Fɪɢ. 1. Fluorescence recovery after photobleaching (FRAP) and molecular kinetics. (A) A schematic illustration of FRAP. Part of a nucleus expressing a GFP-tagged protein (e.g., a histone) is bleached with an intense laser pulse. Bleaching irreversibly damages the GFP fluorophor, and recovery of fluorescence in the bleached area depends on new fluorescent molecules diffusing from the unbleached area into the bleached area. Analysis of the rate of recovery allows the rates of diffusion and association/dissociation to be uncovered. (B) If all molecules are fixed, the intensity of the bleached area never recovers. (C) If all molecules are free to diffuse, the fluorescence in the bleached area quickly recovers almost to the original level; the diffusion rate can be determined by analyzing the recovery curve. (D) If both mobile and immobile fractions exist, the fluorescence recovers with diffusion kinetics but reaches a plateau at a level that reflects the size of the two fractions. (E) Sometimes, an "immobile" fraction defined by short-term analysis is seen to be mobile over the longer term; here, the recovery curve contains two components—rapid diffusion and the slower exchange. (F) The recovery of histone–GFP, where there are three fractions with differing mobilities.

protein is irradiated with an intense laser pulse to bleach the fluorophor (Fig. 1). Different microscopes have different mechanisms for defining the region of interest (ROI) to be bleached; one is an acoustical optical tunable filter (AOTF), which allows almost any area with any shape to be bleached. [Even if the microscope does not feature special ROI tools, a small area can still be bleached, using a high zoom (e.g., ×100); in this case, the middle of the field is usually bleached.] Because the GFP fluorophor is bleached irreversibly, the recovery of fluorescence in the bleached area depends on the influx of unbleached molecules. If all molecules are fixed, the bleached area remains bleached (Fig. 1B); in contrast, if all molecules are free to diffuse, the fluorescence in the bleached area recovers quickly almost to the original level as unbleached molecules equilibrate throughout both bleached and unbleached areas (Fig. 1C). Then, the diffusion rate can be determined by analyzing the recovery curve. When both immobile and mobile fractions coexist, the recovery curve reaches a plateau well below the original level, depending on the relative size of the two fractions (Fig. 1D).

FLIP is complementary to FRAP, and the two should usually be used in conjunction so that the results obtained by one method can be confirmed by the other. In a typical FLIP experiment, a field containing two cells—or in our case, two nuclei—is selected, and raster scanned repeatedly with the laser. (One nucleus is used for reference.) For most of each scan, a low laser power sufficient for imaging is used; then, power is increased for bleaching whenever the laser scans through the ROI (in our case, a rectangle) containing the bottom half of the lower nucleus. This process is then repeated until most fluorescence disappears from the top half of the bleached nucleus. Now the intensity in the unbleached (top) half of the bleached nucleus is expressed relative to its original (unbleached) intensity, and values are further corrected for the slight effects of bleaching during imaging (using the reduction in fluorescence seen in the other unbleached nucleus). Whereas bleaching precedes image collection in FRAP, here the two processes are interspersed. If all the GFP-tagged molecules are freely diffusible, bleaching the bottom half should progressively reduce the (relative) intensity in the top half to zero because unbleached molecules have plenty of time to diffuse into the target area and be bleached; this is the result obtained in control cells expressing GFP. If all are immobile (as in fixed cells), the relative intensity remains at unity because immobile molecules in the top half can never enter the bleaching zone. The results obtained with GFP–Pol lay between these two extremes, and were consistent with the existence of a large "mobile" (diffusing) population, and a smaller "immobile" (engaged) population that could be eliminated by incubation with the transcriptional inhibitor

5,6-dichloro-1-β-D-ribofuranosylbenzimidazole (DRB). As many of the same techniques are used for both FRAP and FLIP, only those used for FRAP are described.

Most studies involving FRAP and FLIP monitor the kinetics over several seconds to a few minutes. Then, most of the population of core histones tagged with GFP appear as immobile because the pool of free protein is so small and the others are so tightly bound to DNA. However, if the kinetics are monitored for longer, the intensity of fluorescence in the bleached area gradually becomes more intense as bound molecules in the unbleached area dissociate, diffuse into the bleached area, and rebind to the DNA (Fig. 1E). Because diffusion is so fast compared with the rate at which a molecule dissociates and rebinds, the recovery curve over the long term can be modeled by first-order dissociation/association kinetics. Unlike the tagged histones, there proves to be a large pool of free GFP–Pol that rapidly enters the bleached area, plus a second fraction that enters more slowly; the latter is probably the engaged polymerase as it is sensitive to the transcriptional inhibitor DRB.

Microscopy

The details of live cell microscopy have been described in this series.[36,37] For our studies, cells expressing GFP-tagged histone are plated a few days before analysis in a 35-mm glass-bottomed dish [e.g., from MatTek (Ashland, MA) or Matsunami (Osaka, Japan)]. It is important to transfer cells quickly from the incubator used for growth to the heated stage, as otherwise the temperature falls and it can take a significant time (usually ~0.5 h) for reequilibration, which is seen as a stabilization of the focal plane.

It is important to keep cells alive during image collection, and the issue of phototoxicity has been discussed elsewhere.[36] It is also difficult to prove formally that a cell does remain alive and physiologically intact during imaging—it often depends on how one defines "alive." However, passage through mitosis provides one good practical indication that the cell is alive enough to pass through the necessary checkpoints. Morphological changes such as the subsequent appearance of membrane blebs can also be used as another practical and sensitive indicator of the health of the cell.[38] Therefore, it is essential to image the cells (usually using differential interference, or phase-contrast optics, which are less damaging than the light used

---

[36] P. M. Conn, ed., *Methods Enzymol.* **302** (1999).
[37] P. M. Conn, ed., *Methods Enzymol.* **309,** (1999).
[38] D. Zink, *Cytometry* **45,** 214 (2001).

for GFP imaging) after the FRAP (or FLIP) data have been collected to check that the cells remain more alive than dead. In our experiments using HeLa cells expressing H2B–GFP,[18] most cells go through mitosis after having been scanned >60 times over 8 h [0.3% power of 25-mW argon laser; pinhole aperture, 4; fast scan mode; ×7 zoom; ×63 PlanApo object-ive, NA 1.4; Bio-Rad (Hercules, CA) μRadiance]. Locating a cell over such long periods is aided by using a glass-bottomed dish with a grid (MatTek or Matsunami). If cells suffer after imaging, survival can be improved by reducing the laser power, scan number, zoom factor, resolution, or by increasing the rate of scanning.

Often only one optical plane is scanned to reduce photodamage. The problem then arises of ensuring that the same optical section is imaged over time, especially over long periods during which the focal plane may shift as the temperature changes or the cell moves. Therefore, the focus must be readjusted manually every time an image is collected, and—for ease of refocusing—an equatorial plane or some specific cellular feature can be selected.

In a typical FRAP experiment, several images are collected at low laser power before bleaching, and the intensity of the zone to be bleached is averaged to provide the initial intensity. Then, scanning the bleaching zone one to four times with full laser power is usually enough to reduce GFP fluorescence significantly. In general, the size of the bleached area should remain constant throughout a series of experiments that are going to be averaged, because the bleaching period and efficiency are directly affected by the number of scan lines and the zoom factor. Note that the frequency of imaging depends on the kinetics to be analyzed. If monitoring the kinetics of a diffusible fraction, many images are usually collected over the next few seconds; if monitoring the transcription cycle of GFP–Pol or the reassociation of histone–GFPs, they are monitored for minutes or even hours.

Measurements of Relative Intensity

For FRAP, the intensity of the bleached area, the whole nucleus, and the background should be measured with an image analysis tool. We use imageJ (provided by W. Rasband, National Institutes of Health, Bethesda, MD; http://rsb.info.nih.gov/ij/) for analysis; however, others may find it easier to use the built-in software associated with the microscope or a com-mercial package like MetaMorph (Universal Imaging, Downingtown, PA; http://www.image1.com). For each time point, the net intensity is obtained by subtracting the background intensity from the intensity of the bleached area or the whole nucleus, and normalizing it relative to the intensity found before bleaching. Then, the relative intensity is determined by the following equation: Relative intensity $[I_{relative}(n)] = [I_{bleach}(n)/I_{nucleus}(n)]/[I_{bleach}(O)/I_{nucleus}(O)]$, where $I_{bleach}(n)$ is the net intensity of the bleached area at time n, $I_{nucleus}(n)$ is the net intensity of the whole nucleus at time n, $I_{bleach}(O)$ is the average net intensity of the bleached area before bleaching, and $I_{nucleus}(O)$ is the average net intensity of the whole nucleus before bleaching.

Curve Fitting

An example involving the slow recovery during FRAP of histone–GFP is given here (Fig. 1F); the bleached molecules dissociate from DNA to be replaced by unbleached molecules, and first-order kinetics apply. If only one exchanging fraction is present, the recovery should occur exponentially and be governed by $I_{relative} = C + P(1 - \exp^{-kt})$, where C is a constant value at time 0, P is the plateau value, k is an association constant, and t is time. Curves can be analyzed by nonlinear regression using software such as Prism (GraphPad, San Diego, CA; http://www.graphpad.com), Origin (OriginLab, Northampton, MA; http://www.originlab.com), or SAAM II (SAAM Institute, Seattle, WA; http://www.saam.com). The constant value (C) represents the sum of unbleached and diffusible fractions. The unbleached fraction, or bleaching efficiency, can be determined by bleaching cells that have been fixed with paraformaldehyde. The plateau value essentially represents the exchanging fraction; considering the unbleached and diffusible fractions, the exchanging fraction $= (P - C)/(1 - C)$. The association half-time of this exchanging fraction is calculated from the association constant k $[t_{1/2} = -\ln(1/2)/k]$. Then, the immobile fraction (which has a $t_{1/2}$ of more than the observation period) $= 1 - $ (exchanging fraction + unbleached fraction). If there are two exchanging fractions with distinct kinetics, recovery is governed by $I_{relative} = C + P_1(1 - \exp^{-k_1 t}) + P_2(1 - \exp^{-k_2 t})$, where P_1 and k_1 refer to population 1 and P_2 and k_2 refer to population 2. When the curve is fitted to a model including more than two different fractions, the different properties of these fractions should be demonstrated (e.g., in the case of H2B–GFP by a differential sensitivity to an inhibitor such as DRB, or by different kinetics in different compartments[18]).

Problems Associated with Analysis over Long Periods

Many FRAP experiments can be completed in a few seconds or minutes, but analysis of histone–GFPs required analysis over many hours, and then nuclear movements made it difficult to identify the bleached area precisely. It turns out that HeLa cells do not move as much as some other cells (e.g., human fibroblasts, CHO cells, and BHK cells). When using rapidly moving cells, a larger area (e.g., half of the nucleus) can be bleached to facilitate identification of the bleached region.[39] Another problem is caused by the rotation of the nucleus in the xy axis, which also makes identification of the bleached area difficult; in our case, this kind of rotation did not occur frequently in HeLa cells.[18] Another problem is caused by the synthesis of histone–GFP during the long period required for imaging, as the newly made protein may contribute to the recovery of fluorescence seen. Although a protein synthesis inhibitor such as cyclo-heximide can be added,[18] the accompanying side effects (e.g., inhibition of DNA replication, inhibition of cell cycle progression, etc.) can complicate analysis. In such cases, the use of an inducible expression vector should be considered.

Present Results and Future Directions

FRAP and FLIP have now been used by different groups to analyze the exchange of various histone–GFP constructs. The linker histone H1 exchanges rapidly even when it is in heterochromatin or a mitotic chromosome,[21,25,31,40] whereas the core histones exchange much more slowly.[18] For example, H2B exchanges more rapidly than H3 and H4, even though all form part of the same structure, the nucleosome. About 3% of H2B (probably the transcriptionally active fraction) exchanges within minutes ($t_{1/2}$, ~6 min), ~40% (probably the euchromatic fraction) more slowly ($t_{1/2}$, ~130 min), and another ~50% (probably in heterochromatin) remains bound permanently ($t_{1/2}$, >8.5 h). More than 80% of H3 and H4 is also bound permanently. These results are consistent with the mobile components—H1, H2B (and perhaps H2A)—facilitating immediate access of transcription factors and polymerases to the DNA, whereas the immobile components—H3 and H4—act as stable epigenetic markers. Although this global view of histone exchange in living cells has emerged, the underlying control mechanisms that govern it remain largely unknown. We are also only beginning to analyze the kinetics of the different histone variants; for example, the *Drosophila* H3.3 variant—which differs from the major from of H3 in only a few amino acids—is incorporated into transcriptionally active chromatin independently of DNA replication whereas H3 is not.[30] And we have yet to analyze how the histone code modifies those

[39] N. Daigle, J. Beaudouin, L. Hartnell, G. Imreh, E. Hallberg, J. Lippincott-Schwartz, and J. Ellenberg, *J. Cell Biol.* **154**, 71 (2001).
[40] F. Catez, D. T. Brown, T. Misteli, and M. Bustin, *EMBO Rep.* **3**, 760 (2002).

kinetics. Similar studies have also revealed two kinetic fractions of RNA polymerase II in living nuclei: most diffuses freely, but a small but significant fraction becomes transiently immobile (association $t_{1/2}$, ~20 min) during transcription.[20,32] One challenge now is to analyze what happens as a single tagged polymerase transcribes a specific nucleosomal template in a living cell.

Acknowledgments

The authors' work was supported by the Wellcome Trust. Miki Hieda is a Research Fellow of the Japanese Society for the Promotion of Science.

[25] Measurement of Dynamic Protein Binding to Chromatin *In Vivo*, Using Photobleaching Microscopy

By Robert D. Phair, Stanislaw A. Gorski, and Tom Misteli

Introduction

Chromatin-binding proteins play a crucial part in every aspect of chromatin structure and gene expression.[1] Direct binding of proteins to chromatin maintains and regulates higher order chromatin structure, and leads to histone modifications and transcriptional activation. Once a gene is activated, components of the RNA polymerase machinery directly contact DNA and mediate transcription. Despite the crucial importance of chromatin proteins, most of what we know about the interaction of these proteins with DNA comes from *in vitro* experiments. Regardless of whether the DNA used in *in vitro* assays consists of naked DNA or reconstituted chromatin, it is unlikely that these templates reflect the physiological binding substrates that are found in a cell nucleus or that the buffer conditions accurately reproduce the ionic environment in a cell. Methods are required to probe the binding of proteins to native, unperturbed chromatin in intact cells.

An experimental approach to studying the binding of protein to chromatin in living cells is the use of photobleaching methods.[2–9] In these

[1] G. Felsenfeld and M. Groudine, *Nature* **421**, 448 (2003).
[2] R. D. Phair and T. Misteli, *Nature* **404**, 604 (2000).
[3] T. Misteli, A. Gunjan, R. Hock, M. Bustin, and D. T. Brown, *Nature* **408**, 877 (2000).
[4] J. G. McNally, W. G. Muller, D. Walker, R. Wolford, and G. L. Hager, *Science* **287**, 1262 (2000).

experiments a fluorescently tagged protein of interest is introduced into cells and its apparent mobility is measured as an indicator of its dynamic properties. Because nuclear proteins move passively by rapid diffusion through the nuclear space,[2,10] binding of a protein dramatically affects its overall mobility and therefore the measured mobility contains information about the *in vivo* binding properties of a protein.[2,8,9,11] Qualitative analysis of photobleaching data gives an impression of whether a protein binds stably or transiently to chromatin *in vivo*.[2–9]

In addition to the standard qualitative analysis of photobleaching experiments, kinetic modeling methods can be applied for data analysis to permit extraction of quantitative information about simple biophysical properties of chromatin proteins.[11] This method can be used to determine the residence time of a protein on chromatin, the size of the bound and free pools, and the number of kinetically distinct fractions of a protein. These parameters are important in quantitatively describing the *in vivo* behavior and function of a protein, and they are crucial as constraints in models of large-scale biological systems. In this chapter, we describe the basics of photobleaching microscopy, the rationale for using photobleaching methods to obtain information about binding properties, and we describe methods for the qualitative and quantitative analysis of the binding properties of proteins to chromatin *in vivo* by photobleaching methods.

Photobleaching Microscopy

Photobleaching methods were first applied to biological samples in the 1970s.[12,13] The early studies were limited to the investigation of lipids and proteins in the plasma membrane because these types of molecules could easily be fluorescently labeled. Subsequently, fluorescently labeled components of the cytoskeleton were microinjected into cells to study cytoskeleton dynamics by photobleaching methods.[14] With the development of the green fluorescent protein, photobleaching of expressed proteins in the cell

[5] M. A. Lever, J. P. H. Th'ng, X. Sun, and M. J. Hendzel, *Nature* **408,** 873 (2000).
[6] D. L. Stenoien *et al., Nat. Cell Biol.* **3,** 15 (2001).
[7] H. Kimura and P. R. Cook, *J. Cell Biol.* **153,** 1341 (2001).
[8] J. Lippincott-Schwartz, E. Snapp, and A. Kenworthy, *Nat. Rev. Mol. Cell. Biol.* **2,** 444 (2001).
[9] A. B. Houtsmuller and W. Vermeulen, *Histochem. Cell. Biol.* **115,** 13 (2001).
[10] O. Seksek, J. Biwersi, and A. S. Verkman, *J. Cell Biol.* **138,** 131 (1997).
[11] R. D. Phair and T. Misteli, *Nat. Rev. Mol. Cell. Biol.* **2,** 898 (2001).
[12] D. Axelrod *et al., Proc. Natl. Acad. Sci. USA* **73,** 4594 (1976).
[13] M. Edidin, Y. Zagyansky, and T. J. Lardner, *Science* **191,** 466 (1976).
[14] P. A. Amato and D. L. Taylor, *J. Cell Biol.* **102,** 1074 (1986).

interior became feasible.[8] Nowadays, photobleaching is rapidly becoming a standard technique in cell biology laboratories in virtually all fields of research.[8,11]

The most commonly used photobleaching method is fluorescence recovery after photobleaching (FRAP) (Fig. 1). Like all photobleaching methods, FRAP is based on the ability of an intense laser pulse to irreversibly bleach the emission of light from fluorescent molecules. In a typical FRAP experiment, a region of interest containing the fluorescent protein of interest is bleached, using a short laser pulse. The fluorescence intensity in the bleached region is then monitored by standard time-lapse microscopy (Fig. 1A). Because the recovery of fluorescence in the bleached region is due to the influx of unbleached molecules from regions surrounding the bleached area, the rate of fluorescence recovery is a reflection of the overall apparent mobility of the observed protein. If the protein is highly mobile, recovery will be rapid; if the protein is immobile, no recovery will occur (Fig. 1B). It is important to realize that photobleaching does not deplete the bleached region of the protein nor does it change the chemistry of protein–DNA interactions, but rather simply makes the protein molecules invisible. As a consequence, the recovery kinetics are not due to the movement of proteins along an artificial generated gradient, but represent the dynamic behavior of proteins in their undisturbed state.

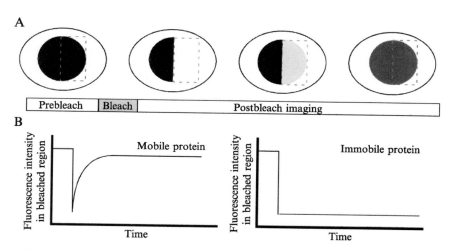

FIG. 1. The principle of fluorescence recovery after photobleaching. (A) In FRAP, a region containing a fluorescently labeled protein of interest is bleached, using a short (250 ms) laser pulse. (B) The recovery of fluorescent signal in the bleached region is then monitored by time-lapse microscopy. The recovery kinetics reflect the apparent mobility of a protein of interest. A highly mobile protein recovers quickly; an immobile protein does not recover.

FRAP to Study Protein Binding

FRAP experiments primarily measure the apparent overall mobility of a protein. In an aqueous environment, the recovery kinetics are a direct reflection of diffusional mobility (Fig. 2A). The case of chromatin proteins is different. Chromatin proteins bind periodically to chromatin and reside on the chromatin fiber. The association of a protein with chromatin slows down its overall mobility (Fig. 2B). This complicates the assessment of the absolute mobility of a protein. However, because the FRAP curve contains information about binding, it can be used to probe the binding of a protein to chromatin *in vivo*.

Proof of this principle of behavior comes from the analysis of pairs of wild-type and mutant proteins, in which the mutant is defective in chromatin binding (Fig. 2C–E). In the case of the linker histone H1, FRAP recovery of the wild-type protein takes several minutes.[3,5] In contrast, an H1 mutant lacking the globular domain, responsible for nucleosomes binding, shows dramatically faster recovery kinetics.[3] Altered FRAP recovery kinetics due to interference with the ability of a protein to bind DNA has been reported for several chromatin proteins.[3,15,16] FRAP thus can be used to obtain information about the binding properties of a protein to native chromatin in living cells.

Collecting FRAP Data

For an FRAP experiment, adherent cultured cells stably or transiently expressing a green fluorescent protein (GFP) fusion protein of interest are plated at ∼50% density into glass-bottom dishes [MatTek (Ashland, MA) or LabTek II chambers (Nalge Nunc, Rochester, NY)] and are grown in their regular growth medium for at least 1 day before the experiment. For imaging, the incubation chamber is placed onto the stage of an inverted confocal microscope capable of executing bleaching routines. Optimal growth temperature can be maintained by use of a Nevus ASI 400 air stream incubator (Nevtek, Burnsville, VA) or a commercially available live cell-imaging chamber (Bioptechs, Butler, PA). The microscope settings should be adjusted such that at least 50–100 images can be acquired without bleaching of more than 10% of the initial signal during imaging. Optimal imaging conditions should be established in pilot experiments. Typical laser settings for the commonly used Zeiss 510 (Carl Zeiss, Thornwood,

[15] T. Cheutin, A. J. McNairn, T. Jenuwein, D. M. Gilbert, P. B. Singh, and T. Misteli, *Science* **299,** 721 (2003).
[16] M. Dundr et al., *Science* **298,** 1623 (2002).

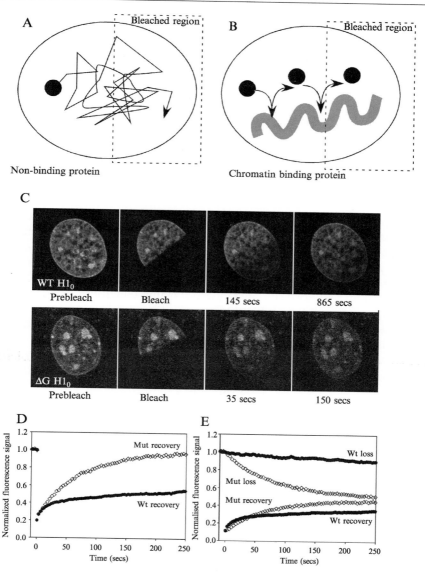

FIG. 2. Analysis of protein binding by FRAP. (A) A freely diffusing protein moves rapidly throughout the nucleus. (B) Binding of a protein to chromatin slows down the overall mobility. (C) Slowed mobility due to chromatin binding can be demonstrated experimentally by FRAP analysis of a wild-type protein and a DNA-binding impaired mutant. For example, half a cell nucleus is bleached in cells stably expressing linker histone $H1_0$ wild-type protein or an $H1_0$ mutant that lacks the globular domain involved in DNA binding. Recovery of the

NY) and imaging of GFP using the 488-nm laser line are 50% output laser power (from a 40-mW laser), 0.1 attenuation, and scan times on the order of 400 ms or less. The bleaching routine is then set up according to the specific instructions for the instrument.

A bleaching routine consists of three phases: prebleach images, bleach pulse, and postbleach images (Fig. 1). Typically, 5–10 prebleach images acquired under standard imaging conditions should be collected to determine the steady state prebleach value of the fluorescence signal in the bleached region and for the operator to ensure that the focal plane is maintained during imaging. The bleach region, bleach duration, and laser intensity during bleaching are then defined using the instrument's software. For analysis of chromatin proteins, we have found it most convenient to use relatively large bleach areas that cover about half the nucleus. In this way, potential heterogeneities from heterochromatin and euchromatin regions or subnuclear structures such as the nucleolus are eliminated. Irrespective of the bleach size chosen, the regions of interest used in experiments in which direct qualitative comparison of FRAP data will be made must be identical. For the postbleach imaging routine, at least 50–100 images should be acquired at standard imaging conditions. If recovery is slow, appropriate intervals between acquired images should be used. Once the entire imaging routine is set up, an appropriate cell is selected and the imaging routine is applied to it.

Several points should be considered when setting up FRAP experiments. First, before use of a fluorescently labeled protein in FRAP procedures, it is critical to establish that addition of the fluorescent tag does not affect the functioning of the protein. Ideally, the fusion protein is introduced into a cell line lacking the protein of interest and the fluorescent protein is assayed for its ability to rescue the knockout phenotype. Whereas this type of complementation assay is easily feasible in yeast systems, it is rarely convenient in mammalian systems. Suitable indicators of protein functionality are correct subcellular localization or the ability to bind to known interaction partners. In the case of chromatin proteins, DNA binding and transcriptional regulation can often be assayed in reporter assays. Second, the bleach duration should be minimized. We have found that depending on the size of the bleached region, bleach durations of 150–400 ms are ideal. Bleaching should occur at full laser intensity. The

wild-type protein is slow, whereas recovery of the mutant protein is fast. Note the different time scales of recovery. Measurements of the average intensity of the fluorescent signal are carried out in the bleached region and the unbleached region. The former provides information about the recovery kinetics, the latter about the loss kinetics of the fluorescent signal. (D) Double normalized recovery kinetics. (E) Single normalized recovery and loss kinetics.

efficiency of bleaching can easily be assessed and normalized between experiments by bleaching of a sample fixed in 4% paraformaldehyde in phosphate-buffered saline (PBS) (15 min, at room temperature). The depth and precision of the bleached region can be determined by a simple linescan measurement across the bleached region. It is advisable to routinely check the bleaching efficiency in this way as laser intensities fluctuate and typically decline with laser age. Bleaching should reduce the fluorescent signal in the bleach region by more than 70% in a fixed sample. Typically, two to five bleach iterations are sufficient to achieve this level of reduction. Third, a frequent observation in FRAP experiments on cellular proteins is a relatively moderate bleach depth (40–50%) in living cells. The most common reason for moderate bleaching is the presence of a rapidly diffusing fraction of the fluorescent protein into the bleached region during acquisition of the first postbleach image. Because the time scale of image acquisition after bleaching is generally slow compared with diffusion, a significant number of unbleached molecules will enter the bleached region while the first image is acquired. The size of this rapidly moving fraction can easily be estimated by comparing the fluorescence intensity in the first postbleach image in a living cell with that in a chemically fixed cell.

FRAP Data Processing

Even qualitative analysis of FRAP data requires computation of normalized recovery curves. To generate a recovery curve, the average intensity in several regions of the cell is measured. Data sets can generally be directly exported from the microscope software and can be manipulated in general spreadsheet software packages (Excel, SigmaPlot, etc.). There are various methods to normalize the recovery curves. We describe two of the commonly used methods here.

Double Normalization

Double normalization involves normalization of the recovery signal to the average prebleach signal and, at the same time, takes into account the loss of total signal due to the bleach pulse and bleaching during postbleach imaging.[2] For double normalization, the average intensity at each imaging time point must be measured for three regions of interest: the bleached region (I_t), the total cell nucleus (T_t), and a random region outside of the cell for background subtraction (BG). When bleaching half the nucleus, the measured regions should be large (at least 50% of the bleached and unbleached regions to account for the heterogeneity in nuclear structure). Normalization occurs in four steps.

1. *Background subtraction:* Calculate for each time point t,

$$I_t - \text{BG} \quad \text{and} \quad T_t - \text{BG}$$

2. *Determination of lost fluorescence signal:* Calculate for each time point t,

$$\frac{(T_{\text{prebleach}} - \text{BG})}{(T_t - \text{BG})}$$

3. *Correction for lost signal due to bleach pulse and bleaching during imaging:* Calculate for each time point t,

$$(I_t - \text{BG}) \frac{(T_{\text{prebleach}} - \text{BG})}{(T_t - \text{BG})}$$

4. *Determination of the relative fluorescence signal in the bleached region:* Calculate for each time point t,

$$\frac{(T_{\text{prebleach}} - \text{BG})(I_t - \text{BG})}{(T_t - \text{BG})(I_{\text{prebleach}} - \text{BG})}$$

Because double normalization takes into account the affect of bleaching, full recovery is expected (Fig. 2D). If the recovery curve does not reach the prebleach value after double normalization, the measured protein is likely present in an "immobile fraction," which is stably bound to a cellular structure. The term "immobile fraction" is somewhat misleading because it describes the behavior of the protein in question only over the period of observation. For example, if an immobile fraction is observed in an experiment in which the postbleach imaging covers a time span of 60 s, a significantly longer observation period of, say, 10 min may reveal that the immobile fraction is simply a slowly exchanging fraction (Fig. 2D).

Single Normalization

An alternative normalization method is single normalization (Fig. 2E). For single normalization, the average intensity of the fluorescence signal in three regions of interest is measured for each time point: The bleached region (I_t), the unbleached region (U_t), and a region outside the cell to determine the background signal (BG). When half the nucleus is bleached, single normalization has the advantage that both the loss of fluorescence signal in the unbleached region and the recovery of fluorescence in the bleached region can be recorded. The two curves should be largely reciprocal. Normalization occurs in two steps.

1. *Background subtraction:* Calculate for each time point t,

$$I_t - BG \quad \text{and} \quad U_t - BG$$

2. *Relative recovery and loss:* Calculate for each time point t,

$$\frac{(I_t - BG)}{(I_{\text{prebleach}} - BG)}$$

$$\frac{(U_t - BG)}{(U_{\text{prebleach}} - BG)}$$

In experiments in which a small region is bleached, it may also be desirable to perform only a single normalization of the data. Single normalization of the data retains information about the fraction of molecules bleached; this adds an important constraint when applying kinetic models to the data (see below). However, this approach may be used only on data sets in which complete recovery has been demonstrated by double normalization.

Regardless of the normalization method applied, several problems may complicate data analysis. A frequent problem is noise in the measurements. The most trivial source of data noise is cell movement or changes in the focal plane during the postbleach imaging phase. Cell movement is highly dependent on the cell type and there is little that can be done to minimize it. One possibility is to use large regions of bleaching. The measurement of the signal recovery can then be carried out in a somewhat smaller region within the bleached region. Changes in the focal plane are often due to temperature effects on the stage. To minimize these, cells should be left for a few minutes to equilibrate at the appropriate temperature on the microscope stage before starting the imaging routine. A further measure to minimize focal drift is the use of a larger pinhole when imaging. Opening the pinhole results in reduction of the confocal effect, but because light is collected from a thicker optical section, the effect of small changes in the focal plane is minimized. Also, the loss of total signal during postbleach imaging before normalization should not be more than 10% of the prebleach signal. A further common observation is that the reproducibility of recovery curves between single cells in the same population is poor. A common cause of this is differential binding properties of proteins during the cell cycle. This can easily be excluded by analysis of synchronized cell populations.

Qualitative Analysis of FRAP Data

Inspection of the recovery curves allows qualitative conclusions regarding the binding of proteins to chromatin. Using this type of approach, it has been demonstrated that core histones H2B, H3, and H4 are virtually immobile and are thus largely statically bound to chromatin.[7] In contrast, the linker histone H1 exchanges dynamically and recovery is on the order of minutes.[3,5] Several chromatin proteins including steroid receptors, transcriptional coactivators, and components of the transcription machinery show recovery kinetics on the order of tens of seconds, suggesting that these proteins bind only transiently to chromatin.[4,6,17–19] These analyses are qualitative and do not provide information about the actual residence time of a protein on chromatin. A more serious limitation is the difficulty of interpreting recovery curves that are multiphasic. In these cases, it is problematic to make accurate estimates as to the residence time of a protein, the number of steady state fractions on chromatin, or the size of the fractions. Simple quantitative analysis is often attempted by fitting of polyexponentials to the recovery curves. Although this gives a reasonable estimate of the minimal number of kinetic fractions present (see below), because of the possibility of multiple competing processes occurring simultaneously, this method does not easily provide accurate residence times and binding rate constants. To be certain these crucial properties of chromatin proteins are extracted accurately, a somewhat more sophisticated quantitative analysis using kinetic modeling methods is required.

Quantitative FRAP Analysis

The goal in the following sections is to detail the process of FRAP data analysis in the context of a hypothesized kinetic model of chromatin protein binding in the nucleus.[11] A kinetic model is a quantitative statement, based on biochemical and biophysical principles, of a working mechanistic hypothesis about a biological system. The process of extracting quantitative parameters from imaging data, using kinetic models, consists of four steps. First, the mechanistic hypothesis is translated into a mathematical description of a kinetic model. Second, a simulation of the experimental protocol is applied to the kinetic model. This yields quantitative predictions that can be compared with experimental data. Third, in order to give each hypothesis its best chance to account for the observed data, numerical optimization tools are used for parameter estimation. Hypotheses that are inconsistent with the data even after optimization are rejected. Fourth, parameter values consistent with the model are extracted. This provides useful additional information of biological interest. The extracted parameter

[17] M. Becker et al., EMBO Rep. **3,** 1188 (2002).
[18] J. Essers et al., EMBO J. **21,** 2030 (2002).
[19] A. B. Houtsmuller, S. Rademakers, A. L. Nigg, D. Hoogstraten, J. H. Hoeijmakers, and W. Vermeulen, Science **284,** 958 (1999).

values are inherently dependent on the particular model used. If this model is later rejected because it is found inconsistent with newer experimental data, then the parameters, too, must be reevaluated.

Approximation of Well-Mixed Nucleoplasm

The apparent mobility of a protein as measured by FRAP is affected by two major components: diffusional mobility between binding events and the binding events themselves. The simplest kinetic data analysis approach involves the assumption that the chromatin protein of interest is kinetically well mixed in the nucleoplasm. This simply means that if we were able to measure the free pool of the protein, we would obtain the same numerical value no matter where in the nucleoplasm we chose to sample. Importantly, this must be true not only in steady states but also during transient states such as those created by photobleaching. Although the assumption of a well-mixed compartment is always fulfilled when analyzing FRAP data from relatively small (1–2 μm) spot bleaches, it must be tested on a case-to-case basis when using larger bleach areas. There are two factors that strongly influence this assumption: diffusion and the relative number of binding sites for the protein.

Diffusion of nuclear proteins is fast on the time scale of a typical half-FRAP experiment. Nonbinding proteins diffuse rapidly in the nucleoplasm, with reported diffusion coefficients on the order of 10–40 μm^2 s^{-1}.[2,10,19] This means that if the protein is free to diffuse, it will require on average less than 2 s to cross a nucleus 8 μm in diameter. Half-FRAP experiments for chromatin proteins display recoveries over time frames of tens or hundreds of seconds. Consequently, for these proteins, diffusion is not rate limiting. As far as diffusion is concerned, the approximation of a well-mixed nucleoplasmic pool is valid.

The second determinant of nucleoplasmic mixing is the relative abundance of binding sites. If a protein of interest is confronted with a large number of available binding sites, and if those binding sites are in large excess over the total number of chromatin protein molecules, a photobleaching experiment can yield long-lasting gradients of fluorescent molecules that are the hallmark of incomplete mixing.

An approach to evaluate the validity of the assumption of a well-mixed compartment is to bleach half the cell nucleus and to record recovery in three regions of interest in the bleached half of the nucleus, spaced evenly on a line approximately perpendicular to the bleach boundary (Figs. 3 and 4). Region 1 is chosen adjacent to the bleach boundary; region 3 is adjacent to the nuclear envelope, as far as possible from the boundary; and region 2 lies between these extremes. Figure 3 illustrates two simulated

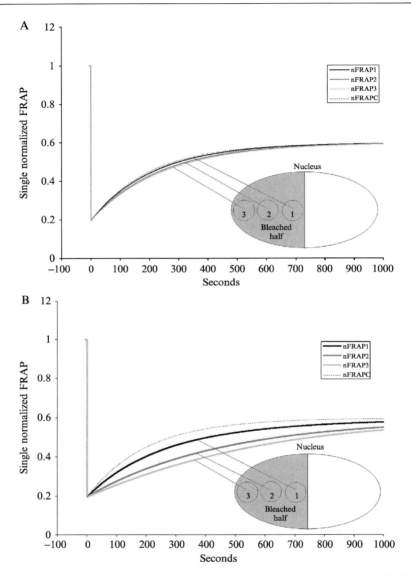

FIG. 3. Full binding-diffusion simulation of half-FRAP compared with a well-mixed approximation. (A) Simulated recoveries (solid lines) in three regions, indicated in the legend, for the case in which the number of binding sites is 50% greater than the number of chromatin proteins. Also shown is a corresponding simulation for the well-mixed approximation (dotted line), using the same value for binding site residence time (200 s). Inset diagram shows the half-FRAP protocol and the distribution of measured regions in the bleached half of the nucleus. (B) Corresponding simulations for the case in which the number of binding sites is 5-fold greater than the number of chromatin protein molecules.

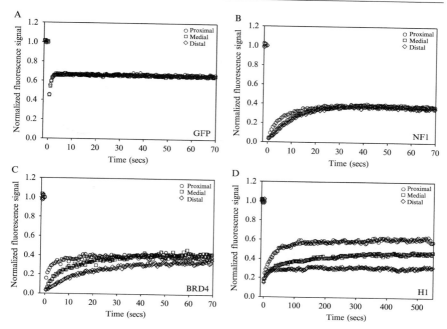

FIG. 4. FRAP recovery kinetics are measured in three regions at increasing distances from the bleach boundary. If the nucleoplasmic pool of the protein is well mixed, the recovery curves are superimposable as for (A) GFP and (B) NF1. Differences among the recovery curves such as those observed for (C) BRD4 and (D) H1 indicate that the protein is not well mixed in the nucleoplasm.

outcomes for this protocol based on a full computational model of diffusion and binding where $D = 20\ \mu m^2\ s^{-1}$, and binding is characterized by a mean residence time of 200 s. The difference between these two simulations is the relative abundance of chromatin proteins and binding sites. In Fig. 3A, there are 50% more binding sites than chromatin proteins; in Fig. 3B, there are five times as many binding sites as chromatin proteins. Notice that as the number of binding sites increases relative to potential ligands, the predicted FRAPs in the three regions are more widely separated. Also plotted in Fig. 3 is a dotted line showing the corresponding simulation for the well-mixed approximation. These simulations indicate that if the number of binding sites is less than or approximately the same as the number of chromatin protein molecules, the well-mixed approximation is a valid simplification. This is clear because the dotted curve in Fig. 3A is nearly identical to the three curves representing FRAP responses at increasing distances from the bleach boundary. In Fig. 3B, however, the dotted curve falls substantially above the family of FRAP responses and consequently the well-mixed approximation is not ideal.

The above-described sampling of multiple regions in a bleached area serves as a simple experimental test for the validity of the assumption of a well-mixed nuclear compartment. If comparison of the three half-FRAP recovery curves measured at increasing distances from the bleach boundary shows they are approximately superimposable, then the well-mixed assumption is sound and the analysis in the following sections can be confidently applied. If, on the other hand, the data appear more like the plot in Fig. 3B, then the well-mixed approximation will underestimate k_{off} and will overestimate residence time on the binding site. Shown in Fig. 4 are experimental data for GFP alone and three chromatin proteins displaying a range of possible outcomes. Although the assumption of a well-mixed compartment is valid for GFP alone and for NF1, BRD4 represents a borderline case with some deviation from the similarity of the three curves, whereas for a protein with a long residence time on the order of minutes and a large excess of free binding sites, such as the linker histone H1, the assumption of a well-mixed compartment should not be made (Fig. 4). Gradients like those in Fig. 3A result in a 13% overestimation of residence time, whereas applying the approximation in a case like Fig. 3B results in a 2.1-fold overestimate.

The finding that, for many chromatin proteins, the nucleoplasmic free pool can be approximated as kinetically homogeneous or well mixed greatly simplifies and speeds the kinetic analysis. It also allows information about binding to be extracted directly from photobleaching data by permitting us to adopt the tools of chemical kinetics for chromatin protein-binding data.

Kinetic Analysis of Photobleaching Experiments for
  Chromatin Proteins

Estimation of Number of Binding Site Classes

A first step in generating a kinetic model of chromatin protein binding is to determine how many classes of binding sites can be resolved from the data. This can be done with software tools capable of fitting data to sums of exponentials and providing statistical information about the parameter estimates. We recommend SAAM II[20,21] (SAAM Institute, Seattle WA; www.saam.com), but other tools, such as ADAPT (USC Biomedical Simulations Resource; http://bmsr.usc.edu/Software/Adapt/overview.html),

[20] P. H. Barrett *et al.*, *Metabolism* **47,** 484 (1998).
[21] C. Cobelli and D. M. Foster, *Adv. Exp. Med. Biol.* **445,** 79 (1998).

have similar features. Single- or double-normalized FRAP data can be fitted to a sum of exponentials to determine how many classes of binding sites exist. For example, the normalized data are fitted to $1 - Ae^{-at}$ to test the hypothesis that one class of binding sites can be resolved. Discussion of judging goodness of fit is outside the scope of this review, but the following criteria can be used (see Refs. 22–24 for more details). First, the experimental data and the exponential fit are plotted on the same axes and inspected for systematic deviations—areas where the fitted line is consistently above or below the data. The goal is to obtain a fit for which the residuals are randomly distributed about zero. A statistical measure of this is the classic runs test.[23] If a systematic deviation is observed, a further exponential term is added and the data are refit to $1 - Ae^{-at} - Be^{-bt}$. This is repeated until a good fit is obtained. Once systematic deviations are eliminated, the coefficients of variation (standard deviation divided by the mean) for the estimated values of A, a, B, and b are examined. If these are less than 30%, the fit is accepted. For FRAP data averaged over 10 or more cells, it is generally possible to obtain coefficients of variation less than 20%.

A frequently asked question concerns how one can be sure that no further classes of binding sites—additional components in an exponential fit—are needed. First, it may well be that there is a continuous distribution of binding affinities for the protein of interest in the nucleus. A fit to a sum of exponentials aims to identify the minimum model that can account for the available experimental data. It does not rule out greater complexity. Instead, it yields a model that is just complex enough. If more binding sites are postulated than are necessary, the coefficients of variation for the added parameters will be greater than 50%, perhaps even greater than 100%, because the optimizer will be unable to determine which of most possible fits is the correct one. In our experience, FRAP experiments for most chromatin proteins can be resolved into two exponential components. For this reason, in the section below we assume that two classes of binding sites are to be resolved. Analogous approaches are applied to proteins with fewer or more binding sites.

Developing Kinetic Model

For each chromatin protein, we know, on thermodynamic grounds, that in addition to the bound pools of protein, there must also exist a free pool

[22] J. A. Jacquez, "Modeling with Compartments." Biomedware, Ann Arbor, MI, 1999.

[23] W. J. Dixon and F. J. Massey, "Introduction to Statistical Analysis," 4th Ed. McGraw-Hill, New York, 1983.

[24] R. B. D'Agostino and M. A. Stephens, "Goodness-of-Fit Techniques." Marcel Dekker, New York, 1986.

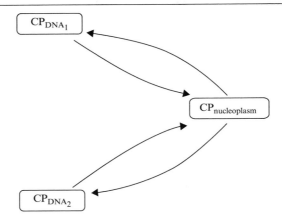

FIG. 5. Initial model. Chromatin proteins (CP) bound to two classes of DNA-binding sites and exchanging with a free nucleoplasmic pool of chromatin protein are shown. Arrows represent processes of binding and unbinding.

of the protein in the nucleoplasm. Consequently, an initial kinetic model is the one shown in Fig. 5.

Evaluation of the rate constants in this model will provide answers to the principal biological questions: what is the mean residence time of a chromatin protein on a particular site, and what is the relative abundance of the two classes of binding sites? Dissociation of a chromatin protein from a DNA-binding site is a first-order process. This means that the dissociation flux (molecules per second) is proportional to the amount bound. The rate law for this process is thus $k_{off_1} \times CP_{DNA_1}$, where k_{off_1} is the rate constant, and CP_{DNA_1} is the mass of chromatin protein bound to the first class of DNA-binding sites. We recommend expressing compartmental masses in units of abundance (molecules per cell). When formulated in this way, unknown volumes of distribution are combined into the rate constants. If volumes are known to be changing on the time scale of the experiment, then a separate differential equation for volume is required and the volume must be factored into the rate laws appropriately.

Things are not so simple for the "on" process. Because this is a binding event, chemical kinetics dictate that the process is second order. Its rate law would thus be $k_{on_{1}so} \times DNA_1 \times CP_{nucleoplasm}$, where $k_{on_{1}so}$ is the second-order rate constant, DNA_1 is the mass of available binding sites, and $CP_{nucleoplasm}$ is the mass of free chromatin protein in the nucleoplasm. If we planned to examine a non-steady state situation, we would have to adopt this rate law because the masses would be changing with time, but GFP-tagged proteins are tracers and photobleaching is assumed not to modify the underlying steady state. This means that for the GFP-tagged chromatin protein, the rate law for binding is also first order: $k_{on_1} \times$ $CP_{nucleoplasm}$. The constant amount of DNA1 has simply been combined with $k_{on_1 so}$ to give an effective first-order rate constant, k_{on_1}.

Modeling Experimental Bleaching Protocol

To apply photobleaching protocols to the model described in the previous section, we must account separately for molecules that are in the bleached area and for molecules that are in the unbleached area of the nucleus. This is accomplished as shown in Fig. 6. Rate constants k_{off_1} and k_{off_2} are the same for bleached and unbleached areas because they characterize exactly the same process in two different places. k_{on_1} is apportioned between the bleached and unbleached areas by multiplying by the fraction of the nucleus that is bleached and the fraction that is unbleached, respectively. k_{on_2} is apportioned similarly. This accounts for the relative areas and molecular abundances of the bleached and unbleached regions.

One step remains to complete the model so that it faithfully represents the experimental protocol. Photobleaching effectively provides a new loss pathway for GFP-labeled proteins, but only in areas subjected to the bleaching laser pulse. This feature of the protocol is included in the model by adding the three new process arrows that yield bleached CP. Photobleaching is an exponential process,[25] so the bleaching arrows in Fig. 6 are characterized by rate constants that are proportional to bleach intensity

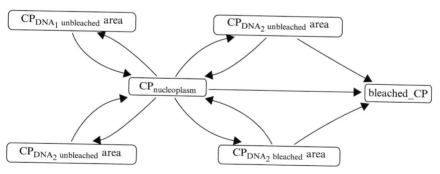

FIG. 6. Complete model of chromatin protein binding to two classes of DNA-binding sites with superimposed bleaching protocol. This model can be used to analyze data from FRAPs of small areas and half-FRAPs by changing the apportionment of k_{on_1} and k_{on_2} to reflect the size of the bleached area. The experimental data are fitted to the sum of the two "bleached area" compartments.

[25] D. Axelrod, D. E. Koppel, J. Schlessinger, E. Elson, and W. W. Webb, *Bio Phys. J.* **16,** 1055 (1976).

or laser power. These rate constants (k_{bleach}) are zero when the bleaching laser is off and large when it is on. As long as $k_{bleach} \gg k_{off_1} > k_{off_2}$, virtually every GFP that is present in the bleached area or enters the bleached area during the bleaching pulse will be bleached and will no longer contribute to the observed fluorescence in the nucleus. Because there is free protein in nucleoplasm within the bleached area, we also include a bleaching rate constant for the nucleoplasmic pool. k_{bleach} for the nucleoplasmic pool is reduced by the factor (bleached area/nuclear area) because only nucleoplasm within the bleached area is exposed directly to the bleaching pulse. However, because the nucleoplasm is treated as well mixed, it will be emptied relatively quickly by all but the shortest bleach pulses. Moreover, for most chromatin proteins, only 10% or less of the total protein is free in the nucleoplasm, so this pool makes only a small contribution to the observed signal.

Required Equations

Having formulated a model for two DNA-binding sites and modified that model to include photobleaching protocols, it remains to indicate the equations that have been used to solve this data analysis problem. Differential equations for the number of fluorescent GFP-tagged protein molecules in each compartment are as follows.

$$\frac{dCP_{DNA_1 unbleached}}{dt} = f_{unbleached} k_{on_1} CP_{nucleoplasm} - k_{off_1} CP_{DNA_1 unbleached}$$

$$\frac{dCP_{DNA_1 bleached}}{dt} = f_{bleached} k_{on_1} CP_{nucleoplasm} - k_{off_1} CP_{DNA_1 bleached}$$
$$- k_{bleach} CP_{DNA_1 bleached}$$

$$\frac{dCP_{DNA_2 unbleached}}{dt} = f_{unbleached} k_{on_2} CP_{nucleoplasm} - k_{off_2} CP_{DNA_2 unbleached}$$

$$\frac{dCP_{DNA_2 bleached}}{dt} = f_{bleached} k_{on_2} CP_{nucleoplasm} - k_{off_2} CP_{DNA_2 bleached}$$
$$- k_{bleach} CP_{DNA_2 bleached}$$

$$\frac{dCP_{nucleoplasm}}{dt} = - k_{off_1}(CP_{DNA_1 bleached} + CP_{DNA_1 unbleached})$$
$$+ k_{off_2}(CP_{DNA_2 bleached} + CP_{DNA_2 unbleached})$$
$$- k_{on_1} CP_{nucleoplasm} - k_{on_2} CP_{nucleoplasm}$$
$$- f_{bleached} k_{bleach} CP_{nucleoplasm}$$

$$\frac{dCP_{bleached}}{dt} = k_{bleach}(CP_{DNA_1 bleached} + CP_{DNA_2 bleached})$$
$$+ f_{bleached} CP_{nucleoplasm})$$

where CP is chromatin protein, subscripts DNA_1bound and DNA_2bound indicate the two classes of chromatin protein-binding sites, subscripts *bleached* and *unbleached* indicate the regions exposed to and not exposed to the bleaching laser pulse, $f_{bleached}$ and $f_{unbleached}$ are the corresponding fractions of nuclear area, subscript *nucleoplasm* indicates the free pool of chromatin protein in the nucleoplasm, k_{on_1} and k_{on_2} are the effective first-order rate constants for chromatin protein binding to sites 1 and 2, respectively, k_{off_1} and k_{off_2} are the rate constants for dissociation of chromatin protein from sites 1 and 2, respectively, and k_{bleach} is the rate constant characterizing the photobleaching process.

Photobleaching protocols are implemented by first measuring $f_{bleached}$ and $f_{unbleached}$. For a typical spot FRAP, these might be 0.05 and 0.95, respectively. For the half-bleach experiments, both are 0.5. For a bleaching protocol that begins at t_{start} and is complete at t_{end}, the SAAM II "change conditions" facility is used to implement the following definition of k_{bleach}:

$$k_{bleach} = \begin{cases} 0 & t < t_{start} \\ k_{bleach}^{on} & t_{start} \le t \le t_{end} \\ 0 & t > t_{end} \end{cases}$$

For any given set of parameter values, relative steady state abundances for each chromatin protein in each compartment are calculated by setting $k_{bleach} = 0$ and including both constant synthesis and protein turnover ($CP_{synth} - k_{deg} CP_{nucleoplasm}$) on the right-hand side of the $\frac{dCP_{nucleoplasm}}{dt}$ equation, and then solving the algebraic equations that result when all the derivatives are set to zero. This procedure is automatic in the SAAM II software, and is a widely available numerical tool.

Some software tools, including SAAM II, have the ability to initialize each state variable with its precalculated steady state value. This is useful, but is not an absolute requirement. An alternative is to initialize just one compartment with the total abundance and solve the differential equation system until it relaxes to a steady state. With either approach, the beginning of the bleach, t_{start}, can then be set to any time after steady state is achieved because we assume that the bleach takes place in a steady state nucleus.

Experimental Data Constraints and Parameter Optimization

To extract useful information from the photobleaching data, we must find parameters that simultaneously account for both the photobleaching kinetics and the initial steady state distribution between bound and free chromatin protein. On our assumption that the nucleoplasm is well mixed, a large fraction of the unbound fluorescent chromatin proteins will be bleached. This permits estimation of the free fraction, f_{free}, from the difference between total prebleach fluorescence and fluorescence in the unbleached region in the first postbleach image. Numerical values of f_{bleached}, $k^{\text{on}}_{\text{bleach}}$, k_{on_1}, k_{off_1}, k_{on_2}, and k_{off_2} are adjusted using standard numerical optimization procedures to account for the experimental data. Normalized recovery kinetics are fitted to

$$F(t) = \frac{\text{CP}_{\text{DNA}_1\text{bleached}} + \text{CP}_{\text{DNA}_2\text{bleached}} + f_{\text{bleached}}\text{CP}_{\text{nucleoplasm}}}{\text{CP}^{ss}_{\text{DNA}_1\text{bleached}} + \text{CP}^{ss}_{\text{DNA}_2\text{bleached}} + f_{\text{bleached}}\text{CP}^{ss}_{\text{nucleoplasm}}}$$

where the superscript ss indicates the steady state value. Notice that absolute values of these steady state abundances are not required, so we do not need to know either CP_{synth} or k_{deg}. If desired, the decline of fluorescence in the unbleached area can be fitted simultaneously to the same function, where all the "bleached" subscripts are changed to "unbleached."

The function $F(t)$ above is used when the experimental data have been single normalized, but investigators often prefer double normalization because this accounts for loss of bleached molecules and avoids the appearance of an artifactual immobile fraction. This normalization complicates data analysis because it requires knowledge of a normalization factor that cannot be calculated until the simulation is complete and the extent of the bleach is known. This difficulty can be overcome by defining

$$f_{\text{remaining}}(t) = \frac{\text{CP}_{\text{tot}} - \text{CP}_{\text{bleached}}(t)}{\text{CP}_{\text{tot}}}$$

Double-normalized data can then be fitted to $\dfrac{F(t)}{f_{\text{remaining}}(t)}$.

The steady state fraction free is simultaneously fitted to

$$f_{\text{free}} = \frac{\text{CP}^{ss}_{\text{nucleoplasm}}}{\text{CP}^{ss}_{\text{DNA}_1\text{unbleached}} + \text{CP}^{ss}_{\text{DNA}_2\text{unbleached}} + \text{CP}^{ss}_{\text{DNA}_1\text{bleached}} + \text{CP}^{ss}_{\text{DNA}_2\text{bleached}} + \text{CP}^{ss}_{\text{nucleoplasm}}}$$

When the data permit estimation of the parameters with coefficients of variation less than 30%, biologically interesting information can be extracted. Mean residence times for fast and slow binding sites are

$$T_{\text{res}_1} = 1/k_{\text{off}_1}$$
$$T_{\text{res}_2} = 1/k_{\text{off}_2}$$

Fractions of total binding which is associated with fast (1) or slow (2) binding sites are

$$f_{\text{bound}_1} = \frac{\text{CP}^{ss}_{\text{DNA}_1\text{unbleached}} + \text{CP}^{ss}_{\text{DNA}_1\text{bleached}}}{\text{CP}^{ss}_{\text{DNA}_1\text{unbleached}} + \text{CP}^{ss}_{\text{DNA}_1\text{bleached}} + \text{CP}^{ss}_{\text{DNA}_2\text{unbleached}} + \text{CP}^{ss}_{\text{DNA}_2\text{bleached}}}$$

$$f_{bound_2} = \frac{CP^{ss}_{DNA_2 unbleached} + CP^{ss}_{DNA_2 bleached}}{CP^{ss}_{DNA_1 unbleached} + CP^{ss}_{DNA_1 bleached} + CP^{ss}_{DNA_2 unbleached} + CP^{ss}_{DNA_2 bleached}}$$

Detailed exposition of the optimization process is a topic in numerical analysis, and the interested reader is referred to a standard textbook[26] and to a full-length article on the use of SAAM II.[21] Briefly, optimization is a numerical procedure that systematically adjusts model parameter values in order to optimize (often minimize) an objective function. This objective is a function of the differences between the experimental data points and the corresponding simulated values. The simplest objective function is the sum of squares of the errors (residuals) between the experimental data and the simulated values. Data weighting is an important issue in optimization, and there are many approaches to this subject. We have found that weighting a data point proportional to its numerical value (i.e., assigning a fractional standard deviation) is an effective approach. This choice will become even more important if methods are developed for measuring GFP fluorescence intensities over several orders of magnitude despite the presence of cellular autofluorescence.

Optimization can be seen as a procedure for determining the noise that is present in a parameter estimate because of noise in the experimental data. The noisier the data, the noisier the parameter estimates. This is why better data yield more robust parameter estimates; it is also why smaller coefficients of variation are better than larger ones. Data analysis aims at coefficients of variation less than 20%, but this is not always attainable for a given model. If, after optimization, the coefficients of variation are 100 or even 500%, it is apparent that the model is too complex or that the data being fitted do not contain sufficient information about all the parameters of the model. Those parameters must either be fixed on the basis of *a priori* knowledge or removed from the model. As an example of an optimized fit and parameter extraction, Fig. 7 shows the optimized fit of a typical half-bleach data set for a fusion protein between GFP and the transcription factor Jun.

Parameters extracted from this fit were $f_{bleached} = 0.705$, $k_{off_1} = 0.168 \text{ s}^{-1}$, $k_{on_1} = 0.624 \text{ s}^{-1}$, $k^{on}_{bleach} = 7.67 \text{ s}^{-1}$, $k_{off_2} = 0.0366 \text{ s}^{-1}$, and $k_{on_2} = 0.524 \text{ s}^{-1}$. Coefficients of variation for these parameter estimates were 0.3, 6.5, 12.1, 0.6, 1.2, and 8.2%, respectively. Using the equations for residence times and fractions bound, and substituting the steady state abundances that correspond to these rate constants, we obtain $T_{res_1} = 5.94 \text{ s}$, $T_{res_2} = 27.3 \text{ s}$, $f_{bound_1} = 0.206$, and $f_{bound_2} = 0.794$. Coefficients of variation on these estimates were 6.5, 1.2, 4.5, and 1.2%, respectively.

[26] J. Nocedal and S. J. Wright, "Numerical Optimization." Springer-Verlag, New York, 1999.

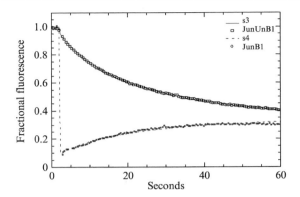

FIG. 7. Optimized fit of half-FRAP experimental data for Jun–GFP. Half-bleach of a nucleus expressing Jun–GFP begins at $t = 2$. The rapid decline in fluorescence in the bleached region (small circles) is followed by a recovery as unbleached molecules move from the unbleached area. The corresponding decline in the unbleached region (open squares) has similar kinetics. Simultaneously optimized model fits are shown for both bleached (dash line) and unbleached (solid line) regions of the nucleus. Plot is from SAAM II.

Summary

We have described procedures for collecting, processing, and analyzing kinetic data obtained by photobleaching microscopy of GFP-tagged chromatin proteins in nuclei of cultured living cells. These procedures are useful for characterizing the *in vivo* binding of chromatin proteins to their natural template—unperturbed, native chromatin in an intact cell nucleus. These techniques have revealed several generalizations that significantly change our view of the nucleus. At the qualitative level, it has become clear that almost all chromatin proteins bind only transiently to their targets. More importantly, the combined use of *in vivo* microscopy and kinetic, computational analysis allows analysis of the kinetics of protein binding *in vivo*. These methods should prove useful in the further *in vivo* investigation of the molecular mechanisms involved in genome organization and expression.

[26] Quantification of Protein–Protein and Protein–DNA Interactions *In Vivo,* Using Fluorescence Recovery after Photobleaching

By Gustavo Carrero, Ellen Crawford, John Th'ng, Gerda de Vries, and Michael J. Hendzel

Introduction

Fluorescence recovery after photobleaching (FRAP) is a fluorescence microscopy method that enables the quantification of protein movement over time in living cells.[1–6] With the addition of spatial information afforded by the fluorescence microscopy platform for FRAP, these measurements can be used to distinguish between a protein that diffuses through the cytoplasm as a monomer and one that is incorporated into a more massive protein complex, binding and dissociation constants, and binding specificity, all within the native environment of the living cell. It is an important complementary tool to techniques such as chromatin immunoprecipitation and biochemical purification, and, when properly controlled, provides a "gold standard" for the interpretation of the biochemistry of the genomic chromatin.

The FRAP technique was developed during the 1970s for studying molecular diffusion. Its first success in a living cell setting was in the study of the diffusion of proteins within cellular membranes.[7–10] More recently, FRAP has emerged as a powerful approach to study the behavior of nuclear proteins *in vivo*. The development of fluorescent protein technology and the general availability of laser scanning confocal microscopes allow widespread access to the specialized instrumentation required. As we discuss in this chapter, the vast potential of FRAP to address biochemical and

[1] A. S. Verkman, *Trends Biochem. Sci.* **27,** 27 (2002).

[2] N. Klonis, M. Rug, I. Harper, M. Wickham, A. Cowman, and L. Tilley, *Eur. Biophys. J.* **31,** 36 (2002).

[3] J. Lippincott-Schwartz, E. Snapp, and A. Kenworthy, *Nat. Rev. Mol. Cell. Biol.* **2,** 444 (2001).

[4] A. B. Houtsmuller and W. Vermeulen, *Histochem. Cell Biol.* **115,** 13 (2001).

[5] M. J. Hendzel, M. J. Kruhlak, N. A. MacLean, F. Boisvert, M. A. Lever, and D. P. Bazett-Jones, *J. Steroid Biochem. Mol. Biol.* **76,** 9 (2001).

[6] R. D. Phair and T. Misteli, *Nat. Rev. Mol. Cell. Biol.* **2,** 898 (2001).

[7] W. W. Webb, L. S. Barak, D. W. Tank, and E. S. Wu, *Biochem. Soc. Symp.* **46,** 191 (1981).

[8] K. Jacobson, *Cell Motil.* **3,** 367 (1983).

[9] N. O. Petersen, *Can. J. Biochem. Cell Biol.* **62,** 1158 (1984).

[10] M. Schindler, J. F. Holland, and M. Hogan, *J. Cell Biol.* **100,** 1408 (1985).

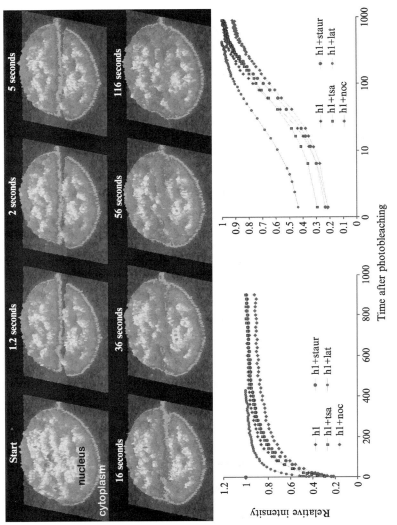

Fig. 1. (continued)

functional questions about proteins in their native environment, the living nucleus, makes this a valuable research tool in the study of chromatin structure, function, and dynamics.

Because FRAP is new to investigators of chromatin function, the method has initially been employed to ask simple questions, such as, are proteins stably bound to chromatin and if not, how long do they remain bound before dissociating? Using this approach, we[11] and others[12–14] have demonstrated that histone H1 moves on and off chromatin quite rapidly (Fig. 1), whereas the histones that compose the nucleosome are much more stable components of chromatin.[11,14,15] While some of these conclusions could be gathered from radiolabeling experiments performed during the 1970s and 1980s,[16] FRAP allowed, for the first time, quantitative estimates of binding dynamics in living cells. Thus, kinetic analysis of proteins in living cells is no longer restricted to the analysis of protein synthesis and turnover. In this chapter, we outline the important principles in designing, analyzing, and interpreting FRAP experiments. We restrict our focus to designing and analyzing experiments performed on proteins that associate with chromatin. The remainder of the introduction provides a brief description of the nuclear environment and the scale, in molecular terms, of chromatin organization within living interphase nuclei. This provides the background for the design and analysis of a FRAP experiment. Because the power of this approach resides in the combination of experimentation

[11] M. A. Lever, J. P. Th'ng, X. Sun, and M. J. Hendzel, *Nature* **408**, 873 (2000).
[12] Y. Dou, J. Bowen, Y. Liu, and M. A. Gorovsky, *J. Cell Biol.* **158**, 1161 (2002).
[13] M. Gong, J. H. Ni, and H. T. Jia, *Cell Res.* **12**, 395 (2002).
[14] T. Misteli, A. Gunjan, R. Hock, M. Bustin, and D. T. Brown, *Nature* **408**, 877 (2000).
[15] H. Kimura and P. R. Cook, *J. Cell Biol.* **153**, 1341 (2001).
[16] M. J. Hendzel and J. R. Davie, *Biochem. J.* **271**, 67 (1990).

FIG. 1. Fluorescence recovery after photobleaching of chromatin-associated histone H1.1. *Top:* SK-N-SH neuroblastoma cells expressing eGFP–histone H1.1 were examined by photobleaching a 1-μm-wide band across the nucleus (see 1.2-s image). As time elapses, this band of photobleached chromatin can be seen to increase in intensity (reflected in the dark blue being replaced by green and yellow). This occurs through the exchange of fluorescence with the remainder of the nuclear chromatin. *Bottom:* Plots of histone H1 recovery under control conditions or after treatment with a histone deacetylase inhibitor (tsa), an inhibitor of protein kinases (staur), an inhibitor of actin polymerization (lat), or an inhibitor of microtubule polymerization (noc). In the left-hand plot, time is shown on a linear scale. This allows for the easy identification of differences in the time required to reach equilibrium. In the right-hand plot, time is plotted on a log scale. This allows the resolution of differences that are occurring in the first few seconds of the experiment. These are primarily differences in diffusion. (See color insert.)

and mathematical modeling, particular attention will be paid to the design and analysis of the experimental data.

Topology of Living Cell Nucleus

Chromatin is a biological polymer with properties distinct from the well-known biological polymers of the cytoskeleton.[17–21] Actin filaments, microtubules, and intermediate filaments assemble end-to-end to establish helical arrays of individual units. Chromatin is based on a polymeric backbone composed of two antiparallel strands, each a complete and independent polymer of deoxyribonucleotides. These polymers are extraordinarily long, comprising millions of base pairs of DNA. At this level, chromatin is not unlike the biopolymers of the cytoskeleton. It differs, however, by the incorporation of an additional layer of material, the chromatin proteins. The chromatin proteins are primarily deposited as a repeating octamer of histones that, when complexed with 147 bp of DNA, is termed the nucleosome. These units are repeated at intervals of 200 base pairs or less and are connected by short intervening sequences of DNA termed linker DNA. Hence, chromatin is also a polymer of nucleosomes that is based on a deoxyribonucleotide polymer. It is the protein component of the nucleosome polymer that allows chromatin to adopt complex and regulated states of folding. This requires contacts between sites that are linearly distant within the nucleotide sequence, something that is not observed in the biopolymers of the cytoskeleton.

The genome is composed of several distinct and independent polymers of nucleosomes. These are termed chromosomes and may vary in number and size between eukaryotic species. Regardless of size or number, each chromosome forms a spatially independent territory within the nucleoplasm.[22–24] The chromosome territory reflects the unique property of the nucleosome polymer to engage in long-range (nonlinear) interactions that allow for the establishment of a complex three-dimensional topology within both the chromosome and the nucleus. The ability of nucleosomes

[17] J. Ostashevsky, *Mol. Biol. Cell* **13,** 2157 (2002).

[18] G. Wedemann and J. Langowski, *Biophys. J.* **82,** 2847 (2002).

[19] K. Rippe, *Trends Biochem. Sci.* **26,** 733 (2001).

[20] D. A. Beard and T. Schlick, *Structure (Camb.)* **9,** 105 (2001).

[21] Y. Cui and C. Bustamante, *Proc. Natl. Acad. Sci. USA* **97,** 127 (2000).

[22] T. Cremer and C. Cremer, *Nat. Rev. Genet.* **2,** 292 (2001).

[23] H. Tanabe, S. Muller, M. Neusser, J. von Hase, E. Calcagno, M. Cremer, I. Solovei, C. Cremer, and T. Cremer, *Proc. Natl. Acad. Sci. USA* **99,** 4424 (2002).

[24] P. Edelmann, H. Bornfleth, D. Zink, T. Cremer, and C. Cremer, *Biochim. Biophys. Acta* **1551,** M29 (2001).

to interact with other nucleosomes over megabase pair distances is a reflection of the propensity of nucleosomes to self-associate into higher order organizations of the basic nucleosome repeat. The principal players in defining these long-range interactions between regions of the chromosome are the amino-terminal domains of the nucleosomal histones.[25,26] Posttranslational modification of these domains provides a regulated mechanism for modulating the affinity and number of interactions between nucleosomes.[25,26]

The individual chromosomes are composed of individual segments, cytogenetically defined as chromosome bands, where the chromatin structure is similar over hundreds of kilobase pairs of sequence. These can be generally classed into euchromatin (R bands) and heterochromatin (G bands).[27] During interphase, alternating segments of the chromosome segregate differently within the nuclear volume.[5,22,27,28] The heterochromatic regions are found in two specific regions of the nucleoplasm: the perinucleolar space and lining the nuclear envelope. The interior of the nucleoplasm is predominated by structures that are best known for their property of being enriched in splicing factors and polyadenylated RNAs.[5,29,30] Surrounding these regions of the nucleoplasm are the R bands, containing the euchromatin component of the genome.[5,29–31] Because of the spatial stability of both the chromatin and these splicing factor compartments during the course of interphase, these distributions can be considered to persist for extended periods of time rather than represent snapshots of dynamic spatial relationships.

It has become apparent that virtually all nuclear proteins show some degree of spatial compartmentalization within the nucleoplasm.[5] There are several spatial features that are important to be aware of when performing FRAP experiments on nonhistone chromatin proteins, particularly transcriptional machinery. First, the euchromatin is enriched on the periphery of intranuclear domains that exclude chromatin. In contrast, the heterochromatin is enriched on the inner surface of the nucleus and on the outer surfaces of the nucleoli. This is illustrated in Fig. 2A, where

[25] J. C. Hansen, *Annu. Rev. Biophys. Biomol. Struct.* **31,** 361 (2002).

[26] J. C. Hansen, C. Tse, and A. P. Wolffe, *Biochemistry* **37,** 17637 (1998).

[27] G. P. Holmquist, *Am. J. Hum. Genet.* **51,** 17 (1992).

[28] N. Sadoni, S. Langer, C. Fauth, G. Bernardi, T. Cremer, B. M. Turner, and D. Zink, *J. Cell Biol.* **146,** 1211 (1999).

[29] K. C. Carter, D. Bowman, W. Carrington, K. Fogarty, J. A. McNeil, F. S. Fay, and J. B. Lawrence, *Science* **259,** 1330 (1993).

[30] C. M. Clemson and J. B. Lawrence, *J. Cell. Biochem.* **62,** 181 (1996).

[31] K. P. Smith, P. T. Moen, K. L. Wydner, J. R. Coleman, and J. B. Lawrence, *J. Cell Biol.* **144,** 617 (1999).

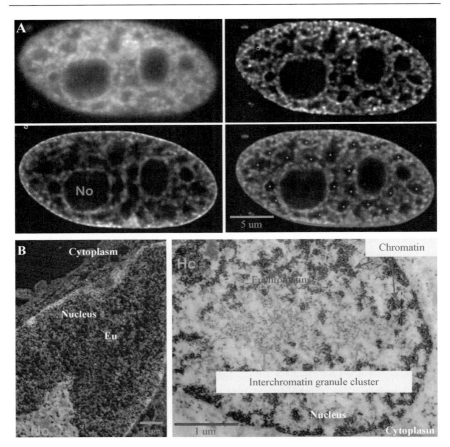

FIG. 2. General organization of euchromatin and heterochromatin within the vertebrate fibroblast cell nucleus. (A) Indian muntjac fibroblasts were fixed with paraformaldehyde and then stained with an antibody recognizing trimethylated Lys-4 of histone H3 (Abcam) and DAPI. The top panels show optical sections of trimethylated Lys-4. *Top left:* The original optical section. *Top right:* The same image after deconvolution to remove the out-of-focus information. *Bottom left:* The deconvolved optical slice from the DAPI image. *Bottom right:* The DAPI is false-colored red and the trimethylated Lys-4 is false-colored green. (B) Transmission electron microscopy images of Indian muntjac fibroblast cells. *Left:* A composite image obtained from a quantitative spatial map of the phosphorus distribution (green) and a quantitative spatial map of the nitrogen distribution (red). This results in a composite image in which the chromatin and ribonucleoproteins are yellow and protein-rich structures are red. The blue dots indicate the positions of interchromatin granule clusters. *Right:* A higher magnification image of an Indian muntjac fibroblast nucleus containing prominent interchromatin granule clusters (green arrows). Eu, euchromatin; Hc, heterochromatin. (See color insert.)

4', 6-diamidino-2-phenylindole (DAPI) is used to stain heterochromatin and an antibody recognizing trimethylated Lys-4 of histone H3 is used to stain euchromatin. Second, when visualized by transmission electron microscopy, the chromatin (Fig. 2B, yellow domains in left panel, dark fibrillar structures in right panel) is excluded from these regions (blue dots in Fig. 2B, left panel). The euchromatin (Eu; Fig. 2B) that surrounds these regions is organized into fibers approaching 200 nm in diameter. Between the chromatin fibers is the interchromatin space with prominent interchromatin granule clusters (blue dots in Fig. 2B, left panel, green arrows in right panel). The chromatin found in these regions represents the euchromatin portion of the genome. The more dense-appearing chromatin that lies under the nuclear lamina (yellow in Fig. 2B) and the dense region (Hc, dark area in Fig. 2B, right panel) on the surface of the nucleolus (not visible in image) represent heterochromatic regions of the genome. Third, this polarized organization of interphase chromosomes and subchromosomal regions is also observed for proteins that regulate or participate in the regulation of RNA polymerase II-transcribed genes. This is illustrated in Fig. 3. In this image, a human MRC-5 lung fibroblast was stained with antibodies recognizing the CREB-binding protein (CBP) and the related protein P300. Both proteins contain a domain that encodes lysine

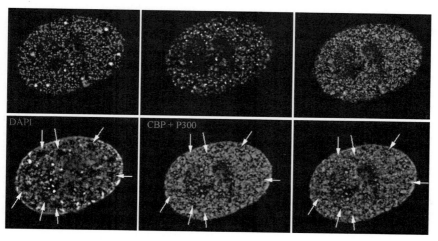

Fig. 3. Typical organization of eukaryotic transcriptional regulators within the vertebrate fibroblast nucleus. Human MRC-5 lung fibroblast cells were fixed and stained with antibodies recognizing CBP and P300. The cells were then counterstained with DAPI to identify the chromatin and imaged by deconvolution microscopy. Each panel shows a projection of the entire 3D volume of the nucleus. The arrows indicate regions that exclude the DAPI stain and are typically associated with splicing factor compartments.

acetyltransferase enzymatic activity. Each protein localizes to several hundred foci scattered throughout the nucleoplasm. In addition, the CBP protein is highly enriched in a smaller number of larger foci. These correspond to promyelocytic leukemia (PML) bodies. When the three-dimensional (3D) projection of each of these proteins is superimposed on the 3D projection of the DNA distribution (DAPI), the foci enriched in CBP and P300 can be seen to be excluded from the heterochromatic regions and the nucleolar domains, but enriched within the DNA-depleted regions where the splicing factor compartments/interchromatin granule clusters reside (arrows).

Scalar Relationship between Cells and Molecules

It has not been uncommon to assume molecular scales to observations made by fluorescence microscopy. Hence, the scale of the events is often misinterpreted when investigators lack familiarity with its use. The relative scales of organization are illustrated in a mouse embryonic fibroblast cell nucleus. The left and right panels in Fig. 4A were acquired from a living mouse fibroblast cotransfected with PML-dsRed2 (blue spheres in Fig. 4A, left; red spheres in Fig. 4A, right) and SC35-GFP (green in Fig. 4A, left). The cells were also counterstained with Hoechst 33258 (Fig. 4A, green in right panel, red in left panel) in order to allow visualization of the chromatin under live cell conditions. Although the distribution of chromatin is often dismissed as "diffuse" when general DNA-binding dyes are used to visualize chromatin in living cells, careful examination of the DNA organization clearly reveals a structured DNA distribution with domains and fibers large enough for visualization by fluorescence microscopy being evident features of the nucleoplasm (Fig. 4A, green in right panel). These are visible even when using standard epifluorescence in living cells (Fig. 4A). The splicing factor compartments identify the regions of the nucleus that lie between chromatin fibers. This region has been classically referred to as the interchromatin space and is visible as a ribonucleoprotein and protein-rich region between higher order chromatin fibers (Fig. 4B). The splicing factor (green) is present in both "speckles" and a more diffuse nucleoplasmic signal. The latter is depleted in the interior of nucleoli, indicating that the nucleolus may be a barrier to diffusion. The PML bodies are approximately 500 nm in diameter and represent an abundant nuclear body present within the nucleoplasm (blue spheres in Fig. 4A, left; red spheres in Fig. 4A, right). The small white bar superimposed on each of the two images in Fig. 4A represents the same length as the entire length of the high-magnification electron microscopy image in Fig. 4C. The yellow fibrillar material is chromatin. Careful examination will reveal that the

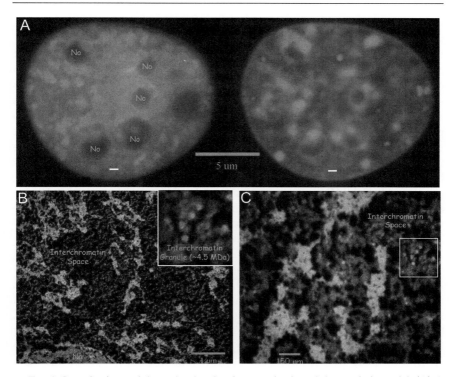

FIG. 4. Organization and the scale of molecular organization of chromatin in nuclei. (A) A mouse 10T1/2 fibroblast cell was cotransfected with SC35-GFP (green in left-hand image, absent in right-hand image) and PML-dsRed2 (blue in left-hand image, red in right-hand image). The DNA was visualized by adding Hoechst 33258 to the culture medium of the living cells and is shown in red (*left*) and green (*right*). The small white bars superimposed on each image are shown to allow comparison with (C). The length of these bars is equal to the entire width of the electron microscopy image shown in (C). (B) Transmission electron microscopy image of a mouse 10T1/2 cell nucleus. The specimen was fixed with paraformaldehyde, cut in 30-nm-thick sections after embedding, and then visualized with an energy filter to acquire quantitative maps of nitrogen and phosphorus. This procedure allows the easy discrimination of nucleoprotein from protein structure. Nucleoprotein structures are yellow-orange, depending on their relative abundance of RNA or DNA versus protein. The protein structures are red or red-orange. (C) Higher magnification of subregion in (B). The boxed region of (C) contains a region enriched in interchromatin granules [this is shown at ×4 magnification in the inset of (B)]. An individual granule, which we have previously demonstrated is approximately 4.5 MDa, is highlighted by showing only its phosphorus signal (green). (See color insert.)

larger chromatin domains have a thin "spaghetti-like" substructure. It is the 30-nm fibers that provide the basis for this appearance. The euchromatin fibers observed in these cells are in the 100- to 200-nm diameter range and are composed of 30-nm chromatin fibers in tight association with each other. Chromatin is not abundant in less condensed conformations outside of these larger fibers. Instead, the space between these chromatin fibers is dominated by a network of protein fibers (red) and ribonucleoprotein particles. The small box in Fig. 4C encloses a small cluster of one class of ribonucleoprotein particles found within the interchromatin space, the interchromatin granules. These provide a reference for molecular scale. We have previously demonstrated that these particles contain, on average, approximately 3.1 MDa of protein and 1.7 MDa of RNA for a total mass approaching 5 MDa. This is more than twice the mass of the largest chromatin-remodeling complexes that have been isolated and is approximately 25 times the mass of the nucleosome.

From this introduction to scale and organization within the nucleoplasm, we can conclude that the FRAP approach does not have the spatial resolution to study individual protein–DNA, protein–RNA, or protein–protein interactions. In the case of chromatin studies, even resolution-limited spot photobleaching will sample a range of chromatin structures. Similarly, the approximately 200-nm foci evident when chromatin-remodeling machinery and transcription factors are visualized by fluorescence microscopy constitute tens to hundreds of individual binding events rather than a gene-specific association at the molecular level. Despite this difference, because FRAP provides a measurement of molecular behavior and the capacity to measure the behavior of binding sites in different regions or compartments (e.g., heterochromatin versus euchromatin), FRAP provides a link between the molecular and cellular scales of chromatin organization.

Study of Chromatin Dynamics, Using FRAP

FRAP has been successfully applied to study the dynamics of both histone and nonhistone protein associations with chromatin. The principal conclusion is that most proteins associate with chromatin only transiently, for durations of seconds to minutes, before dissociating, diffusing, and then binding to another site within the chromatin.[13,32–38] The one notable

[32] J. G. McNally, W. G. Muller, D. Walker, R. Wolford, and G. L. Hager, *Science* **287,** 1262 (2000).

[33] D. Hoogstraten, A. L. Nigg, H. Heath, L. H. Mullenders, R. van Driel, J. H. Hoeijmakers, W. Vermeulen, and A. B. Houtsmuller, *Mol. Cell* **10,** 1163 (2002).

[34] W. Rodgers, S. J. Jordan, and J. D. Capra, *J. Immunol.* **168,** 2348 (2002).

[35] P. Maruvada, C. T. Baumann, G. L. Hager, and P. M. Yen, *J. Biol. Chem.* **278,** 12425 (2003).

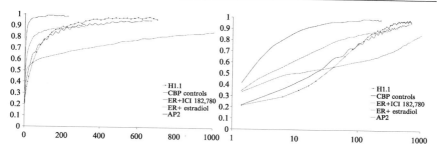

FIG. 5. Recovery profiles of example nuclear proteins that function within chromatin. The recovery profiles of histone H1.1 are compared with a transcriptional coactivator (CBP), a sequence-specific DNA-binding protein and transcriptional regulator (AP2), and an inducible sequence-specific regulator of transcription (the estrogen receptor α, ER). In the case of the ER protein, its mobility is compared in the presence of estradiol (its natural ligand) or in the presence of a potent inhibitor of estradiol, ICI 182,780. (See color insert.)

exception is the exchange of nucleosomal histones, which is a relatively slow process.[15,39] Although the exchange of histone H1 and the slow exchange of the nucleosomal histones were revealed many years earlier by radiolabeling techniques,[16] the *in vivo* exchange rates for H1 histones could not be defined because they undergo multiple associations with chromatin before being degraded. Histone H1, and the HP1 protein, have surprisingly labile associations within heterochromatic domains.[38,40]

FRAP is a powerful approach for investigating the assembly of transcription complexes in living cells. The approach complements the steady state analysis that is possible using either chromatin immunoprecipitation or indirect immunofluorescence approaches. Several sequence-specific DNA-binding proteins and coactivators have been characterized by FRAP. With the exception of the association of the TATA-binding protein (TBP) with metaphase chromosomes,[41] these factors were generally completely replaced within seconds to minutes. This is illustrated in Fig. 5, which shows

[36] D. L. Stenoien, K. Patel, M. G. Mancini, M. Dutertre, C. L. Smith, B. W. O'Malley, and M. A. Mancini, *Nat. Cell Biol.* **3,** 15 (2001).

[37] M. J. Schaaf and J. A. Cidlowski, *Mol. Cell. Biol.* **23,** 1922 (2003).

[38] R. Festenstein, S. N. Pagakis, K. Hiragami, D. Lyon, A. Verreault, B. Sekkali, and D. Kioussis, *Science* **299,** 719 (2003).

[39] J. S. Siino, I. B. Nazarov, M. P. Svetlova, L. V. Solovjeva, R. H. Adamson, I. A. Zalenskaya, P. M. Yau, E. M. Bradbury, and N. V. Tomilin, *Biochem. Biophys. Res. Commun.* **297,** 1318 (2002).

[40] T. Cheutin, A. J. McNairn, T. Jenuwein, D. M. Gilbert, P. B. Singh, and T. Misteli, *Science* **299,** 721 (2003).

[41] D. Chen, C. S. Hinkley, R. W. Henry, and S. Huang, *Mol. Biol. Cell* **13,** 276 (2002).

the recovery kinetics of several different types of chromatin-associated proteins. Only the estrogen receptor, when treated with the direct antagonist ICI 182,780, requires more than 10 min to reach equilibrium. It is interesting to note that steady state changes in loading of individual components of the regulatory machinery during gene activation occur over the course of minutes to hours whereas the actual binding times of individual proteins is typically seconds.

Biological Principles Underlying FRAP of Nuclear Proteins

When molecules are coupled to a fluorescent tag, they can be measured with respect to both position and concentration, using fluorescence microscopy. With the development of fluorescent protein technology, proteins can now be visualized in living cells by fluorescent microscopy. Under typical experimental conditions, the fluorescent proteins are at steady state and the fluorescent molecules are too abundant to resolve the movement of individual fluorescent proteins. This allows the tracking of steady state structures by direct fluorescent visualization but it does not reveal the movement of molecules into and out of these steady state structures. FRAP is a method that enables the measurement of molecular dynamics. It does this by introducing a rapid shift away from steady state without disrupting the actual concentrations of the molecule under study. This is achieved by exposing a small portion of the molecules with a sufficiently intense laser pulse to irreversibly photobleach their fluorescence emission. This creates a condition in which "all" the molecules outside of the photobleached region are fluorescent whereas all the molecules within the photobleached region are nonfluorescent and, hence, resolved by fluorescence microscopy. If the underlying steady state distribution reflects molecules that are in constant flux, the fluorescent and nonfluorescent pools will mix over time and the distinction between the region photobleached and the surrounding environment will reduce over time until a steady state distribution of the photobleached and fluorescent copies of the protein is achieved. The time required to reach steady state is a reflection of the rate that the fluorescently tagged protein normally moves through its environment.

The movement of fluorescent molecules is dependent on its molecular weight and whether or not it binds to other molecules that either diffuse slower or are immobilized over the time scale of the experiment.[6] Although both properties are of interest to investigators interested in chromatin structure and function, the binding events are of particular interest because they reflect the period of time that a molecule functionally engages the genome. When binding events are of significant duration, the time scale of the binding event is typically two or more orders of magnitude greater than the time scale of diffusion across the same nuclear space.

For inert molecules that do not undergo specific interactions, diffusion will be the principal determinant of the measured rate of movement. These types of movements occur rapidly on the time scale required to photobleach a region of the nucleoplasm. Hence, FRAP is not a technique that is well suited to resolving differences in diffusion between biomolecules that differ in size by 2-fold or less. Fluorescence correlation spectroscopy, although not as widely accessible, is an approach that is better suited for measuring differences in diffusion. FRAP is more appropriately applied for studying molecules that interact with structures sufficiently large to slow the migration of the protein by an order of magnitude or more. In the case of proteins that function by binding to the DNA or chromatin, the bound state is essentially immobile on the scale of both the diffusion and the binding–unbinding events.[11,14] This property of chromatin makes processes occurring on chromatin particularly well suited for investigation by FRAP.

Designing and Characterizing Fluorescent Reporter for Live Cell-Imaging Experiments

The fluorescent protein tags used for studying protein behavior in living cells are relatively bulky and can influence the function and properties of the tagged protein. For example, the 27-kDa green fluorescent protein isolated from a jellyfish is a barrel-shaped structure just over 4 nm in length and approximately 2.5 nm in diameter. This is sufficiently large to potentially disrupt protein–protein interactions. Although GFP is fluorescent as a monomer, the red fluorescent protein must dimerize or multimerize, dependent on the variant, in order to fluoresce. Hence, red fluorescent protein constructs will drive multimerization of fusion proteins and this could alter the ability to assemble into macromolecular complexes, affinity and dissociation constants, and function. The purpose of the experiment will determine what properties are essential to maintain if the fusion with the fluorescent protein alters some properties of the native protein. When the purpose of an FRAP experiment is to extract information about protein dynamics as it pertains to the execution of a biological function, the maintenance of all key molecular interactions at or near native affinities is critical. Consequently, it is important to evaluate the influence of the fluorescent protein tag on the behavior of the tagged protein.

Although simply demonstrating that a fluorescently tagged protein is incorporated into the same immunocomplexes and with the same efficiency as the endogenous protein in immunoprecipitation experiments performed on nuclear extracts provides some information, it is not conclusive data. Because of the different conditions between the living cell nucleus and any assays of functionality outside of the living cell, *in vitro* characterization by immunoprecipitation or general assays of protein binding may not be sufficiently sensitive to detect differences that may be physiologically relevant. For example, extensive biochemical characterization of chromatin binding revealed no differences between histone H1 with GFP fused to the C-terminus and the normal H1 counterparts.[11,14,42] Nonetheless, when we compared N-terminal and C-terminal fusion proteins, we found that when the GFP was fused to the amino terminus of histone H1, the fusion protein had a reduced exchange rate. Deletion analysis further identified the C terminus as being critical for binding to chromatin in living cells. Thus, the C-terminal fusion proteins may exchange more readily than endogenous histone H1.1.[42a]

Because of the potential influence of fluorescent tags on protein behavior, we recommend that, minimally, both C-terminal and N-terminal fusion proteins should be generated and the kinetics of the two fusion proteins compared with each other. Ideally, these results should also be verified by microinjecting purified protein that has been conjugated with a small fluorescent dye [e.g., fluorescein isothiocyanate (FITC)] *in vitro*. Morphology can also provide rigorous criteria for exclusion or inclusion in analysis. In our experience, two types of artifactual distributions are commonly observed for the fusion protein but not the endogenous counterpart. In many cases, the expressed protein is aggregated, probably by misfolding, into large nuclear bodies that, in at least some instances, are associated with proteosome activity.[43] These large nuclear inclusions are not present in normal healthy nuclei.[44] A second common observation is that the protein is no longer capable of binding with noticeable affinity to any of its nuclear substrates.[45] Consequently, these fusion proteins have a truly diffuse distribution (something not common to nuclear proteins) and move at rates expected of freely diffusing proteins in their size range.

[42] A. Gunjan, B. T. Alexander, D. B. Sittman, and D. T. Brown, *J. Biol. Chem.* **274**, 37950 (1999).
[42a] M. J. Hendzel *et al.*, in preparation (2003).
[43] C. J. Cummings, M. A. Mancini, B. Antalffy, D. B. DeFranco, H. T. Orr, and H. Y. Zoghbi, *Nat. Genet.* **19**, 148 (1998).
[44] D. L. Stenoien, M. Mielke, and M. A. Mancini, *Nat. Cell Biol.* **4**, 806 (2002).
[45] G. Carrero, D. McDonald, E. Crawford, G. de Vries, and M. J. Hendzel, *Methods* **29**, 14 (2003).

Performing the Experiment: Optimizing Data Collection

The precise experimental design of an FRAP experiment is determined by the experimental question and the analytical approach. For example, when the experimental question is to determine the *effective* diffusion coefficient, the most commonly employed approach is based on that of Axelrod and colleagues[46] and requires photobleaching a diffraction-limited spot to initiate the experiment. Although this experimental design has also been applied to studying proteins undergoing binding events, photobleaching a band across the nucleoplasm has advantages for *in vivo* binding assays because it simplifies the region in which the events are taking place (see later).

In most instances, one does not know at the onset of the experiment whether or not the protein is primarily diffusing or bound with residency times of seconds or greater. Hence, it is best to perform pilot experiments before designing the quantitative experimental procedure. Pilot experiments can be performed simply by initially photobleaching a 1-μm-diameter circle near the middle of the nucleoplasm (but away from the nucleolar boundaries if a nucleoplasmic protein is under study). The photobleaching should be able to be achieved with approximately 20 iterations of a 25-mW 488-nm laser line at full power in well under 1 s. Images can then be collected at 1% or less laser power at a rate of one per second for 60 s. Typically, most of the recovery will occur in the first 10 s. The purpose of the pilot experiment is to determine (1) how frequently images should ideally be collected during the first 10 s of the recovery phase and (2) whether the entire experiment reaches equilibrium within the 60-s time scale. From this pilot experiment, the experimental setup of the quantitative experiment can be rapidly determined (see Table I for details on optimizing the experimental setup for quantitative data collection). An important note of caution is that if the raw data are examined simply by plotting the intensity of the photobleached region without normalizing the data (see later for details on normalization), the small amount of photobleaching that occurs during the collection of the approximately 60 images of the recovery phase will usually be significant enough to lead to an underestimation of the time required to reach equilibrium. Hence, it is important to normalize the data from the pilot experiment before defining the initial time course for quantitative analysis. The remainder of this review focuses on extracting quantitative information on binding events *in vivo*, using mathematical models of protein–protein interactions to fit experimental data.

[46] D. Axelrod, D. E. Koppel, J. Schlessinger, E. Elson, and W. W. Webb, *Biophys. J.* **16**, 1055 (1976).

TABLE I
OPTIMIZATION OF FRAP DATA COLLECTION

Parameter	Optimization
Laser settings during photobleaching	The number of iterations required for photobleaching should be determined by defining the minimum number of iterations of the laser pulse necessary to photobleach the desired region to background fluorescence levels. The goal is to minimize the time required for the photobleaching process. This is to minimize the exchange of molecules between the surrounding nucleoplasm and the photobleached region during the photobleaching process
Pinhole aperture (confocal microscopes only)	When using a laser scanning confocal microscope to perform FRAP experiments, the goal is to optimize image sampling of both intensity and time. Since these two parameters are inversely related (faster collection means lower dwelling times on each point in the image and consequently fewer photons collected), gains can be made on intensity sampling by opening the pinhole aperture. Optical section thickness is only important if it can be used to resolve potentially kinetically different cellular compartments (e.g., nucleus from cytoplasm) based on resolution in the z axis of the specimen
Size of photobleached region	While the size of the photobleached region does not influence the movement of the fluorescent molecules under study, it does influence the amount of time required to complete the photobleaching process. For proteins that are freely diffusing, the smaller the region photobleached, the more accurate the data. For proteins that are undergoing binding events, their rates of movement through the nucleoplasm are typically "slow" relative to the period of time required to photobleach. Small photobleached regions (up to 2 μm in width or diameter) should be able to be photobleached in well under 1 s. Photobleaching half of a typical fibroblast nucleus, in contrast, would typically take several seconds using current confocal microscope configurations
Frequency and number of images collected during recovery	The key principles to balance with respect to image collection are (1) scan speed and interval, (2) adequate sampling of the different kinetic phases of the recovery curve, and (3) to minimize photobleaching during acquisition of the recovery curve. To accomplish this, we collect images at different intervals at different parts of the recovery curve. Typically, we sample at rates of 1 or more images per second for the first 10 s of recovery

(continued)

TABLE I   *(continued)*

Parameter	Optimization
	(using a 2-μm wide photobleached band across the width of the nucleus). The first few seconds contain the molecules that approach free diffusion on their path from outside to inside the photobleached region. Binding events happen on longer time scales. We determine the interval of scanning based upon the period required for equilibrium to be reached after photobleaching and then dividing this period into equal intervals such that the number of scans in the recovery phase of the experiment totals 60
Spatial sampling (pixel dimensions)	The resolution of the data collection is principally determined by the experimental question. For example, we have previously pushed the resolution limits of the procedure to photobleach a band across the center of a nuclear body (the PML body), which itself measures only about 500 nm in diameter. This allowed us to demonstrate the immobility of the PML protein not only in exchanging with the surrounding nucleoplasm but also with respect to movement within individual PML bodies. If the goal is to resolve subregions of the nucleus in order to define the specific kinetics of these domains (e.g., heterochromatin versus euchromatin), then sampling frequencies of 70 nm per pixel are appropriate in order to achieve the resolution capabilities of the instrument. This also requires the use of a high numerical aperture objective (e.g., ×63 1.4 N A). Note, however, that the higher the spatial resolution, the lower the signal that is present for each pixel in the image. Often, experiments performed under these conditions must balance sampling with photobleaching and are limited by signal:noise ratios of the digital acquisition equipment. To avoid this, we typically sample at spatial resolutions between 100 and 150 nm per pixel. This is appropriate for most FRAP studies of nuclear proteins
Effect of protein expression levels	It is widely appreciated that the functional consequences of expressing a particular protein are often concentration dependent. Moreover, it is obvious that unless a cell upregulates the number of binding sites for a particular protein, expression above a certain level will saturate the capacity of the cell to form specific complexes that require this protein. In these instances, the freely diffusing pool that predominates the first few seconds of recovery within the FRAP experiment will be overrepresented. By using morphological criteria to

(continued)

TABLE I *(continued)*

Parameter	Optimization
	evaluate the "health" of the cell and by intentionally biasing the experiment toward the lower levels of expression observed within the culture, the effects and requirements of introducing a protein expression construct to express alongside the endogenous protein can be minimized
Transient versus stable transfection	While it is natural to assume that generating clones or by using cells following drug selection for stable expression, the transfectants may adapt to stable overexpression of a protein through mechanisms other than attenuating the expression of the stably integrated expression vector and/or the endogenous protein expression levels. For example, we have found that the CREB-binding protein (CBP) is almost invariably altered in its binding characteristics over the course of selection for stable expression. We recommend (1) determining the subnuclear distribution of the endogenous protein using indirect immunofluorescence prior to initiating kinetic studies of GFP fusion proteins, (2) restricting analysis to cells within the culture that are not expressing the transfected protein at "high" levels, (3) restricting analysis to cells that have a "normal" morphology, and (4) restricting analysis to cells that maintain the endogenous subnuclear distribution
Mathematical analysis	The methods that may be used to model the experimental data in order to obtain quantitative information from these studies depend on the experimental question. For example, if the question is simply to determine whether or not drugs that alter specific aspects of nuclear function have an effect on the binding of a protein to chromatin, it is sufficient merely to compare recovery curves in the presence and absence of the drug. This allows the extraction of a half-time of recovery and the proportion of molecules that readily exchange with the surrounding nucleoplasm. The section, "Mathematical Analysis of Molecular Binding Events" in this chapter is devoted to more complex mathematical modeling approaches that can be used to define the binding and diffusion properties of the molecules under study. These methods are most appropriate when it is desirable to quantify binding events and should be considered when the goal of the experiment is to define the dynamics of binding (e.g., to DNA or chromatin) within the living cell system

Preparing Data for Analysis

As mentioned in the previous section, to accurately quantify the results of the FRAP experiment, the data must be corrected to account for changes in the fluorescence pool throughout the course of the experiment. Normalization of the data is done to account for (1) the loss of the total cellular fluorescence during the photobleaching portion of the experiment and (2) changes in fluorescence intensity that occur during the collection of the recovery curve. During the initial photobleaching process, a significant proportion of the total nuclear pool is rendered nonfluorescent. This alters the theoretical maximum intensity that the photobleached region may recover to from its initial intensity before photobleaching to its initial intensity multiplied by the ratio of the total cellular fluorescence at the first time point/total cellular fluorescence at the onset of the experiment. This correction will allow for an accurate estimate of the percentage of the population that is mobile over the time scale of the experiment. During the collection of the recovery curve, the cell is exposed to many iterations of laser illumination. This may also result in some photobleaching. There may also be minor fluctuations in total laser output per time point. Both of these processes can influence the profile of the curve. To normalize for these changes, we set the initial intensity of the cell at the first time point post photobleaching as 1.0. The total fluorescence intensity of the cell (or cell nucleus) is measured at each time point and then normalized to the value of the first time point. The value obtained for normalization of the total fluorescence pool is also used to correct the fluorescence values obtained for the photobleached region at each time point. When the curve is normalized in this manner, the mobile and immobile fractions can be read directly from the curve. If the experimental goal is only to define half-times of recovery for the mobile fraction and no measurement of the immobile portion of the population is required, it is useful to apply an additional normalization. If the absolute value of the photobleached region at the first time point postbleaching is set to zero and the maximum value reached during the recovery phase is set to 1.0, then the half-time of recovery can be acquired directly from the plot.

Mathematical Analysis of Molecular Binding Events

When analyzing the overall mobility of nuclear proteins by FRAP measurements, one could simply fit the diffusion equation to estimate an effective diffusion coefficient.[45,46] Because most functional nuclear proteins interact with structures (e.g., chromatin, speckles, and nucleoli), it would be more informative biologically to obtain both measurements of diffusion and measurements of binding from FRAP experiments. During FRAP experiments, biomolecules redistribute into the photobleached regions by first dissociating from their binding site outside of the photobleached region and then, through a random walk, eventually encountering a binding site within the photobleached region. The durations of binding events are often one or more orders of magnitude greater than the time required to diffuse through the same space. This results in the binding–unbinding turnover as the rate-limiting event in the recovery of fluorescence.

In this section, we discuss the use of two different mathematical models to quantify FRAP measurements when molecular binding events are taking place. We focus on biomolecules that undergo binding and unbinding events with an approximately spatially homogeneous structure that is considered immobile on the time scale of molecular movement. For example, chromatin-associated proteins interact with interphase chromatin that approaches a homogeneous distribution in human cell lines and is immobile on the time scale of a typical FRAP experiment.

The first model that we consider consists of a system of reaction-diffusion equations, that is, a system of partial differential equations, which has been applied in the context of reversible chemical reactions[47] and in the context of cytoplasmic actin dynamics.[48] The other model is a compartmental model that consists of a system of ordinary differential equations. We introduced this model to study the kinetics of nuclear proteins.[45] We summarize both models. On the basis of their solutions, we obtain explicit expressions for the theoretical fluorescence recovery curves that can be used to fit the FRAP data in order to obtain a quantification of the interactions occurring *in vivo*. A simple parameter estimation methodology is given along with an application on the dynamics of H1 histones.

Models

To describe the dynamics after photobleaching of fluorescent proteins that undergo a reversible binding–unbinding process with a structure that is assumed to be immobile on the time scale of molecular movement and spatially homogeneously distributed (e.g., chromatin), two models have been proposed.[45]

The first model is the well-known linear system of reaction-diffusion equations for reversible reactions subject to Neumann (no-flux) boundary conditions:

[47] J. Crank, "The Mathematics of Diffusion." Oxford University Press, London, 1975.
[48] Y. Tardy, J. L. McGrath, J. H. Hartwig, and C. F. Dewey, *Biophys. J.* **69,** 1674 (1995).

$$\frac{\partial}{\partial t}u(x,t) = D\frac{\partial^2}{\partial x^2}u(x,t) - k_b u(x,t) + k_u v(x,t), \quad x \in (0,l), \quad t > 0$$

$$\frac{\partial}{\partial t}v(x,t) = k_b u(x,t) - k_u v(x,t), \qquad\qquad x \in (0,l), \quad t > 0 \qquad (1)$$

$$\frac{\partial u}{\partial x} = \frac{\partial v}{\partial x} = 0 \qquad\qquad\qquad\qquad x = 0,l, \quad t > 0$$

$$u(x,0) = f(x), \quad v(x,0) = g(x), \qquad\qquad x \in (0,l)$$

where u and v represent the populations of biomolecules free to diffuse and bound to the immobile structure, respectively; D is the diffusion coefficient; k_b and k_u represent the binding and unbinding rates, respectively; t represents time; x is the spatial coordinate of position; $(0, l)$ is the spatial domain, and $f(x)$ and $g(x)$ are the initial conditions of fluorescent unbound and bound species immediately after photobleaching, respectively.

Note that the spatial domain in Eq. (1) is one-dimensional. Therefore, it is appropriate to approximate the shape of the domain (e.g., the cell nucleus) with a rectangle of length l, and assume a narrow band of width $2h$, centered at c as a photobleaching profile (Fig. 6A). Assuming that the fluorescence intensity in a region is proportional to its size, the length l is estimated in terms of the fluorescence intensity of the nucleus before photobleaching, F_0, and immediately after photobleaching, F_a, as follows:

$$l = 2h\frac{F_0}{F_0 - F_a} \qquad (2)$$

A similar procedure was followed by Tardy et al.[48] and McGrath et al.[49] when they applied Eq. (1) in photoactivated fluorescence (PAF) studies in the context of cytoplasmic actin dynamics.

Therefore, assuming a narrow band as the photobleaching profile, and that this photobleaching is performed in a region that has reached a homogeneous steady state solution (u_0, v_0) that satisfies $u_0 + v_0 = 1$, we can express, without loss of generality, the initial conditions of Eq. (1) as

$$f(x) = \begin{cases} 0, & |x - c| \leq h \\ P_u = \dfrac{k_u}{k_b + k_u}, & |x - c| > h \end{cases} \quad \text{and} \quad g(x) = \begin{cases} 0, & |x - c| \leq h \\ P_b = \dfrac{k_b}{k_b + k_u} & |x - c| > h \end{cases}$$

$$(3)$$

where $P_u = k_u/(k_b + k_u)$ and $P_b = k_b/(k_b + k_u)$ represent the proportions of unbound and bound populations, respectively.

The second model is a linear compartmental model. The model is based on the fact that photobleaching a narrow band causes the two populations

[49] J. L. McGrath, Y. Tardy, C. F. Dewey, Jr., J. J. Meister, and J. H. Hartwig, *Biophys. J.* **75,** 2070 (1998).

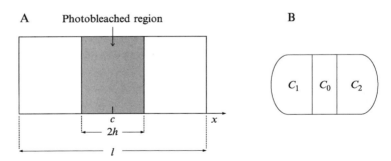

A   Photobleached region   B

FIG. 6. (A) Geometric approximation of the domain. The shape of the domain (e.g., the cell nucleus) is approximated with a rectangle of length l. The photobleached region is a narrow band of width $2h$, centered on the x axis at c. (B) The three physical compartments of the cell nucleus caused by the photobleaching of a narrow band. C_0 represents the photobleached region, and C_1 and C_2 are the left and right unbleached regions, respectively.

(bound and unbound) to occupy three physical compartments within the cell nucleus, namely, the photobleached band C_0, the left unbleached region C_1, and the right unbleached region C_2 (Fig. 6B).

The compartmental model can be written as the following system of ordinary differential equations [Eq. (16) in Ref. 45]:

$$
\begin{aligned}
\dot{u}_0 &= -2D_1 u_0 + D_2 u_1 + D_2 u_2 - k_b u_0 + k_u v_0 \\
\dot{u}_1 &= D_1 u_0 - D_2 u_1 - k_b u_1 + k_u v_1 \\
\dot{u}_2 &= D_1 u_0 - D_2 u_2 - k_b u_2 + k_u v_2 (4) \\
\dot{v}_0 &= k_b u_0 - k_u v_0 \\
\dot{v}_1 &= k_b u_1 - k_u v_1 \\
\dot{v}_2 &= k_b u_2 - k_u v_2
\end{aligned}
\tag{4}
$$

where the dot denotes the derivative with respect to time t; u_0, u_1, and u_2 represent the population of diffusing fluorescent molecules in C_0, C_1, and C_2, respectively; v_0, v_1, and v_2 represent the population of fluorescent molecules bound to the immobile structure in C_0, C_1, and C_2, respectively; k_b is the binding rate of molecules to the immobile structure; k_u is the unbinding rate of molecules from the structure; D_1 is the fractional diffusional transfer coefficient from compartment C_0 to compartment C_1 or C_2; and D_2 is the fractional diffusional transfer coefficient from compartments C_1 and C_2 to compartment C_0. It is known that the fractional diffusional transfer coefficients D_1 and D_2 can be described in terms of only one parameter D_1, called the diffusional transfer coefficient, as follows [Eq. (17) in Ref. 45]:

$$D_1 = \frac{r}{2-r} D_t, \quad D_2 = 2 \frac{1-r}{2-r} D_t \qquad (5)$$

where

$$r = \frac{F_a}{F_0} \qquad (6)$$

and F_0 and F_a denote the total fluorescence in the nucleus before and immediately after photobleaching, respectively.

To obtain an initial condition for Eq. (4), we assume that the photobleaching is performed on an equilibrium state $(\boldsymbol{u}, \boldsymbol{v}) = (u_0, u_1, u_2, v_0, v_1, v_2)$ that satisfies $u_0 + u_1 + u_2 + v_0 + v_1 + v_2 = 1$. Thus, the initial condition that reflects the experimental setting is given by Eq. (18) in Ref. 45:

$$(\boldsymbol{u_0}, \boldsymbol{v_0}) = \left(0, \frac{k_u}{k_b + k_u} \frac{1}{2} r, \frac{k_u}{k_b + k_u} \frac{1}{2} r, 0, \frac{k_b}{k_b + k_u} \frac{1}{2} r, \frac{k_b}{k_b + k_u} \frac{1}{2} r \right) \qquad (7)$$

Model Solutions and Explicit Theoretical Recovery Curves. To obtain an explicit solution for reaction-diffusion Eq. (1) subject to initial condition (3), we applied the method of Laplace transform.[50] This technique also was used in Ref. 47 when solving Eq. (1) in the context of reversible chemical reactions, and in Ref. 48 in the context of cytoplasmic actin dynamics. By doing so, we can obtain an explicit solution for the total fluorescence concentration of biomolecules $u(x, t) + v(x, t)$, which is integrated over the photobleached region to obtain the following normalized theoretical recovery curve:

$$R(t) = 1 - \frac{l^2}{2h(l - 2h)(k_b - k_u)} \sum_{n=1}^{\infty} (A_{1n} F_{1n} e^{r_{1n} t} + A_{2n} F_{2n} e^{r_{2n} t}) S_n^2 \qquad (8)$$

where

$$S_n = \frac{1}{n\pi} \left[\sin \left(\frac{n\pi(c - h)}{l} \right) - \sin \left(\frac{n\pi(c + h)}{l} \right) \right] \qquad (9)$$

$$r_{jn} = -\frac{k_b + k_u + D_t \left(\frac{n\pi}{l} \right)^2}{2} + (-1)^{j+1} \frac{Q_n}{2}, \quad j = 1, 2 \qquad (10)$$

$$Q_n = \sqrt{\left[k_b + k_u + D_t \left(\frac{n\pi}{l} \right)^2 \right]^2 - 4 k_u D_t \left(\frac{n\pi}{l} \right)^2} \qquad (11)$$

[50] E. Zauderer, "Partial Differential Equations of Applied Mathematics." John Wiley & Sons, New York, 1998.

$$F_{jn} = \frac{(-1)^{j+1}\left(r_{jn} + k_b + k_u\right)}{Q_n}, \quad j = 1, 2 \tag{12}$$

and

$$A_{jn} = k_b + k_u + r_{jn} + D_t\left(\frac{n\pi}{l}\right)^2, \quad j = 1, 2 \tag{13}$$

To obtain the solution of compartmental model (4) subject to initial condition (7), we used the fact that the eigenvalues of the Jacobian matrix generated by linear system (4) are distinct and real, and therefore the solution is given by a sum of exponentials [Eq. (2.51) in Ref. 51]. By doing so, we obtain an explicit solution for the total fluorescent population $u_0 + v_0$ in the photobleached region C_0, which is normalized to obtain the following theoretical fluorescence recovery curve [Eq. (19) in Ref. 45]:

$$R(t) = 1 - \gamma \exp\left(\alpha t\right) - (1 - \gamma) \exp\left(\beta t\right) \tag{14}$$

where α and γ are negative eigenvalues generated by linear system (4), given by

$$\begin{aligned} \alpha &= -S_1 + S_2 \\ \beta &= -S_1 - S_2 \end{aligned} \tag{15}$$

with

$$S_1 = \frac{k_b + k_u}{2} + \frac{D_t}{2 - r}$$

$$S_2 = \sqrt{\left(\frac{k_b + k_u}{2} + \frac{D_t}{2 - r}\right)^2 - 2k_u\frac{D_t}{2 - r}} \tag{16}$$

and

$$\gamma = \frac{1}{2}\frac{k_u}{k_b + k_u}\Gamma_1\Gamma_2 \tag{17}$$

where

$$\Gamma_1 = 1 + \frac{S_1 - S_2}{k_u} + \frac{k_b}{k_u} + \frac{2D_t}{k_u(2 - r)}$$

$$\Gamma_2 = \frac{S_1 - S_2}{S_2} - \frac{2D_t}{S_2(2 - r)} \tag{18}$$

and r is given by Eq. (6).

[51] K. Godfrey, "Compartmental Models and Their Applications." Academic Press, London, 1983.

Parameter Estimation Methodology: Application to Type H1 Histone

After having collected the experimental raw fluorescence intensity data, and normalized it to correct for the loss of total fluorescence that arises from photobleaching a narrow band region, one is ready to use either the theoretical recovery curve given by Eq. (8) or the theoretical recovery curve given by Eq. (14) to fit the data. Fitting the experimental FRAP data with the theoretical recovery curve (14) obtained from the solution of the reaction-diffusion model will allow estimation of diffusion coefficient D, whereas fitting the data with theoretical recovery curve (8) will provide diffusional transfer coefficient D_t instead. Moreover, the fitting of either equation allows for the estimation of the binding and unbinding rates, which can be used to infer the following biological information:

$\tau_r = 1/k_u$: the average residency time of biomolecules in bound form
$\tau_w = 1/k_b$: the average wandering time of biomolecules between binding events
$P_b = \dfrac{k_b}{k_b + k_u}$: the steady state proportion of biomolecules in bound form
$P_u = \dfrac{k_u}{k_b + k_u}$: the steady state proportion of biomolecules in unbound form

In both cases, the fitting procedure consists of a multiple parameter estimation problem, in which three parameters are to be estimated.

The procedure for parameter estimation using the reaction-diffusion equation can be summarized as follows.

1. Estimate the length l of the domain (cell nucleus), using Eq. (2).
2. Truncate the series given by Eq. (8) to a finite number of terms. For typical time scales of recovery and diffusion coefficients of nuclear proteins, a truncation using the first 20 or more terms provides an accurate approximation.
3. Incorporate the expressions given by Eqs. (9)–(13) into the truncated series to obtain an expression depending on the diffusion coefficient, and binding and unbinding rates.
4. Using the method of nonlinear least squares,[52] fit the truncated series to the normalized fluorescence recovery data to obtain estimates for the three dynamic parameters (D, k_b, and k_u).

[52] R. H. Myers, "Classical and Modern Regression with Applications." Duxbury Press, Boston, 1986.

In the case of the compartmental model, note first that the theoretical recovery curve given by Eq. (14) is expressed in terms of three new parameters, namely, α, β, and γ. Instead of expressing these new parameters in terms of the diffusional transfer coefficient and the binding and unbinding rates, using Eqs. (15)–(18), we have proposed the following simple procedure that requires less computer capability than using the reaction-diffusion equation.

1. Fit the theoretical recovery curve given by Eq. (14), using the method of nonlinear least squares,[52] to obtain estimates for α, β, and γ.

2. Obtain estimates for S_1 and S_2 by solving the linear system given by Eq. (15). According to Eq. (16), these estimates must be positive. If they are not, it is necessary to interchange the values of α and β, and reassign the value $1 - \gamma$ to γ.

3. Substitute Eq. (18) into Eq. (17).

4. Solve, with any numerical scheme, the nonlinear system given by Eq. (16) and Eq. (17) to obtain estimates for the three dynamical parameters D, k_b, and k_u.

To illustrate a particular application of the results in the context of chromatin-associated proteins, we estimate the dynamical parameters for the type H1 histone, which is assumed to be in either of two states in the cell nucleus, that is, bound to the chromatin structure (assumed to be spatially homogeneously distributed), or unbound and therefore free to diffuse throughout the nucleus.

The FRAP data are obtained by photobleaching a narrow band of width 1.5 μm in a cell nucleus of estimated length $l = 5.6$ μm [see Eq. (2)]. Fitting Eq. (8) to the FRAP data, using the procedure explained above, we obtain a proportion $P_u = k_u/(k_b + k_u) \approx 0.88$ for the population moving freely with a diffusion coefficient $D = 0.071$ μm^{2}s^{-1}, whereas for the population bound to the chromatin structure proportion $P_b = k_b/(k_b + k_u) \approx 0.12$ (Fig. 2). The estimated diffusion coefficient is smaller than expected for a diffusing biomolecule of the molecular weight of a histone. Also, the proportion of bound population of histone was smaller than biologically expected. To interpret this apparent discrepancy, we hypothesize that the 12% obtained for the bound population actually corresponds to a high-affinity (strongly) bound subpopulation, and the apparent high percentage of free molecules is constituted by the actual diffusing subpopulation and a low-affinity (weakly) bound subpopulation. Therefore, the diffusion coefficient is indeed an effective diffusion coefficient D_{eff} that accounts for the actual diffusing population and for the weakly bound population (Fig. 7).

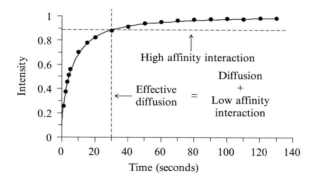

FIG. 7. Experimental fluorescence recovery data (solid circles) of type H1 histone, obtained after photobleaching a band of width 1.5 μm in a cell nucleus of estimated length $l = 5.6$ μm [see Eq. (2)]. The experimental data were fitted with the theoretical recovery curve given by Eq. (8), using the method of nonlinear least squares (solid curve). The estimated values for the binding and unbinding rates are $k_b = 0.0024$ s^{-1} and $k_u = 0.0183$ s^{-1}, respectively. The first phase of the recovery represents the proportion $p_u = k_u/(k_b + k_u) \approx 0.88$ of the population moving with an effective diffusion coefficient $D_{eff} = 0.071$ μm^{2}s^{-1}. This proportion accounts for the actual diffusing subpopulation and for a subpopulation weakly bound to the chromatin structure. The slow rate of recovery in the last phase is caused by the proportion $p_b = k_b/(k_b + k_u) \approx 0.12$ of the population strongly bound to the chromatin structure.

From this section, we can conclude that the specific theoretical recovery curve to be used for the fitting depends entirely on the criteria and needs of the particular experiment. If the main concern is the estimation of the binding and unbinding rates, then one could take advantage of the simplicity of Eq. (14) and use the compartmental model for the fitting. On the other hand, if one is also interested in the estimation of a diffusion coefficient, then Eq. (8) would be more appropriate, and the reaction-diffusion model can be used. However, the theoretical recovery curve given by Eq. (8) is not as simple as the exponential sum obtained for the solution of the compartmental model given by Eq. (14). The simplicity of the latter solution allows for a straightforward parameter estimation procedure. However, the apparent tradeoff for this simplicity is that the compartmental model does not provide as straightforward an estimation of a diffusion coefficient as the reaction-diffusion equation does. Finally, both explicit solutions given by Eq. (8) and Eq. (14) offer little in terms of explaining the different qualitative behaviors exhibited by the recovery curves. As a work in progress, we are currently analyzing the spectrum of possible qualitative behaviors of both the reaction-diffusion model and the compartmental model. This study will enable us to find a relationship between the models, and also to obtain simpler theoretical recovery curves that can reflect the behavior of the experimental FRAP data and simplify the task of parameter estimation even further.

Conclusions

Although obtaining estimates of recovery times for individual proteins is an important contribution to the understanding of chromatin structure and function in living cells, the realization of the analytical power of FRAP and related live-cell approaches requires the complementation of experiments with mathematical modeling. Because there may be more than one way to fit the experimental data, mathematical modeling alone cannot establish the underlying biological processes involved. It can, however, exclude biological hypotheses to explain the observed dynamics in living cells.[6] In other words, the reliability of the mathematical model in quantifying the biological mechanisms is determined by the extent that the biological hypothesis matches the *in vivo* process. When the mathematical models are based on valid biological hypotheses and closely reproduce the experimental data, FRAP provides an elegant quantitative approach to begin to ask relatively complex questions, such as the order of assembly of proteins on a gene promoter.[53]

Acknowledgments

Original research related to this technical review was supported by operating grants from the Canadian Institutes of Health Research (J. T. and M. H.) and the Natural Sciences and Engineering Research Council (NSERC). G. C. was supported by funding from MITACS. M. H. is a scholar of the Canadian Institutes of Health Research and the Alberta Heritage Foundation for Medical Research.

[53] M. Dundr, U. Hoffmann-Rohrer, Q. Hu, I. Grummt, L. I. Rothblum, R. D. Phair, and T. Misteli, *Science* **298**, 1623 (2002).

[27]   Fluorescence Recovery after Photobleaching (FRAP) Methods for Visualizing Protein Dynamics in Living Mammalian Cell Nuclei

By Diana A. Stavreva and James G. McNally

Introduction

Fluorescence recovery after photobleaching (FRAP) is an optical technique used to measure the temporal dynamics of fluorescently tagged molecules. FRAP is a relatively old technique,[1-5] but its application to the study of proteins in living cells is more recent. This renaissance is driven largely by the advent of confocal microscopy and the introduction of green fluorescent protein (GFP) as an endogenous protein marker.

After GFP (or another fluorophore) is covalently attached to the protein of interest, its cellular distribution can be visualized at low light intensities that do not damage cellular processes. Photobleaching is the irreversible destruction of fluorescence in a region within the sample by brief exposure to high light intensities. After bleaching a region, the recovery of fluorescence over time there can be recorded to measure the rate of redistribution of fluorescent molecules.

This rate of fluorescence redistribution provides information about the processes involved in the movement of the molecule. The movement reflects diffusion, which can be retarded if the fluorescently tagged molecule binds to other molecules. In the latter case, the rate of redistribution of bleached and unbleached molecules will contain information about the strength of the binding interaction. Thus, FRAP is a valuable technique to study molecular dynamics in live cells.

Designing a FRAP Experiment

The most commonly available instrumentation for FRAP is the laser scanning confocal microscope, the scanning capability of which provides

[1] M. Poo and R. A. Cone, *Nature* **247,** 438 (1974).
[2] P. A. Liebman and G. Entine, *Science* **185,** 457 (1974).
[3] M. Edidin, Y. Agyansky, and T. J. Lardner, *Science* **191,** 466 (1976).
[4] J. Schlessinger, D. E. Koppel, D. Axelrod, K. Jacobson, W. W. Webb, and E. L. Elson, *Proc. Natl. Acad. Sci. USA* **73,** 2409 (1976).
[5] D. Axelrod, D. E. Koppel, J. Schlessinger, E. Elson, and W. W. Webb, *Biophys. J.* **16,** 1055 (1976).

METHODS IN ENZYMOLOGY, VOL. 375

flexibility in defining the bleach and imaging regions. The confocal microscope must also have the ability to rapidly change the laser illumination intensity and collect a time-lapse series of images. Conventional microscope configurations with charge-coupled device (CCD) cameras can also be used for FRAP.[6]

After identifying a region of interest, the strategy in a typical experiment is to (1) collect a set of images before the bleach (to be used for normalization of the collected intensities after the bleach), (2) bleach the selected area by scanning it with high laser power, and (3) monitor the recovery of fluorescence by scanning with low laser power over time until the fluorescent and photobleached molecules redistribute to equilibrium. The time course of intensity changes reflects the fluorescence recovery, and this generates a FRAP curve that provides quantitative information about the mobility of the fluorescent molecules (Fig. 1). Analyses of the FRAP curve can show the presence of an immobile or a transiently immobilized fraction, and also, in principle, provide information about both the diffusion coefficient and the strength of binding interactions occurring within the bleached region. Here we focus on FRAP, because it is the most widely used photobleaching strategy, but other photobleaching approaches are sometimes valuable. The scanning capabilities of the confocal microscope enable two relatively new approaches known as fluorescence loss in photobleaching (FLIP) and inverse FRAP (IFRAP). Here we briefly describe these two approaches. For more information, consult Cole et al.[7] and McNally and Smith.[8]

FLIP is a technique in which fluorescence is bleached at one site in the cell, and loss of fluorescence is monitored at another site in the cell. Its chief value is to assess whether there is exchange of molecules between two compartments. For example, if a protein is localized to nucleoli, it may be of interest to know whether there is exchange of the protein between individual nucleoli, or whether once targeted to the nucleolus, the protein remains there. FLIP can be used to assess this by bleaching fluorescence in one nucleolus and then monitoring whether there is loss of fluorescence in other nucleoli. If there is no exchange between different nucleoli, then bleaching one nucleolus will not affect the fluorescence in other nucleoli. On the other hand, if there is some exchange between nucleoli, then bleaching one nucleolus will eventually reduce fluorescence in other nucleoli. To detect this loss of fluorescence in other nucleoli, it is often necessary to

[6] A. S. Verkman, *Methods Enzymol.* **360,** 635 (2003).

[7] N. B. Cole, C. L. Smith, N. Sciaky, M. Terasaki, M. Edidin, and J. Lippincott Schwartz, *Science* **273,** 797 (1996).

[8] J. G. McNally and C. L. Smith, *in* "Confocal and Two-Photon Microscopy: Foundations, Applications, and Advances" (A. Diaspro, ed.), p. 525. Wiley-Liss, New York, 2002.

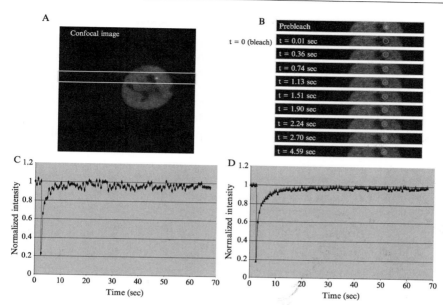

Fig. 1. FRAP of the GFP-tagged glucocorticoid receptor (GFP–GR) at its regulatory sites. Cell line 3617, stably containing 200 tandem repeats of a 9-kb element composed of the MMTV promoter followed by *ras* and BPV genes stably expressing GFP-tagged GR under the control of a tetracycline-off system [D. Walker, H. Htun, and G. L. Hager, *Methods* **19,** 386 (1999)], was used. (A) After hormone induction, a population of the cells shows a visible bright spot in the nucleus corresponding to the MMTV array. (B) Series of images showing the recovery of the GFP–GR after photobleaching. For faster imaging, only a strip of the nucleus including the structure of interest was imaged. Resulting individual FRAP curves are rather noisy (C) but the average of 12 individual curves (D) gives a good representative FRAP curve. (See color insert.)

repeat the photobleach in the initial nucleolus as it recovers fluorescence. This gradually depletes the shared pool of fluorescent molecules, finally leading to a clear loss of fluorescence in all of the other nucleoli.

IFRAP is a photobleaching strategy in which all regions of the cell except the region of interest are bleached. Loss of fluorescence from the region of interest is then monitored. This approach is especially useful as a control to test whether loss of fluorescence in FRAP is an artifact due to bleaching the molecules there. In IFRAP, these molecules are not bleached, so if they still leave the site, this loss cannot arise from damage due to photobleaching. As long as the distribution of fluorescence is at equilibrium before bleaching, the loss of fluorescence from the site in IFRAP should occur at the same rate as the recovery of fluorescence at the site in FRAP.

Cells and Cell Culture Conditions

The first step in a FRAP experiment is to demonstrate that the fluorescently tagged protein of interest is functional and associates with the appropriate compartments within the cell. Either cell lines stably expressing the fluorescently tagged protein or transiently transfected cells can be used.

It is preferable to use stable cell lines in which the protein under study is not significantly overexpressed. Overexpression can lead to an unnatural situation in which nonspecific binding reactions may arise that alter the mobility of the protein. Thus, stable cell lines should be selected for moderate expression levels. Alternatively, expression of the fluorescently tagged protein can be placed under the control of a regulated promoter, such as the tetracycline-regulated system.

To prepare cells for a FRAP experiment, they must be grown in a chamber suitable for live cell imaging. The basic requirements are that the cells grow on a glass surface equivalent to a standard no. 1.5 coverslip for which all microscope objectives are designed. In addition, the chamber ideally should enable temperature and pH control by CO_2. A number of such systems are commercially available [Bioptechs (Butler, PA); Carl Zeiss (Thornwood, NY); binomics chamber [20/20 Technology, Wilmington, NC); Harvard Apparatus (Holliston, MA)], ranging in cost from $3000 to $20,000. A less expensive alternative that offers less precise control of environmental conditions is a live cell coverslip chamber (chambered coverglass system; Nalge Nunc International, Naperville, IL). By heating the microscope stage with an air stream stage incubator (Nevtek, Burnsville, VA) the temperature in these chambers can be regulated, while pH can be controlled by using a buffer that does not require CO_2.

For any of these chambers, cells must first be transferred to the chamber in phenol red-free growth medium. Phenol red is fluorescent; thus, removing it from the medium reduces background fluorescence. If the GFP-tagged protein is under tetracycline control (positive or negative), then the antibiotic must be added or removed to induce expression. Cells are typically incubated overnight in the appropriate chamber to permit attachment. If the cells are not adherent, then it is essential to induce attachment by growth on a modified surface such as polylysine, because fluorescence recovery must be measured from a stationary cell. It is recommended that the concentration of suspended cells giving 60 to 70% cell confluence the next day be experimentally determined. If the cells are not confluent, it takes more time to find them and to perform enough FRAP measurements. When the cells are too nearly confluent, it becomes difficult to distinguish individual cells, and overgrowth may also lead to unhealthy conditions that could alter the function and mobility of the protein.

If transiently transfected cells are used, they can be examined 18–24 h posttransfection. Again, it is essential to perform FRAP experiments on cells with low expression levels that mirror the endogenous distribution of the protein. Immunofluorescence of the endogenous protein can be used as a control.

For reproducibility and reliability of collected data, optimal growth conditions should be maintained during the FRAP experiment. FRAP is a sensitive technique and for some proteins even small deviations from the optimal growth state can lead to slower fluorescence recovery. Keeping cells at 37° by using stage and objective heaters and maintaining the proper pH (using 5% CO_2 when necessary) is especially critical when the mobility of the protein is energy dependent. It is also a useful control to compare FRAP curves measured in the first few minutes on the microscope stage with those measured at later times. If these differ, then conditions for maintaining the cells on the microscope become suspect.

Setting up the Microscope

Different proteins have specific functions, binding partners, and inter-actions of different strength. As a result, mobilities differ from protein to protein, and so for each protein under investigation, specific experimental conditions (frame rate, bleach intensity, and duration of measurement) must be established.

Controlling Laser Illumination Intensity

There are three general ways to change the illumination intensity: by varying the output of the laser, by inserting the neutral density filter in front of the laser, or by using an acousto-optical modulator (AOM) or acousto-optic tunable filter (AOTF). Acousto-optical devices are included with more recent confocal microscopes from Zeiss, Leica (Bannockburn, IL), and Bio-Rad (Hercules, CA). These devices allow 1000-fold variation in illumination intensity within a few seconds, as well as graded control over laser illumination. The rapid switch in intensity levels is critical, espe-cially for studying the mobility of rapidly exchanging molecules to prevent their redistribution into the bleached region before the first postbleach image is taken.

To determine the conditions for photobleaching, fixed cells can be used. Cells are grown on coverslips and fixed for 15 min with 3.5% paraformal-dehyde in phosphate-buffered saline (PBS) with shaking, washed in PBS (three times, 5–10 min each), and mounted on a slide for microscopy. Ideally, conditions should be set to bleach to background levels. In practice, this may require multiple iterations of the bleach, which takes longer to perform. This is a drawback in the case of a fast-moving protein that may redistribute extensively during the bleach. Naturally, laser power should be set at high levels, because this leads to a deeper bleach in less time, and also because the laser is more stable in this mode than at low output levels. Another important factor in the rate of bleaching is the zoom setting on the confocal microscope Photobleaching increases with the square of the zoom,[9] so the scan area at zoom 5 is 1/25th the area at zoom 1 and the rate of bleaching is 25 times faster.

However, if the laser power or zoom is set too high, it may lead to substantial bleaching as the recovery is monitored, even if the AOTF is set at low levels. Too low a transmission setting for the AOTF (<1%) may also lead to unstable transmission levels, so a compromise must be found between laser power levels, AOTF levels, and zoom settings that leads to rapid and deep bleaching during the photobleach but minimal bleaching while monitoring the recovery. In the actual FRAP experiment, a bleach depth that is 50% of starting intensity is often sufficient.

Defining Scan and Bleach Regions

The most sophisticated microscopes allow the user to define regions to be bleached that can be any shape or size. This is a considerable advantage because many biological structures have unusual shapes and therefore bleaching can be restricted to precisely that structure, or to specific subdomains of the structure. This feature also enables IFRAP, in which the bleach region invariably has an irregular shape encompassing the whole fluorescent pool except a small region of interest. These arbitrary bleaching patterns are achieved by using an AOM or AOTF to blank the beam at defined times during the scan.

In selecting a region to be imaged, several considerations arise. For rapidly moving fluorescent molecules, it is advisable to scan as rapidly as possible to enable high temporal resolution for the recovery. Rapid bleaching is achieved by using a high transmission objective (e.g., a Fluar or Planapo objective) with high numerical aperture (NA), by using only one iteration of the bleach, and by scanning a narrow section from the cell nucleus including the region or structure of interest, rather than scanning the whole cell. The changes in fluorescence outside the bleached region within the narrow section are used later for correcting for the loss of intensity due to the scanning (Fig. 1B).

[9] V. Centonze and J. B. Pawley, in "Handbook of Biological Confocal Microscopy" (J. B. Pawley, ed.), p. 549. Plenum Press, New York, 1995.

For both better reproducibility and easier comparison of FRAP curves generated on different days or under different conditions (mutant protein or drug treatment), it is critical to use the same parameters for all experiments (magnification, bleach spot size, laser power, scan speed, number of iterations, time lapse intervals, and duration). Despite this, in our experience, some small day-to-day differences between the FRAP curves arise when using well-standardized experimental conditions. As a consequence, to permit valid comparisons for each experimental treatment, control experiments must always be performed in parallel.

Collecting Data

Before starting the experiments, it is necessary to empirically determine the frame rate and duration of measurement. Most nuclear proteins recover more slowly than expected for free diffusion, that is, with a total recovery period on the order of tens of seconds, to minutes. However, there is usually a substantial change in the first few seconds often accounting for 50–80% of the total recovery. Thus, it is usually necessary to obtain high temporal resolution in the first few seconds of the recovery phase. This can be achieved by using the AOM or AOTF and high-throughput objectives as described in Controlling Laser Illumination Intensity (see previously). In addition, a short interval between the bleach and the first time point is important. This interval is a function of the instrumentation and software in use, and should be considered when selecting a specific microscope. Rapid bleaching is important to provide a more accurate determination of the first time point's intensity. This is achieved by using high laser power, using a high transmission objective, selecting a small bleach area, and using only one iteration of the bleach.

To determine how much time is needed to record the recovery, a pilot experiment should be performed to measure the recovery at later time points (after at least 5–10 s or even minutes). The recovery curve will flatten at some time point, so enough data must be collected to observe this flattening. The time point at which flattening occurs may change with different mutants or drug treatments, so different experiments will require different durations.

In our experiments, we typically acquire ~100 data points during the recovery phase. The number of collected data points is a compromise between acquiring enough time points to generate a smooth recovery curve and minimizing the photobleaching during recovery. As a rule, this photobleaching should not be greater than 5–10%; otherwise, any errors in correcting for this bleaching will contribute substantially to the final FRAP curve. When both steep and shallow components are present in a curve, it is preferable to acquire images at highest scan speed during the steep phase, and then increase the time interval between images during the shallow phase. This reduces bleaching during the recovery.

The parameters used for imaging the recovery are important. All recorded intensity values should be within the detector range (no saturation). For a 12-bit photomultiplier tube, this means that no pixel value should be at 4095. This ensures that the FRAP curve is quantitative. Another parameter that needs to be set is the pinhole diameter, which impacts the signal-to-noise ratio. A larger pinhole improves the signal-to-noise ratio, but also increases the optical section thickness. Thicker optical sections may yield recovery data from undesired regions above or below the structure of interest.

In our experiments, the area monitored during recovery is typically larger than the bleached region. This makes it possible to simultaneously collect data from the same cell to correct for bleaching due to imaging and also makes it possible to verify that the specimen remains in focus during the experiment. When data are collected in a strip image that extends beyond the bleach spot along the scanning axis, data collection time is not increased on a Zeiss 510 confocal microscope. This is because the microscope always scans the full length of the axis anyway, whether or not a subregion has been selected.

To generate a FRAP recovery curve from the image data, average intensities at each time point must be calculated for the bleached region (**on** the bleach spot), for the region to be used for the bleach correction due to monitoring (**off** the bleach spot), and for the background signal in the image (**bgd**). Care should be exercised in selecting the region for bleach correction so that it is far enough away from the bleach spot so as not to be influenced by the intensity changes occurring near the bleach spot. Most confocal microscope software packages enable calculation of these three values by drawing the three regions of interest on the first time point of the image sequence. The software then computes the average intensity at every time point for each of the three regions. For these measurements to be accurate, it is critical to ensure that the cell has remained in focus and not drifted during the course of the experiment. Thus, the image sequence should be examined carefully with particular attention paid to whether the bleach spot region appears to be shifting in any way. The numbers generated from the region measurements can then be imported into a spreadsheet program. After the data are tabulated, the next step is to subtract the background level intensity from all of the measurements (**on** − **bgd** and **off** − **bgd**). The background-corrected data are further corrected for bleaching due to monitoring (I = **on** − **bgd** divided by **off** − **bgd**). Computed intensities (I) are further normalized to the average intensity of the prebleached region [I_{norm} = I divided by a **constant** (average intensity of prebleached *I* values)]. Typically, we acquire at least 5–10 prebleach images to provide a good estimate of the prebleach values.

Once the corrected and normalized recovery data are generated, they should be plotted as a function of time, and the resultant FRAP curve examined. In some cases, there may be large spikes in the curve due to a fluctuation in laser intensity or a sudden movement of the cell. Such data should be discarded. In addition, sometimes the FRAP curve will rise steadily, but then start to gradually decline in intensity at later time points. This is physically impossible, and indicates either that the cell is drifting out of focus, or that there is some problem with the bleach correction. In other cases, the recovery curve will increase steadily and become significantly larger than one. This is also physically impossible, and again indicates problems with either cell focus or the bleach correction. All such anomalous data should be discarded. Individual FRAP curves are often rather noisy (Fig. 1C) due to the low light levels used for imaging, and possibly to small jiggling motions in the live cell under study. In addition, there may be substantial cell-to-cell variability in FRAP curves, reflecting not only this noise, but also possibly subtle physiological differences among cells. To obtain a good, representative FRAP curve, about 10 individual FRAP measurements should be averaged and plotted as a function of time (Fig. 1D). When comparing FRAP curves obtained from controls with those from mutant proteins or drug-treated cells, differences may be small but real. This reflects the fact that diffusion often contributes significantly to the FRAP recovery, and does not change even though binding interactions do. To detect small differences, the experiment should be performed on at least three separate days to determine whether the shift between the two curves is reproducible.

Experiments with FRAP

The FRAP technique is not limited only to studies of protein mobility under normal physiological conditions. It can also be used in different experiments in which the GFP fusion protein is altered either by mutations or specific drug treatments. By comparing the resultant fluorescence recovery curve with the control recovery curve, useful information about changes in protein mobility and binding affinity can be obtained. Using FRAP, it is possible to determine whether protein mobility is an energy-dependent process. To answer that question, FRAP experiments must be performed at 37° to generate a control FRAP curve and compared with curves produced either at 22° or in the presence of an ATP inhibitor (e.g., sodium azide). If these conditions do not alter the FRAP recovery, then protein

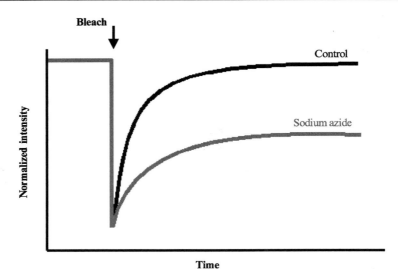

FIG. 2. After treatment with an ATP inhibitor (sodium azide), the resulting FRAP curve is slower, which indicates involvement of energy-dependent factors in the mobility of the protein. (See color insert.)

mobility is not energy dependent. If changes are observed, then energy dependence is indicated. Typically, for energy dependence, fluorescence recovery will be much slower than the control and exhibit an immobile fraction (Fig. 2).

When performing FRAP experiments, it is advisable to limit the time that a specimen remains on the microscope stage. For energy-independent protein mobility, 2 h is the recommended maximum; 1 h or less is recommended for either energy-dependent mobility or after any form of drug treatment. These limits need to be assessed by several controls. The simplest is a test for cell viability after the FRAP experiment, for example, by trypan blue exclusion. In addition, it is critical to compare FRAP recovery curves obtained immediately after the cells are placed on the microscope with those obtained later. If there are differences, this can indicate problems with temperature or pH control on the stage, or simply nonspecific effects arising from prolonged imaging of the sample.

A second control is needed to assure that a drug treatment or other disruption has not changed the mobility of all proteins in the nucleus. A simple approach here is to transfect cells with GFP only, and then treat with the same concentration of test drug or with vehicle. GFP is uniformly distributed in the whole cell and its dynamics are governed by pure

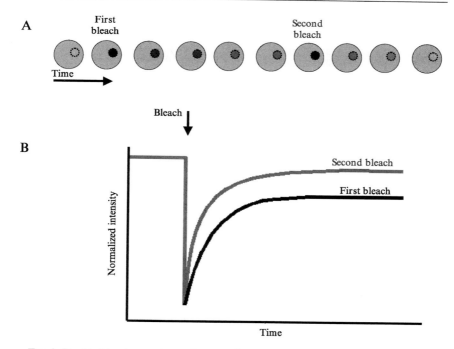

FIG. 3. Double-bleach experimental approach. After the first bleach, the same region of interest is bleached again (A) and the resulting FRAP curves are compared (B). In the case of an immobile fraction, the second bleach leads to a complete recovery because the immobile molecules are now invisible. (See color insert.)

diffusion.[10] The recovery of the GFP-expressing, drug-treated cells should be compared with the recovery of the control cells (treated only with vehicle). No change in GFP recovery rates indicates that the drug treatment has not generically changed nuclear mobilities. It is also possible to look at another functional protein unrelated to the protein under study, and show that its mobility is unaffected by the drug treatment.

A variation of FRAP known as a double bleach is useful to confirm the presence of an immobile fraction. In this approach, a region of interest is bleached and, after a defined recovery time, a second bleach is performed at the same place (Fig. 3A). Recovery of fluorescence after the second bleach is compared with recovery after the first bleach (Fig. 3B). In the case of an immobile fraction, the first bleach renders the bound molecules "invisible" in the recovery after the second bleach. Thus, the fluorescence

[10] R. Swaminathan, C. P. Hoang, and A. S. Verkman, *Biophys. J.* **72**, 1900 (1997).

available for the second bleach should be fully mobile and recover to 100%. If it does not, then the fraction is not truly immobile. The time delay between the first and second bleach can be varied to determine how long the molecules remain "immobilized."

Quantitative Analysis of FRAP Data

At a minimum, FRAP recovery data contain information about the diffusion rate of the protein under study. However, in many cases, the protein not only diffuses but also interacts with other molecules. These binding interactions slow the rate of fluorescence recovery.

As a first step in obtaining estimates of diffusion and binding constants, the recovery of the protein should be compared with the recovery of the fluorescent protein tag alone (e.g., GFP only for a GFP-tagged protein). If under the same bleach conditions the recoveries are nearly identical, then this indicates little or no retardation due to binding, and so the recovery reflects principally diffusion of the tagged protein. (The larger size of the tagged protein compared with GFP alone has a minimal effect on its diffusion constant, because this goes as the cube root of the mass.) Thus, if the protein is not binding, then its recovery will be rapid, because diffusion is fast. For a 1-μm-diameter bleach spot, recovery is complete within ~1 s for a freely diffusing protein.

If the fluorescence recovery is slower than that of GFP alone, then this indicates that binding interactions occur. The longer the interaction with other molecules, the slower is the resulting FRAP curve. In the extreme case, when the molecules are completely immobile, the fluorescence of the bleached spot does not recover at all and the resulting FRAP curve resembles the curves from a fixed specimen.

For extraction of on and off rates of binding, it is critical to determine whether diffusion contributes to the recovery curve. In some cases, diffusion and binding are uncoupled. This occurs when diffusion is fast compared with the binding interactions. In this scenario, diffusion can be ignored because it occurs so rapidly, and thus the resultant FRAP curve reflects only binding interactions. In all other cases, diffusion and binding are inextricably coupled, and this must be considered in extracting estimates of the on and off rates.

To determine whether diffusion contributes to the FRAP recovery, bleaches of different spot sizes are performed (Fig. 4). Two different spots are used: one smaller and one larger than the typical spot used for a FRAP experiment. If diffusion is extremely rapid compared with binding, then the FRAP recovery will be identical for different spot sizes. However, to detect a difference, it is important to determine an average value over at least 10

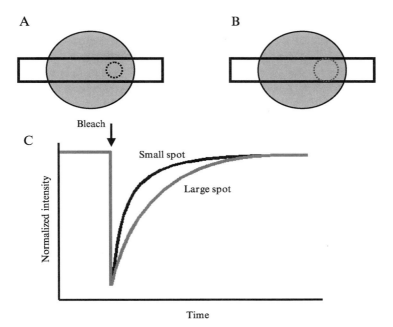

FIG. 4. Spot size dependence of fluorescence recovery. When diffusion plays a role in the mobility of the protein under study, a larger spot size [compare (A) and (B)] will lead to a slower recovery curve (C). (See color insert.)

bleaches for each spot size. If there is a difference in recovery as a function of spot size, then diffusion contributes to the recovery and must be included in the model for fluorescence recovery. Because the diffusion constant can be estimated using GFP only, this value can be substituted into the model describing the FRAP recovery, which will depend on this diffusion constant and the on and off rates of binding.

We are developing a systematic protocol for fitting FRAP data, including a set of model equations for the range of possible behaviors observed with either a single binding reaction or multiple, independent binding reactions.[11] These protocols can be consulted for detailed instructions, and a Web version of this software will soon be available.

[11] B. L. Sprague, R. L. Pego, D. A. Stavreva, and J. G. McNally, submitted (2003).

[28] Correlative Light and Electron Spectroscopic Imaging of Chromatin *In Situ*

By Graham Dellaire, Rozalia Nisman, and David P. Bazett-Jones

Introduction

Perhaps more than for any other cellular organelle, models of the nucleus and its ultrastructure have been continually debated and refined. Until relatively recently, only the nucleolus was recognized as a subdomain where a defined nuclear process was carried out, the transcription of rRNA and the assembly of ribosomal components. We now know that many other subcompartments exist within the nucleus,[1] such as interchromatin granule clusters (IGCs), associated with the regulation of pre-mRNA splicing; Cajal bodies, implicated in small nuclear ribonucleoprotein (snRNP) biogenesis; and promyelocytic leukemia (PML) bodies, variously implicated in cell senescence, apoptosis, and transcriptional regulation. DNA within the eukaryotic nucleus is complexed with histones to form chromatin. Chromatin structure plays an important role in regulation of the transcription, repair, and replication of DNA within the nucleus.[2] Chromosomes, once thought to be randomly distributed in a "spaghetti-like" manner within the mammalian nucleus, have been convincingly shown to reside in distinct territories[3,4] and their relative positions with respect to each other, and within the nuclear volume, may be important for proper nuclear function.[5,6] Advances in microscopy have been central to our understanding of chromatin organization, in the definition of novel subnuclear compartments, and subsequently in the refinement of models describing nuclear structure and function.

Several influential microscopy techniques, aimed at determining structure–function relationships within the nucleus, have been described that either map specific molecules *in situ,* in fixed cells by indirect

---

[1] A. I. Lamond and W. C. Earnshaw, *Science* **280,** 5363 (1998).

[2] A. P. Wolffe, "Chromatin Structure and Function," 1st Ed. Academic Press, San Diego, CA, 1995.

[3] T. Cremer, P. Lichter, J. Borden, D. C. Ward, and L. Manuelidis, *Hum. Genet.* **80,** 3 (1988).

[4] T. Cremer, A. Kurz, R. Zirbel, S. Dietzel, B. Rinke, E. Schrock, M. R. Speicher, U. Mathieu, A. Jauch, P. Emmerich, T. Reid, C. Cremer, and P. Lichter, *Cold Spring Harb. Symp. Quant. Biol.* **58,** 777 (1993).

[5] N. L. Mahy, P. E. Perry, S. Gilchrist, R. A. Baldock, and W. A. Bickmore, *J. Cell Biol.* **157,** 4 (2002).

[6] J. R. Chubb and W. A. Bickmore, *Cell* **112,** 4 (2003).

immunofluorescence, or in living cells by expressing specific proteins fused with fluorescent tags such as green fluorescent protein (GFP) and its variants.[7-9] These and other light microscopy techniques continue to impact molecular and cellular biology, as new insights into nuclear structure and function have emerged from learning how, when, and where two or more factors interact with each other, or how certain factors move from one domain to another, and the rate at which individual protein molecules move through the nucleoplasm.[10,11] Despite the power and the enormous value of these light microscopy methods, it is recognized that higher resolution techniques are required to define the ultrastructure responsible for the cellular complexity observed with the light microscope.

Electron microscopy has played a key role in defining nuclear ultrastructure. In particular, the discovery of the nucleosomal subunit by both biochemical and electron microscopy data,[12,13] the observed reversible folding of 10-nm into 30-nm chromatin fibers,[14] and the imaging of actively transcribing ribosomal genes[15] have together greatly affected our concept of how DNA is organized within the nucleus. Further insight into the structure and behavior of the chromatin fiber has come from scanning force and electron cryomicroscopy.[16,17] More recently, our laboratory has applied correlative light microscopy and electron spectroscopic imaging (ESI) to study DNA–protein complexes including chromatin, visualized both *in vitro* and *in situ*.[18] ESI or energy-filtered transmission electron microscopy is based on the principle of electron energy loss spectroscopy. When a specimen is bombarded with electrons, the elements within the specimen can become ionized, and the energy for that ionization event is equal to the energy lost by the incident electron responsible for that event. An imaging electron spectrometer produces an energy loss spectrum reflecting the elemental content of the sample and is capable of reconstituting an image with electrons that have lost a particular amount of energy. In so doing, an

[7] M. Chalfie, Y. Tu, G. Euskirchen, W. W. Ward, and D. C. Prasher, *Science* **263**, 5148 (1994).
[8] D. M. Miller, III, N. S. Desai, D. C. Hardin, D. W. Piston, G. H. Patterson, J. Fleenor, S. Xu, and A. Fire, *Biotechniques* **26**, 5 (1999).
[9] M. J. Kruhlak, M. A. Lever, W. Fischle, E. Verdin, D. P. Bazett-Jones, and M. J. Hendzel, *J. Cell Biol.* **150**, 1 (2000).
[10] T. Misteli, *Science* **291**, 5505 (2001).
[11] R. D. Phair and T. Misteli, *Nat. Rev. Mol. Cell. Biol.* **2**, 12 (2001).
[12] C. L. Woodcock, *J. Cell Biol.* **59**, 368a (1973).
[13] A. L. Olins and D. E. Olins, *Science* **183**, 122 (1974).
[14] F. Thoma, T. Koller, and A. Klug, *J. Cell Biol.* **83**, 2 (1979).
[15] O. L. Miller, Jr. and B. R. Beatty, *Science* **164**, 882 (1969).
[16] C. Bustamante, G. Zuccheri, S. H. Leuba, G. Yang, and B. Samori, *Methods* **12**, 1 (1997).
[17] J. Bednar, R. A. Horowitz, J. Dubochet, and C. L. Woodcock, *J. Cell Biol.* **131**, 6 (1995).
[18] D. P. Bazett-Jones and M. J. Hendzel, *Methods* **17**, 2 (1999).

image can be formed with electrons that have interacted with a particular element. Nitrogen in the nucleus can be used as an "endogenous stain" for both protein and nucleic acid, and phosphorus can be used as a specific probe for nucleic acids. Thus, chromatin can be imaged within the nucleus without the use of heavy atom stains that might obscure ultrastructural detail or otherwise create ambiguities because of the nonuniform staining characteristics of different biochemical components *in situ*. Because the technique is quantitative, ribonucleoprotein or chromatin structures can be identified in the electron micrographs on the basis of their ratio of nitrogen to phosphorus. Although with this imaging technique there is an associated mass loss from the specimen, ESI remains quantitative for most biological molecules as the rates of mass loss for nitrogen, phosphorus, oxygen, and carbon are similar.[19]

The potential of correlative light and electron microscopy is well recognized and has been applied in numerous studies to determine the structures that underlie a particular cellular behavior[20] (reviewed in Ref. 21), and has been combined with fluorescent probes and immunolabeling methods.[22,23] We describe a technique here that makes significant improvements over previous approaches, so that the technique can readily become a routine method in any electron microscopy laboratory. We combine correlative fluorescence microscopy with ESI, so that elemental maps can be obtained. These permit delineation of specific biochemical features and provide an opportunity for the development of quantum dot (Q-dot)[24,25] and metal-tagged probes[26] for mapping proteins and other features with both high resolution and a high degree of detection sensitivity.

Sample Preparation for Correlative Light and Electron Microscopy

Immunofluorescence Labeling

Cells can be cultured on glass coverslips under the recommended conditions for that cell type (American Type Culture Collection, Manassas, VA). Cells are then fixed with fresh 1–4% paraformaldehyde

[19] D. P. Bazett-Jones, *Microbeam Anal.* **2**, 69 (1993).
[20] G. Sluder and C. L. Rieder, *J. Cell Biol.* **100**, 3 (1985).
[21] C. L. Rieder and G. Cassels, *Methods Cell Biol.* **61**, 297 (1999).
[22] T. J. Mitchison, L. Evans, E. Schultz, and M. W. Kirschner, *Cell* **45**, 4 (1986).
[23] C. C. Robinett, A. Straight, G. Li, C. Willhelm, G. Sudlow, A. Murray, and A. S. Belmont, *J. Cell Biol.* **135**, 6 (1996).
[24] W. C. Chan and S. Nie, *Science* **281**, 5385 (1998).
[25] J. K. Jaiswal, H. Mattoussi, J. M. Mauro, and S. M. Simon, *Nat. Biotechnol.* **21**, 1 (2003).
[26] M. Malecki, A. Hsu, L. Truong, and S. Sanchez, *Proc. Natl. Acad. Sci. USA* **99**, 1 (2002).

in phosphate-buffered saline (PBS, pH 7.5) for 5–10 min at room temperature, and rinsed twice with PBS. Subsequently, cells are permeabilized in
PBS containing 0.5% Triton X-100 for 5 min at room temperature, and
rinsed twice with PBS. By varying the fixation conditions, it is possible to
vary the contrast of certain subnuclear structures visualized in net nitrogen
images. For instance, high contrast images of chromatin can be obtained by
gentle fixation with 1% paraformaldehyde followed by extraction of
soluble nuclear proteins and nucleic acids by the Triton X-100 treatment.
The choice of fixation conditions is even more important for immuno-
electron microscopy (EM), as a balance must always be found between
the preservation of ultrastructure and the preservation of antigenicity.
Individual conditions for a particular antibody–protein antigen pair must
be determined empirically.

Cells on coverslips once fixed and permeabilized are then labeled with
primary and secondary antibodies against nuclear proteins, using standard
immunodetection conditions.[27] For preembedding detection of nuclear
antigens, using Nanogold (Nanoprobes, Yaphank, NY) or Aurion Ultra
Small ImmunoGold (Electron Microscopy Sciences, Fort Washington,
PA) secondary antibodies, a silver enhancement step is required before
embedding (R-Gent silver enhancement kit for EM: Electron Microscopy
Sciences). Briefly, after immunodetection of primary antibodies with
nanogold-coupled secondary reagents for 1 h to overnight, the cells on coverslips are fixed with 2% glutaraldehyde for 5–10 min and then washed five
times in ultrapure, deionized water for 10–15 min. The coverslips are then
subjected to silver enhancement, according to the standard protocol provided by Aurion, for 25–30 min, followed immediately by several quick
washes in distilled water to stop the silver reaction. The coverslips can then
be mounted onto slides with a glycerol-based mounting medium, 90% (v/v)
glycerol in 1 × PBS, before being sealed with transparent nail polish or
rubber cement. The sealant is applied to avoid contamination of the
mounting medium with cleaning solutions used later and to prevent the
coverslip from drying out.

Identification of Cell of Interest

In some instances, it is important to find a region of interest (e.g., a
single cell at a particular stage in the cell cycle) before embedding. Fluorescence imaging is performed on an epifluorescence microscope using
filters optimized for the fluors being used (e.g., Cy3; Molecular Probes,

[27] G. Dellaire, E. M. Makarov, J. J. Cowger, D. Longman, H. G. Sutherland, R. Luhrmann,
J. Torchia, and W. A. Bickmore, *Mol. Cell. Biol.* **22**, 14 (2002).

Eugene, OR). After immuno labeling, cells are examined with a high-magnification oil immersion objective lens ($\times$ 63; 1.4 NA) and imaged with a 12-bit cooled charge-coupled device (CCD) camera (Hamamatsu Orca-ER; Hamamatsu, Bridgewater, NJ). When a cell or a region containing cells of interest has been imaged at high magnification, the region is centered with the aid of the eyepiece reticule and imaged with a low-magnification objective lens ($\times$ 10). Despite the presence of aberrations caused by imaging cells through the added layer of immersion oil while using a low-magnification air lens, an acceptable image map of the distribution of the cells of interest can be obtained. After an image record of the cell(s) is obtained, the immersion oil is removed from the coverslip with ethanol, using a cotton tip, and a 50 mesh copper grid (Electron Microscopy Sciences) is positioned over the region of interest (Fig. 1A). The slide is then removed gently from the microscope, and the edges of the grid are fixed to the coverslip with transparent tape. The slide is subsequently returned to the microscope and examined with the low-magnification lens to reconfirm that the region of interest is represented, and the precise location with respect to the 16 squares of the grid is noted. Other regions of interest within the area of the first grid can also be identified, or additional grids can be used in different regions of the coverslip. The recorded fluorescence images are then processed for future reference (see below).

Resin Selection, Coverslip Embedding, and Sample Sectioning

Unlike imaging *in vitro*-purified nucleoprotein complexes, *in situ* imaging is complicated by a higher background generated by the embedding medium. The signal generated by the resin is particularly important in the imaging of protein-rich structures of relatively low mass density. Optimally, the embedding medium preserves good physical stability and good sectioning properties, produces a low background, and provides good ultrastructure. The commonly used Epon 812 resin produces a signal greater than that of the protein structures of the cell. In this resin, protein structures appear with negative contrast in images collected below the carbon K edge at 284 eV. Thus, Epon is generally not the resin of choice for ESI. Alternatively, the acrylic resin LR White produces a low background in ESI and is suitable for most postembedding immunodetection protocols such as those using 10-nm immunogold. A disadvantage of LR White is its reduced stability in the electron beam as compared with epoxy resins. Taking into account the need for stability and low background, we have found that Quetol 651, an epoxy resin, satisfies these criteria for ESI.

The protein components in cells embedded in Quetol 651 are moderately contrasted relative to the embedding medium at all energy losses

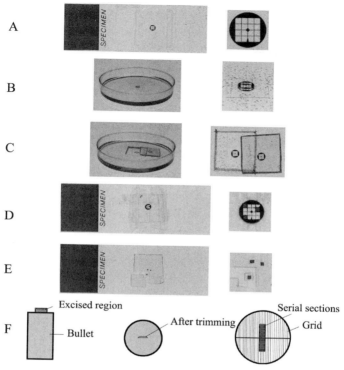

Fig. 1. Schematic diagram outlining the major preparation steps to identify a region of interest by correlative fluorescence and energy-filtering transmission electron microscopy. Cells of interest are first identified by fluorescence microscopy. (A) A 50 mesh electron microscope grid is attached to the coverslip on the side opposite from the cells and covering the region of interest. (B) The cells are refixed and embedded in resin along with the coverslip and initial grid, and a second grid is attached, on top of the embedding plastic, in register with the first grid to maintain identification of the region of interest. (C) After embedding, the resin is cut around the edges of the coverslip and carefully peeled off the coverslip. (D) A third grid is placed on the cell surface side, in register with the second grid. With the aid of the fluorescence microscope, the regions of interest are marked with a scalpel and felt pen, and these regions are cut out of the resin block (E), and (F) glued onto a bullet for sectioning with an ultramicrotome. A ribbon of sections is picked up on EM grids of 400 horizontal mesh. (See color insert.)

below the carbon K edge. The phosphorus-rich chromatin and ribonucleo-protein structures show high contrast at energy losses between 150 and 260 eV. The contrast of phosphorus-rich structures is similar to that of structures that contain no or little phosphorus seen in nitrogen-enhanced images recorded at an energy loss of 415 eV.

Embedding and Relocating Cell of Interest

Before dehydration and embedding, it is important to relocate the cell of interest (Fig. 1A). Another grid can be attached to the stage of a dissecting microscope with transparent tape. This grid is then aligned with the grid on the top face of the coverslip. The slide is then fixed to the stage of the dissecting microscope with tape. The grid on the coverslip is removed and the coverslip is washed with ethanol. A new grid is then placed in the same registration and fixed with transparent tape. This step is required to fix the grid to the coverslip so that it will not wash off during the dehydration and embedding steps that follow. The coverslip is removed gently from the slide and placed into a 40-mm petri dish. The coverslip is washed three times, each for 60 s, in 1 × PBS to remove the mounting medium. The samples that have not been labeled with nanogold can then be postfixed with 2% glutaraldehyde in PBS for 10 min at room temperature. After the postfixation step, the cells are dehydrated with a series of graded ethanol steps of 30, 50, 70, and 95%, with incubations on a shaker for 30–60 min at each step.

Once dehydration of the cells has been accomplished, we can proceed to embedding. One or 2 ml of Quetol 651 resin [Electron Microscopy Sciences (EMS)] is then placed on the coverslip and incubated with shaking for 2–3 h. The resin is removed and replaced with Quetol mix (Quetol 651-NSA kit; EMS) and incubated with shaking for 2–3 h. This step is repeated one more time before the coverslip is placed in a petri dish (Permanox; Nalge Nunc International, Rochester, NY) or a glass petri dish and covered with a thin layer, 0.5 to 2 mm, of Quetol mix. Polymerization of the resin is then accomplished by incubating the dish at 65° for 24 h. A thin layer of Quetol mix is preferable if low fluorescence signal is expected (Fig. 1B).

After polymerization, a second grid is glued onto the resin with the aid of the dissecting microscope, directly above and in register with the original grid on the coverslip. The petri dish is placed on a hot plate at approximately 70° for a few minutes. When the resin becomes soft, it is possible to cut around the edges of the coverslip, just within the perimeter of the coverslip area. The resin containing the cells is then carefully peeled off the coverslip (Fig. 1C). The block is then placed onto a glass slide, cell side up, and the edges taped down with transparent tape. The region of interest is again confirmed with a low-magnification lens by epifluorescence. A third grid in perfect register with the second grid is attached to the cell side of the Quetol block with transparent tape (Fig. 1D). The point of a sharp scalpel is used to mark the edges of the region to be trimmed, and the region is also marked with a felt-tip pen. The grid on the cell side is then removed; the block is again placed on a hot plate at approximately 70°. A large selected area is cut with a razor blade on the hot plate (Fig. 1E). The cut fragment is then glued onto a bullet of Quetol resin and mounted into a ultramicrotome for sectioning (Fig. 1F). If the block surface is lightly wiped with 95% ethanol, the cells can be seen on the surface of the trimmed block with the ultramicrotome binoculars. The region of interest can then be compared with images captured earlier with the fluorescence microscope, so that the cell of interest is kept in the center of a keystone formed by scalpel trimming. The shape of the keystone should have a high length:width ratio, which will permit long ribbons of serial sections to remain intact as they come off the diamond knife during the ultramicrotome sectioning (Fig. 1F).

Serial Sectioning

The region of interest should be seen in the trimmed block being cut by the ultramicrotome. We use an Ultracut UCT ultramicrotome (Leica Microsystems, Richmond Hill, ON, Canada). When a ribbon of 5–15 serial sections, generally in the 60- to 90-nm thickness range, has been obtained on the surface of the water, the ribbon can be manipulated so that the sections are perpendicular to the 400 mesh parallel grid bars. As the ribbon is picked up, the region of interest must be carefully positioned so that it is located over the grid openings and not on a grid bar.

Preparation of Grids and Carbon Films for ESI

Generally, 400 and 600 mesh copper grids containing a center bar (Gilder; EMS) can be used. If postsection immunodetection is required, then it is necessary to use nonreactive nickel grids.

For *in situ* analysis of chromatin, sections between 30 and 80 nm thick, embedded in Quetol 651, become resistant to deformation in the electron beam if coated with carbon films that are 3–4 nm thick. The acceptable range of carbon film thickness can be determined by eye when the films are floated on the surface of water. Initially, however, thickness is estimated by comparisons of the optical density of a double layer of the film compared with an internal mass standard such as DNA or tobacco mosaic virus (TMV) on such a film and imaged by dark-field energy loss.[19] The carbon films are prepared by evaporating carbon onto the surface of freshly cleaved mica (Cressington Scientific, Watford, UK) by electron beam evaporation in a vacuum evaporator operated at approximately 10^{-6} torr. The films are then floated onto distilled water [GIBCO tissue culture grade (0.1-μm pore size filtered); Invitrogen, Carlsbad, CA] and picked up onto the specimen sections with the section side facing up.

Imaging Conditions

Correlative Fluorescence and Electron Imaging of Serial Sections

Before the addition of a carbon film, the grids are checked for integrity of the sections by epifluorescence microscopy and images are recorded for correlative fluorescence and electron microscopy. A grid with serial sections is placed on a glass slide, covered with a glass coverslip, and secured with transparent tape (Fig. 1A) and fluorescence images are recorded at both low and high magnification. Low-magnification images reveal both the integrity of the section as well as the relative orientation and position of the section with respect to the grid center, and high magnification ($\times$63, NA 1.4) is used to obtain the highest resolution spatial map of the fluorescence labeling of the sectioned specimen. Because the sections are typically in the 60- to 90-nm range of thickness, the images are truly confocal. The absence of out-of-focus haze results in enhanced z-resolution with a signal-to-noise ratio that supersedes both the laser scanning confocal microscope and digital deconvolution of serial optical sections.

After addition of a carbon film, the regions of interest are imaged with a Tecnai 20 (FEI, Eindhoven, The Netherlands) transmission electron microscope, equipped with an electron imaging spectrometer (Gatan, Pleasanton, CA) and operated at 200 kV.[18] Serial sections are first stabilized to the beam by imaging with a low flux of electrons at low magnification. Stabilization of the sections helps to prevent deformation, stretching, or rupture of thin sections during imaging. Net phosphorus ratio maps are produced from pre- and postedge images recorded at 120 and 155 eV ($L_{II,III}$ edge), and net nitrogen ratio maps are produced from pre- and postedge images recorded at 385 and 415 eV (K edge). Elemental maps are generated by dividing the element-enhanced postedge image by the preedge image after alignment by cross-correlation. The recording times required to obtain the preedge and postedge images are in the range of 10 to 30 s, and result in electron exposures of the specimen to approximately 10^5 electrons/nm^2.

Phosphorus and Nitrogen Imaging of Chromatin *In Situ*

Analysis of Centromeric Heterochromatin *In Situ*

As discussed above, one major advantage of ESI is the ability to visualize nucleic acid and protein structures in nuclei without stains that might obscure detail or lead to ambiguities that complicate the interpretation and analysis of electron micrographs. Furthermore, by using correlative light microscopy and ESI, we are able to observe substructures within chromatin beyond that possible by conventional transmission electron microscopy (TEM). Unlike *in situ* detection protocols for DNA, such as fluorescence or electron *in situ* hybridization, which require denaturation of DNA and proteinase digestion of nuclear proteins, antibody detection of chromatin components provides a way of imaging chromatin substructures without destroying the underlying nuclear architecture we wish to observe. In Fig. 2A, we have used an anti-centromere immune serum (CREST anti-serum; gift of W. C. Earnshaw, University of Edinburgh) for fluorescence microscopy to define and correlate the position of centromeres in the interphase nucleus for analysis by ESI. This immune serum detects the centromere proteins CENP-A, -B, and -C.[28] Two centromeres were selected for further analysis and high-magnification net phosphorus (red) and nitrogen (green) ESI micrographs were obtained. By overlaying the nitrogen and phosphorus images, one can distinguish highly condensed chromatin (yellow) corresponding to the region detected by the anti-centromere immune serum (Fig. 2B-1 and 2B-2). The first centromere (Fig. 2B-1) is found adjacent to a cytoplasmic invagination (cyt) and on the edge of this invagination is a nucleopore (np). In contrast to the centromeric heterochromatin, adjacent chromatin is much less condensed as demonstrated by loosely organized 30-nm chromatin fibers in the vicinity of the centromere (white arrow, Fig. 2B-1). Interestingly, in this centromere, an array of tightly packed 30-nm chromatin fibers can be seen (white arrowheads, Fig. 2B-1, enlarged region i), suggesting that centromeric, heterochromatin *in situ* consists of ordered arrays of 30-nm fibers. Similarly, analysis of scanning and transmission electron micrographs of mouse centromeres has led to the suggestion that substructures corresponding to 30-nm chromatin fibers exist within the mammalian centromere.[29] Whether these 30-nm chromatin fibers are, in turn, organized in higher ordered arrays *in situ* remains to be determined.

Last, regions containing nitrogen-dense (and phosphorus-poor) non-chromatin protein complexes can also be seen in the composite images of the net phosphorus and nitrogen electron micrographs (bright green regions, highlighted by arrows in Fig. 2B-1, enlarged region ii; and Fig. 2B-2, enlarged region iii). The first centromere is found in close proximity to a nitrogen-rich region with a concentration of phosphorus that is intermediate between those of chromatin and pure protein (Fig. 2B-1, enlarged region ii). This most likely corresponds to nucleolar material, as it

[28] W. C. Earnshaw and N. Rothfield, *Chromosoma* **91,** 313 (1985).
[29] J. B. Rattner and C. C. Lin, *Genome* **29,** 4 (1987).

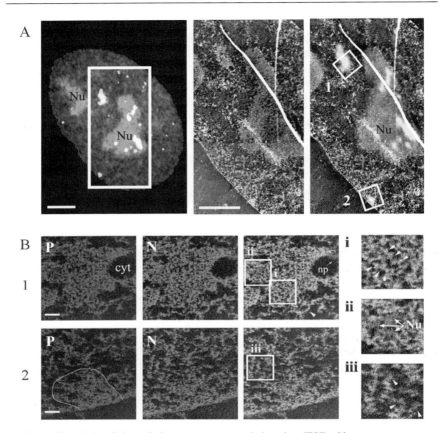

FIG. 2. Correlative light and electron spectroscopic imaging (ESI) of human centromeres. Centromeres were detected in paraformaldehyde-fixed human SK-N-SH cells by indirect immunofluorescence using human CREST antisera (Earnshaw reference) and anti-human secondary antibodies coupled to Cy3 (yellow signal), before being processed for ESI. (A) The immunofluorescence image of a single section of a cell is presented, followed by a low-magnification electron micrograph at 155 eV by ESI of a portion (white rectangle) of the same cell containing several centromeres and a third panel showing the correlative LM/EM image of that region. Two prominent centromeres are apparent in the correlative overlay of the fluorescence and electron micrograph images, one between two nucleoli (Nu) and another adjacent to the nuclear lamina (white box 1 and 2, respectively). (B) High-magnification ESI micrographs of the phosphorus (P, red) and nitrogen (N, green) content of centromeres 1 and 2. The position of a cytoplasmic invagination in the nucleus is shown in B1 (cyt) as well as a nucleopore (np) on the edge of this invagination in the combined net phosphorus and nitrogen overlay shown in B1. The approximate boundaries of the centromere in B2, determined from the fluorescence image, are indicated by a white line in the P image. The combined overlays also indicate a high packing of centromeric chromatin for both centromeres (B1 and B2 and enlarged region of B1-i) with respect to adjacent 30-nm fibers (e.g., white arrow in the B1 overlay). In addition, the chromatin within the first centromeres (B1 and enlargement in B1-i)

aligns, in the correlative fluorescence and low-magnification electron micrograph in Fig. 2A, with the nucleolar signal found near the centromere signal. Nitrogen-rich structures with no detectable phosphorus content correspond to protein-based components. Such structures are found throughout the nucleoplasm, examples of which can be seen in the vicinity of the second centromere (white arrowheads, Fig. 2B-2, enlarged region iii). These nitrogen-rich accumulations may correspond to transcription or splicing complexes as they are located directly adjacent to loosely packed 30-nm chromatin fibers consistent with euchromatin. It should be possible to define the components of these complexes, and thus find clues to their function, using immunogold labeling in combination with correlative LM and ESI. Examples of these techniques are demonstrated later in this chapter for the ultrastructural analysis of both PML bodies and IGCs (see Figs. 4 and 5).

Ultrastructural Analysis of Chromatin during Mitosis

Changes in chromatin and nuclear architecture during mitosis have been extensively studied at the resolution of the light microscope. However, the low spatial resolution of light microscopy does not lend itself to the analysis of cell cycle-dependent changes in the ultrastructure of chromatin or the various subnuclear compartments. Using correlative light microscopy and ESI, we have been able to identify and image at high resolution the nuclear ultrastructure of cells within various stages of the cell cycle. We used both differential interference contrast (DIC) microscopy and immunofluorescence to identify cells of interest. Because a variety of antibodies are commercially available for many chromatin proteins, including histone variants and antihistone modifications, it is possible to identify cells, or regions within a cell, enriched in these proteins for structure and function analysis of chromatin *in situ* during the cell cycle. We have successfully used antibodies directed against several histone modifications including anti-phosphohistone H3 (Ser10) (Upstate Cell Signaling Solutions, Lake Placid, NY), which can be used to identify cells in mitosis in serial sections.[30] For example, in Fig. 3, low-magnification ESI micrographs are shown for various stages of mitosis. The mitotic cells were first

---

appears to contain an ordered array of 30-nm fibers (as indicated by white arrowheads). Another enlarged region of the first centromeres (B1-ii) also shows protein-dense, nonchromatin material (white arrows) that corresponds with nucleolar signal (Nu) in the correlative fluorescence and low-magnification electron micrograph in Fig. 1A. Similarly, large protein complexes can be found in association with 30-nm fibers in the vicinity of the second centromere (white arrows, enlarged region B2-iii), which could represent transcription and/or splicing complexes. Scale bars: (A) 2 μm; (B) 0.2 μm. (See color insert.)

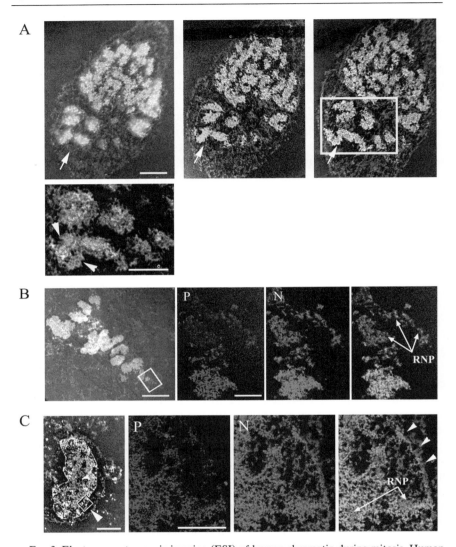

FIG. 3. Electron spectroscopic imaging (ESI) of human chromatin during mitosis. Human SK-N-SH cells grown on coverslips were paraformaldehyde fixed before processing for ESI, and mitotic cells were identified by immunofluorence detection using an anti-phosphohistone H3 (Ser10) antibody (Upstate Cell Signaling Solutions) in serial sections. Low-magnification ESI electron micrographs were taken at 155 eV for cells in late prophase, metaphase, and late anaphase of the cell cycle (A–C). (A) Three serial sections of a late prophase cell are shown, in which large compact chromatin fibers are found that appear in all three sections (white arrow), corresponding to 210–240 nm. A composite of the anti-phosphohistone (Ser10) immunostaining (blue signal) is shown for the first serial section electron micrograph. Two

identified by anti-phosphohistone H3 (Ser10) immunofluorescence of fixed cells before embedding and processing for ESI. Histone H3 is phosphorylated on Ser-10 during the early G_2 phase of the cell cycle within pericentromeric heterochromatin and spreads along chromatin at the onset of chromatin condensation in mitosis.[30] Three serial sections of a cell in late prophase are shown in Fig. 3A, in which we can follow a large condensed chromatin fiber (representing one or more chromosomes) through all three sections (white arrow). A three-dimensional (3D) volume representation of this chromatin fiber (white box, third panel of Fig. 3A) was produced from these serial sections (using Volocity software version 2.0.1; Improvision, Lexington, MA). In the 3D representation, two sister chromatid arms can be seen (white arrows) of what appears to be a single mitotic chromosome. This additional figure demonstrates the power of serial section reconstruction to reveal ultrastructural details that would have otherwise remained unobserved. In Fig. 3B, a low-magnification ESI micrograph is shown of a cell in metaphase. High-magnification net phosphorus and net nitrogen ESI micrographs are also shown for a region of a mitotic chromosome in the low-magnification image (white box, Fig. 3B). Interestingly, the level of condensation seen in the high-magnification ESI micrographs of mitotic chromatin is similar in appearance to centromeric chromatin in interphase nuclei (see Fig. 2B-1 and 2B-2). In the vicinity of this large condensed chromosome region are a number of ribonucleoprotein (RNP) fibers and a large RNP complex, as indicated by white arrows in the net phosphorus and nitrogen composite electron micrograph (Fig. 3B). The larger RNP complex, in particular, is similar in appearance to remnants of interchromatin granule clusters (IGCs) our laboratory has previously

[30] M. J. Hendzel, Y. Wei, M. A. Mancini, A. Van Hooser, T. Ranalli, B. R. Brinkley, D. P. Bazett-Jones, and C. D. Allis, *Chromosoma* **106,** 6 (1997).

sister chromatid arms of a single chromosome (white arrows) can be seen in the three-dimensional volume representation of a selected region of these sections (white box). (B) An electron micrograph of metaphase chromatin at low magnification and high-magnification net phosphorus (P), nitrogen (N), and combined ESI micrographs for a portion of a mitotic chromosome (white box) in which ribonucleoprotein complexes and fibers can be seen adjacent to the chromatin (white arrows, RNP). (C) A low-magnification electron micrograph of a cell in late anaphase and corresponding high-magnification net phosphorus (P), nitrogen (N), and combined ESI micrographs of a selected region of the cell (white box). Chromatin in this cell has begun to decondense in late anaphase and a number of nitrogen-rich RNP complexes can be seen surrounded by chromatin (white arrows, RNP). In addition, the newly formed nuclear envelope can be seen as a green rim in the combined electron micrograph (three white arrows). Scale bars in low- and high-magnification electron micrographs represent 2 and 0.5 μm, respectively. (See color insert.)

observed in metaphase (our unpublished observations). In Fig. 3C, one of two reforming nuclei is shown for a cell in late anaphase, in which the chromatin has begun to decondense. By overlaying the high-magnification net nitrogen and net phosphorus ESI micrographs for a portion of the reforming nucleus (white box, Fig. 3C), two large RNP complexes surrounded by chromatin become apparent (white arrows, RNP). These structures are nitrogen rich but contain intermediate levels of phosphorus. As well, we can see nitrogen-rich (and phosphorus-poor) protein-based fibers (green) at the nuclear rim that represent proteins such as the lamins in the nascent nuclear envelope (three white arrowheads, Fig. 3C). Using conventional transmission electron microscopy, it would impossible to distinguish subnuclear structures such as the nuclear lamina or RNP complexes from chromatin without the use of a specific heavy metal stain that might obscure ultrastructural detail. This ability of ESI to distinguish between various macromolecular complexes on the basis of their phosphorus and nitrogen content can be used to great advantage to study various subnuclear compartments in relation to chromatin, as we see in the next section.

Analysis of Subnuclear Compartments in Relation to Chromatin

ESI of Chromatin in Vicinity of PML Nuclear Bodies and IGCs

Many subcompartments or domains within the nucleus have been defined.[1] Using antibodies directed against proteins within these compartments, we can probe both the ultrastructure of these domains and their physical relationship to chromatin, using correlative light and ESI. In Fig. 4, we have used antibodies directed against the CREB-binding protein (CBP) to define a region in a human SK-N-SH cell nucleus containing PML nuclear bodies (NBs). The transcription coactivator and histone acetyltransferase CBP is a dynamic component of PML NBs in several cell types including SK-N-SH cells.[31] Thus, not only can we define which regions of the nucleus contain PML NBs, we can use secondary antibodies labeled with gold to mark the relative position of CBP within these bodies [white dots in the black and white image (Au) at the far left of Fig. 4A and in Fig. 4B]. Two serial sections of one PML NB are shown that are representative of the morphology of the majority of PML bodies in human cells. These PML bodies are completely surrounded by chromatin (Fig. 4A-1 and 4A-2; yellow signal in the overlay net phosphorus and net nitrogen ESI

---

[31] F. M. Boisvert, M. J. Kruhlak, A. K. Box, M. J. Hendzel, and D. P. Bazett-Jones, *J. Cell Biol.* **152,** 5 (2001).

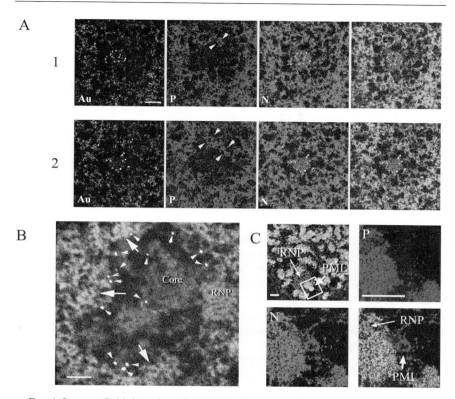

FIG. 4. ImmunoGold detection of CREB-binding protein (CBP) in PML nuclear bodies imaged by ESI. Before processing for ESI, human SK-N-SH cells grown on coverslips were fixed in paraformaldehyde before detection of PML protein with a mouse anti-CBP monoclonal antibody (Santa Cruz Biotechnology, Santa Cruz, CA) and an anti-mouse secondary antibody coupled to Aurion Nanogold particles (EMS) either alone or in conjunction with a Cy3-labeled secondary antibody. (A and B) Interphase structure of PML nuclear bodies. ESI micrographs for gold (Au, white), phosphorus (P, red), or nitrogen (N, green) were taken for serial sections at high magnification for PML nuclear bodies (NBs). The gold particles are silver (Ag) enhanced, and thus the ESI map of the particles is obtained via Ag detection. The combined phosphorus and nitrogen overlay of these micrographs helps distinguish PML bodies from RNP structures such as IGCs [see RNP in (B)] because of their nitrogen-dense protein core (green), which can be easily distinguished from the surrounding chromatin and RNP fibers (yellow). Two serial sections of one PML NB surrounded by chromatin are shown in (A1) and (A2). A single chromatin fiber can be followed in the two sections (white arrows in net P), demonstrating the close contact of these NBs to chromatin. The Ag-enhanced gold particles (white dots in the overlay) appear across the top of the PML NB in the first section (A1) and around the periphery of the second section (A2), indicating that the CBP is found at the periphery of PML NBs and not at their core. (B) A high-magnification ESI micrograph of a second PML NB illustrating the peripheral localization of the PML protein within these structures. The core of the PML NB and an adjacent

micrograph at the far right). One 30-nm chromatin fiber can be seen beginning in one serial section and continuing in the next [Fig. 4A-1 and 4A-2, net phosphorus image (red), white arrowheads]. An enlargement of a second PML NB in Fig. 4B shows the surrounding chromatin fibers in greater detail (yellow fibers indicated by the white arrows). The central core of a PML body is nitrogen dense (Fig. 4, green signal in the center of the composite images) and contains no detectable phosphorus above featureless background regions in the nucleoplasm, indicating that both chromatin and RNPs (red signal) are excluded from its protein-rich core. CBP localizes on the surface "cap" of the PML NB in the first serial section (Fig. 4A-1) and around the periphery in the next serial section (Fig. 4A-2). Similarly, CBP can be localized to the periphery of a second PML NB shown in Fig. 4B (white dots and arrowheads). Thus, it appears that CBP localizes primarily to the interface between the surrounding chromatin and the periphery of the protein-based core of the PML NB. The limited resolution of light microscopy would preclude the observation of such detail by coimmunofluorescence. Previously, it has been suggested that the PML NBs are involved in transcriptional regulation.[32,33] The localization of the CBP protein to the periphery of PML NBs, at the interface between the body and the surrounding chromatin, would support a role for these NBs in chromatin transactions, including transcriptional activation and/or repression.

During cell division, the entire contents of the cell, including the nucleus and its various subcompartments, must be broken down and segregated faithfully to produce two daughter cells. The manner in which subnuclear compartments disassemble and reassemble, the mechanisms governing their relative position to chromatin, and to what extent they might retain their function during mitosis are not completely understood.

---

[32] S. Zhong, P. Salomoni, and P. P. Pandolfi, *Nat. Cell Biol.* **2**, 5 (2000).
[33] C. H. Eskiw and D. P. Bazett-Jones, *Biochem. Cell Biol.* **80**, 3 (2002).

---

accumulation of RNPs are indicated. Surrounding chromatin (yellow fibers) is indicated by large arrows and Ag-enhanced gold particles corresponding to the CBP protein are indicated by small arrows (white dots). (C) Ultrastructure of a PML NB during prophase. First, a PML NB was located by overlaying the light microscopy image of a single section of a prophase cell containing a PML NB (yellow spot, PML) with the low-magnification electron micrograph of the same cell taken at 155 eV by ESI. A large RNP complex is also apparent in the correlative LM/EM micrograph (RNP). High-magnification net phosphorus (P), nitrogen (N), and combined ESI micrographs of the region of the prophase cell containing the PML NB (white box) are shown. At high magnification, the PML NB is less dense than during interphase and has a pronounced donut shape. This magnified region also contains RNP complexes closely associated with chromatin (RNP). Scale bars: (A and B) 0.2 μm; (C) 1 μm. (See color insert.)

Using ESI, we can explore the changing ultrastructure and spatial relationship of subnuclear compartments, such as PML NBs or IGCs relative to chromatin during mitosis. In Fig. 4C, for example, we have imaged a PML NB in a prophase SK-N-SH cell. As chromatin has condensed during prophase, the PML NB is no longer surrounded by chromatin as in the interphase nucleus (see Fig. 4A and B) but is still associated physically with chromatin. The PML NB has also become less dense and has taken on a donut-like shape. A large RNP complex can also be seen in the field of the high-magnification ESI micrograph that may represent a remnant IGC. The ultrastructure of these IGCs, also known as SC35 domains, or "splicing speckles," in relation to chromatin is explored in the next section.

Elemental mapping of other subnuclear structures, such as IGCs, and their points of contact to the surrounding chromatin, can also be visualized with ESI. Figure 5A shows the phosphorus, nitrogen, and composite ESI micrographs, respectively, of a region bordered by the nuclear envelope and containing an IGC, a PML body, and the surrounding chromatin fibers. A nuclear pore (np) within the nuclear membrane is indicated by a depletion of signal in the phosphorus map, illustrating the locations of fused outer and inner membranes. The core of the IGC consists of phosphorus-rich particles, which generate lower phosphorus signal intensity than the surrounding chromatin fibers. Each large particle corresponds to a highly folded RNP fibril.[34] The total mass and stoichiometric relationship between protein and RNA, based on the phosphorus and nitrogen maps, has been calculated[35] to show that IGCs contain heterogeneous nuclear (hn) RNA. The IGC region in the net nitrogen ESI micrograph consists of a large network of protein fibers and, in the composite image, it is clear that the phosphorus-rich nodules (yellow particles) are embedded in this protein architecture. The nitrogen image also depicts protein-based fibrils extending from the core of the IGC into the PML body and into the surrounding chromatin fibers. It will be important to deduce whether this protein structure assembles before the recruitment of the phosphorus-rich granules and the surrounding chromatin, during the postmitotic reestablishment of IGC domains.

The IGCs are readily observed by immunofluorescence microscopy using antibodies directed against splicing factors such as SC-35.[36] We have used Nanogold secondaries with silver enhancement to localize the SC-35 splicing factor within the IGC at the EM level. This can be especially useful for characterization of mitotic IGCs, as the distinctive granules found

[34] M. J. Hendzel, M. J. Kruhlak, and D. P. Bazett-Jones, *Mol. Biol. Cell* **9**, 2491 (1998).

[35] D. P. Bazett-Jones, M. J. Hendzel, and M. J. Kruhlak, *Micron.* **30**, 151 (1999).

[36] X.-D. Fu and T. Maniatis, *Nature* **343**, 437 (1990).

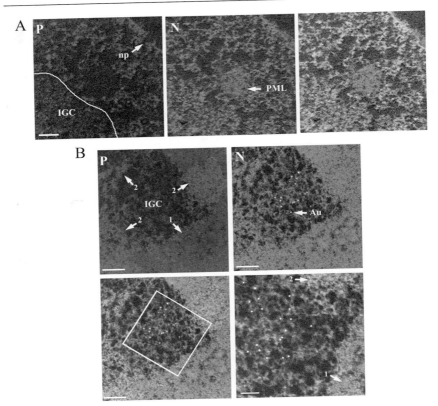

FIG. 5. Organization of subnuclear components. (A) Human SK-N-SH cells were double-labeled with primary anti-PML (Santa Cruz Biotechnology) antibody and primary anti-SC35 monoclonal cell culture supernatant, followed by secondary fluorescent antibodies for correlative microscopy. ESI micrographs of a thin section of an embedded cell show the phosphorus, nitrogen, and composite maps of a region of the cell containing a PML nuclear body, an IGC, and a nuclear pore. The IGC region is defined in the phosphorus map. Scale bar: 0.2 μm. (B) Human SK-N-SH cells were incubated with primary anti-SC35 antibody, followed by secondary labeling with Nanogold and fluorescently labeled antibodies for correlative identification. The low-magnification nitrogen, phosphorus, and composite maps illustrate the position of an IGC with respect to the nucleolus (arrow 1) and the surrounding condensed chromatin (arrow 2). The white particles are silver-enhanced, gold-labeled antibodies and show the location of the SC35 splicing factor within the RNP fibrillar network. The boxed region within the low-magnification composite map is shown at a higher magnification for a detailed view of the IGC and surrounding chromatin structure. Scale bars: 0.5 μm in the low-magnification P, N, and composite maps; 0.2 μm in the high-magnification composite map. (See color insert.)

within phosphorus ESI micrographs of interphase IGCs are no longer available for definitive localization. After incubation with the Nanogold antibody, exposure of the remaining available epitopes to a fluorescent secondary allows us to use correlative microscopy to locate the regions of interest, as described previously. Figure 5B shows the phosphorus, nitrogen, and composite ESI micrographs of an interphase cell in which an IGC region is proximal to the nucleolus (arrow 1) and is surrounded by condensed chromatin (arrows 2). The phosphorus image clearly delineates the regions of chromatin from that of the nucleolus, which has a lower phosphorus signal intensity and, subsequently, a lower phosphorus-to-nitrogen ratio in the composite map. The gold particles (white dots; Fig. 5B, Au) illustrate the location of the SC-35 splicing factor within the RNP scaffolding. The enlarged micrograph of the composite map, as marked by the box within the image, shows a detailed view of the IGC region and the surrounding chromatin, with the corresponding increase in ultrastructure and compositional information.

The presence of condensed chromatin at the periphery of IGCs is of significance, as we have previously determined[34] that the chromatin surrounding these domains is enriched in the most highly and dynamically acetylated forms of the core histone H3, and is therefore transcriptionally active. The precise location of transcriptionally active chromatin with respect to the condensed structure and the core of the IGC remains to be determined. We are presently employing markers such as Nanogold antibodies against acetylated histones to address this question.

Lower Limit of Detection of Nucleic Acid-Based Fibers in Sections by ESI

Because transmission electron microscopic images are 2D projections, the interpretation of structural features is facilitated by using sections that are as thin as possible, unless tomographic or serial sectioning methods are used. The thin sections required for elemental detection by ESI are well suited to the observation of structures that are not visible in thicker stained sections imaged by conventional EM techniques. In the overlay image of the net phosphorus and nitrogen in Fig. 2B-1, chromatin fibers that are in the 30-nm range are easily observed (white arrow pointing to a fiber). Regions where these chromatin fibers decondense to 10-nm diameters are also frequently observed. In Fig. 6, we are able to see such fibers in a region of decondensed chromatin surrounding a PML nuclear body. The individual nucleosomes are clearly resolved within the phosphorus and composite maps as red and yellow nodules, respectively. Enlarged views of the boxed areas within the phosphorus map are presented in Fig. 6B,

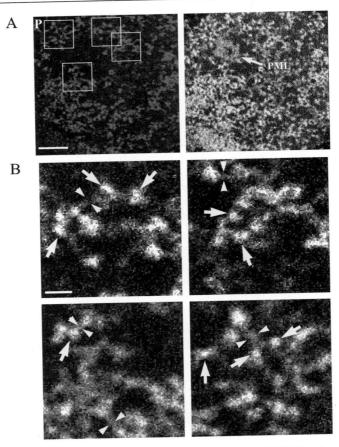

FIG. 6. Nucleosomes in decondensed chromatin fibers. Human SK-N-SH cells were labeled with an anti-PML antibody, embedded, and sectioned. (A) Low-magnification phosphorus and composite maps of a region of decondensed chromatin surrounding a PML nuclear body. The locations of clearly resolved nucleosomes are highlighted with boxes in the phosphorus image. Scale bar: 0.2 μm. (B) A series of magnified views of the highlighted regions within the phosphorus map. Individual nucleosomes are marked with single arrows, and the linking DNA is indicated with double arrowheads. Scale bar: 20 nm. (See color insert.)

and show individual nucleosomes (large arrows) linked by strands of phosphorus-rich material (double arrowheads). This material is most likely linker DNA, and at 2 nm in diameter, represents the limit of resolution for nucleic acids by ESI within embedded thin sections. In contrast, the lower limit of fluorescence *in situ* hybridization (FISH) analysis of chromatin is approximately 170 nm. It is also apparent from the high-magnification images that the nucleosomal spacing is highly variable within decondensed chromatin *in situ*. Therefore, ESI can be routinely used to analyze chromatin *in situ* both in its most condensed state, that of mitotic chromosomes, and in the most decondensed form of chromatin, the 10-nm nucleosome fiber.

Summary and Future Prospects

We have illustrated the power of ESI in providing structural and analytical information about both chromatin and subnuclear domains imaged *in situ*. The simultaneous resolution of structural and analytical information provided by ESI has and will reveal important insights into chromatin organization and other nucleoprotein structures *in vitro* and *in situ*. As with any technique, the information that is provided should be supplemented with that obtained by other means. Whenever possible, either biochemical or additional imaging techniques should be used to corroborate observations made using ESI. This will help to overcome some of the drawbacks or potential artifacts that are inherent with electron and other microscopy techniques. An obvious drawback for conventional EM techniques is that samples are generally fixed with aldehyde cross-linkers before imaging. Precautions to minimize fixation artifacts, such as optimization of fixation for ultrastructural preservation, and the use of cryo-EM techniques, can be taken. Nonetheless, one must always be aware of the potential for distortion of structures. A particular disadvantage of ESI is the high electron beam exposure that is required to form element-specific images. For sections in the 50-nm-thickness range, the radiation exposure can be as high as 100 electrons/Å^2. Although this level of radiation does result in some mass loss, mass loss and specimen distortion are much less of a concern with the stabilization provided by the use of thin carbon films.

Future of ESI: EM Tomography and Multiplex Detection of Nucleic Acids and Protein In Situ

Understanding the function of particular nuclear domains will be enhanced by imaging approaches. These will include ESI because of its ability to image the structures themselves, rather than the contrast agents that bind to them in a nonlinear and biochemically unpredictable manner, and because of the ability to distinguish biochemical entities on the basis of elemental content. We have shown that ESI can be used in conjunction with immunolabeling with antibodies or streptavidin tagged with metal complexes such as colloidal gold. Quantum dots are also particularly interesting for correlative LM/EM in that they can be distinguished by their autofluorescence, using the light microscope, and by their cadmium or zinc content, using ESI.[37] Similarly, a range of Fluoro-Nanogold secondary antibodies that are both fluorescent and electron dense are available (Nanoprobes). It may also be possible to directly tag proteins with metals through the ability of certain protein domains to coordinate metal ions, such as the zinc finger (zinc, magnesium), metallothionein (zinc, copper, mercury, lead, and manganese), polyhistidine (nickel), and calmodulin (calcium). Nickel tagging of recombinant single-chain, variable fragment antibodies (scFv) has already been used for ESI of protein complexes imaged *in vitro*.[26] In addition, antibodies coupled to other elements such as boron[38,39] have been used for the immunodetection of molecules by ESI. Currently, we are investigating the possibility of detecting halogenated nucleic acids such as RNA by pulse labeling cells with fluoro- or iodouridine before processing for ESI. Therefore, it may be possible in the near future to distinguish DNA from RNA and image nucleic acids while identifying multiple protein components within the nucleus or within isolated supramolecular assemblies of protein imaged *in vitro*. Last, correlative imaging of serial thin sections or thick sections by EM tomography can be used to reconstruct three-dimensional representations of chromatin and other subnuclear domains *in situ*. As the gene products that play a role in the organization of nuclear domains are identified, the mechanisms by which they function will require structural analysis, using imaging approaches. In this regard, ESI has the potential to play a key role in the continued refinement of nuclear structure and function models.

Acknowledgments

We acknowledge the skilled work of Ying Ren and Ren Li for aid in preparing and imaging serial sections for the ESI micrographs in this manuscript. G.D. is a Senior Postdoctoral Fellow of the Canadian Institutes of Health Research (CIHR). These studies are supported by operating grants to D.P.B.-J. from the CIHR, the Natural Sciences and Engineering Research Council, and the Cancer Research Society, Inc.

[37] R. Nisman, G. Dellaire, Y. Ren, R. Li, and D. P. Bazett-Jones, *J. Histochem. Cytochem.* in press (2004).

[38] M. M. Kessels, B. Qualmann, and W. D. Sierralta, *Scanning Microsc.* Suppl, **10,** 327 (1996).

[39] B. Qualmann, M. M. Kessels, F. Klobasa, P. W. Jungblut, and W. D. Sierralta, *J. Microsc.* **183,** 1 (1996).

[29] RNA Fluorescence *In Situ* Hybridization Tagging and Recovery of Associated Proteins to Analyze *In Vivo* Chromatin Interactions

By Lyubomira Chakalova, David Carter, and Peter Fraser

Introduction

The emerging picture of the vertebrate cell nucleus suggests a high degree of dynamic, structural organization of chromosomes and genes that may reflect or influence function.[1–3] It is generally recognized that individual chromosomes occupy distinct positions or territories within the cell nucleus. During interphase, there appears to be little net movement of chromosomes from their territories; however, subchromosomal regions have been observed to undergo rapid, locally constrained movements with an average range of 0.5 μm.[4] Similar rapid movements of chromatin have also been observed in yeast[5] and *Drosophila*.[6] Such movements may play a role in the nuclear repositioning of some genes, which has been correlated with transcriptional activation, silencing, and replication timing.[7–9] Rapid diffusional movements of chromatin may also play a role in long-range interactions between distal *cis* elements as has been proposed for a number of gene loci, perhaps most notably the β-globin genes and β-globin locus control region (LCR).[10,11]

Light microscopy (LM) and electron microscopy (EM) have been used to great effect in the study of nuclear and chromosome architecture and chromatin structure. However, a large gap exists, both in the practical application of these techniques and in their effective resolving powers. For example, LM in conjunction with DNA fluorescence *in situ* hybridization

[1] T. Cremer and C. Cremer, *Nat. Rev. Genet.* **2**, 292 (2001).
[2] L. Parada and T. Misteli, *Trends Cell Biol.* **12**, 425 (2002).
[3] J. R. Chubb and W. A. Bickmore, *Cell* **112**, 403 (2003).
[4] J. R. Chubb, S. Boyle, P. Perry, and W. A. Bickmore, *Curr. Biol.* **12**, 439 (2002).
[5] S. M. Gasser, *Science* **296**, 1412 (2002).
[6] J. Vazquez, A. S. Belmont, and J. W. Sedat, *Curr. Biol.* **11**, 1227 (2001).
[7] K. E. Brown, S. Amoils, J. M. Horn, V. J. Buckle, D. R. Higgs, M. Merkenschlager, and A. G. Fisher, *Nat. Cell. Biol.* **3**, 602 (2001).
[8] S. T. Kosak, J. A. Skok, K. L. Medina, R. Riblet, M. M. Le Beau, A. G. Fisher, and H. Singh, *Science* **296**, 158 (2002).
[9] J. Gribnau, K. Hochedlinger, K. Hata, E. Li, and R. Jaenisch, *Genes Dev.* **17**, 759 (2003).
[10] M. Wijgerde, F. Grosveld, and P. Fraser, *Nature* **377**, 209 (1995).
[11] D. Carter, L. Chakalova, C. S. Osborne, Y. F. Dai, and P. Fraser, *Nat. Genet.* **32**, 623 (2002).

(FISH) has been used to assess relatively large changes in nuclear localization of specific gene loci and to study large-scale structure of chromosomes. However, the resolution of LM is limited to approximately half the wavelength of light, which means that any two objects less than 250 nm apart will appear to colocalize. This limits the usefulness of LM in the analysis of chromatin fiber conformations of specific gene loci and in detecting interactions between defined genetic loci. EM has extraordinary resolving power and has been used to visualize primary, secondary, and tertiary chromatin fiber structures[12] in chromatin spreads but its use in the analysis of specific loci in three-dimensionally preserved nuclear specimens is lacking.

We have developed a method designed to detect tertiary chromatin interactions between specific DNA sequences *in vivo* (Fig. 1). We have used this method to show that the distal β-globin locus control region is intimately associated with an actively transcribed β-globin gene in erythroid cells.[11] The technique, called RNA FISH TRAP (fluorescence *in situ* hybridization tagging and recovery of associated proteins), utilizes the targeting power of *in situ* hybridization to tag proteins near a specific, transcriptionally active gene locus. Briefly, a hapten-labeled, antisense DNA probe is hybridized to the intron sequences of a nascent primary transcript associated with an actively transcribed gene in formaldehyde-fixed cells. A hapten-specific antibody conjugated with horseradish peroxidase (HRP) is then directed to the DNA probe, thereby localizing HRP activity to the specific gene locus. The HRP is then used to catalyze the covalent attachment of a tag (in this case, a biotinylated tyramide) to proteins in the immediate vicinity of the gene. This tag can then be used in affinity purification procedures to recover proteins and chromatin complexes near the site of transcription. This technique helps to bridge the gap between LM and EM in that it allows the recovery and analysis of sequences that are engaged in functional interactions or specifically juxtaposed to a defined gene locus *in vivo*. Permutations of this technique will undoubtedly aid in our understanding of higher order chromatin structure and the role of chromatin interactions in various nuclear processes.

Probes for RNA FISH TRAP

Principles

RNA TRAP is based on a modified RNA FISH technique. A successful FISH experiment is highly dependent on the choice and quality of probe. The sequence of the probe, the length, and the method of labeling can all

---

[12] C. L. Woodcock and S. Dimitrov, *Curr. Opin. Genet. Dev.* **11,** 130 (2001).

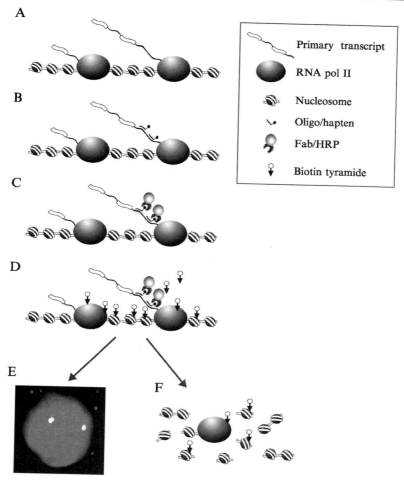

Fig. 1. Schematic of the RNA FISH TRAP technique. (A) Schematic of a transcribed gene in the nucleus of a fixed cell; RNA Pol II synthesizes primary transcripts. (B) *In situ* hybridization of hapten-labeled, antisense oligonucleotide probes complementary to intron sequence of the primary transcript. (C) Addition of HRP-conjugated Fab fragment localizes HRP activity to the site of transcription. (D) Addition of biotin–tyramide. HRP catalyzes the activation of the tyramide ring into a radical intermediate that covalently attaches to proteins in the immediate vicinity. (E) One slide is stained with an avidin–Texas red conjugate to visualize biotin–tyramide deposition by epifluorescence microscopy. (F) Cells are scraped off the remaining slides and sonicated to yield chromatin fragments with an average DNA size of 400 bp. Biotinylated chromatin fragments can then be purified by affinity chromatography or ChIP.

affect the outcome. Obviously, probes containing unique sequences are desirable; however, probes with minor amounts of repetitive sequences can be used if hybridization of the repetitive sequences is suppressed by preannealing with excess C_0t-1 DNA. In conventional *in situ* experiments, probe sensitivity generally increases with the length of the labeled probe. This is especially useful in low copy number target applications such as DNA FISH of single gene loci. Good-quality signals are obtained with longer probes as more label concentrates at the site of hybridization and therefore detection is more efficient. When the target of interest is the nascent RNA of a transcribed gene, the number of target molecules per gene locus may vary greatly depending on the gene. Longer probes should, in theory, permit the visualization of genes with a low abundance of primary transcripts, whereas short probes can be used for genes with abundant target RNAs. For RNA TRAP on the β-globin genes, we used four 30-mer, antisense oligonucleotides complementary to the large intron of the β-globin gene. Each oligonucleotide is triple-labeled on the 5' end with digoxygenin (Dig) tethered by hexacetylglycerol spacers. These commercially available modified oligos provide a stable and consistent source of probe, which can be stored frozen in aliquots for several years. Although triple-labeled oligonucleotides have a high specific activity, they are not sensitive enough for genes that are often described as having a "low transcription rate." Detection of genes with a low abundance of target RNAs at the site of transcription requires larger probes. The most commonly used protocols for generating relatively large labeled probes for *in situ* hybridization applications are nick translation, random priming, and hapten incorporation during polymerase chain reaction (PCR). One intrinsic disadvantage of all these methods is that the labeled probes are double-stranded, leading to probe reannealing in solution and reduced hybridization to the desired target. Single-stranded RNA probes are often used to circumvent this problem, but we have found them to yield unacceptably high levels of background in RNA FISH. To overcome this problem, we have developed a procedure for generating large, single-stranded DNA probes with high levels of incorporated hapten. The method involves *in vitro* transcription of the probe region from a plasmid template, followed by incorporation of a hapten-labeled nucleotide into a single-stranded cDNA probe through random-primed reverse transcription. This method generates single-stranded DNA probes with high specific activity, although the probes can cover a large region (we have made RNAs up to 5.5 kb; J. Miles, unpublished data, 2002). Random priming ensures that the labeled DNA fragments are under 400 bp, making them ideal for *in situ* hybridization. A comparison between the oligonucleotide probes mentioned above and double-stranded and single-stranded DNA probes clearly shows that the RT probes have much greater sensitivity. We have successfully used such probes covering nongene regions of 1–4 kb to detect rare intergenic transcripts in the β-globin locus and gene primary transcripts in a number of other gene loci with low primary transcript levels. Although double-stranded probes can be used effectively for RNA TRAP, they are less sensitive, necessitating the use of larger probes and possibly decreasing resolution. A summary of the single-stranded RT method is given below.

Method: Synthesis of Single-Stranded DNA Probes

Preparation of Template for In Vitro Transcription

The sequence to be used as a probe should be cloned in a plasmid vector with prokaryotic promoter sequences flanking the polycloning site (e.g., T3 and T7 promoters of pBluescript; Stratagene, La Jolla, CA). This allows labeling of either strand regardless of insert orientation. A standard *in vitro* transcription procedure from a plasmid template is carried out. *In vitro* transcription kits are commercially available. Standard procedures for avoiding RNase contamination should be followed.

1. Linearize 10–30 μg of plasmid DNA with an appropriate restriction enzyme that cuts downstream of the insert with respect to the bacterial promoter.
2. To destroy traces of RNase in the DNA sample, adjust to 0.5% sodium dodecyl sulfate (SDS), proteinase K (100 μg/ml), and incubate for 30 min at 37°.
3. Phenol–chloroform extract, and then chloroform extract.
4. Ethanol precipitate.
5. Dissolve the pellet in RNase-free water.

In Vitro Transcription Reaction

1. Mix in the following order at room temperature (to prevent precipitation of spermidine, a component of the enzyme buffer):

RNase-free water	100 μl (final volume)
DNA template (linearized plasmid DNA)	10 μg
NTP mix (25 mM each; final concentration, 2.5 mM each)	10 μl
10× RNA polymerase buffer	10 μl
RNA polymerase (T3, T7, or SP6)	200 U

2. Add 40 units of RNasin (optional).
3. Incubate at 37° for 2–4 h.

4. Add 1 μl of DNase I (1 mg/ml).
5. Incubate at 37° for 15–20 min.
6. Add 40 μl of 6 M ammonium acetate (final concentration, 1.7 M).
7. Phenol–chloroform extract.
8. Ethanol precipitate with 3 volumes of ethanol, for 30 min on ice.
9. Dissolve the pellet in RNase-free water.
10. Make sure the RNA concentration is about 2 μg/μl. The expected yield is tens of micrograms.

Reverse Transcription

A hapten-labeled nucleotide is present in the reverse transcription reaction and is incorporated into the DNA product. The example below describes how to prepare a Dig-labeled probe. DNP-11-dUTP or biotin-16-dUTP can be used as alternatives. The reaction can be scaled up. Adjust all volumes accordingly.

1. Mix the following on ice:
   RNase-free water   20 μl (final volume)
   RNA template (from step 7 above)   4 μg
   Random hexanucleotide mix (10 μg/μl)   1.5 μl
2. Denature at 65° for 5 min; quickly chill on ice.
3. Add the following:
   5× SuperScript first-strand buffer   4.0 μl
   0.1 M dithiothreitol (DTT; final concentration, 10 mM)   2.0 μl
   dNTP-Dig mix (2 mM dATP, 2 mM dCTP, 2 mM dGTP,
   1.3 mM dTTP, 0.7 mM Dig-11-dUTP   7.5 μl
4. Add RNasin (optional).
5. Mix gently and incubate at 42° for 2 min.
6. Add 400 U of SuperScript II.
7. Incubate at 42° for 90 min. Use an incubator (rather than a hot block) to prevent condensation.
8. Add 2 μl of 4 M NaOH and incubate at 37° for 30 min to degrade RNA.
9. Add 2 μl of 4 M HCl to neutralize pH.
10. *Optional:* Add 5 μl of 1 M Tris-HCl, pH 7.5, to stabilize the pH.
11. Purify the probe, using a QIAquick nucleotide removal kit (Qiagen, Hilden, Germany). Elute in 50 μl. If the reaction is scaled up, the elution volume can still be 50 μl as having concentrated probe preparations provides certain advantages. Alternatively, the labeled probe can be ethanol precipitated for 1 h at 80°.

RNA FISH TRAP

Principles of Fixation

There are two main requirements in choosing cross-linking reagents for fixation of the cells in an RNA TRAP experiment. First, cellular or nuclear structure must be preserved. This requirement must be balanced with the need for permeability to allow probe and antibodies to penetrate to their targets. Second, the cross-links must be reversible so that DNA and/or proteins can be separately isolated after affinity purification. Formaldehyde–acetic acid fixation is known to provide good preservation of cellular structure,[13] and the chemical cross-links are readily reversible under conditions that protect chromatin. After fixation, great care should be taken to avoid nuclease contamination through all subsequent steps. Recommendations for minimizing RNase activity should be strictly followed. All solutions should be made with the highest grade reagents and autoclaved if possible. Use only nuclease-free water [reverse osmosis water treated with diethyl pyrocarbonate (DEPC) and then autoclaved] to make solutions. Maintain a set of dedicated RNase-free glassware for RNA FISH. It should be cleaned thoroughly and baked overnight at 80° after each use. Wear clean gloves throughout the RNA FISH procedure. Handle slides with clean forceps to minimize contamination risks.

Start with approximately 1–2×10^8 cells if possible. The method of allowing the cells to attach to poly-L-lysine slides that we used is gentle but inefficient. Approximately 90% of the cells are washed away when the slides are immersed in fixative. More efficient methods of cell attachment are possible; hence, fewer starting cells may be needed. However, Cytospin preparations or other methods that may disrupt cell structure or flatten cells should be avoided. Aim for a single layer of cells on the slides, with few cells in direct contact with other cells. For cells in culture, collect the cells by gentle centrifugation at 4° and wash and resuspend in ice-cold phosphate-buffered saline (PBS). Pipette up and down to ensure a single-cell suspension. Single-cell suspensions can also be obtained from some tissues such as fetal liver by repeated pipetting through a yellow tip or small-bore syringe in ice-cold PBS. We find that RNA FISH on primary cells is much more robust than on cell lines. Work quickly and avoid extensive processing of the cells because long incubation times (>30–40 min) in PBS at room temperature result in noticeable decreases in the number and intensity of transcription foci.

---

[13] R. C. Curran and J. Gregory, *J. Clin. Pathol.* **33,** 1047 (1980).

Method

1. Resuspend cells at a density of 1–2×10^7 cells/ml.

2. Pipette 150 μl of the single-cell suspension onto poly-L-lysine-coated slides (prepare at least 40 slides). Spread the cell suspension evenly over the entire surface (use a horizontal yellow tip to gently spread the droplet). Allow the cells to settle onto the slides for 2 min.

3. Fix cells by laying the slides gently into a large shallow tray containing 4% formaldehyde, 5% acetic acid, 0.14 M NaCl for 18 min at room temperature. The fixative should be made fresh. The depth of the fixative in the tray should be about 0.7 cm (just enough to cover the slide and cells). To ensure consistent incubation times, divide the slides into groups of five to eight, set up a timer to count up and note the time each group goes into fixative, and so on.

4. Wash the slides in PBS at room temperature three times, 5 min each, by immersion in PBS. Use Coplin/staining jars containing 100 ml of PBS. All subsequent incubations are done in Coplin jars with 100 ml of solution unless stated otherwise.

Slides can be stored for up to 2 weeks in racks immersed in 70% ethanol in sealed boxes at $-20°$. However, best results are obtained by proceeding directly to hybridization—step 2 in Nascent RNA *in Situ* Hybridization (the next section).

Nascent RNA *In Situ* Hybridization

The precise amount of each probe used in the hybridization should be optimized using conventional RNA FISH.[10,14,15] This should be followed by a modified RNA FISH[11] in which tyramide signal amplification is used. The aim is to find hybridization and chromatin-labeling conditions that produce a robust, localized signal with a good signal-to-noise ratio as monitored by epifluorescence microscopy. In our experience, good signals with low background provide the best TRAP results.

Prehybridization Treatment

The following steps increase permeability of the cells to facilitate probe access. We use limited pepsin digestion; however, it may be possible to substitute a 10-min incubation with 0.5% Triton X-100.

[14] J. Gribnau, E. de Boer, T. Trimborn, M. Wijgerde, E. Milot, F. Grosveld, and P. Fraser, *EMBO J.* **17,** 6020 (1998).

[15] J. Gribnau, K. Diderich, S. Pruzina, R. Calzolari, and P. Fraser *Mol. Cell* **5,** 377 (2000).

1. Equilibrate the slides in 70% ethanol at room temperature for 5 min. Omit this step if slides have not been stored in 70% ethanol before use.

2. Equilibrate the slides in 0.1 M Tris-HCl (pH 7.5), 0.15 M NaCl (TS) at room temperature for 5 min.

3. Digest with 0.01% pepsin in 0.01 M HCl for 5 min at 37°.

4. Rinse in distilled, DEPC-treated H_2O briefly to wash away the pepsin, and fix again for 5 min in 3.7% formaldehyde in PBS at room temperature.

5. Wash the slides in PBS for 5 min at room temperature.

6. Dehydrate in 70, 90, and 100% ethanol steps (3 min each) and place in clean racks to air dry.

Hybridization

1. Probe is added to hybridization mix at a concentration of 1–5 ng/µl [hybridization mix: 50% formamide, 2× saline–sodium citrate (SSC), 5× Denhardt's, sheared denatured salmon sperm DNA (200 µg/ml), 1 mM EDTA, NaH_2PO_4/Na_2HPO_4, pH 7.0).

Note: For the β-globin primary transcripts, we use a mixture of four 30-mer oligonucleotides triple-labeled with Dig at the 5′ end at a final concentration of 1 ng/µl each.

2. For large probes, add C_0t-1 DNA to a final concentration of 25 ng/µl.

3. Denature the probe mix for 5 min at 85°.

4. Preanneal to C_0t-1 DNA, if present, for 15 min at 37°.

5. With slides lying flat, apply 40–50 µl of hybridization mix with denatured probe to each of the dried slides.

6. Gently cover with a 24 × 60 mm coverslip, avoiding air bubbles, and incubate at 37° overnight in a humidified chamber. Chamber should be humidified with 25% formamide, 2 × SSC (use soaked paper towels).

Washing after Hybridization

1. Remove the coverslips by dipping the slides briefly in room temperature 2× SSC and gently letting the coverslips slide off.

2. Wash the slides three times for 10 min each in 2× SSC at 37°.

3. Quench endogenous peroxidases in 0.15% H_2O_2 in PBS for 10 min.

4. Wash for 5 min in 0.1 M Tris-HCl (pH 7.5), 0.15 M NaCl, 0.05% Tween 20 (TST) at room temperature.

Tagging Chromatin Proteins *In Situ*

Principle of Method

After hybridization, horseradish peroxidase (HRP) activity is directed to the site of the labeled probe by the binding of a hapten-specific antibody conjugated to HRP. Biotin–tyramide is then applied in the presence of a low concentration of peroxide. Under such conditions, HRP catalyzes the conversion of the tyramide into a highly reactive, short-lived radical intermediate.[16,17] The radical can then either decay spontaneously or bind covalently to electron-dense residues (primarily tyrosines but also histidine and phenylalanine) in the immediate vicinity of the enzyme. As a result, biotin is deposited on chromatin and other proteins in the vicinity of the target RNA—in this case, the site of transcription.

The extent or spread of biotin–tyramide deposition is a major concern in the design of a TRAP experiment. Several steps can be taken to control or limit the spread of deposition, which will ultimately affect the resolution. In most cases, a limited degree of diffusion of the activated tyramide is desirable in order to label nearby proteins. The simplest way to control the sphere of deposition is to vary the tyramide concentration and/or the reaction time. We used this strategy in our own experiments with the β-globin genes. Biotin deposition, measured by epifluorescence detection, indicated that the majority of deposition was limited to within 450 nm of the gene, with concentric shells of decreasing intensity with increased distance from the epicenter.[11,18] Other results, which assessed diffusion via EM, suggest that resolution can be improved dramatically, perhaps greater than 10-fold, by limiting activated tyramide diffusion through the use of larger tyramide conjugates or addition of dextran sulfate to the labeling reaction.[18,19] Thus, a controllable and wide range of resolution is possible, which greatly increases the utility of the technique. These parameters should be considered carefully and optimized in pilot experiments to suit the aims of the experiment.

---

[16] M. N. Bobrow, T. D. Harris, K. J. Shaughnessy, and G. J. Litt, *J. Immunol. Methods* **125,** 279 (1989).

[17] R. P. van Gijlswijk, H. J. Zijlmans, J. Wiegant, M. N. Bobrow, T. J. Erickson, K. E. Adler, H. J. Tanke, and A. K. Raap, *J. Histochem. Cytochem.* **45,** 375 (1997).

[18] G. Mayer and M. Bendayan, *J. Histochem. Cytochem.* **45,** 1449 (1997).

[19] R. P. van Gijlswijk, J. Wiegant, A. K. Raap, and H. J. Tanke, *J. Histochem. Cytochem.* **44,** 389 (1996).

Method

Blocking and Antibody Binding

1. Block by preincubating the slides in TSB [0.1 *M* Tris-HCl (pH 7.5), 0.15 *M* NaCl, with 1× blocking reagent (Roche Molecular Biochemicals, Indianapolis, IN)]. Pipette 100 μl of TSB directly onto the slides while they are lying flat and cover each with a 24 × 60 mm coverslip.

2. Incubate for 30 min in a humidified chamber at room temperature. The chamber should be humidified with TS (or TST).

3. Lift each slide and let the coverslip slide off.

4. Lay each slide flat and apply 100 μl of anti-Dig Fab fragment–HRP conjugate (Roche Molecular Biochemicals 1207733) diluted 1:100 in TSB.

5. Gently apply a coverslip and incubate in humidified chamber for 30 min at room temperature.

6. Lift each slide and let the coverslip slide off.

7. Wash twice for 5 min each in TST at room temperature.

Labeling with Biotinyl Tyramide

1. Remove the slides individually from the TST and wipe off excess liquid from the back and sides of each slide with a tissue. Remove as much liquid as possible from the front of the slide by soaking it up by capillary action with a tissue placed along the side of the slide; be careful not to disturb the cells or to let them dry out.

2. Lay the slides flat. Work in batches of five slides at a time to avoid long periods on the bench that may allow them to dry out. TST washes can be extended if necessary with no effect on results. Apply the biotinyl tyramide reagent diluted in 100 μl of amplification reagent [0.1 *M* boric acid (pH 8.5), 0.003% H_2O_2; made fresh].

3. Gently apply a coverslip avoiding bubble formation, and incubate for 1–10 min at room temperature. For the β-globin gene TRAP, we incubated for 1 min with a 150-fold dilution of biotinyl tyramide stock [PerkinElmer Life Sciences (Boston, MA) kit].

4. Lift each slide and let the coverslip slide off.

5. Immediately dip each slide in fresh 0.15% H_2O_2 in PBS to stop the enzymatic reaction. Incubate for 10 min at room temperature.

6. Wash two times for 5 min in TST at room temperature.

FISH Detection: The efficiency of hybridization and targeting of biotin–tyramide labeling is assessed by staining one slide with an avidin D–Texas red conjugate. Good quality hybridization and tagging are represented by a high percentage of cells with primary transcript signals; staining should be localized to the site of RNA synthesis with low background.

1. Apply avidin D–Texas red (at a dilution of 1:250; Vector Laboratories, Burlingame, CA) in 100 μl of TSB. Store fluorescent reagents protected from light. In this step and all subsequent steps, minimize exposure to light (incubate slides in a "light-tight" humidified chamber and when washing, keep the staining jars containing the slides covered. We use an inverted coffee can).

2. Wash twice for 5 min each in TST at room temperature.

3. Wash once for 5 min in TS at room temperature.

4. Dehydrate the slides in 70, 90, and 100% ethanol steps (3 min each) and air dry.

5. Apply 50–100 μl of mounting medium with antifading agents directly onto each dried slide and cover with a coverslip. If it is not possible to proceed with observation, the slide can be stored flat at 4°.

Disruption of Cells and Recovery of Labeled Complexes

The cells are harvested from the slides and sonicated to fragment the chromatin before affinity purification. The resolution of the technique depends primarily on the spread of biotin deposition *in situ,* but also on the size of the fragmented chromatin. There are many published protocols for sonication of fixed cells that may be substituted; the conditions that we use should reduce the chromatin to fragments with an average DNA length of 400 bp.

Method

Sonication

1. Apply 100 μl of PBS to the surface of the slide and scrape the cells off into a petri dish, using a straight-edge razor blade. Typical recovery is in the range of 1–2 × 10^7 cells.

2. Pool the scrapings and centrifuge at 2900g for 25 min at 4°.

3. Resuspend the cell pellet in 700 μl of sonication buffer (5 *M* urea, 2 *M* NaCl, 10 m*M* EDTA) and transfer to a 1.5-ml Eppendorf tube.

4. Sonicate for eight 25-s bursts with 2 min between each burst, using a Microson sonicator (Kimble/Kontes, Vineland, NJ) set to power level 5. Keep the sample chilled on ice throughout. Optimization may be required to obtain the desired chromatin size.

5. Centrifuge for 15 min at 13,000 rpm (on a conventional benchtop microfuge) to separate soluble from insoluble material.

6. Set aside the supernatant containing the soluble material and resuspend the pellet in another 700 μl of sonication buffer and repeat the sonication.

7. Collect the second soluble chromatin fraction after centrifugation at 13,000 rpm for 15 min and combine with the first.

This usually represents more than 90% of the DNA. Take a small aliquot to check the size of the chromatin. Adjust the volume of the aliquot to 1 ml with TE, reverse the cross-links as described below, and run on a 1.2% agarose gel. If the chromatin is of the desired size, then dialyze the rest of the sample overnight against 4 liters of PBS at 4°.

Recovery of Tagged Chromatin. The sample now consists of chromatin fragments that are mostly one or two nucleosomes in length. Protein complexes that were in proximity to the targeted HRP (in this case, the active gene) should be extensively biotinylated, and separable by chromatography using a streptavidin–agarose conjugate. The streptavidin–agarose is packed into a column and the sample is passed over the column in aliquots, allowing each aliquot ample time to bind. After binding, the column is washed extensively to remove nonbiotinylated components. The streptavidin–agarose is then removed from the column and the DNA is released by reversal of the cross-links and protease digestion. The DNA can then be purified and analyzed for enrichment of various sequences. With other hapten systems, standard chromatin immunoprecipitation (ChIP) protocols could be used for chromatin protein recovery. It should be pointed out that the purification strategy described below is relatively crude, but allows recovery of greater than 50% of the biotinylated material (D. Carter, unpublished, 2002), facilitating detection of enrichment of specific DNA sequences via PCR. Analysis of other components will probably require greater amounts of starting material and more stringent purification strategies.

1. Recover the sample from the dialysis tubing and set aside 10% as the input fraction.

2. Prepare a streptavidin column: take 500 μl of a 50% slurry of streptavidin–agarose (Molecular Probes, Eugene, OR) in 25 mM Tris (pH 8.0), 140 mM NaCl, 1% Triton X-100, 0.1% SDS, 3 mM EDTA and add to the column.

3. Wash the column with 3 ml of PBS and then remove the void volume by centrifuging at 1000g for 2 min.

4. Apply the sample to the column in 200-μl aliquots (i.e., enough to replace the void volume).

5. Cap the bottom of the column and incubate each aliquot at room temperature for 45–60 min.

6. After all the chromatin has been passed over the column, wash with 2 ml of PBS, followed by 2 ml of TSE 150 [20 mM Tris (pH 8.0), 1% Triton, 0.1% SDS, 2 mM EDTA, 150 mM NaCl], 2 ml of TSE 500 [20 mM

Tris (pH 8.0), 1% Triton, 0.1% SDS, 2 mM EDTA, 500 mM NaCl], and 2 ml of TE.

7. Remove the streptavidin–agarose beads from the column in TE [this is the affinity-purified (AP) fraction].

Reversal of Cross-Links and Purification of DNA

1. Incubate beads with the affinity-purified chromatin at 65° overnight with shaking to reverse the formaldehyde cross-links.

2. Do the same with the input chromatin.

3. Add RNase A (2 μg/ml, final concentration) and incubate for 30 min at 37°.

4. Add proteinase K (100 μg/ml, final concentration) and incubate for 6 h at 37°.

5. Phenol extract and ethanol precipitate, using glycogen as carrier (20 μg/ml).

6. Resuspend the pellets in molecular biology-grade water. Quantitation of the input is achieved by measuring the optical density at 260 nm, using a conventional spectrophotometer.

The affinity-purified fraction typically contains levels of DNA that are below the threshold of accurate quantitation by a spectrophotometer. A more sensitive method to measure low quantities of DNA is by incubation with the dye Picogreen[20] (Molecular Probes). Make serial dilutions of the input DNA to 1.6, 0.8, 0.4, 0.2, and 0.1 ng/μl and incubate 50 μl of each with 50 μl of Picogreen. Use this as a standard curve to quantitate 50 μl of a 1:5 or 1:10 dilution of the AP DNA (also incubated with 50 μl of Picogreen). Accurate quantitation is achieved by measuring emission at 530 nm after excitation at 485 nm in a standard plate reader.

Quantitation of Enrichment

The enrichment of a given sequence can be assessed in several ways as in standard ChIP experiments. We have principally used quantitative real-time PCR with SYBR green[21] to measure enrichment of different sequences. Primers are designed and optimized to minimize primer–dimer effects and to ensure there is a single product. Typically, the amplicons are approximately 100–300 bp in size, a requirement imposed by the size of the DNA after sonication. To quantitatively assess enrichment, the affinity-purified DNA is compared with the input DNA. The absolute

[20] V. L. Singer, L. J. Jones, S. T. Yue, and R. P. Haugland, *Anal. Biochem.* **249**, 228 (1997).
[21] C. T. Wittwer, M. G. Hermann, A. A. Moss, and R. P. Rasmussen, *Biotechniques* **22**, 134 (1997).

level of enrichment of any particular sequence is then normalized to the enrichment of an unrelated gene on another chromosome.

Perspectives

With the increased sensitivity of large single-stranded DNA probes, we believe the RNA FISH TRAP technique will be widely applicable to many other genes to identify and study long-range interacting elements. Improved recovery of the *in situ*-labeled chromatin and methods for rapid detection of enrichment over large genomic regions, such as high-density oligonucleotide arrays,[22] will greatly facilitate the speed and utility of the technique. We have used the technique to assay sequence elements in the immediate vicinity of an actively transcribed gene; however, DNA elements interacting with other RNAs, such as noncoding RNAs that may function at remote locations, could also be assessed. We are also developing DNA FISH TRAP in order to detect specific interactions between distal chromatin regions that are not directly associated with an RNA or transcription site. Together, these techniques may help to bridge some of the gaps in our understanding of the relationships between chromatin and chromosomes' structure and function *in vivo*.

[22] P. Kapranov, S. E. Cawley, J. Drenkow, S. Bekiranov, R. L. Strausberg, S. P. Fodor, and T. R. Gingeras, *Science* **296,** 916 (2002).

[30] 3C Technology: Analyzing the Spatial Organization of Genomic Loci *In Vivo*

By Erik Splinter, Frank Grosveld, and Wouter de Laat

Introduction

Spectacular advances in light microscopy have reestablished the technique at the forefront of cell biological research. Among a wealth of knowledge in many areas of interest, it has provided new insight into how chromosomes and chromosomal regions are organized in the context of the nucleus. For example, chromosomes were found to occupy distinct but fluctuating nuclear territories, while loci on these chromosomes move rapidly within a restricted small volume of this territory. Changes in the transcriptional status of a locus sometimes, but not always, alter its nuclear positioning, as measured against large nuclear landmarks such as centromeres and the nuclear membrane. However, optical constraints set limits to what can be resolved by light/fluorescence microscopy. Thus, it is not possible as yet to visualize the structural organization of a single gene locus that spans, for example, 200 kilobases of genomic DNA. However, intricate structural organizations are to be expected at this level of resolution, for example, when enhancers or other transcriptional regulatory elements communicate with distant promoters located *in cis*. Two novel, independently developed assays have allowed insight concerning the spatial organization of such genomic loci *in vivo*. One assay, called RNA-TRAP, was developed by Carter *et al.*[1] and involves targeting of horseradish peroxidase (HRP) to nascent RNA transcripts, followed by quantitation of HRP-catalyzed biotin deposition on chromatin nearby. The other technique, developed by Dekker *et al.*,[2] was named 3C technology (chromosome conformation capture) and involves quantitation of cross-linking frequencies between two DNA restriction fragments as a measure of their frequency of interaction in the nuclear space. Originally applied to the structural organization of yeast chromosomes,[2] Tolhuis *et al.*[3] adapted 3C technology to analyze the conformation of a 200-kb region spanning the mouse β-globin gene cluster in its active and inactive transcriptional state. Here, we discuss a detailed protocol for 3C analysis in mammalian cells, evaluate interpretation of results, and discuss the advantages and disadvantages over RNA-TRAP.

Principles of 3C Technology

An outline of 3C technology is provided in Fig. 1. In brief, cells (or isolated nuclei) are treated with formaldehyde to cross-link proteins to other, neighboring proteins and DNA. The resulting DNA–protein network is then subjected to cleavage by a restriction enzyme, which is followed by ligation at low DNA concentration. Under such conditions, ligation between cross-linked DNA fragments, which is intramolecular, is strongly favored over ligation between random fragments, which is intermolecular. After ligation, the cross-links are reversed and ligation products are detected and quantified by polymerase chain reaction (PCR) across the newly ligated ends of fragments. The cross-linking frequency of two specific restriction fragments, as measured by the amount of corresponding ligation product, is proportional to the frequency with which these two genomic sites are close to each other in space. Thus, 3C analysis provides information about

---

[1] D. Carter, L. Chakalova, C. S. Osborne, Y. F. Dai, and P. Fraser, *Nat. Genet.* **32,** 623 (2002).

[2] J. Dekker, K. Rippe, M. Dekker, and N. Kleckner, *Science* **295,** 1306 (2002).

[3] B. Tolhuis, R. J. Palstra, E. Splinter, F. Grosveld, and W. de Laat, *Mol. Cell* **10,** 1453 (2002).

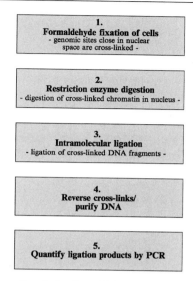

FIG. 1. Outline of 3C technology.

the spatial organization of chromosomes or chromosomal regions *in vivo*. Below, we discuss each step of this procedure in detail.

Description of 3C Technology

In Vivo *Formaldehyde Fixation of Cells*

Formaldehyde is an excellent cross-linking agent to study the composition and structure of chromatin *in vivo:* it reacts with amino and imino groups of proteins and nucleic acids to form protein–protein and protein–nucleic acid cross-links, but does not react with free double-stranded DNA. Formaldehyde cross-links bridge relatively short distances (2 Å), selecting intimate interactions, and can easily be reversed under mild conditions.[4–6]

1. Fixation is performed by incubating (tumbling) 1×10^7 cells in 10 ml of DMEM–10% FCS supplemented with 2% formaldehyde for 10 min at room temperature. The reaction is stopped by the addition of glycine (0.125 M final concentration) and transferred to 4°.

[4] M. J. Solomon and A. Varshavsky, *Proc. Natl. Acad. Sci. USA* **82,** 6470 (1985).
[5] V. Orlando, H. Strutt, and R. Paro, *Methods* **11,** 205 (1997).
[6] V. Jackson, *Methods* **17,** 125 (1999).

All amounts and volumes mentioned in this protocol refer to 1×10^7 cells of starting material. In our hands, this number of cells yields enough template for at least 600 PCRs, which should be sufficient to estimate the structural organization of, for example, a 200-kb genomic region. We find that cross-linking with 2% formaldehyde for 10 min at room temperature is sufficiently stringent to detect long-range interactions between β-globin genomic sites separated by more than 100 kb. However, the stringency of fixation required to detect a given interaction will depend on the frequency and stability of this interaction, and therefore other loci may require different fixation conditions for 3C analysis. In the original protocol, Dekker and co-workers[2] reported a loss of PCR signal when formaldehyde fixation is carried out on intact *Saccharomyces cerevisiae* cells and therefore they perform fixation on isolated nuclei. We prefer to keep manipulation of mammalian cells before fixation to a minimum and add formaldehyde directly to intact living cells; the PCR signals we obtain are reproducible and quantifiable. When working with a mixed cell population, it is important to remember that 3C technology provides an estimate of the average conformation of a genomic region as it is present in all cells included in the analysis. This implies that if the structure of a locus is to be correlated to, for example, its transcriptional activity, the percentage of nonexpressing cells present in the sample needs to be sufficiently low to not obscure the detection of possible transcription-specific interactions. We routinely use populations in which at least 80% of the cells display the phenotype of interest. Cell sorting may be required to obtain a good representation of a single cell type, particularly when working with mixtures of cell types in a tissue. Formaldehyde-fixed cells can easily be sorted on the basis of cellular markers. We have good experience with sorting fixed erythroid cells from mouse adult bone marrow, using Ter-119 as an erythroid-specific cellular marker. Sorting is done either by fluorescence-activated cell sorting (FACS) or magnetic beads, both increasing the representation of Ter-119$^+$ cells from ~20 to >90%, but with magnetic beads being much faster in processing large amounts of cells.

If working with tissues, it is important to have a homogeneous single-cell suspension before carrying out fixation, because the presence of cell aggregates will seriously affect the efficiency and reproducibility of cross-linking. Soft tissues such as 14.5 dpc mouse fetal liver and brain can initially be disrupted with a yellow tip in 100 μl of DMEM–10% FCS, whereas tissues such as bone marrow first need to be crushed with a pestle and mortar (or cut and flushed) before taking up cells in medium. Next, all cells should be passed through a cell strainer to obtain homogeneous single-cell suspensions.

Lysis of Cells

2. Cells are pelleted by centrifugation at 320g for 8 min at 4° and washed once in 10 ml of cold PBS. The pellet is resuspended in 5 ml of cold lysis buffer [10 mM Tris (pH 8), 10 mM NaCl, 0.2% NP-40, protease inhibitors], lysis is allowed to proceed for 10 min at 4°, and nuclei are collected by centrifugation at 600g for 5 min at 4°. Pellets of nuclei are either frozen via liquid nitrogen and stored at −80°, or directly processed for digestion.

To confirm the presence of isolated nuclei after lysis, one can stain proteins and DNA with, for example, Methylgrün-Pyronin and analyze the cells under a light microscope; all cells should be lysed.

DNA Digestion in Intact Nuclei

3. Take up the pellet of nuclei in 0.5 ml of 1.2 × standard restriction buffer (will be 1 × after addition of SDS and Triton X-100; the appropriate buffer depends on the restriction enzyme of choice). Transfer the suspension to a 1.5-ml tube and add 7.5 μl of 20% SDS to a final concentration of 0.3%; shake for 1 h at 37°. Add 50 μl of 20% Triton X-100 to a final concentration of 1.8% and shake for another 1 h at 37°. Add 400 U of highly concentrated restriction enzyme and digest while shaking overnight at 37°.

The conditions in step 3 were designed by Dekker *et al.*[2] to allow restriction enzyme digestion of DNA in the context of the cross-linked nucleus, a critical event in the assay. SDS serves to remove any non-cross-linked proteins from the DNA, Triton X-100 is added to sequester SDS and allow subsequent digestion. We also find that addition of both substances is necessary for any digestion to occur, but others have reported conditions not requiring SDS and Triton X-100 for *Pst*I to digest DNA in nuclei.[7] The percentages of SDS and Triton X-100 given here were obtained after carefully titrating each component. They represent optimal conditions for *Bgl*II, *Hin*dIII, and *Eco*RI digestion, but other enzymes, such as *Bam*HI, *Spe*I, *Pst*I, and *Nde*I, do not work (optimally) under these conditions. Dialyzing the lysed nuclei to reduce the concentration of SDS and then adding Triton-X100 to complex what is left may be an option to achieve full activity of these and other enzymes (M. Spivakov and N. Dillon, personal communication, 2003).

The restriction enzyme of choice depends on the locus to be analyzed. It preferably isolates potentially interesting DNA sequences as small (<2 kb)

[7] K. M. Leach, K. F. Vieira, S. H. Kang, A. Aslanian, M. Teichmann, R. G. Roeder, and J. Bungert, *Nucleic Acids Res.* **31**, 1292 (2003).

discrete fragments, because larger fragments tend to display more background cross-linking. Preliminary experiments should be done to optimize digestion conditions. De-cross-linking DNA immediately after exposure to restriction enzyme, and running it on an ethidium bromide-stained agarose gel, show whether an enzyme has cut or not. Southern blot analysis and quantification needs to be done to determine the percentage of cleavage. An example is given in Fig. 2. Digestion efficiencies of fetal liver cells subjected to different concentrations of formaldehyde (0.5, 1, and 2%) are compared. Efficiency of cleavage by *Hin*dIII is already high (~85%)

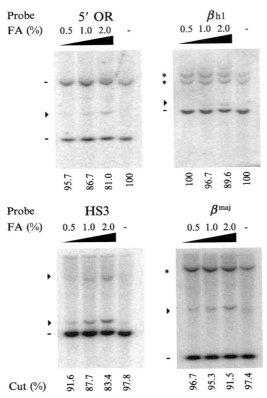

FIG. 2. Digestion efficiency depends on fixation and is uniform throughout a locus. Southern blots show that in 14.5dpc fetal livers, digestion efficiency of cross-linked chromatin depends on formaldehyde concentration and is equal near nontranscribed genes (5′ OR and βh1), a transcribed gene (βmajor, with DNase I-hypersensitive promoter), and a DNase I-hypersensitive site (HS3). Yield of specifically cut fragments is shown (percentages). Arrowheads depict partial digests and asterisks (expected) cross-hybridization signals with other β-globin-like genes.

at 2% formaldehyde, and increases to nearly 100% with decreasing amounts of formaldehyde. Thus, formaldehyde fixation accounts for the fact that DNA is not digested to completion. Digestion efficiency is independent of local chromatin configuration, because cutting works equally well at an inactive olfactory receptor gene (5' OR), an inactive globin gene (βh1), a DNase I-hypersensitive site (HS3), and an active globin gene (with a DNase I-hypersensitive promoter: βmajor) (Fig. 2). In brain cells, which do not express any of these genes, we find similar digestion efficiency at all of these sites (data not shown). These important controls show that there is no bias in the assay due to preferred cleavage of one site over another. Once conditions have been defined for a given enzyme in a given cell type, they appear to be applicable to other cell types as well. Thus, in our hands, BglII, HindIII, and EcoRI cleave equally well in mouse fetal liver and brain cells, mouse adult bone marrow cells, mouse erythroleukemia (MEL) cells (an established tissue culture line), and primary rat Schwann cells (M. Ghazvini, personal communication) (data not shown).

Intramolecular Ligation

4. The next day (day 2), 40 μl of 20% SDS is added to a final concentration of 1.3% and the solution is incubated for 20 min at 65° to inactivate the restriction enzyme. The solution is transferred to a 50-ml tube and diluted by adding 7 ml of 1× standard ligation buffer. A 375-μl volume of 20% TX-100 is added to a final concentration of 1% and the solution is incubated for 1 h at 37° to sequester SDS. T4 ligase (100 U, from highly concentrated stock) is added and DNA is ligated for 4–5 h at 16°, followed by a 30-min ligation at room temperature.

At low DNA concentration (here, $\sim$3.7 ng/μl), ligation between cross-linked DNA fragments (i.e., intramolecular ligation) is highly favored over ligation between non-cross-linked DNA fragments (i.e., intermolecular or random ligation), because the kinetics of essentially a monomolecular reaction is much faster than that of a bimolecular reaction. Indeed, when we fix cells with 2% formaldehyde, we never detect ligation products between random restriction fragments (e.g., coming from different chromosomes), whereas DNA fragments normally located *in cis* within 20 kilobases (and often more) on a chromosome can easily be observed (data not shown).

Reversal of Cross-Links and Purification of DNA

5. Cross-links are reversed by overnight incubation at 65° in the presence of proteinase K (300 μg total). The next day (day 3) 300 μg

of RNase is added and RNA is degraded for 30–45 min at 37°, followed by phenol extraction and ethanol precipitation of DNA (including a 70% ethanol wash). DNA pellets are resolved in 150 μl of 10 mM Tris (pH 7.5). DNA is now ready to be analyzed by PCR.

Quantitative PCR Analysis of Cross-Linking Frequencies

Primer Design for PCR

A number of issues must be considered for the design of primers for the quantitative PCR analysis of ligation products. First, primers should hybridize to unique DNA sequences and one therefore needs to carefully check for the presence of repetitive DNA sequences near restriction sites of interest to avoid using a repetitive DNA primer. Second, for reproducible and quantifiable PCR signals, the size of the PCR products should be kept small. We try to design all our PCR primers within 50–100 bp from the restriction sites analyzed, yielding PCR products of 100–200 bp in size. Third, differences in the T_m of primers should be kept to a minimum ($< 2°$) to allow simultaneous analysis of all possible primer combinations. Primers can be designed at either end of a given restriction fragment.

Quantification of Cross-Linking Frequencies

3C technology involves quantifying ligation frequencies of restriction fragments by PCR, which gives a measure of their cross-linking frequencies. We quantify the formation of PCR products by scanning their signal intensities after separation on ethidium bromide-stained agarose gels, using a Typhoon 9200 imager (Molecular Dynamics, Sunnyvale, CA). For this analysis to be quantitative, the amount of DNA template added per PCR should be in the range that shows linear PCR product formation. This needs to be determined for each new template by testing PCR product formation on serial dilutions of template, using multiple primer combinations. We typically find that, for most primer combinations, ∼20 ng of DNA template is in the range of linear PCR product formation, using 36 cycles of PCR amplification (Fig. 3). The fact that such a large number of PCR cycles is needed shows that only a little ligation product is present in (and amplified from) a vast excess of genomic DNA. This may explain why quantitation of SYBR green incorporation by real-time PCR has so far failed to provide us with satisfying results. Measuring incorporation of ^{32}P-labeled nucleotides in PCR products, a third method to quantify PCR products, generally requires fewer cycles of amplification but otherwise involves the same principles as measuring EtBr incorporation. Both techniques allow quantification of the correctly sized fragment, while real-time

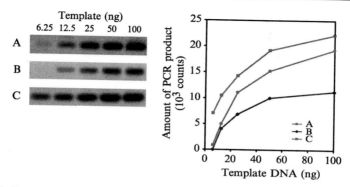

FIG. 3. Determination of the linear range of PCR amplification of cross-linked DNA template. Examples are shown of titrations of the cross-linked DNA template cut with *Eco*RI (A) and *Hin*dIII (B and C) that was obtained from 14.5dpc fetal livers. PCR products were separated on 2% agarose gels stained with ethidium bromide and scanned with a Typhoon 9200 imager (Molecular Dynamics). Graph shows the quantitation of the PCR products. Primer sets analyze ligation between restriction fragments containing the active βmajor gene and HS2 (A), the inactive εy gene and HS-60 (B), and two sites ~7 kb apart in the XPB (component of the basal transcription factor TFIIH) locus, which is taken as a control (C).

PCR also measures the incorporation of dye in background products, primer–dimers, and so on. Given the number of PCRs often involved in 3C analysis, we prefer the use of the nonradioactive method that measures intercalated ethidium bromide.

Controls

Correct interpretation of data obtained by 3C technology depends critically on several controls. We mentioned previously that it is important to exclude a bias in the assay due to preferred restriction enzyme digestion of one site over the other; this can be checked by Southern analysis of DNA directly after digestion.

A quantitative comparison of signal intensities of PCR products obtained with different primer sets is valid only after correction for PCR amplification efficiency of each primer set. Thus, a control template is required in which all possible ligation products are present in equimolar amounts. In yeast, this template was obtained by digesting and randomly ligating non-cross-linked genomic DNA.[2] For mammalian cells, with a genome 100 times the size of the yeast genome, we found that random ligation of two specific loci is too rare an event to be detected by PCR. One therefore needs to enrich for ligation products of interest. This can be done by taking a BAC, PAC, or YAC carrying an artificial chromosome that covers the genomic region to be analyzed (Fig. 4A). Digestion and ligation

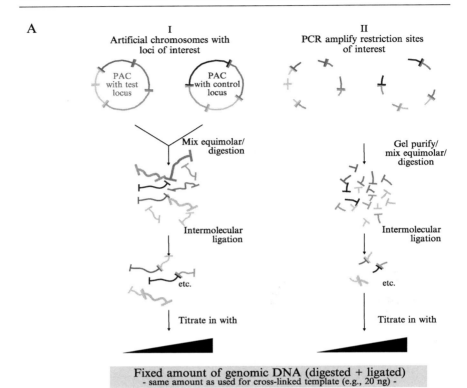

FIG. 4. Controls in 3C technology. (A) Two methods (I and II) to prepare the random ligation control template required to correct for differences in PCR amplification efficiency. Perpendicular bars on PACs (or YACs, BACs) and on DNA fragments indicate restriction sites. Different shades indicate different restriction fragments. PCR analysis is done with

of the artificial chromosome yield a mixture that contains all possible ligation products of interest in equimolar amounts. An alternative method is to
mix equimolar amounts of DNA fragments that span each of the restriction
sites to be analyzed (Fig. 4A). Such fragments can often most easily be
obtained by PCR. The approach involves gel purifying these fragments
and carefully determining their concentration before mixing them in equimolar amounts, followed by restriction enzyme digestion and ligation.
Clearly, the latter is much more laborious and intrinsically less accurate
than working with artificial chromosomes.

The PCR amplification efficiency of a primer set is also influenced by
the amount of genomic DNA present in the PCR. Thus, for correct comparison, the random ligation template described above should be mixed
with an amount of genomic DNA similar to that used for the test samples
($\sim$20 ng; see above). To completely mimic conditions, we mix the random
ligation template with digested and religated genomic DNA. A serial dilution of this template in a fixed amount of genomic DNA is carried out to
determine the proper control template concentration. By taking the ratio
of signal obtained by quantitative PCR on cross-linked template versus
control template, one corrects for differences in amplification efficiency
between primer sets and also for differences in signal intensities due to
the size of PCR products. These controls are absolutely essential for a
quantitative, or even a qualitative, interpretation of the PCR data and thus
for any conclusion on the spatial organization of a locus.

If different cross-linking frequencies must be measured in different
tissues, an internal standard is required that accounts for variations in
efficiency of cross-linking and ligation. This is done by analyzing the
cross-linking frequency of a locus that is unrelated and that can reasonably
be assumed to adopt a similar spatial organization in the different tissues.
Gene loci that express at similar levels in the tissues analyzed are thought
to meet this criterion. Previously, we analyzed cross-linking frequencies
between two fragments in the transcribed part of the calreticulin locus
(CalR) with the restriction sites analyzed $\sim$1.5 kilobases apart. The CalR
locus is embedded in an area of ubiquitously expressed genes and expresses
at similar levels in the tissues analyzed in that particular study (14.5 dpc
fetal brain and liver).[3] However, other loci with well-defined expression
levels and patterns may be equally suitable. By normalizing, within a tissue,
each cross-linking frequency to the cross-linking frequency observed

---

primers that amplify across the ligated sites. (B) Equation used to calculate relative cross-
linking frequency. A, Measured signal intensity of PCR product. The relative cross-linking
value of 1 arbitrarily corresponds to the cross-linking frequency between two control
fragments ($c_1 + c_2$) (fragments, e.g., present in XPB or CalR; see text).

between the control fragments, one can correct for differences in quality and amount of template.

Finally, adding the control locus to the random ligation mix (either as an artificial chromosome or as DNA fragments, in equimolar amounts to the locus that is analyzed), and routinely analyzing its ligation efficiency, provides a loading control to correct for differences in the amount of control template between experiments.

The equation used to calculate the relative cross-linking frequency is given in Fig. 4B. As a result of this normalization, the "cross-linking frequency" value 1 arbitrarily corresponds to the cross-linking frequency between the control fragments (e.g., CalR fragments).

Interpretation of Data Obtained by 3C Technology

As pointed out originally by Dekker et al.,[2] measuring cross-linking efficiency by the formation of ligation products largely depends on the frequency with which two genomic sites interact. They showed that contributions of other parameters, such as local protein concentrations or a favorable geometry of the cross-linked intermediate, are minor. By fitting their data to polymer models,[8] the authors interpreted the relationship between cross-linking frequency and genomic site separation to give an estimate of the three-dimensional organization of chromosome III of *S. cerevisiae*. For a number of reasons, we are still reluctant to interpret data obtained by 3C analysis in a strictly quantitative manner. In the first place, we find that additional parameters, for example, the fragment size, notably affect the cross-linking efficiency. Particularly difficult to interpret are cross-linking frequencies measured for large fragments carrying multiple *cis*-regulatory elements, which are each likely to be engaged in unique and different interactions. Clearly, the specificity of measuring interactions increases with smaller fragments containing such regulatory sites as isolated entities. Second, changes in fixation conditions differentially affect cross-linking frequencies. Figure 5 shows the cross-linking frequencies obtained with 2, 1, and 0.5% formaldehyde for three neighboring restriction fragments. In each PCR, the same amount of template was used and all products are in the linear range of PCR amplification. The data show that nearby fragments require a lower percentage of formaldehyde to reach maximum cross-linking efficiency than more distal fragments. Third, interactions between distal DNA elements are thought to be dynamic, while the measurements represent steady state average levels. Thus, short-lived but important interactions may score much lower than more long-lived

[8] K. Rippe, *Trends Biochem. Sci.* **26**, 733 (2001).

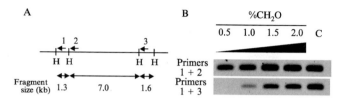

FIG. 5. Changes in fixation conditions differentially affect the efficiency of cross-linking between different sites. (A) Schematic presentation of part of the XPB locus, showing *Hind*III restriction sites (H) at that site. Primers used to analyze ligation products are indicated by arrows and are numbered. (B) PCR analysis of ligation products formed after fixation with 0.5, 1, 1.5, and 2% formaldehyde. In all cases, 20 ng of template was used (in linear range of amplification; see text and Fig. 3). Under the conditions used, ligation between neighboring fragments 1 and 2 is saturated at 1.5 and 2%, whereas ligation between fragments 1 and 3 is not. Control (C) was obtained with the random ligation control template (see text); intensities of these signals are a measure of the efficiency of amplification of each primer set.

interactions. Finally, proper modeling would require incorporation of an estimate of the packing ratio of the chromatin fiber. However, even for the extensively studied β-globin locus, it is impossible to say whether chromatin is folded as a 10-nm fiber or a 30-nm fiber, or whether it adopts yet another conformation. Moreover, chromatin folding is certainly not uniform along the entire locus. Thus, for many loci in higher eukaryotes, it is presently difficult to give a reliable estimate of the packing ratio of the chromatin fiber. We currently prefer to describe data obtained by 3C technology in a qualitative manner,[3] rather than interpreting the relationship between cross-linking frequency and genomic site separation in terms of real distances. Measured interactions become particularly meaningful if they can be correlated to a phenotype, for example, if they occur in a transcriptionally active locus and not in the inactive one.

3C Technology and RNA TRAP

RNA TRAP was designed to identify DNA sequences that are in close proximity to actively transcribed genes.[1] It applies fluorescence *in situ* hybridization (FISH) technology to target horseradish peroxidase (HRP) to nascent RNA transcripts, followed by HRP-catalyzed biotin deposition on chromatin nearby. After affinity purification, the relative abundance of biotinylated DNA sequences is determined by quantitative PCR analysis and is taken as a measure of proximity to the labeled nascent transcript. Although both 3C technology and RNA TRAP have been developed recently and undoubtedly will evolve further, it is useful to summarize the advantages and disadvantages of the two techniques.

A disadvantage of the 3C technology is that the resolution is restricted by the occurrence of restriction sites. Ideally, (potential) regulatory DNA sequences are analyzed by 3C as isolated entities present on small DNA fragments. Frequently cutting enzymes such as *Sau*3AI or *Nla*III, which on average cleave once every 256 base pairs, often yield such fragments, but inevitably will also cut at unfavorable positions. Multiple restriction enzymes can be included in the 3C analysis of a locus, but each enzyme requires its own PCR primers.

Another drawback of 3C technology is the difficulty in discriminating between directly and indirectly bound DNA fragments. Whereas in RNA TRAP the DNA fragment closest to the nascent transcript may shield other chromatin from the deposition of biotin, in 3C technology a nearby cross-linked DNA fragment will bring in other cross-linked DNA fragments, the DNA ends of which can all be ligated to the restriction site of interest. RNA TRAP may therefore be better to find directly interacting genomic sites, while 3C technology gives a more general picture of DNA fragments nearby the site of interest. To optimize the specificity of ligation in 3C technology, one must find cross-linking conditions that minimize the average number of DNA fragments present per cross-linked protein–DNA aggregate without losing the interactions of interest.

RNA TRAP is dependent on nascent transcripts coming from actively transcribed genes, which is both an advantage and a disadvantage. The advantage is that nonexpressing cells do not contribute to the measured values, meaning that genomic site interactions with the transcriptionally active locus can be picked up even in mixed cell populations. However, to determine whether such interactions are specific for the transcriptionally active status of this locus, one needs to know its silent conformation, which is an important control that cannot be checked by RNA TRAP. Moreover, it may be difficult to obtain satisfying results by RNA TRAP for genes with moderate transcription rates. The globin genes, which were used to develop RNA TRAP, are transcribed efficiently, and these genes are exceptionally good targets for visualizing nascent transcripts.[9] Most other genes, however, are not transcribed so actively, and visualizing ongoing transcription here is more troublesome. The minimal density of nascent transcripts that is required for RNA TRAP remains to be determined. Another disadvantage of RNA TRAP is that, even in a transcriptionally active locus, it does not allow analysis of the spatial organization near sites other than the transcribed gene itself.

[9] M. Wijgerde, F. Grosveld, and P. Fraser, *Nature* **377,** 209 (1995).

DNA TRAP, which would involve hybridization to specific DNA sequences rather than to nascent RNA transcripts,[1] may solve the above-described issues, but will undoubtedly be more difficult to develop. It requires specific binding of probes to chromatinized double-stranded DNA without denaturing and changing its higher order structure and hybridization to single-copy genomic DNA is much less sensitive than to a multicopy RNA transcript. DNA TRAP will also circumvent the problem of RNA transcripts being dispersed along the transcribed region of the gene. This dispersed localization of label could be particularly inconvenient when applying RNA TRAP to genes much larger than the globin genes, which are only 1.5 kb in size. Also, DNA TRAP, like the 3C methodology, would have the potential to detect specific spatial interactions between transcriptional regulatory elements such as promoters and enhancers, which cannot be done by RNA TRAP.

Despite their current limitations, both 3C technology and RNA/DNA TRAP clearly are important new tools that will undoubtedly lead to exciting new insights concerning the structural and functional organization of chromatin in the living nucleus, at a level presently not possible with microscopes.

Acknowledgments

We thank Bas Tolhuis and Robert-Jan Palstra for contributions to setting up the 3C technology in our laboratory. We thank Mhernaz Ghazvini, M. Spivakov, and Niall Dillon for sharing unpublished results. This work is supported by NWO (the Netherlands Organization for Scientific Research) to W.d.L. as part of the Innovational Research Incentives Scheme and by NWO and EC grants to F.G.

Author Index

Numbers in parentheses are footnote reference numbers and indicate that an author's work is referred to although the name is not cited in the text.

A

Abbondanzo, S. J., 240
Abbott, D. W., 10
Abrams, M., 293
Adamson, R. H., 425
Adcock-Downey, L., 60
Adler, K. E., 488
Agard, D. A., 346(15), 347, 363, 364(30)
Agyansky, T., 443
Ahmad, K., 18, 118, 253, 385, 386(29), 392(30)
Ajiro, K., 293(80), 294
Alami, R., 293
Albig, W., 278
Aldridge, T. C., 213
Alexander, B. T., 292, 293, 428
Alexander, C., 44
Alfieri, J. A., 170
Allan, J., 179, 180(10), 281, 292
Allis, C. D., 132, 179, 285, 287, 293, 294, 296, 385, 469
Almouzni, G., 117, 117(4), 118, 119, 120, 123(4; 17), 126, 126(24), 127(25)
Alpert, A. J., 284, 285(34), 286(34)
Althaus, F. R., 295
Altschul, S. F., 4
Alvarez, P., 371
Amato, P. A., 394
Amoils, S., 479
Ando, S., 253, 254, 264(11), 269(11), 270, 277(11)
Ando, T., 267
Andrulis, A. D., 346
Annan, R. S., 309
Annunziato, A. T., 294
Ansari, A., 171, 175(1)
Antalffy, B., 428
Aoyagi, S., 212

Apweiler, R., 3
Aragon-Alcaide, L., 350
Aravind, L., 317
Archer, T. K., 132, 293, 295(75)
Arminski, L., 3
Asahara, T., 66
Aslanian, A., 497
Aurias, A., 118
Ausio, J., 10, 48, 145, 179, 180, 180(8), 236, 281, 327
Ausubel, F. M., 299, 315(22), 316(22), 317(22)
Axelrod, D., 394, 409, 429, 433(46), 443

B

Bac, Y., 89
Bains, W. A., 213
Baird, G. S., 350
Bairoch, A., 3
Baldock, R. A., 456
Baldwin, J. P., 52
Balhorn, R., 287, 294(43)
Ballantine, J. E., 213
Bandiera, A., 282, 294(27)
Banks, G. C., 293, 295(75), 305, 307(32), 308(32), 312(32), 318, 320(64), 321(64)
Bannister, A. J., 296
Bao, Y., 23
Barak, L. S., 415
Barham, S. S., 293(79), 294
Barker, W. C., 3
Barlos, K., 67
Barrett, P. H., 406
Bartholomew, B., 89, 94, 94(8), 97(8), 99(8), 100(8), 102(8), 193, 194, 196, 203, 203(9), 206, 209(13), 210(13), 218
Barton, S. C., 240
Bartsch, J., 132

Subject Index

A

Acid-urea-Triton gel electrophoresis
  histone H1 variant resolution, 287–288
  histone two-dimensional gel
    electrophoresis with AUT/AUC gels
    DNA double strand break
      studies, 87–88
    first dimension AUT gels
      casting, 81–82
      running conditions, 82
      staining and destaining, 82–83
    histone preparation
      sample preparation, 80
      tissue cells, 78–79
      tissue culture cells, 78–79
      trichloroacetic acid precipitation, 80
      yeast histones, 79
    overview, 76–77
    radiolabeled histones, 86–87
    second dimension AUC gels
      casting, 83–85
      running conditions, 85
      staining and destaining, 85–86
      storage, 86
    Western blotting, 86–87
APB, *see* 4-Azidophenacyl bromide
A/T hook peptide motif
  distribution in chromatin remodeling
    complexes, 322
  domain swapping for B-box peptide in
    HMGB1
    DNA and nucleosome binding studies,
      321–322
    overview, 318, 320
    recombinant protein production,
      320–321
  nucleosome binding, 317–318, 320
AUT gel electrophoresis, *see* Acid-urea-
  Triton gel electrophoresis
4-Azidophenacyl bromide
  histone tail–DNA interaction studies using
    model dinucleosome system

dinucleosome DNA template
  construction, 187–189
dinucleosome preparation, 189–191
histone cysteine substitution mutants
  4-azidophenacyl bromide
    modification, 187
  design and expression, 180–181
overview, 179–180
recombinant histone H3/H4 tetramer
  preparation
  cation-exchange chromatography,
    183, 185
  dialysis, 183
  DNA removal, 185–186
  expression in *Escherichia coli,* 182–183
  lysate preparation, 183
  storage, 186
  yield, 187
ultraviolet-induced cross-linking,
  191–193
site-directed histone–DNA contact
  mapping for nucleosome dynamics
  analysis
  4-azidophenacyl bromide modification,
    199–200
  cross-linked nucleotide
    identification, 203
  cysteine replacement site engineering,
    195–197
  DNA cleavage, 202–203
  histone overexpression and purification,
    197–198
  nucleosome reconstitution, 198–199
  octamer refolding, 198
  principles, 193–195
  remodeled nucleosomal state
    characterization
    gel-purified remodeled nucleosomes,
      209–210
    histone octamer displacement beyond
      DNA ends, 204–205
    nucleosomal conformation
      implications, 205, 207–208

527

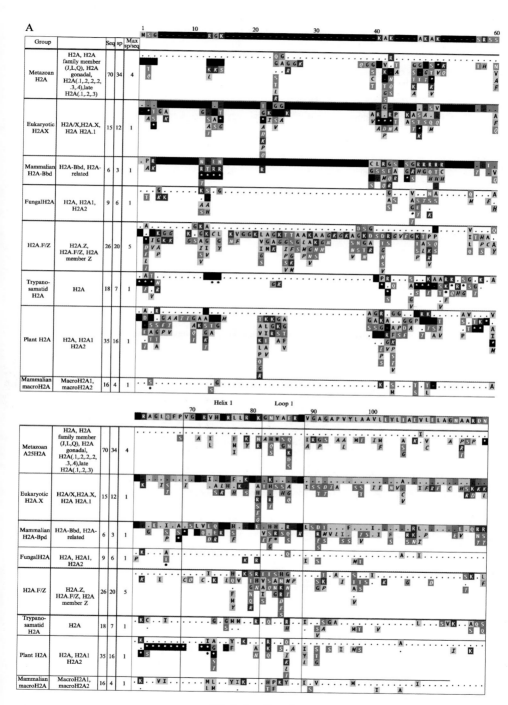

FIG. 1. *(continued)*

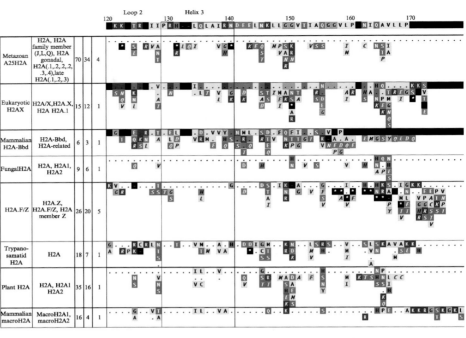

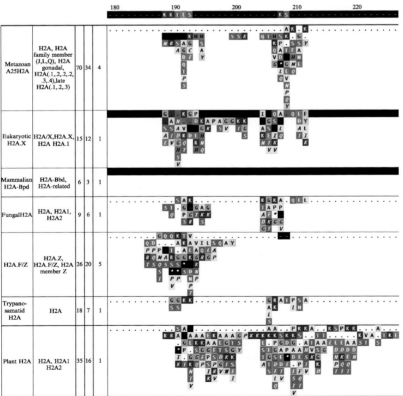

FIG. 1. (continued)

B

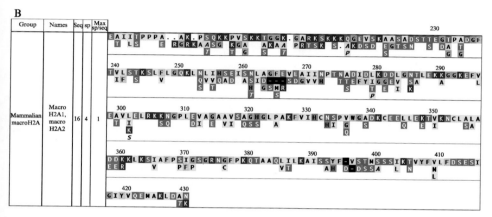

SULLIVAN AND LANDSMAN, CHAPTER 1, FIG. 1. Summary of H2A subclasses and variants. (A) A consensus sequence of all aligned H2A sequences is shown at the top. Dots in the sequences below indicate identity to the consensus. Groups are named on the basis of clustering patterns observed in neighbor-joining trees of aligned H2A sequences (not shown). *Names,* a selection of sequence descriptors found in the definition lines of the sequence records; *seq,* number of unique sequences in the group; *sp,* number of species in the group; *max sp/seq,* the greatest number of species having the same sequence in the group. For each group the first line is the consensus sequence for that group. Variations from the group consensus are indicated below it. Italic indicates a "singleton," i.e., the residue was found in only one sequence from one species in the group. An asterisk (*) indicates singleton identity or a gap. Background color key: white, identity to the anchored consensus; black, gap; orange, aromatic; yellow, aliphatic/hydrophobic; light green, glycine; green, hydrophilic; light blue, histidine; blue, basic; red, acidic. (B) C-terminal section of macroH2A.

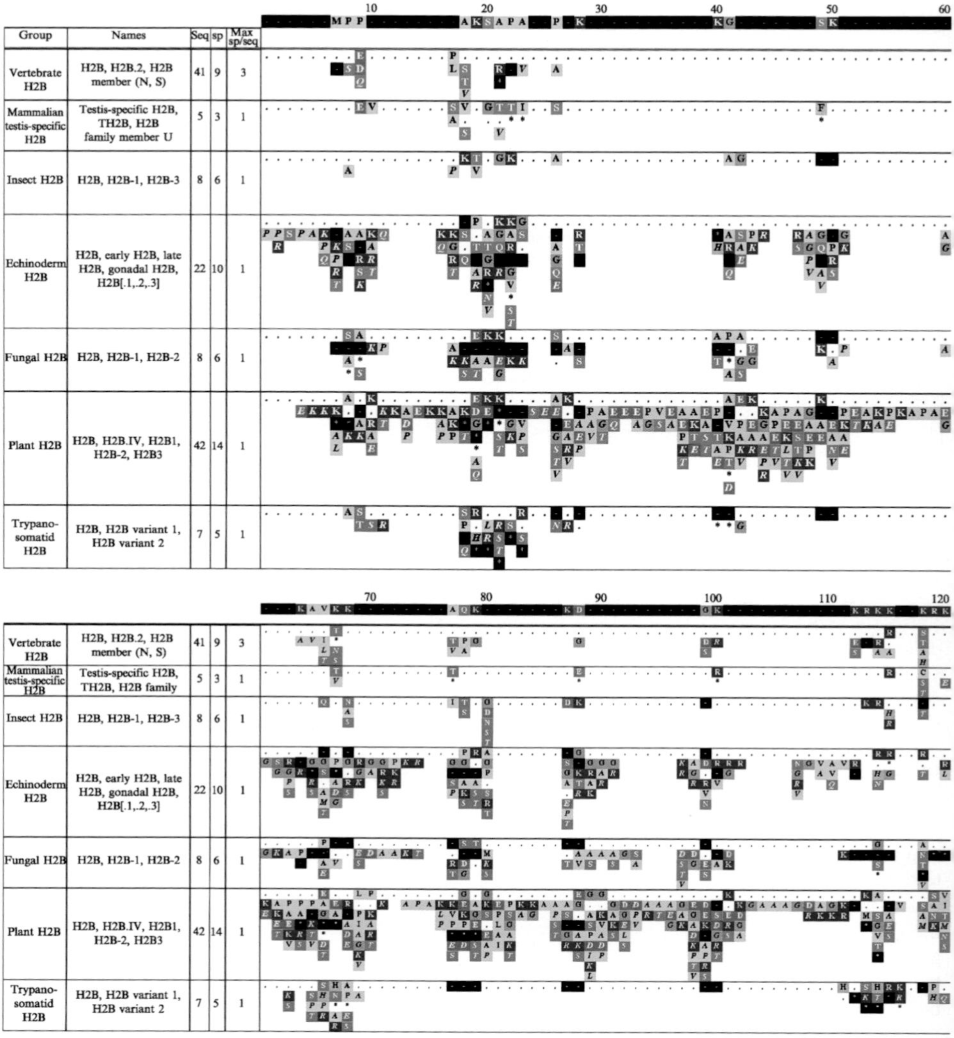

This is a multiple sequence alignment figure (histone H2B) showing protein sequences across different taxonomic groups with position rulers at positions 10–60 (top block) and 70–120 (bottom block). The sequence columns contain letter symbols positioned individually along the alignment and cannot be meaningfully rendered as a grid; the stable tabular portion (row labels and counts) is transcribed below.

Top block (positions 1–60)

Consensus ruler: `MPP ... AKSAPA P K ... KG ... SK`

Group	Names	Seq	sp	Max sp/seq
Vertebrate H2B	H2B, H2B.2, H2B member (N, S)	41	9	3
Mammalian testis-specific H2B	Testis-specific H2B, TH2B, H2B family member U	5	3	1
Insect H2B	H2B, H2B-1, H2B-3	8	6	1
Echinoderm H2B	H2B, early H2B, late H2B, gonadal H2B, H2B[.1,.2,.3]	22	10	1
Fungal H2B	H2B, H2B-1, H2B-2	8	6	1
Plant H2B	H2B, H2B.IV, H2B1, H2B-2, H2B3	42	14	1
Trypanosomatid H2B	H2B, H2B variant 1, H2B variant 2	7	5	1

Bottom block (positions 70–120)

Consensus ruler: `KAVKK ... AQK ... KD ... GK ... KRKK KRK`

Group	Names	Seq	sp	Max sp/seq
Vertebrate H2B	H2B, H2B.2, H2B member (N, S)	41	9	3
Mammalian testis-specific H2B	Testis-specific H2B, TH2B, H2B family	5	3	1
Insect H2B	H2B, H2B-1, H2B-3	8	6	1
Echinoderm H2B	H2B, early H2B, late H2B, gonadal H2B, H2B[.1,.2,.3]	22	10	1
Fungal H2B	H2B, H2B-1, H2B-2	8	6	1
Plant H2B	H2B, H2B.IV, H2B1, H2B-2, H2B3	42	14	1
Trypanosomatid H2B	H2B, H2B variant 1, H2B variant 2	7	5	1

Fig. 2. (*continued*)

SULLIVAN AND LANDSMAN, CHAPTER 1, FIG. 2. Summary of H2B subclasses and variants. A consensus sequence of all aligned H2B sequences is shown at the top. Dots in the sequences below indicate identity to the consensus. Groups are named on the basis of clustering patterns observed in neighbor-joining trees of aligned H2B sequences (not shown). *Names*, a selection of sequence descriptors found in the definition lines of the sequence records; *seq*, number of unique sequences in the group; *sp*, number of species in the group; *max sp/seq*, the greatest number of species having the same sequence in the group. For each group the first line is the consensus sequence for that group. Variations from the group consensus are indicated below it. Italic indicates a "singleton," i.e., the residue was found in only one sequence from one species in the group. An asterisk (*) indicates singleton identity or a gap. Background color key: white, identity to the anchored consensus; black, gap; orange, aromatic; yellow, aliphatic/hydrophobic; light green, glycine; green, hydrophilic; light blue, histidine; blue, basic; red, acidic.

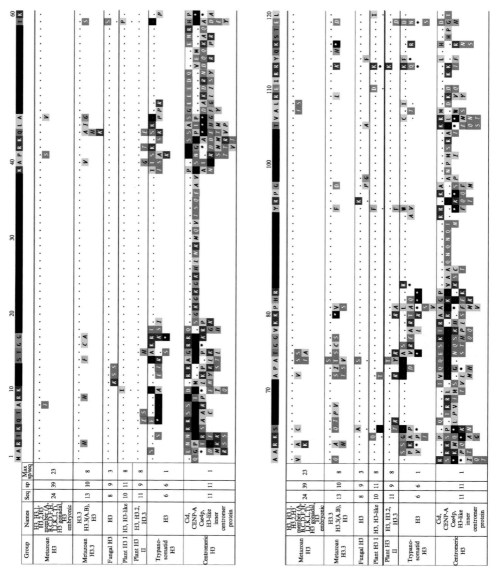

SULLIVAN AND LANDSMAN, CHAPTER 1, FIG. 3. Summary of H3 subclasses and variants. A consensus sequence of all aligned H3 sequences is shown at the top. Dots in the sequences below indicate identity to the consensus. Groups are named on the basis of clustering patterns observed in neighbor-joining trees of aligned H3 sequences (not shown). *Names,* a selection of sequence descriptors found in the definition lines of the sequence records; *seq,* number of unique sequences in the group; *sp,* number of species in the group; *max sp/seq,*

FIG. 3. *(continued)*

the greatest number of species having the same sequence in the group. For each group the first line is the consensus sequence for that group. Variations from the group consensus are indicated below it. Italic indicates a "singleton," i.e., the residue was found in only one sequence from one species in the group. An asterisk (*) indicates singleton identity or a gap. Background color key: white, identity to the anchored consensus; black, gap; orange, aromatic; yellow, aliphatic/hydrophobic; light green, glycine; green, hydrophilic; light blue, histidine; blue, basic; red, acidic.

Multiple sequence alignment of histone H4 across eukaryotic groups.

Top panel (residues 1–~55, helix 1):

Consensus/maximum sequence:
`M S · G R G K G G · · K G L · · · · G K G G A · K R · H · R K V L · R D N I Q G I T K P A I R R L A R R`

Position markers: 10   20   30   40   50 · helix 1

Group	Names	Seq sp	Max sp seq
Metazoan H4	H4	13	15
Insect H4	H4	7	9
Fungal H4	H4	7	2
Slime mold H4	H4.1, H4.2		1
Plant H4	H4		9
Ciliophoran H4	H4, H4 major, H4minor	8	3
Kinetoplastid H4	H4	4	1
Parabasilid H4	H4	1	1
Pelobiont H4	H4	1	1
Entamoebid H4	H4	1	1
Apicomplexan H4	H4	1	1
Diplomonad H4	H4	1	1
Cryptomonad H4	H4	2	1

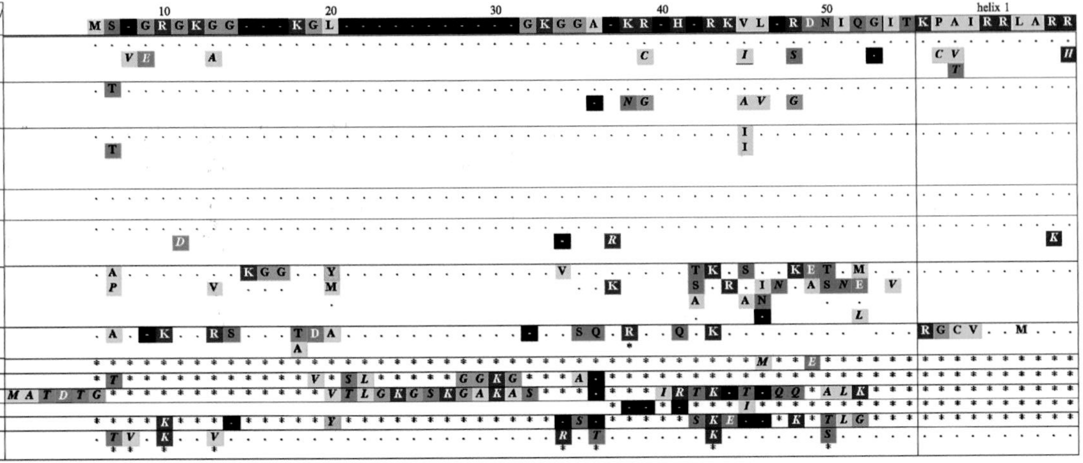

Bottom panel (residues ~56–120):

Consensus/maximum sequence:
`G G V K R · I S G L I Y E E T R G V L K V F L E N V I R D A V T Y T E H A K R K T V T A M D V V Y A L K R Q G R T L Y G F G G`

Position markers: 70 · Loop 1 · 80 · Helix 2 · 90 · Helix 3 · 100 · Loop 2 · 110 · Helix 3 · 120

Group	Names	Seq sp	Max sp seq
Metazoan H4	H4	13	15
Insect H4	H4	7	9
Fungal H4	H4	7	2
Slime mold H4	H4.1, H4.2		1
Plant H4	H4		9
Ciliophoran H4	H4, H4 major, H4minor	8	3
Kinetoplastid H4	H4	4	1
Parabasilid H4	H4	1	1
Pelobiont H4	H4	1	1
Entamoebid H4	H4	1	1
Apicomplexan H4	H4	1	1
Diplomonad H4	H4	1	1
Cryptomonad H4	H4	2	1

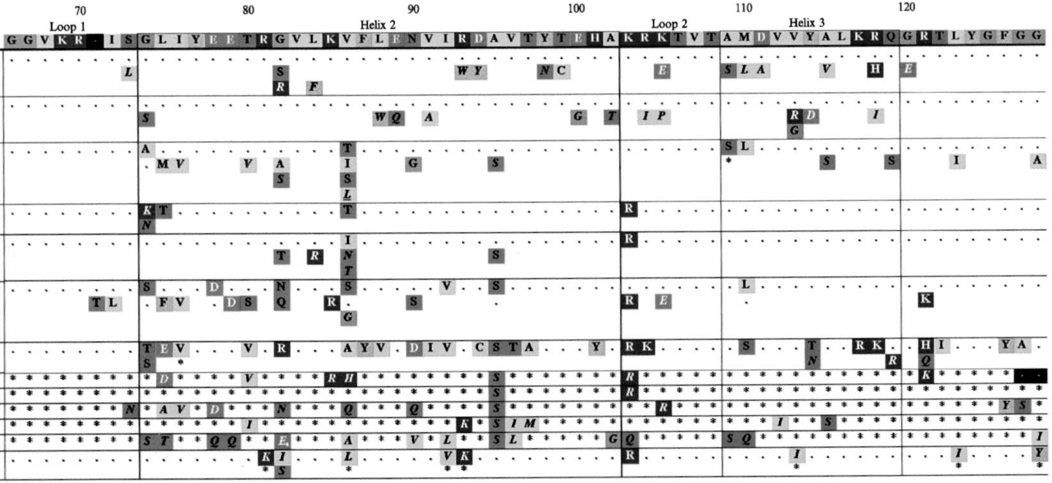

Sᴜʟʟɪᴠᴀɴ ᴀɴᴅ Lᴀɴᴅsᴍᴀɴ, Cʜᴀᴘᴛᴇʀ 1, Fɪɢ. 4. Summary of H4 subclasses and variants. A consensus sequence of all aligned H4 sequences is shown at the top. Dots in the sequences below indicate identity to the consensus. Groups are named on the basis of clustering patterns observed in neighbor-joining trees of aligned H4 sequences (not shown). *Names,* a selection of sequence descriptors found in the definition lines of the sequence records; *seq,* number of unique sequences in the group; *sp,* number of species in the group; *max sp/seq,* the greatest number of species having the same sequence in the group. For each group the first line is the consensus sequence for that group. Variations from the group consensus are indicated below it. Italic indicates a "singleton," i.e., the residue was found in only one sequence from one species in the group. An asterisk (*) indicates singleton identity or a gap. Background color key: white, identity to the anchored consensus; black, gap; orange, aromatic; yellow, aliphatic/hydrophobic; light green, glycine; green, hydrophilic; light blue, histidine; blue, basic; red, acidic.

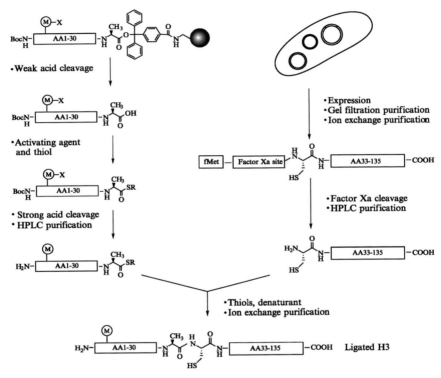

SHOGREN-KNAAK AND PETERSON, CHAPTER 4, FIG. 1. Native chemical ligation strategy for generating histone H3 proteins containing specifically modified N-terminal residues. An N-terminal peptide fragment of histone H3 that contains specifically modified amino acid residues (in this example, a methylated lysine residue denoted by an encircled "M"), and a C-terminal thioester moiety (COSR), is produced by standard solid-phase peptide synthesis on an acid-hypersensitive support *(left)*. A C-terminal protein fragment of histone H3 containing an N-terminal cysteine residue is generated by proteolytic trimming of recombinant protein *(right)*. Reaction of these two fragments in the presence of thiol reagents produces native full-length histone H3 containing the modifications of interest.

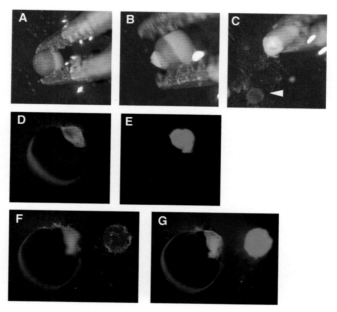

Ridgway *ET AL.*, Chapter 16, Fig. 3. Nuclear localization of overexpressed H2A.Z–EGFP in *Xenopus laevis* oocytes. (A–C) Manual isolation of nuclei from *Xenopus laevis* oocytes. (A) Oocyte is held in place with forceps. (B) Animal pole of an oocyte is cut, using the point of a needle to make an incision for nuclear removal. (C) Oocyte is squeezed with forceps to remove nucleus. Clear nucleus is indicated by the arrowhead. Nucleus is immediately transferred to a tube on ice. (D–G) Oocytes were injected into the cytoplasm with H2A.Z–EGFP synthetic mRNA (10 ng/oocyte) and incubated for 18 h for *in vivo* protein expression and accumulation as described. (D) Bright-field color image of incision in animal pole of oocyte. (E) Fluorescence image, using Leica GFP Plus filter set, corresponding to (D). Brilliant green nucleus can be seen through the incision. (F) Bright-field color image of clear extracted nucleus. (G) Fluorescence image, using Leica GFP filter set (with some bright-field light to show location of remainder of oocyte), corresponding to (F). Brilliant green of H2A.Z–EGFP fusion protein accumulated in nucleus is illustrated.

HMGA proteins bind to nucleosome core particles

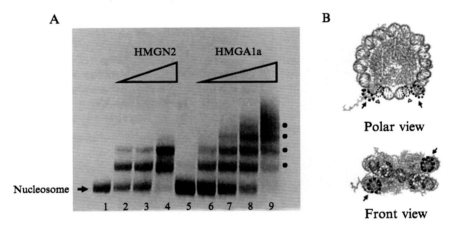

REEVES, CHAPTER 20, FIG. 3. Nonmodified, recombinant HMGA1 proteins bind to four regions of random sequence nucleosome core particles. (A) The results of EMSA gel assays in which increasing concentrations of either purified nonhistone HMGN2 (which binds to two sites on nucleosome core particles) or recombinant human HMGA1a protein were bound to nucleosome core particles isolated from chicken erythrocytes. (B) Two different views (polar and front) of the X-ray structure of the nucleosome cores particle, showing the sites of binding of HMGA proteins (dashed circles) determined by DNA footprinting analyses and other techniques (see text for details).

A

HMGA "A/T Hook" II DNA binding motif

P T *P* K R *P* R **G** R *P*

K D *P* N A *P* K R P *P*

HMGB "B-box" DNA binding motif

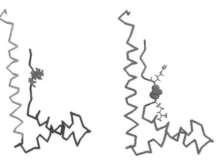

Extended N-terminus

Hydrophobic hinge region

HMGB "B-box"

Pro   Arg
Gly
Pro   Arg

I   II   III

N+ ▬▬■▬▬▨▬▬■▬▬ COO-

Full-length HMGA1a

B

HMGB "B-box"
peptide

Hybrid "B-Box:A/T"
DNA binding peptide

REEVES, CHAPTER 20, FIG. 6. (A) Schematic diagram of a domain-swap experiment in which a single A/T-hook motif (DNA-binding domain 2 from the human HMGA1 protein) is exchanged for the extended N-terminal peptide segment of the second B box of the HMGB1 protein. Also shown is a sequence comparison between the two exchanged peptide segments. (B) Computer-generated models of the wild-type B-box peptide and the hybrid B-box:A/T-hook peptides, showing the relative orientation of the side chains of the P-K-R-P-P residues in the wild-type B box and the P-R-G-R-P residues in the hybrid protein. See text for discussion. Redrawn from Reeves (2000), with modifications.

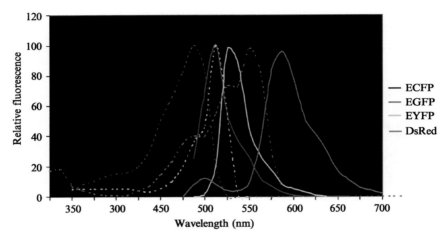

HEDIGER *ET AL.*, CHAPTER 22, FIG. 2. Excitation and emission spectra of enhanced cyan, green, yellow, and red fluorescent proteins. Excitation spectra are represented as dashed lines, excitation spectra as unbroken lines. ECFP, EGFP, and EYFP are all variants of the green fluorescent protein from *Aequorea victoria*. DsRed is derived from a coral of the *Discosoma* genus (for more information, see www.clontech.com).

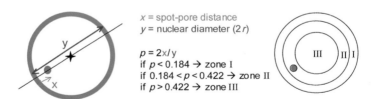

x = spot-pore distance
y = nuclear diameter ($2r$)

$p = 2x/y$
if $p < 0.184$ → zone I
if $0.184 < p < 0.422$ → zone II
if $p > 0.422$ → zone III

HEDIGER *ET AL.*, CHAPTER 22, FIG. 4. Schematic representation of the Nup49–GFP and DNA tagged locus fluorescent signals. For hundreds of different cells, the distance from the middle of the spot to the middle of the envelope signal (x), and the nuclear diameter (y), are measured. By dividing x by $y/2$ ($p = 2x/y$), we can classify the spot position into three concentric zones of equal surface. The outer-most zone (I) contains peripheral spots ($p < 0.184$). Zone II regroups intermediate positioned spots ($0.184 < p < 0.422$). Zone III contains internal spots ($p > 0.422$).

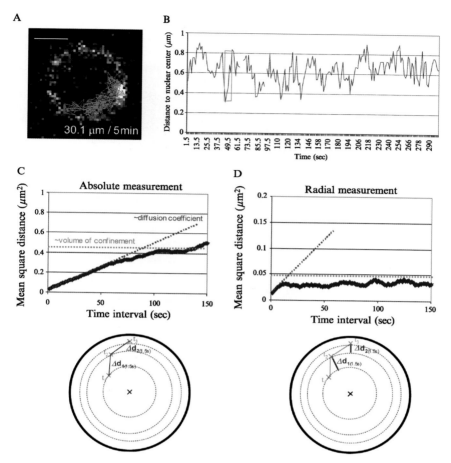

HEDIGER *ET AL.*, CHAPTER 22, FIG. 6. (A) A representative trace of a tagged telomere on Chr 14L in a G_1-phase yeast cell is shown. Images were captured over 5 min with frames taken every 1.5 s. The position of the spot was tracked after alignment of the nuclei on the basis of their nuclear envelope fluorescence, using the tracking algorithm. The trajectory of the locus is projected in red on a single focal section of the nucleus. The mean length of the path in micrometers for a 5-min movie (200 frames) averaged over 8 movies is indicated. Scale bar: 1 μm. (B) Radial movement of the GFP-tagged telomere 14L relative to the nuclear envelope (NE) was monitored by measuring the distance from the middle of the spot to the middle of the pore signal in each frame. The red box indicates radial movement > 0.5 μm within 10.5 s or 7-frame intervals. (C) Mean square difference (MSD) of actual point-to-point distances is shown for the time-lapse series pictured in (A) and (B). For each time interval (1.5 s to 150 s), the mean of the absolute distance (Δd) covered by the spot is calculated and plotted against time intervals. (D) The MSD of radial distances is shown for the same time-lapse series. For each time interval (1.5 s to 150 s), the mean of the radial distance (Δd) covered by the spot is calculated and plotted against time intervals. Telomere 14L moves significantly along the nuclear envelope and thus appears less constrained when the MSD is calculated from absolute distance as compared with radial distances. The fact that telomere 14L is restricted to a small volume close to the nuclear periphery is shown by the low plateau in the radial MSD graph.

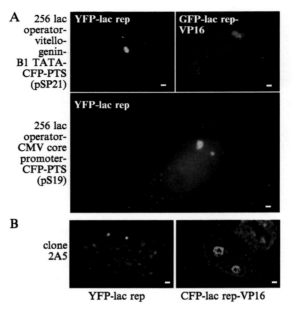

CARPENTER AND BELMONT, CHAPTER 23, FIG. 1. Testing pSP plasmids in transient transfections. (A) CFP-PTS observed by microscopy. pSP plasmids were transfected into wild-type CHO-K1 cells along with YFP–lac rep or GFP–lac rep–VP16. GFP bleeds through in the CFP channel, so the GFP–lac rep–VP16 protein is not distinctly visible. Single optical sections are shown. Scale bar: 1 μm. (B) Stable cell lines containing pSP plasmids. Chromatin unfolding and reporter activation in response to VP16 activator in stable clone 2A5 with the 256 lac op–vitellogenin B1 TATA–CFP-PTS plasmid (pSP21) integrated into the genome. DNA stained with DAPI is shown in red, the YFP channel is shown in green, and the CFP channel is shown in blue. A single optical section from a deconvolved image series is shown. Scale bar: 1 μm.

CARRERO ET AL., CHAPTER 26, FIG. 1. Fluorescence recovery after photobleaching of chromatin-associated histone H1.1. *Top:* SK-N-SH neuroblastoma cells expressing eGFP–histone H1.1 were examined by photobleaching a 1-μm-wide band across the nucleus (see 1.2-s image). As time elapses, this band of photobleached chromatin can be seen to increase in intensity (reflected in the dark blue being replaced by green and yellow). This occurs through the exchange of fluorescence with the remainder of the nuclear chromatin. *Bottom:* Plots of histone H1 recovery under control conditions or after treatment with a histone deacetylase inhibitor (tsa), an inhibitor of protein kinases (staur), an inhibitor of actin polymerization (lat), or an inhibitor of microtubule polymerization (noc). In the left-hand plot, time is shown on a linear scale. This allows for the easy identification of differences in the time required to reach equilibrium. In the right-hand plot, time is plotted on a log scale. This allows the resolution of differences that are occurring in the first few seconds of the experiment. These are primarily differences in diffusion.

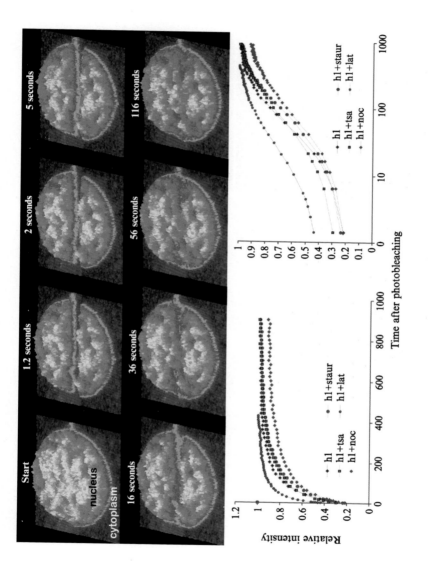

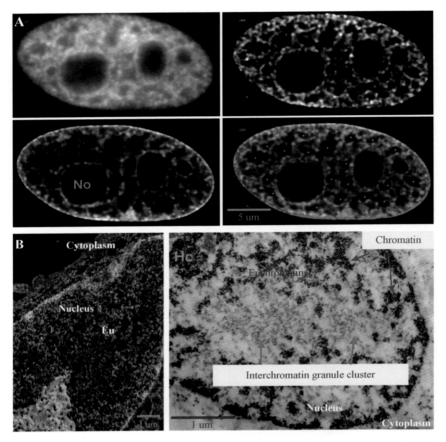

CARRERO *ET AL.*, CHAPTER 26, FIG. 2. General organization of euchromatin and heterochromatin within the vertebrate fibroblast cell nucleus. (A) Indian muntjac fibroblasts were fixed with paraformaldehyde and then stained with an antibody recognizing trimethylated Lys-4 of histone H3 (Abcam) and DAPI. The top panels show optical sections of trimethylated Lys-4. *Top left:* The original optical section. *Top right:* The same image after deconvolution to remove the out-of-focus information. *Bottom left:* The deconvolved optical slice from the DAPI image. *Bottom right:* The DAPI is false-colored red and the trimethylated Lys-4 is false-colored green. (B) Transmission electron microscopy images of Indian muntjac fibroblast cells. *Left:* A composite image obtained from a quantitative spatial map of the phosphorus distribution (green) and a quantitative spatial map of the nitrogen distribution (red). This results in a composite image in which the chromatin and ribonucleoproteins are yellow and protein-rich structures are red. The blue dots indicate the positions of interchromatin granule clusters. *Right:* A higher magnification image of an Indian muntjac fibroblast nucleus containing prominent interchromatin granule clusters (green arrows). Eu, euchromatin; Hc, heterochromatin.

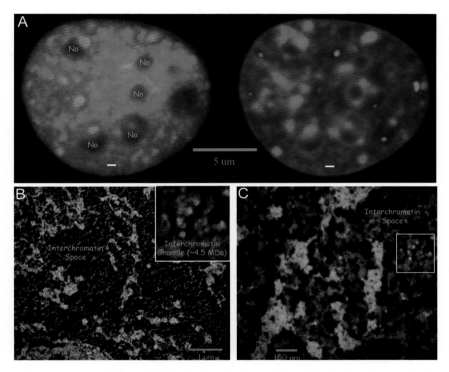

CARRERO *ET AL.*, CHAPTER 26, FIG. 4. Organization and the scale of molecular organization of chromatin in nuclei. (A) A mouse 10T1/2 fibroblast cell was cotransfected with SC35-GFP (green in left-hand image, absent in right-hand image) and PML-dsRed2 (blue in left-hand image, red in right-hand image). The DNA was visualized by adding Hoechst 33258 to the culture medium of the living cells and is shown in red *(left)* and green *(right)*. The small white bars superimposed on each image are shown to allow comparison with (C). The length of these bars is equal to the entire width of the electron microscopy image shown in (C). (B) Transmission electron microscopy image of a mouse 10T1/2 cell nucleus. The specimen was fixed with paraformaldehyde, cut in 30-nm-thick sections after embedding, and then visualized with an energy filter to acquire quantitative maps of nitrogen and phosphorus. This procedure allows the easy discrimination of nucleoprotein from protein structure. Nucleoprotein structures are yellow-orange, depending on their relative abundance of RNA or DNA versus protein. The protein structures are red or red-orange. (C) Higher magnification of subregion in (B). The boxed region of (C) contains a region enriched in interchromatin granules [this is shown at ×4 magnification in the inset of (B)]. An individual granule, which we have previously demonstrated is approximately 4.5 MDa, is highlighted by showing only its phosphorus signal (green).

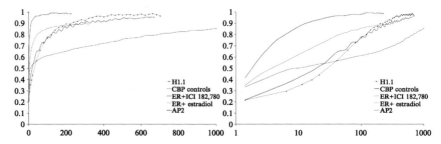

CARRERO *ET AL.*, CHAPTER 26, FIG. 5. Recovery profiles of example nuclear proteins that function within chromatin. The recovery profiles of histone H1.1 are compared with a transcriptional coactivator (CBP), a sequence-specific DNA-binding protein and transcriptional regulator (AP2), and an inducible sequence-specific regulator of transcription (the estrogen receptor α, ER). In the case of the ER protein, its mobility is compared in the presence of estradiol (its natural ligand) or in the presence of a potent inhibitor of estradiol, ICI 182,780.

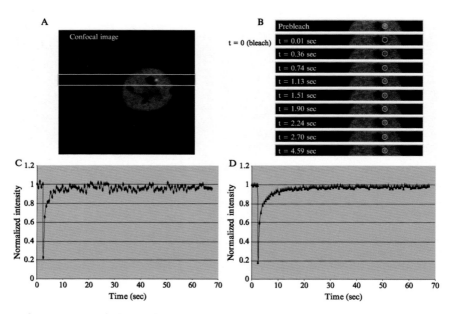

STAVREVA AND MCNALLY, CHAPTER 27, FIG. 1. FRAP of the GFP-tagged glucocorticoid receptor (GFP–GR) at its regulatory sites. Cell line 3617, stably containing 200 tandem repeats of a 9-kb element composed of the MMTV promoter followed by *ras* and BPV genes stably expressing GFP-tagged GR under the control of a tetracycline-off system [D. Walker, H. Htun, and G. L. Hager, *Methods* **19,** 386 (1999)], was used. (A) After hormone induction, a population of the cells shows a visible bright spot in the nucleus corresponding to the MMTV array. (B) Series of images showing the recovery of the GFP–GR after photobleaching. For faster imaging, only a strip of the nucleus including the structure of interest was imaged. Resulting individual FRAP curves are rather noisy (C) but the average of 12 individual curves (D) gives a good representative FRAP curve.

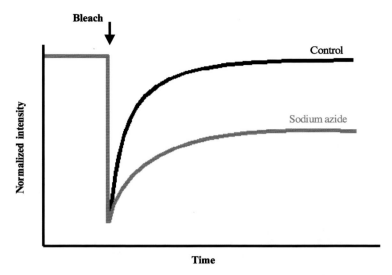

STAVREVA AND MCNALLY, CHAPTER 27, FIG. 2. After treatment with an ATP inhibitor (sodium azide), the resulting FRAP curve is slower, which indicates involvement of energy-dependent factors in the mobility of the protein.

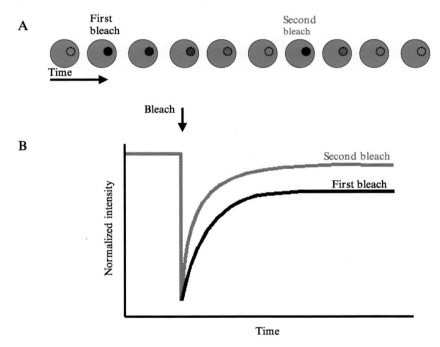

STAVREVA AND MCNALLY, CHAPTER 27, FIG. 3. Double-bleach experimental approach. After the first bleach, the same region of interest is bleached again (A) and the resulting FRAP curves are compared (B). In the case of an immobile fraction, the second bleach leads to a complete recovery because the immobile molecules are now invisible.

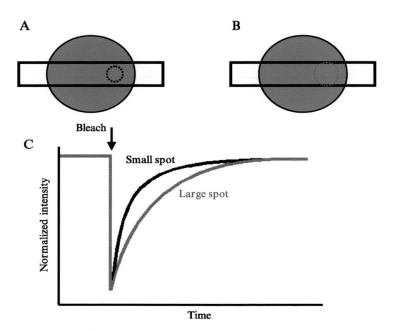

STAVREVA AND MCNALLY, CHAPTER 27, FIG. 4. Spot size dependence of fluorescence recovery. When diffusion plays a role in the mobility of the protein under study, a larger spot size [compare (A) and (B)] will lead to a slower recovery curve (C).

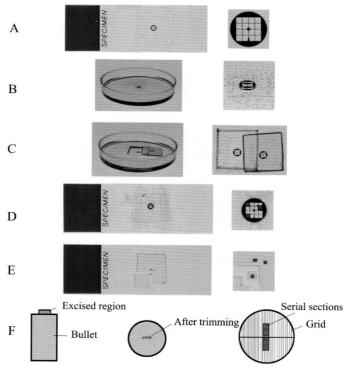

Dellaire *et al.*, Chapter 28, Fig. 1. Schematic diagram outlining the major preparation steps to identify a region of interest by correlative fluorescence and energy-filtering transmission electron microscopy. Cells of interest are first identified by fluorescence microscopy. (A) A 50 mesh electron microscope grid is attached to the coverslip on the side opposite from the cells and covering the region of interest. (B) The cells are refixed and embedded in resin along with the coverslip and initial grid, and a second grid is attached, on top of the embedding plastic, in register with the first grid to maintain identification of the region of interest. (C) After embedding, the resin is cut around the edges of the coverslip and carefully peeled off the coverslip. (D) A third grid is placed on the cell surface side, in register with the second grid. With the aid of the fluorescence microscope, the regions of interest are marked with a scalpel and felt pen, and these regions are cut out of the resin block (E), and (F) glued onto a bullet for sectioning with an ultramicrotome. A ribbon of sections is picked up on EM grids of 400 horizontal mesh.

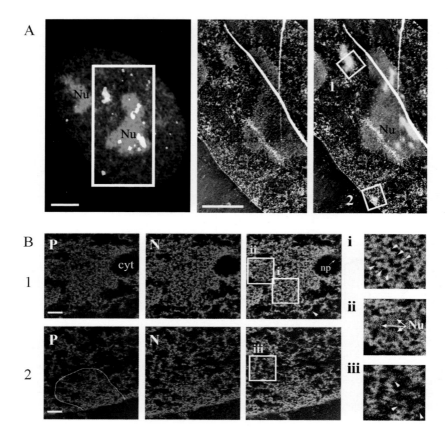

A

Nu

Nu

1

Nu

2

B

P   cyt   N   ii   np   i

1   i

P   N   iii   ii   →Nu

2   iii

DELLAIRE *ET AL.*, CHAPTER 28, FIG. 2. Correlative light and electron spectroscopic imaging (ESI) of human centromeres. Centromeres were detected in paraformaldehyde-fixed human SK-N-SH cells by indirect immunofluorescence using human CREST antisera (Earnshaw reference) and anti-human secondary antibodies coupled to Cy3 (yellow signal), before being processed for ESI. (A) The immunofluorescence image of a single section of a cell is presented, followed by a low-magnification electron micrograph at 155 eV by ESI of a portion (white rectangle) of the same cell containing several centromeres and a third panel showing the correlative LM/EM image of that region. Two prominent centromeres are apparent in the correlative overlay of the fluorescence and electron micrograph images, one between two nucleoli (Nu) and another adjacent to the nuclear lamina (white box 1 and 2, respectively). (B) High-magnification ESI micrographs of the phosphorus (P, red) and nitrogen (N, green) content of centromeres 1 and 2. The position of a cytoplasmic invagination in the nucleus is shown in B1 (cyt) as well as a nucleopore (np) on the edge of this invagination in the combined net phosphorus and nitrogen overlay shown in B1. The approximate boundaries of the centromere in B2, determined from the fluorescence image, are indicated by a white line in the P image. The combined overlays also indicate a high packing of centromeric chromatin for both centromeres (B1 and B2 and enlarged region of B1-i) with respect to adjacent 30-nm fibers (e.g., white arrow in the B1 overlay). In addition, the chromatin within the first centromeres (B1 and enlargement in B1-i) appears to contain an ordered array of 30-nm fibers (as indicated by white arrowheads). Another enlarged region of the first centromeres (B1-ii) also shows protein-dense, nonchromatin material (white arrows) that corresponds with nucleolar signal (Nu) in the correlative fluorescence and low-magnification electron micrograph in Fig. 1A. Similarly, large protein complexes can be found in association with 30-nm fibers in the vicinity of the second centromere (white arrows, enlarged region B2-iii), which could represent transcription and/or splicing complexes. Scale bars: (A) 2 μm; (B) 0.2 μm.

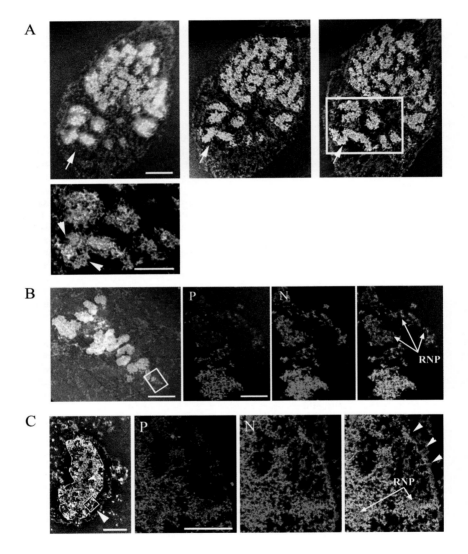

DELLAIRE *ET AL.*, CHAPTER 28, FIG. 3. Electron spectroscopic imaging (ESI) of human chromatin during mitosis. Human SK-N-SH cells grown on coverslips were paraformaldehyde fixed before processing for ESI, and mitotic cells were identified by immunofluorence detection using an anti-phosphohistone H3 (Ser10) antibody (Upstate Cell Signaling Solutions) in serial sections. Low-magnification ESI electron micrographs were taken at 155 eV for cells in late prophase, metaphase, and late anaphase of the cell cycle (A–C). (A) Three serial sections of a late prophase cell are shown, in which large compact chromatin fibers are found that appear in all three sections (white arrow), corresponding to 210–240 nm. A composite of the anti-phosphohistone (Ser10) immunostaining (blue signal) is shown for the first serial section electron micrograph. Two sister chromatid arms of a single chromosome (white arrows) can be seen in the three-dimensional volume representation of a selected region of these sections (white box). (B) An electron micrograph of metaphase chromatin at low magnification and high-magnification net phosphorus (P), nitrogen (N), and combined ESI micrographs for a portion of a mitotic chromosome (white box) in which ribonucleo-protein complexes and fibers can be seen adjacent to the chromatin (white arrows, RNP). (C) A low-magnification electron micrograph of a cell in late anaphase and corresponding high-magnification net phosphorus (P), nitrogen (N), and combined ESI micrographs of a selected region of the cell (white box). Chromatin in this cell has begun to decondense in late anaphase and a number of nitrogen-rich RNP complexes can be seen surrounded by chromatin (white arrows, RNP). In addition, the newly formed nuclear envelope can be seen as a green rim in the combined electron micrograph (three white arrows). Scale bars in low- and high-magnification electron micrographs represent 2 and 0.5 μm, respectively.

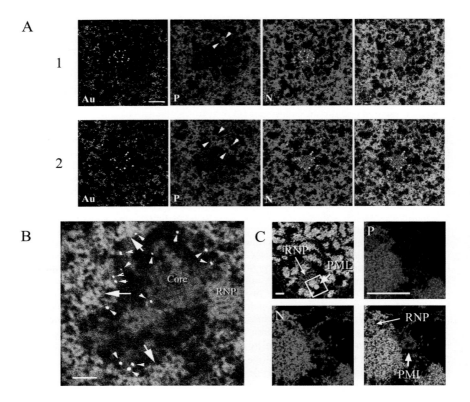

DELLAIRE ET AL., CHAPTER 28, FIG. 4. ImmunoGold detection of CREB-binding protein (CBP) in PML nuclear bodies imaged by ESI. Before processing for ESI, human SK-N-SH cells grown on coverslips were fixed in paraformaldehyde before detection of PML protein with a mouse anti-CBP monoclonal antibody (Santa Cruz Biotechnology, Santa Cruz, CA) and an anti-mouse secondary antibody coupled to Aurion Nanogold particles (EMS) either alone or in conjunction with a Cy3-labeled secondary antibody. (A and B) Interphase structure of PML nuclear bodies. ESI micrographs for gold (Au, white), phosphorus (P, red), or nitrogen (N, green) were taken for serial sections at high magnification for PML nuclear bodies (NBs). The gold particles are silver (Ag) enhanced, and thus the ESI map of the particles is obtained via Ag detection. The combined phosphorus and nitrogen overlay of these micrographs helps distinguish PML bodies from RNP structures such as IGCs [see RNP in (B)] because of their nitrogen-dense protein core (green), which can be easily distinguished from the surrounding chromatin and RNP fibers (yellow). Two serial sections of one PML NB surrounded by chromatin are shown in (A1) and (A2). A single chromatin fiber can be followed in the two sections (white arrows in net P), demonstrating the close contact of these NBs to chromatin. The Ag-enhanced gold particles (white dots in the overlay) appear across the top of the PML NB in the first section (A1) and around the periphery of the second section (A2), indicating that the CBP is found at the periphery of PML NBs and not at their core. (B) A high-magnification ESI micrograph of a second PML NB illustrating the peripheral localization of the PML protein within these structures. The core of the PML NB and an adjacent accumulation of RNPs are indicated. Surrounding chromatin (yellow fibers) is indicated by large arrows and Ag-enhanced gold particles corresponding to the CBP protein are indicated by small arrows (white dots). (C) Ultrastructure of a PML NB during prophase. First, a PML NB was located by overlaying the light microscopy image of a single section of a prophase cell containing a PML NB (yellow spot, PML) with the low-magnification electron micrograph of the same cell taken at 155 eV by ESI. A large RNP complex is also apparent in the correlative LM/EM micrograph (RNP). High-magnification net phosphorus (P), nitrogen (N), and combined ESI micrographs of the region of the prophase cell containing the PML NB (white box) are shown. At high magnification, the PML NB is less dense than during interphase and has a pronounced donut shape. This magnified region also contains RNP complexes closely associated with chromatin (RNP). Scale bars: (A and B) 0.2 μm; (C) 1 μm.

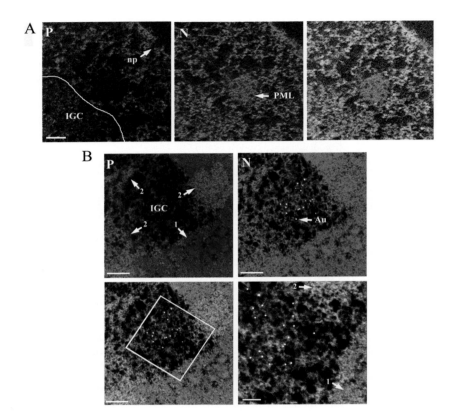

Dellaire *et al.*, Chapter 28, Fig. 5. Organization of subnuclear components. (A) Human SK-N-SH cells were double-labeled with primary anti-PML (Santa Cruz Biotechnology) antibody and primary anti-SC35 monoclonal cell culture supernatant, followed by secondary fluorescent antibodies for correlative microscopy. ESI micrographs of a thin section of an embedded cell show the phosphorus, nitrogen, and composite maps of a region of the cell containing a PML nuclear body, an IGC, and a nuclear pore. The IGC region is defined in the phosphorus map. Scale bar: 0.2 µm. (B) Human SK-N-SH cells were incubated with primary anti-SC35 antibody, followed by secondary labeling with Nanogold and fluorescently labeled antibodies for correlative identification. The low-magnification nitrogen, phosphorus, and composite maps illustrate the position of an IGC with respect to the nucleolus (arrow 1) and the surrounding condensed chromatin (arrow 2). The white particles are silver-enhanced, gold-labeled antibodies and show the location of the SC35 splicing factor within the RNP fibrillar network. The boxed region within the low-magnification composite map is shown at a higher magnification for a detailed view of the IGC and surrounding chromatin structure. Scale bars: 0.5 µm in the low-magnification P, N, and composite maps; 0.2 µm in the high-magnification composite map.

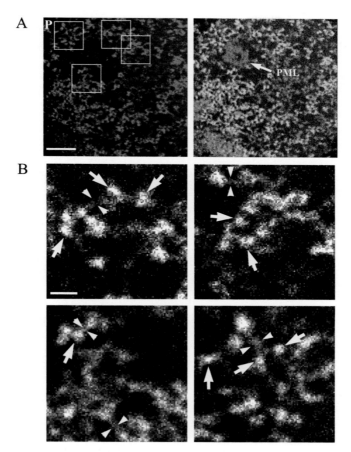

DELLAIRE *ET AL.*, CHAPTER 28, FIG. 6. Nucleosomes in decondensed chromatin fibers. Human SK-N-SH cells were labeled with an anti-PML antibody, embedded, and sectioned. (A) Low-magnification phosphorus and composite maps of a region of decondensed chromatin surrounding a PML nuclear body. The locations of clearly resolved nucleosomes are highlighted with boxes in the phosphorus image. Scale bar: 0.2 μm. (B) A series of magnified views of the highlighted regions within the phosphorus map. Individual nucleosomes are marked with single arrows, and the linking DNA is indicated with double arrowheads. Scale bar: 20 nm.